Leonas Valkunas,
Darius Abramavicius, and
Tomáš Mančal

Molecular Excitation Dynamics and Relaxation

Related Titles

Radons, G., Rumpf, B., Schuster, H. G. (eds.)

Nonlinear Dynamics of Nanosystems

2010

ISBN: 978-3-527-40791-0

Siebert, F., Hildebrandt, P.

Vibrational Spectroscopy in Life Science

2008

ISBN: 978-3-527-40506-0

Schnabel, W.

Polymers and Light
Fundamentals and Technical Applications

2007

ISBN: 978-3-527-31866-7

Reich, S., Thomsen, C., Maultzsch, J.

Carbon Nanotubes
Basic Concepts and Physical Properties

2004

ISBN: 978-3-527-40386-8

May, V., Kühn, O.

Charge and Energy Transfer Dynamics in Molecular Systems

2011

ISBN: 978-3-527-40732-3

Yakushevich, L. V.

Nonlinear Physics of DNA

1998

ISBN: 978-3-527-40417-9

Leonas Valkunas, Darius Abramavicius,
and Tomáš Mančal

Molecular Excitation Dynamics and Relaxation

Quantum Theory and Spectroscopy

WILEY-VCH
Verlag GmbH & Co. KGaA

The Authors

Prof. Leonas Valkunas
Department of Theoretical Physics
Vilnius University
Center for Physical Sciences and Technology
Vilnius, Lithuania
leonas.valkunas@ff.vu.lt

Prof. Darius Abramavicius
Department of Theoretical Physics
Vilnius University
Vilnius, Lithuania
darius.abramavicius@ff.vu.lt

Dr. Tomáš Mančal
Faculty of Mathematics and Physics
Charles University in Prague
Prague, Czech Republic
tomas.mancal@mff.cuni.cz

Cover Picture
Rephasing and non-rephasing two-dimensional spectra calculated for the Fenna–Matthews–Olson complexes at zero population time ($T = 0$), see Chapter 17 for details.

Library of Congress Card No.:
applied for

British Library Cataloguing-in-Publication Data:
A catalogue record for this book is available from the British Library.

Bibliographic information published by the Deutsche Nationalbibliothek
The Deutsche Nationalbibliothek lists this publication in the Deutsche Nationalbibliografie; detailed bibliographic data are available on the Internet at http://dnb.d-nb.de.

Print ISBN 978-3-527-41008-8
ePDF ISBN 978-3-527-65368-3
ePub ISBN 978-3-527-65367-6
mobi ISBN 978-3-527-65366-9
oBook ISBN 978-3-527-65365-2

Composition le-tex publishing services GmbH, Leipzig
Printing and Binding Markono Print Media Pte Ltd, Singapore
Cover Design Grafik-Design Schulz, Fußgönheim

Printed in Singapore

Printed on acid-free paper

Contents

Preface

Classical mechanics is known for its ability to describe the dynamics of macroscopic bodies. Their behavior in the course of time is usually represented by classical trajectories in the real three-dimensional space or in the so-called phase space defined by characteristic coordinates and momenta, which together determine the degrees of freedom of the body under consideration. For the description of the dynamics of a microscopic system, however, quantum mechanics should be used. In this case, the system dynamics is qualified by the time evolution of a complex quantity, the wavefunction, which characterizes the maximum knowledge we can obtain about the quantum system. In terms of the quantum mechanical description, coordinates and momenta cannot be determined simultaneously. Their values should satisfy the Heisenberg uncertainty principle. At the interface between the classical world in which we live and the world of microscopic systems, this type of description is inherently probabilistic. This constitutes the fundamental differences between classical and quantum descriptions of the system dynamics. In principle, however, both classical and quantum mechanics describe a reversible behavior of an isolated system in the course of time.

Irreversibility of time evolution is a property found in the dynamics of open systems. No realistic system is isolated; it is always subjected to coupling to its environment, which in most cases cannot be considered as a negligible factor. The theory of open quantum systems plays a major role in determining the dynamics and relaxation of excitations induced by an external perturbation. A typical external perturbation is caused by the interaction of a system with an electromagnetic field. In resonance conditions, when the characteristic transition frequencies of the system match the frequencies of the electromagnetic field, the energy is transferred from the field to the system and the system becomes excited. The study of the response of material systems to various types of external excitation conditions is the main objective of spectroscopy. Spectroscopy, in general, is an experimental tool to monitor the features and properties of the system based on the measurement of its response. More complicated spectroscopic experiments study the response which mirrors the dynamics of excitation and its relaxation.

Together with the widely used conventional spectroscopic approaches, two-dimensional coherent spectroscopic methods were developed recently, and they have been applied for studies of the excitation dynamics in various molecular systems, such as photosynthetic pigment–protein complexes, molecular aggregates, and polymers. Despite the complexity of the temporal evolution of the two-dimensional spectra, some of these spectra demonstrate the presence of vibrational and electronic coherence on the subpicosecond timescale and even picosecond timescale. Such observations demonstrate the interplay between the coherent behavior of the system, which might be considered in terms of conventional quantum mechanics, and the irreversibility of the excitation dynamics due to the interaction of the system with its environment.

From the general point of view, quantum mechanics is the basic approach for considering various phenomena in molecular systems. However, a typical description must be based on a simplified model, where specific degrees of freedom are taken into consideration, and the rest of them are attributed to an environment or bath. This is the usual approach used for open quantum systems. Thus, complexity of the molecular system caused by some amount of interacting molecules has to be specifically taken into account by describing the quantum behavior of the system. For this purpose the concept of excitons is usually invoked.

As can be anticipated, this area of research covers a very broad range of fields in physics and chemistry. Having this in mind, we have divided this book into two parts. Part One, being more general, describes the basic principles and theoretical approaches which are necessary to describe the excitation dynamics and relaxation in quantum systems interacting with the environment. These theoretical approaches are then used for the description of spectroscopic observables in Part Two.

Consequently, we have many different readers of this book in mind. First of all, the book addresses undergraduate and graduate students in theoretical physics and chemistry, molecular chemical physics, quantum optics and spectroscopy. For this purpose the basic principles of classical physics, quantum mechanics, statistical physics, and stochastic processes are presented in Part One. Special attention is paid to the interface of classical and quantum physics. This includes discussion on the decoherence and entanglement problems, the projection operator, and stochastic classical and quantum problems. These processes are especially relevant in small molecular clusters, often serving as primary natural functioning devices. Therefore, the adiabatic description of molecules, the concept of Frenkel and Wannier–Mott excitons, charge-transfer excitons, and problems of exciton self-trapping and trapping are also presented. This knowledge helps understand other chapters in this book, especially in Part Two, which is more geared toward graduate students and professionals who are interested in spectroscopy. Since different approaches to the problem are widely used to describe the problem of coherence, various methods used for the description are also discussed. Possible modern approaches for observation of the processes determining the excitation dynamics and relaxation in molecular systems are discussed in Part Two, which is mainly devoted to the theoretical description of the spectroscopic observations. For this purpose the response function formalism is introduced. Various spectroscopic methods are discussed, and the results demonstrating the possibility to distinguish the coherent effects on the excitation dynamics are also presented.

We would like to thank our colleagues and students for their contribution. First of all we mention Vytautas Butkus, who produced almost all the figures in this book and who pushed us all the time to proceed with the book. He was also involved in the theoretical analysis of the two-dimensional spectra of molecular aggregates. We are also grateful to our students, Vytautas Balevicius Jr., Jevgenij Chmeliov, Andrius Gelzinis, Jan Olsina, and others, who were involved in solving various theoretical models. We are thankful to our colleagues and collaborators, the discussions with whom were very stimulating and helped in understanding various aspects in this rapidly developing field of science. Especially we would like to express our appreciation to our colleagues Shaul Mukamel and Graham R. Fleming, who were initiators of two-dimensional coherent electronic spectroscopy and who have inspired our research. We also thank our wives and other members of our families for patience, support, and understanding while we were taking precious time during holidays and vacations to write this book.

Vilnius, Prague
November 2012

Leonas Valkunas
Darius Abramavicius
Tomáš Mančal

Part One
Dynamics and Relaxation

1 Introduction

Photoinduced dynamics of excitation in molecular systems are determined by various interactions occurring at different levels of their organization. Depending on the perturbation conditions, the excitation in solids and molecular aggregates may lead to a host of photoinduced dynamics, from coherent and incoherent energy migration to charge generation, charge transfer, crystal lattice deformation, or reorganization of the environmental surroundings. The theoretical description of all these phenomena therefore requires one to treat part of the molecular system as an open system subject to external perturbation. Since perfect insulation of any system from the rest of the world is practically unattainable, the theory of open systems plays a major role in any realistic description of experiments on molecular systems.

In classical physics, the dynamics of an open system is reflected in the temporal evolution of its parameters, leading to a certain fixed point in the corresponding phase space. This fixed point corresponds to a thermodynamic equilibrium, with the unobserved degrees of freedom determining the thermodynamic bath. Many situations in molecular physics allow one to apply a classical or semiclassical description of the evolution of the perturbation-induced excitation in an open system. Often, the influence of the large number of degrees of freedom can be efficiently simulated by stochastic fluctuations of some essential parameters of the system. Such fluctuations may lead to transitions between several stable fixed points in the phase space of the system, or, in a semiclassical situation, to transitions between several states characterized by different energies.

Apart from classical fluctuations, a genuine quantum description might be required when entanglement between constituents of the system has to be considered. This is especially essential for systems with energy gaps larger than the thermal energy, which is an energy characteristics of the bath defined by macroscopic degrees of freedom. Only a full quantum description then leads to proper formation of a thermal equilibrium.

Indeed it is impossible to switch off fluctuations completely. Even if we place a system in a complete vacuum and isolate it from some light sources, there still exist background vacuum fluctuations of the electromagnetic field. Even at zero temperature these fluctuations affect the quantum system, and the resulting spontaneous

Molecular Excitation Dynamics and Relaxation, First Edition. L. Valkunas, D. Abramavicius, and T. Mančal.

emission emerges. All these fluctuations cause decay of excited states and establish thermal equilibrium and stochasticity "in the long run."

The first part of this book presents a coarse-grained review of the knowledge which is needed for a description of excitation dynamics and relaxation in molecular systems. Basic topics of classical physics which are directly related to the main issue of this book are presented in Chapter 2. It is worthwhile mentioning that concepts of classical physics are also needed for better understanding of the basic behavior of quantum systems. The electromagnetic field, which is responsible for electronic excitations, can usually be well described in terms of classical electrodynamics. Thus, the main principles of this theory and the description of the field–matter interaction are also introduced in Chapter 2. The concept and main applicative features of stochastic dynamics are presented in Chapter 3. Markov processes, the Fokker–Planck equation, and diffusive processes together with some relationships between these descriptions and purely stochastic dynamics are also described in Chapter 3. The basic concepts of quantum mechanics, which is the fundamental theory of the microworld, are presented in Chapter 4. Together with its main postulates and equations, some typical model quantum systems with exact solutions are briefly discussed. The density matrix and second quantization of the vibrations and electromagnetic field are briefly introduced as well. Special attention is paid in this book to consideration of molecular aggregates. The adiabatic approximation, the exciton concept, Frenkel excitons, Wannier–Mott excitons, and charge-transfer excitons are described together with vibronic interactions, the self-trapping problem, and the exciton trapping problem in Chapter 5. Chapter 6 is devoted to a discussion of decoherence and entanglement concepts. The problem of measurements in quantum mechanics and the relative state interpretation are also discussed. The basics of statistical physics are then presented in Chapter 7. The relationship between the statistical approach and thermodynamics is briefly outlined, and standard statistics used for descriptions of classical and quantum behavior are presented. The harmonic oscillator model of the system–bath interaction is described in Chapter 8. In Chapter 9 we describe the projection operator technique together with the concept of the reduced density matrix and its master equations. The path integral technique is then discussed in Chapter 10 together with the stochastic Schrödinger equation approach and the so-called hierarchical equations of motions. Excitation dynamics and relaxation in some model systems are discussed in Chapter 11.

2
Overview of Classical Physics

In this chapter we will review some of the most important concepts of classical physics. Despite the eminent role played by quantum mechanics in the description of molecular systems, classical physics provides an important conceptual and methodological background to most of the theories presented in later chapters and to quantum mechanics itself. Often classical or semiclassical approximations are indispensable to make a theoretical treatment of problems in molecular physics feasible. In the limited space of this chapter we have no intention to provide a complete review as we assume that the reader is familiar with most of the classical concepts. Specialized textbooks are recommended to the interested reader in which the topics presented in this chapter are treated with full rigor (e.g., [1–4]).

2.1
Classical Mechanics

Classical mechanics, as the oldest discipline of physics, has provided the formal foundation for most of the other branches of physics. Perhaps with the exception of phenomenological thermodynamics, there is no theory with a similar general validity and success that does not owe its foundations to mechanics. Classical mechanics reached its height with its Lagrangian and Hamiltonian formulations. These subsequently played a very important role in the development of statistical and quantum mechanics.

In classical mechanics, the physical system is described by a set of idealized material points (point-sized particles) in space which interact with each other by a specific set of forces. The coordinates and velocities of all particles fully describe the state of the system of the particles. The three laws formulated by Newton fully describe the properties of motion of this system. The first law states that the particle moves at a constant speed in a predefined direction if it is not affected by a force. The second law relates the change of motion of the particle due to the presence of external forces. The third law defines the symmetry of all forces: particle a acts on particle b with the same force as particle b acts on particle a.

Molecular Excitation Dynamics and Relaxation, First Edition. L. Valkunas, D. Abramavicius, and T. Mančal.

The dynamics of the system of N particles is described by a set of differential equations [1, 2, 4]:

$$m_i \ddot{\boldsymbol{r}}_i = \sum_j \boldsymbol{F}_{ij}(\boldsymbol{r}_1 \dots \boldsymbol{r}_N) \,. \tag{2.1}$$

Here m_i is the mass of the ith particle and $\boldsymbol{F}_{ij}$ is the force created by the jth particle acting on the ith particle. The velocity of the ith particle is given by a time derivative of the coordinate $\dot{\boldsymbol{r}}_i$. For a problem formulated in three spatial dimensions the particle momenta $\boldsymbol{p}_i = m_i \dot{\boldsymbol{r}}_i$ together with the coordinates $\boldsymbol{r}_i$ create a $6N$-dimensional *phase space* in the three-dimensional real space.

The real phase space is often smaller due to specific symmetries, resulting in certain conservation laws. For instance, if the points describe some finite body, which is at rest, the center of mass of all points may be fixed. In that case the dimension of the phase space effectively decreases by six (three coordinates and three momenta corresponding to a center of mass equal to zero). If additionally the body is rigid, we are left with three dimensional phase space, characterizing orientation of the body (e.g. three Euler angles).

A single point in the phase space defines an instantaneous state of the system. The notion of the system's state plays an important role in quantum physics; thus, it is also useful to introduce this type of description in classical physics. The motion of the system according to Newton's laws draws a trajectory in the phase space. In the absence of external forces, the energy of the system is conserved, and the trajectory therefore corresponds to a particular energy value. Different initial conditions draw different trajectories in the phase space as shown schematically in Figure 2.1. The phase space trajectories never intersect or disappear. Later in the discussion of statistical mechanics this notion is used to describe the microcanonical ensemble of an isolated system.

Note that in Newton's equation, (2.1), we can replace t by $-t$ and the equation remains the same. Thus, the Newtonian dynamics is invariant to an inversion of the time axis, and the dynamics of the whole system is reversible. This means that Newton's equation for a finite isolated system with coordinate-related pairwise forces has no preferred direction of the time axis. Because energy is conserved, the whole system does not exhibit any damping effects. The damping is often introduced phenomenologically. In order to achieve irreversible dynamics using a

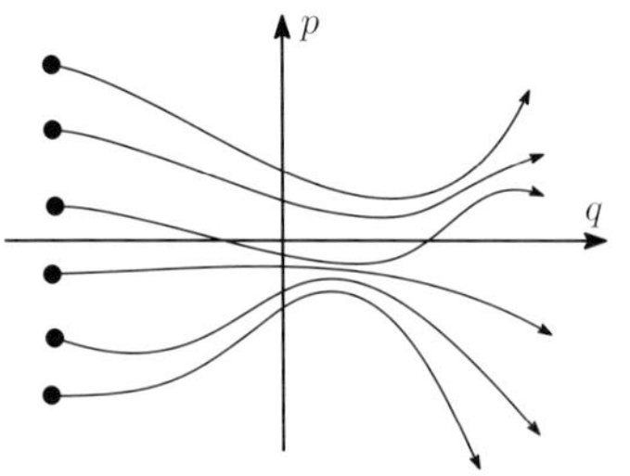

Figure 2.1 Motion of the system in a phase space starting with different initial conditions.

microscopic description, one has to introduce an infinitely large system so that the observable part is a small open subsystem of the whole. In such a subsystem the damping effects occur naturally from statistical arguments. Various treatments of open systems are described in subsequent chapters.

2.1.1 Concepts of Theoretical Mechanics: Action, Lagrangian, and Lagrange Equations

Some problems in mechanics can be solved exactly. The feasibility of such an exact solution often depends crucially on our ability to express the problem in an appropriate coordinate system. Let us find now a more general way of expressing mechanical equations of motion that would have the same general form in an arbitrary system of coordinates, and would therefore allow a straightforward transformation from one coordinate system to another. This new form of the representation of Newton's equations is called the Lagrangian formulation of mechanics.

Let us start with Newton's law, (2.1), in the following form:

$$\sum_i (\boldsymbol{F}_i - m_i \ddot{\boldsymbol{r}}_i) = 0 . \tag{2.2}$$

Here we sum up over all particles in the system, and $\boldsymbol{F}_i = \sum_j \boldsymbol{F}_{ij}$ is the total force acting on the ith particle. With a given initial condition, the whole trajectory $\boldsymbol{r}_i(t)$ of the ith particle satisfies (2.2). At every point of the trajectory, we can imagine a small displacement of the trajectory $\delta \boldsymbol{r}_i(t)$ from $\boldsymbol{r}_i(t)$ to $\boldsymbol{r}_i(t) + \delta \boldsymbol{r}_i(t)$, that is, an infinitesimal variation. We multiply each term of the sum in (2.2) by $\delta \boldsymbol{r}_i(t)$ and integrate it over time from t_1 to t_2. The right-hand side of the equation remains zero. On the left-hand side we assume that the force can be expressed by means of a gradient of the potential V as $\boldsymbol{F}_i = -\partial V / \partial \boldsymbol{r}_i$, so we get

$$\int_{t_1}^{t_2} \mathrm{d}t \left(\sum_i \frac{\partial V}{\partial \boldsymbol{r}_i} + m_i \ddot{\boldsymbol{r}}_i \right) \cdot \delta \boldsymbol{r}_i(t) = 0 . \tag{2.3}$$

The first term on the left-hand side of (2.3) can obviously be written as a variation of an integral over the potential:

$$\int_{t_1}^{t_2} \mathrm{d}t \sum_i \frac{\partial V}{\partial \boldsymbol{r}_i} \cdot \delta \boldsymbol{r}_i(t) = \delta \int_{t_1}^{t_2} V \mathrm{d}t . \tag{2.4}$$

The second term on the left-hand side can be turned into a variation as well. We apply integration by parts and interchange the variation with the derivative to obtain

$$\int_{t_1}^{t_2} \mathrm{d}t m_i \ddot{\boldsymbol{r}}_i \cdot \delta \boldsymbol{r}_i = [m_i \dot{\boldsymbol{r}}_i \cdot \delta \boldsymbol{r}_i]_{t_1}^{t_2} - \int_{t_1}^{t_2} \mathrm{d}t m_i \dot{\boldsymbol{r}}_i \cdot \delta \dot{\boldsymbol{r}}_i . \tag{2.5}$$

By assuming now that variation of the trajectory $\delta \boldsymbol{r}_i(t)$ is zero at times t_1 and t_2, that is, $\delta \boldsymbol{r}_i(t_1) = 0$ and $\delta \boldsymbol{r}_i(t_2) = 0$ for all i, we set the first term on the right-hand side to zero. Therefore, (2.3) reads

$$\delta \int_{t_1}^{t_2} \mathrm{d}t \left(\sum_i \frac{1}{2} m_i |\dot{\boldsymbol{r}}_i|^2 - V \right) = 0 . \tag{2.6}$$

Here we used the rules of variation of a product, and we multiplied the equation obtained by -1. Now, the first term denotes the total kinetic energy of the system. The second term is the full potential energy. Thus, the variation of the kinetic energy must be anticorrelated with the variation of the potential energy. This result is also implied by the conservation of the total energy.

We next denote the kinetic energy term $\sum_i 1/2 m_i |\dot{\boldsymbol{r}}_i|^2$ by T, and introduce two new functions:

$$S = \int_{t_1}^{t_2} L \mathrm{d}t , \tag{2.7}$$

where

$$L = T - V . \tag{2.8}$$

Here, S denotes the *action functional* or simply the *action*. The scalar function L is the *Lagrangian function*, or the *Lagrangian*. The whole mechanics therefore reduces to the variational problem

$$\delta S = 0 , \tag{2.9}$$

also known as the *Hamilton principle*. According to this principle, the trajectories $\boldsymbol{r}_i(t)$, which satisfy Newton's laws of motion, correspond to an extremum of the action functional S. In Chapter 10, we will see that the action functional plays an important role in the path integral representations of quantum mechanics.

This formulation is independent of any specific choice of coordinates. Trajectories $\boldsymbol{r}_i(t)$ can also be expressed in terms of coordinates different from the original Cartesian coordinates $\boldsymbol{r}$. Let us have the Lagrangian expressed in terms of generalized coordinates $\{q_i\} = \{q_1, q_2, \ldots, q_{3N}\}$ and their time derivatives $\{\dot{q}_i\}$, where N is the number of particles. The variational problem, (2.9), then leads to

$$\int_{t_1}^{t_2} \mathrm{d}t \left(\frac{\partial L}{\partial q_i} \delta q_i + \frac{\partial L}{\partial \dot{q}_i} \delta \dot{q}_i \right) = 0 . \tag{2.10}$$

By integrating the second term by parts under the assumption that $\delta q_i(t_1) = \delta q_i(t_2) = 0$ as done for (2.6), we obtain

$$\int_{t_1}^{t_2} \mathrm{d}t \left[\frac{\partial L}{\partial q_i} - \frac{\mathrm{d}}{\mathrm{d}t} \left(\frac{\partial L}{\partial \dot{q}_i} \right) \right] \delta q_i = 0 . \tag{2.11}$$

This can only be satisfied for an arbitrary value of δq_i if

$$\frac{\mathrm{d}}{\mathrm{d}t}\left(\frac{\partial}{\partial \dot{q}_i} L\right) - \frac{\partial}{\partial q_i} L = 0 . \tag{2.12}$$

Equation (2.12) is the famous *Lagrange equation* of classical mechanics in a form independent of the choice of the coordinate system.

There is some flexibility in choosing a particular form of the Lagrangian. If we define a new Lagrangian L' by adding a total time derivative of a function of coordinates,

$$L'(q_i, \dot{q}_i, t) = L(q_i, \dot{q}_i, t) + \frac{\mathrm{d}}{\mathrm{d}t} f(q_i, t) , \tag{2.13}$$

the equations of motion remain unchanged. The corresponding action integral S' is

$$\begin{aligned} S' &= \int_{t_1}^{t_2} \mathrm{d}t L + \int_{t_1}^{t_2} \mathrm{d}t \frac{\mathrm{d}}{\mathrm{d}t} f(q_i, t) \\ &= \int_{t_1}^{t_2} \mathrm{d}t L + f(q_i(t_2), t_2) - f(q_i(t_1), t_1) , \end{aligned} \tag{2.14}$$

where the last two terms do not contribute to a variation with fixed points at times t_1 and t_2. By means of (2.13), the Lagrangian can sometimes be converted into a form more convenient for description of a particular physical situation. We will give an example of such a situation in Section 2.4.3.

2.1.2
Hamilton Equations

A more symmetric formulation of mechanics can be achieved by introducing generalized momenta p_i as conjugate quantities of coordinates q_i. So far the independent variables of the Lagrangian were q_i and $\dot{q}_i$. Now we will define the generalized momentun corresponding to the coordinate q_i as

$$p_i = \frac{\partial}{\partial \dot{q}_i} L . \tag{2.15}$$

It can be easily shown that in Cartesian coordinates the momentum $p_i = m\dot{q}_i$ is conjugate to the coordinate r_i. Let us investigate the variation of the Lagrangian:

$$\delta L = \sum_i \frac{\partial L}{\partial q_i} \delta q_i + \sum_i \frac{\partial L}{\partial \dot{q}_i} \delta \dot{q}_i . \tag{2.16}$$

First, from (2.12) and (2.15) we obtain a very symmetric expression:

$$\delta L = \sum_i \dot{p}_i \delta q_i + \sum_i p_i \delta \dot{q}_i , \tag{2.17}$$

which can also be written as

$$\delta L = \sum_i \dot{p}_i \delta q_i + \delta\left(\sum_i p_i \dot{q}_i\right) - \sum_i \dot{q}_i \delta p_i \,. \tag{2.18}$$

This in turn can be written in such a way that we have a variation of a certain function on the left-hand side and an expression with variations of p_i and q_i only on the right-hand side:

$$\delta\left(\sum_i p_i \dot{q}_i - L\right) = \sum_i \dot{q}_i \delta p_i - \sum_i \dot{p}_i \delta q_i \,. \tag{2.19}$$

The expression on the left-hand side,

$$H = \sum_i p_i \dot{q}_i - L \,, \tag{2.20}$$

must thereqfore be a function of parameters p_i and q_i only, that is, $H = H(p_i, q_i)$. By taking its formal variation and using (2.19), we arrive at

$$\delta H = \sum_i \frac{\partial H}{\partial q_i} \delta q_i + \sum_i \frac{\partial H}{\partial p_i} \delta p_i = \sum_i \dot{q}_i \delta p_i - \sum_i \dot{p}_i \delta q_i \,. \tag{2.21}$$

Comparing the coefficients of variations of δq_i and δp_i, we get two independent equations:

$$\dot{p}_i = -\frac{\partial H}{\partial q_i} \tag{2.22}$$

and

$$\dot{q}_i = \frac{\partial H}{\partial p_i} \,. \tag{2.23}$$

Equations (2.22) and (2.23) are known as the canonical or *Hamilton equations* of classical mechanics. We usually call the momentum p_i the *canonically conjugated momentum* only to the coordinate q_i. The Hamilton equations represent mechanics in a very compact and elegant way by the set of first-order differential equations.

The Hamiltonian or Lagrangian formalism applies to systems with gradient forces, that is, those which are given by derivatives of potentials. This assumption is true when considering gravitational, electromagnetic, and other fundamental forces. However, frictional forces often included phenomenologically in the mechanical description of dynamic systems cannot be given as gradients of some friction potential. Thus, the Hamiltonian description cannot describe friction phenomena. The microscopic relaxation theory and openness of the dynamic system are required to obtain a theory with the relaxation phenomena.

2.1.3
Classical Harmonic Oscillator

Let us consider a one-dimensional case describing the movement of a particle along coordinate x. Correspondingly the potential is defined as $V(x)$. The force acting on the particle is then $F(x) = -\text{grad}\,V = -\partial/\partial x\, V(x)$, and according to Newton's laws we can write the equation of motion as

$$m\ddot{x} = -\frac{\partial}{\partial x} V(x) . \tag{2.24}$$

In the Lagrange formulation we can define the Lagrangian as the difference of kinetic and potential energies, getting for a particle with mass m

$$L = m\frac{\dot{x}^2}{2} - V(x) . \tag{2.25}$$

From (2.12) it follows that $\partial/\partial \dot{x}\, L = m\dot{x}$, $\partial/\partial x\, L = -\partial/\partial x\, V(x)$, and thus

$$\frac{\mathrm{d}}{\mathrm{d}t}(m\dot{x}) + \frac{\partial}{\partial x} V(x) = 0 , \tag{2.26}$$

which is equivalent to the Newton's equation as demonstrated in the previous sections.

Similarly, we can write the Hamiltonian

$$H = \frac{p^2}{2m} + V(x) , \tag{2.27}$$

where the momentum $p = m\dot{x}$. In this case the Hamilton equations of motion read

$$\dot{p} = -\frac{\partial}{\partial x} V(x) , \tag{2.28}$$

$$\dot{x} = \frac{p}{m} . \tag{2.29}$$

Again we get the same set of equations of motion, which means that the dynamics is equivalent whatever type of description is chosen. However, the Hamiltonian formulation gives one clue about the number of independent variables. In this case we obtain two equations for variables x and p, the coordinate and the momentum, respectively. Thus, in the context of dynamic equations, it is a two-dimensional system (two-dimensional phase space).

We can easily solve the equations of motion when the potential surface has a parabolic form as shown in Figure 2.2. In this case the dynamics corresponds to the time evolution of the harmonic oscillator with the potential defined by $V(x) = kx^2/2$. Then the equation of motion is

$$m\ddot{x} + kx = 0 , \tag{2.30}$$

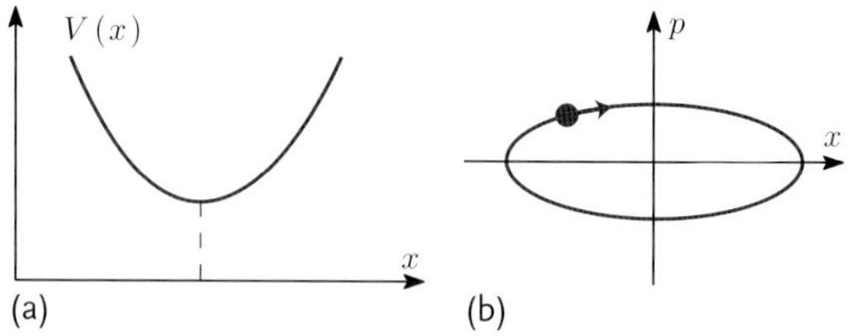

Figure 2.2 Parabolic potential of the harmonic oscillator (a), and the two-dimensional phase space of the oscillator (b). The trajectory is the ellipse or the circle.

and the solution is given by

$$x(t) = A\cos(\omega t) + B\sin(\omega t) , \tag{2.31}$$

which yields

$$\omega^2 = k/m . \tag{2.32}$$

Let us take the initial condition $x(0) = x_0$, $\dot{x}(0) = \dot{x}_0$. We then get $A = x_0$ and $B = \dot{x}_0/\omega$. The final solution is then

$$x(t) = x_0\cos(\omega t) + \frac{\dot{x}_0}{\omega}\sin(\omega t) . \tag{2.33}$$

We thus find that the frequency of the oscillator is described by the stiffness of the force parameter k and the mass of the particle m. Keeping this in mind, we can write the potential energy as

$$V(x) = m\omega^2\frac{x^2}{2} . \tag{2.34}$$

The oscillator equation can be given in somewhat more convenient form by introducing dimensionless parameters. Let us take Hamiltonian (2.27) and denote $m\omega^2 l^2 = \alpha\omega$, where l is some typical length of the oscillation and α is a constant. Denoting $y = x/l$ and $z = p/(m\omega l)$ or $z = \dot{y}/\omega$, we get the Hamiltonian in a symmetric form where the coordinate and the momentum are dimensionless:

$$H = \frac{1}{2}\alpha\omega(y^2 + z^2) . \tag{2.35}$$

Later we will find that this form of the Hamiltonian is equivalent to the Hamiltonian of the quantum harmonic oscillator and the constant α is associated with the reduced Planck constant.

The solution of the dynamic equations can now be written as

$$y(t) = \text{Re}\left(\frac{x_0}{l} + \frac{\dot{x}_0}{\text{i}l\omega}\right)\text{e}^{\text{i}\omega t} , \tag{2.36}$$

$$z(t) = -\mathrm{Im}\left(\frac{x_0}{2l} + \frac{\dot{x}_0}{2\mathrm{i}l\omega}\right)\mathrm{e}^{\mathrm{i}\omega t}\,, \tag{2.37}$$

which shows that the phase space defined by the y and z axes corresponds to the complex plane and a point $x_0/l + \dot{x}_0/(\mathrm{i}l\omega)$ in this space draws a circle. In the following we often face the application of classical or quantum oscillators. The latter is described in Section 4.6.1.

2.2 Classical Electrodynamics

For our introduction to classical electrodynamics, the microscopic Maxwell–Lorentz equations provide a convenient starting point. They enable us to view matter as an ensemble of charged particles, as opposed to the continuum view of macroscopic electrodynamics. The microscopic electric and magnetic fields are usually denoted by $\boldsymbol{E}$ and $\boldsymbol{B}$, respectively. Let us assume that there are particles with charges q_i located at points $\boldsymbol{r}_i$ in space. The density of charge and the density of current can be then defined as

$$\varrho(\boldsymbol{r}) = \sum_i q_i \delta(\boldsymbol{r} - \boldsymbol{r}_i)\,, \quad \boldsymbol{j}(\boldsymbol{r}) = \sum_i q_i \dot{\boldsymbol{r}}_i \delta(\boldsymbol{r} - \boldsymbol{r}_i)\,. \tag{2.38}$$

The Maxwell–Lorentz equations for the fields in a vacuum read [3, 5]

$$\nabla \cdot \boldsymbol{E} = \frac{\varrho(\boldsymbol{r})}{\epsilon_0}\,, \tag{2.39}$$

$$\nabla \cdot \boldsymbol{B} = 0\,, \tag{2.40}$$

$$\nabla \times \boldsymbol{E} = -\frac{\partial}{\partial t}\boldsymbol{B}\,, \tag{2.41}$$

$$\nabla \times \boldsymbol{B} = \frac{1}{c^2}\frac{\partial}{\partial t}\boldsymbol{E} + \mu_0 \frac{\partial}{\partial t}\boldsymbol{j}\,. \tag{2.42}$$

We introduced the usual constants – vacuum permittivity ϵ_0, magnetic permeability μ_0, and the speed of light in a vacuum c, which are all related through $c = 1/\sqrt{\epsilon_0 \mu_0}$. $\nabla\cdot$ denotes divergence, and $\nabla\times$ is the curl operator as described in Appendix A.1.

The same equations are valid for the microscopic and macroscopic cases. The difference is only in the charge and current densities, which in the macroscopic case are assumed to be continuous functions of space, while in the microscopic case the charge and current densities are given as a collection of microscopic points and their velocities.

2.2.1 Electromagnetic Potentials and the Coulomb Gauge

For the subsequent discussion, it is advantageous to introduce the *vector potential* $\boldsymbol{A}$ which determines the magnetic field through the following relation:

$$\boldsymbol{B} = \nabla \times \boldsymbol{A} . \tag{2.43}$$

The magnetic field given by such an expression automatically satisfies the second Maxwell–Lorentz equation, (2.40). Since for any scalar function χ we have the identity $\nabla \times (\nabla \chi) = 0$, the vector potential is defined up to the so-called *gauge function* χ, and the transformation

$$\boldsymbol{A} \to \boldsymbol{A} + \nabla \chi \tag{2.44}$$

does not change the magnetic field.

The same identity allows us to rewrite the third Maxwell–Lorentz equation, (2.41), in a more convenient form. Applying definition (2.43) to (2.41), we obtain

$$\nabla \times \left(\boldsymbol{E} + \frac{\partial}{\partial t} \boldsymbol{A} \right) = 0 , \tag{2.45}$$

which can be satisfied by postulating a *scalar potential* ϕ through

$$-\nabla \phi = \boldsymbol{E} + \frac{\partial}{\partial t} \boldsymbol{A} . \tag{2.46}$$

It is easy to see that if $\boldsymbol{A}$ is transformed by (2.44), the simultaneous transformation

$$\phi \to \phi - \frac{\partial}{\partial t} \chi \tag{2.47}$$

keeps (2.46) satisfied. The transformation composed of (2.44) and (2.47) is known as the *gauge transformation*, and the Maxwell–Lorentz equations are invariant with respect to this transformation. This phenomenon is denoted as gauge invariance.

The freedom in the choice of $\boldsymbol{A}$ and ϕ can be used to transform Maxwell–Lorentz equations into a form convenient for a particular physical situation. Here we will use the well-known *Coulomb gauge*, which is useful for separating the radiation part of the electromagnetic field from the part associated with charges. The Coulomb gauge is defined by the condition

$$\nabla \cdot \boldsymbol{A} = 0 , \tag{2.48}$$

which can always be satisfied [6].

2.2.2 Transverse and Longitudinal Fields

The Maxwell–Lorentz equations provide a complete description of the system of charges and electromagnetic fields, including their mutual interaction. In most of this book we will be interested in treating radiation as a system that interacts weakly with matter represented by charged particles. It would therefore be extremely useful to separate electromagnetic fields into those fields that are associated with the radiation, that is, those that can exist in free space without the presence of charges and currents, and those fields that are directly associated with their sources. Such a separation can be achieved by the so-called Helmholtz theorem [6]. This states that any vector field $\boldsymbol{a}$ can be decomposed into its transverse (divergence-free – denoted by $\perp$) and longitudinal (rotation-free – denoted by $\|$) parts. That is, any vector field $\boldsymbol{a}$ can be written as

$$\boldsymbol{a} = \boldsymbol{a}^{\perp} + \boldsymbol{a}^{\|} . \tag{2.49}$$

The transverse field is defined by

$$\nabla \cdot \boldsymbol{a}^{\perp} = 0 , \tag{2.50}$$

while the longitudinal component satisfies

$$\nabla \times \boldsymbol{a}^{\|} = 0 . \tag{2.51}$$

The magnetic field is purely transverse due to (2.40), and thus the decomposition of electric and magnetic fields reads

$$\boldsymbol{E} = \boldsymbol{E}^{\perp} + \boldsymbol{E}^{\|} , \quad \boldsymbol{B} \equiv \boldsymbol{B}^{\perp} . \tag{2.52}$$

The Maxwell–Lorentz equations for the transverse and longitudinal fields can then be given separately:

$$\nabla \cdot \boldsymbol{E}^{\|} = \frac{\varrho}{\epsilon_0} , \tag{2.53}$$

$$\nabla \times \boldsymbol{E}^{\perp} = -\frac{\partial}{\partial t}\boldsymbol{B} , \tag{2.54}$$

$$\nabla \times \boldsymbol{B} = \frac{1}{c^2}\frac{\partial}{\partial t}\boldsymbol{E}^{\perp} + \mu_0 \boldsymbol{j}^{\perp} , \tag{2.55}$$

and

$$0 = \frac{1}{c^2}\frac{\partial}{\partial t}\boldsymbol{E}^{\|} + \mu_0 \boldsymbol{j}^{\|} . \tag{2.56}$$

The last of these equations can be converted into the well-known *continuity equation* by applying ∇ and using (2.53):

$$\frac{\partial}{\partial t}\varrho = -\nabla \cdot \boldsymbol{j}^{\|} . \tag{2.57}$$

This means that the longitudinal current density is related to the change of the charge density and the charge is conserved in the absence of currents.

From (2.39) and (2.46) we can derive the Poisson equation which relates the scalar potential and the charge density,

$$\nabla^2 \phi(\boldsymbol{r}) = -\frac{\varrho(\boldsymbol{r})}{\epsilon_0}, \tag{2.58}$$

and so in the Coulomb gauge the scalar potential is given by the instantaneous charge distribution. Equation (2.46), which relates the scalar potential to vectors $\boldsymbol{A}$ and $\boldsymbol{E}$, can also be decomposed into transverse and longitudinal parts, yielding

$$\boldsymbol{E}^{\parallel} = -\nabla \phi \tag{2.59}$$

and

$$\boldsymbol{E}^{\perp} = -\frac{\partial}{\partial t}\boldsymbol{A}. \tag{2.60}$$

Equations (2.59) and (2.60) therefore decompose the electric field into the $\boldsymbol{E}^{\parallel}$ part generated by the charge distributions through the scalar potential and the $\boldsymbol{E}^{\perp}$ part associated with the vector potential $\boldsymbol{A}$.

The vector potential, and therefore also the transverse part of the electric field, can exist without charges. This part then naturally represents the radiation part of the electromagnetic field and it is necessarily related to the magnetic field. The other part is all due to charges: the charges create the scalar potential, which generated the longitudinal electric field. Equations (2.54) and (2.55) lead to

$$-\nabla \times (\nabla \times \boldsymbol{A}) - \frac{1}{c^2}\frac{\partial^2}{\partial t^2}\boldsymbol{A} = \left(\nabla^2 - \frac{1}{c^2}\frac{\partial^2}{\partial t^2}\right)\boldsymbol{A} = -\mu_0 \boldsymbol{j}^{\perp}, \tag{2.61}$$

where we used the Coulomb gauge condition, (2.48), and the vector identity, (2.60). The term on the right-hand side is a natural source of the light–matter interaction.

2.3 Radiation in Free Space

In this section we will show that the relationships of electrodynamics also yield to the Lagrangian and Hamiltonian formalisms discussed in Section 2.1. For this purpose we have to identify proper conjugate momenta for the selected "coordinate" variables of the field. For now, we will consider the radiation in a space free of charges.

2.3.1 Lagrangian and Hamiltonian of the Free Radiation

We now consider the case where the charge density $\varrho(\boldsymbol{r})$ is zero and thus in the Coulomb gauge the scalar potential $\phi(\boldsymbol{r})$ is taken to be zero as well. All electric and magnetic fields are then necessarily given by the vector potential $\boldsymbol{A}$. We can therefore choose $\boldsymbol{A}$ as a suitable "coordinate" for the description of the radiation. The equation of motion for the vector potential is given by

$$\frac{1}{\mu_0}\nabla \times (\nabla \times \boldsymbol{A}) + \epsilon_0 \frac{\partial^2}{\partial t^2}\boldsymbol{A} = 0 \,, \tag{2.62}$$

which follows from (2.61) in the case when the current is zero. We multiplied (2.61) by $1/\mu_0$ for later convenience.[1)]

Let us take the Cartesian coordinate system. Equation (2.62) can be understood as the equation of motion for the vector potential. We express the equation in components A_x, A_y, and A_z and multiply the components by their variations (a dot product) to obtain

$$\sum_i \left[\frac{1}{\mu_0} \sum_{jklm} \varepsilon_{ijk} \frac{\partial}{\partial x_j} \left(\varepsilon_{klm} \frac{\partial}{\partial x_l} A_m \right) + \epsilon_0 \frac{\partial^2}{\partial t^2} A_i \right] \delta A_i = 0 \,. \tag{2.63}$$

Here we used the Levi-Civita symbol ε_{ijk} to express the cross product $\boldsymbol{a} \times \boldsymbol{b}$ (see Appendix A.1). In order to convert the expression on the left-hand side into a variation of a functional, we have to integrate it not only over time (as we did in the case of classical mechanics), but also over space. We will use the same trick to treat the double spatial derivative as in the case of the time derivatives – we will integrate it by parts. We also assume that the variations are zero at times t_1 and t_2 (the limits of the time integration) and at the limits of the spatial integration. Under the spatial integration, the first term on the left-hand side of (2.63) yields

$$\begin{aligned} &\int \mathrm{d}^3\boldsymbol{r} \sum_i \sum_{jklm} \varepsilon_{ijk} \frac{\partial}{\partial x_j} \left(\varepsilon_{klm} \frac{\partial}{\partial x_l} A_m \right) \delta A_i \\ &= - \int \mathrm{d}^3\boldsymbol{r} \sum_i \sum_{jklm} \varepsilon_{ijk} \varepsilon_{klm} \left(\frac{\partial}{\partial x_l} A_m \right) \delta \left(\frac{\partial}{\partial x_j} A_i \right) \\ &= \int \mathrm{d}\boldsymbol{r} \sum_k (\nabla \times \boldsymbol{A})_k \, \delta \, (\nabla \times \boldsymbol{A})_k \,, \end{aligned} \tag{2.64}$$

where we used the properties of the Levi-Civita symbol, $\varepsilon_{ijk} = \varepsilon_{kij} = -\varepsilon_{kji}$. The second term in (2.62) is handled in the same way as in classical mechanics, and the resulting variational problem reads

$$\delta \int \mathrm{d}\boldsymbol{r} \int_{t_1}^{t_2} \mathrm{d}t \frac{\epsilon_0}{2} \left[c^2 (\nabla \times \boldsymbol{A})^2 - \left(\frac{\partial}{\partial t} \boldsymbol{A} \right)^2 \right] = 0 \,. \tag{2.65}$$

Consequently, the *Lagrangian density* of the radiation field defined as

$$\mathcal{L}_{\mathrm{rad}} = \frac{\epsilon_0}{2} \left[\dot{\boldsymbol{A}}^2 - c^2 (\nabla \times \boldsymbol{A})^2 \right] \tag{2.66}$$

1) We divided (2.61) by μ_0 in order to obtain the Hamiltonian corresponding to the energy density. We could derive the Lagrangian function and the Hamiltonian function without this step and multiply them by a suitable constant at the end.

leads to correct equations of motion, which can be verified by inserting them into the Lagrange equations, (2.12) [6].

The momentum $\boldsymbol{p}$ conjugate to $\boldsymbol{A}$ is given as

$$\boldsymbol{\Pi}(\boldsymbol{r}) = \frac{\partial \mathcal{L}}{\partial \dot{\boldsymbol{A}}} = \epsilon_0 \dot{\boldsymbol{A}} \,, \tag{2.67}$$

and the *Hamiltonian density* $\mathcal{H}$ (given by $\mathcal{H} = \boldsymbol{\Pi} \cdot \boldsymbol{A} - \mathcal{L}$) of the radiation field is

$$\mathcal{H}_{\text{rad}} = \frac{\epsilon_0}{2} \left[\frac{\boldsymbol{\Pi}^2}{\epsilon_0^2} + c^2 (\nabla \times \boldsymbol{A})^2 \right] . \tag{2.68}$$

Using (2.43) and (2.60), we can recast this result in a more familiar form,

$$H_{\text{rad}} = \frac{\epsilon_0}{2} \int \mathrm{d}^3 r \left[(\boldsymbol{E}^{\perp})^2 + c^2 \boldsymbol{B}^2 \right] , \tag{2.69}$$

with transverse electric and magnetic fields.

We find that the Hamiltonian of the electromagnetic field has a quadratic form reminiscent of a harmonic oscillator described in Section 2.1.3. Note that in the theory of electromagnetic fields we need to distinguish the Lagrangian and Hamiltonian from their densities. The latter are denoted by calligraphic letters $\mathcal{L}$ and $\mathcal{H}$, respectively. We use this distinction throughout this chapter.

2.3.2
Modes of the Electromagnetic Field

It is very useful to introduce the notion of field *modes*. With this concept we will be able to show that the free radiation is formally similar to an ensemble of harmonic oscillators. This idea will be very useful when we turn to field quantization.

A natural way of attacking the solution of (2.62) would be to apply the Fourier transform in time and space to it, that is, to expand the vector potential in terms of exponential functions $\mathrm{e}^{-\mathrm{i}\omega t + \mathrm{i}\boldsymbol{k}\cdot\boldsymbol{r}}$. These exponential functions solve (2.62) if $\omega = ck$, $k = |\boldsymbol{k}|$; ω is now the carrier frequency and $\boldsymbol{k}$ is the wave vector. The expansion of the (real) vector potential can be written as Fourier series

$$\boldsymbol{A}(\boldsymbol{r}, t) = \sum_{\boldsymbol{k}} \left(\boldsymbol{A}_{\boldsymbol{k}} \mathrm{e}^{-\mathrm{i}\omega_{\boldsymbol{k}} t + \mathrm{i}\boldsymbol{k}\cdot\boldsymbol{r}} + \boldsymbol{A}_{\boldsymbol{k}}^{*} \mathrm{e}^{\mathrm{i}\omega_{\boldsymbol{k}} t - \mathrm{i}\boldsymbol{k}\cdot\boldsymbol{r}} \right) . \tag{2.70}$$

The Coulomb gauge requires that for each $\boldsymbol{k}$

$$\nabla \cdot \boldsymbol{A}_{\boldsymbol{k}} = \mathrm{i}\mathrm{e}^{\mathrm{i}\boldsymbol{k}\cdot\boldsymbol{r}} \boldsymbol{k} \cdot \boldsymbol{A}_{\boldsymbol{k}} = 0 \,, \tag{2.71}$$

and consequently $\boldsymbol{A}_{\boldsymbol{k}}$ is a vector in a plane perpendicular to $\boldsymbol{k}$. As $\boldsymbol{k}$ is essentially the propagation direction of the field modes, the vector potential is perpendicular to the propagation direction. Defining unit vectors $\boldsymbol{e}_{1\boldsymbol{k}}$ and $\boldsymbol{e}_{2\boldsymbol{k}}$, which are perpendicular to each other and to $\boldsymbol{k}$, we can write

$$\boldsymbol{A}_{\boldsymbol{k}} = \sum_{\lambda=1,2} \boldsymbol{e}_{\lambda\boldsymbol{k}} A_{\lambda\boldsymbol{k}} \,, \tag{2.72}$$

where $A_{\lambda k} = \boldsymbol{e}_{\lambda k} \cdot \boldsymbol{A}_k$. It is possible to integrate the Hamiltonian density resulting from the discrete sum, (2.70), only when the field is limited to a finite integration volume Ω. We imagine the field to be enclosed in a large cubic box with side of length L, that is, the integration volume is $\Omega = L^3$. We assume a perfectly reflective box, so only the field modes for which $e^{i\boldsymbol{k}\cdot\boldsymbol{r}} = 0$ at the borders of the box are allowed. Such modes can only have components

$$k_1 = \frac{\pi}{L} n_1 \,, \quad k_2 = \frac{\pi}{L} n_2 \,, \quad k_3 = \frac{\pi}{L} n_3 \,, \quad n_{1,2,3} = 1, 2, \ldots \tag{2.73}$$

The exponential factors with components, (2.73), form a Kronecker delta under the integration over space,

$$\frac{1}{\Omega} \int d^3 \boldsymbol{r} e^{i(\boldsymbol{k}-\boldsymbol{k}')\cdot \boldsymbol{r}} = \delta_{\boldsymbol{k}\boldsymbol{k}'} = \begin{cases} 1 \,, & \text{for} \quad \boldsymbol{k} = \boldsymbol{k}' \\ 0 \,, & \text{for} \quad \boldsymbol{k} \neq \boldsymbol{k}' \end{cases} \,, \tag{2.74}$$

as one can verify by direct integration. The Hamiltonian of the radiation, (2.69), then reads

$$H_{\text{rad}} = \int\limits_{\Omega} d\boldsymbol{r} \mathcal{H}_{\text{rad}} = 2\Omega \epsilon_0 \sum_{\lambda \boldsymbol{k}} \omega_k^2 A_{\lambda k}^* A_{\lambda k} \,. \tag{2.75}$$

If we define two real variables,

$$q_{\lambda k} = \sqrt{\Omega \epsilon_0} \left(A_{\lambda k} + A_{\lambda k}^* \right) \tag{2.76}$$

and

$$p_{\lambda k} = -i\omega_k \sqrt{\Omega \epsilon_0} \left(A_{\lambda k} - A_{\lambda k}^* \right) \,, \tag{2.77}$$

it turns into a notoriously well known form:

$$H_{\text{rad}} = \sum_{\lambda k} \frac{1}{2} \left(p_{\lambda k}^2 + \omega_k^2 q_{\lambda k}^2 \right) \,. \tag{2.78}$$

Equation (2.78) represents the radiation as an ensemble of independent harmonic oscillators of unit masses. This makes quantization of the radiation rather straightforward, and enables us to apply to the radiation all sorts of results derived originally for harmonic oscillators.

A very important class of such results are those concerning the statistical thermodynamics of radiation as will be described later in the book. Radiation is an omnipresent thermodynamic bath into which the energy is damped during radiative decays of excited states of molecules. It is therefore important to know how many oscillators (i.e., degrees of freedom) interact with a molecular transition at a given frequency ω. We are thus interested in the *density of modes* $n(\omega)$, which gives the number of modes (per unit volume) in the frequency interval $(\omega, \omega + d\omega)$ by $n(\omega)d\omega$. The easiest way to determine the number of modes is to consider the number of allowed vectors $\boldsymbol{k}$ which correspond to a given interval $(k, k + dk)$ in accord with the dispersion relationship $\omega = ck$ [7]. The endpoints of all vectors $\boldsymbol{k}$ with a given absolute value k form a sphere with surface $4\pi k^2$, and the corresponding interval $(k, k + \mathrm{d}k)$ forms a spherical layer with a volume of $4\pi k^2 \mathrm{d}k$. Equation (2.73) tells us that a cubic volume of π^3/L^3 corresponds to each endpoint of an allowed $\boldsymbol{k}$. In the spherical layer with radius k there are therefore $4k^2 L^3 \mathrm{d}k/\pi^2$ endpoints. Since according to (2.73) all components of $\boldsymbol{k}$ are positive, only 1/8 of the volume is relevant. Finally, for each vector $\boldsymbol{k}$ we can have two orthogonal polarization vectors, and thus the number of endpoints we counted so far has to be multiplied by 2. Dividing by L^3, we obtain the desired density as a function of k and using $\omega = ck$ also as a function of ω:

$$n(k)\mathrm{d}k = \frac{k^2}{\pi^2}\mathrm{d}k\ , \quad n(\omega)\mathrm{d}\omega = \frac{\omega^2}{c^3\pi^2}\mathrm{d}\omega\ . \tag{2.79}$$

This result finds use, for example, in the description of spontaneous emission in Section 4.8, where the radiation forms a bath or environment for an excited emitter.

2.4 Light–Matter Interaction

In the previous section we described the radiation field free of any matter. However, the full description of the system of fields and charged particles by (2.61) contains some matter properties on the right-hand side. We write Lagrangian L_{mat} to describe Newton's laws for the particles and Lagrangian L_{rad} to describe the free radiation. The transverse current $\boldsymbol{j}^{\perp}$ influences the vector potential $\boldsymbol{A}$, and at the same time it depends on $\dot{\boldsymbol{r}}$. It will therefore play a role in the mechanical part of the equations of motion. This opens a way to define the Hamiltonian that will describe the light–matter interaction.

2.4.1 Interaction Lagrangian and Correct Canonical Momentum

In order to find the Hamiltonian formulation corresponding to (2.61) we will be looking for a light–matter interaction Lagrangian,

$$L_{\mathrm{int}} = \int \mathrm{d}\boldsymbol{r}\mathcal{L}_{\mathrm{int}}\ , \tag{2.80}$$

which produces the desired right-hand side term in the equations of motion. The free space Lagrangian density of the radiation field does not depend explicitly on $\boldsymbol{A}$ (it only depends on $\dot{\boldsymbol{A}}$ and $\nabla \boldsymbol{A}$). The term $\partial\mathcal{L}/\partial\boldsymbol{A}$ in the Lagrange equation, (2.12), is therefore equal to zero. This term can be used to obtain the right-hand side of (2.61). Defining the interaction Lagrangian density by

$$\mathcal{L}_{\mathrm{int}} = \boldsymbol{j}^{\perp} \cdot \boldsymbol{A} \tag{2.81}$$

correctly leads to (2.61).

The current $j = j^{\perp} + j^{\parallel}$ explicitly contains $\dot{r}$ (see the definition given by (2.38)), and its presence in the total Lagrangian L therefore complicates the definition of the conjugate momentum:

$$p = \frac{\partial \mathcal{L}}{\partial \dot{r}} \, . \tag{2.82}$$

The Lagrangian of the isolated matter leads to the purely kinetic conjugate momentum $p = m\dot{r}$, and (2.82) gives

$$p = m\dot{r} + \frac{\partial}{\partial \dot{r}} \int \mathrm{d}^3 r \; j^{\perp} \cdot A \, . \tag{2.83}$$

To evaluate (2.83) we have to identify the transverse part of j which gives zero under application of ∇. Using the definition given by (2.82), we can write

$$\begin{aligned} \nabla \cdot j &= \int \frac{\mathrm{d}k}{(2\pi)^3} \sum_{i\alpha} q_i \dot{x}_{i\alpha} \frac{\partial}{\partial x_\alpha} \mathrm{e}^{\mathrm{i}k\cdot(r-r_i)} \\ &= \mathrm{i} \sum_i q_i \int \frac{\mathrm{d}k}{(2\pi)^3} \dot{r}_i \cdot k \mathrm{e}^{\mathrm{i}k\cdot(r-r_i)} \, , \end{aligned} \tag{2.84}$$

where in the first line of (2.84) we use the component of vectors $\dot{r}_i = (\dot{x}_{i1}, \dot{x}_{i2}, \dot{x}_{i3})$. It is now clear that $j^{\perp}$ is formed by the components of $\dot{r}_i$ which are perpendicular to k. The following decomposition of the velocity vector

$$\dot{r}_i = (\dot{r}_i \cdot n) n + \sum_\lambda (\dot{r}_i \cdot e_{\lambda k}) e_{\lambda k} \, , \qquad n = \frac{k}{|k|} \, , \tag{2.85}$$

can be used to identify the longitudinal and transverse parts of j. The decomposition can be written in a tensor manner using the components of the unit vector n as follows:

$$\dot{x}_{i\alpha} = \sum_\beta \delta_{\alpha\beta} \dot{x}_{i\beta} = \sum_\beta [n_\alpha n_\beta + (\delta_{\alpha\beta} - n_\alpha n_\beta)] \dot{x}_{i\beta} \, . \tag{2.86}$$

The components of the transverse part of j are thus defined as

$$j_\alpha^{\perp} = \int \frac{\mathrm{d}k}{(2\pi)^3} \sum_{i\beta} q_i \dot{x}_{i\beta} (\delta_{\alpha\beta} - n_\alpha n_\beta) \mathrm{e}^{\mathrm{i}k\cdot(r-r_i)} = \sum_{i\beta} q_i \dot{x}_{i\beta} \delta_{\alpha\beta}^{\perp}(r - r_i) \, , \tag{2.87}$$

where we defined the decomposition of a unity tensor,

$$\delta_{\alpha\beta} \delta(r) = \delta_{\alpha\beta}^{\perp}(r) + \delta_{\alpha\beta}^{\parallel}(r) \, , \tag{2.88}$$

by two tensors:

$$\delta_{\alpha\beta}^{\perp}(r) = \int \frac{\mathrm{d}k}{(2\pi)^3} (\delta_{\alpha\beta} - n_\alpha n_\beta) \mathrm{e}^{\mathrm{i}k\cdot r} \tag{2.89}$$

and

$$\delta^{\parallel}_{\alpha\beta}(\boldsymbol{r}) = \int \frac{\mathrm{d}\boldsymbol{k}}{(2\pi)^3} n_\alpha n_\beta \mathrm{e}^{\mathrm{i}\boldsymbol{k}\cdot\boldsymbol{r}} . \tag{2.90}$$

Using $\delta^{\perp}_{\alpha\beta}(\boldsymbol{r})$ and $\delta^{\parallel}_{\alpha\beta}(\boldsymbol{r})$, we can obtain the longitudinal and transverse parts of any given vector field $\boldsymbol{a}$ as

$$a^{\parallel}_{\alpha}(\boldsymbol{r}) = \int \mathrm{d}^3\boldsymbol{r} a_\beta \delta^{\parallel}_{\alpha\beta}(\boldsymbol{r}) , \quad a^{\perp}_{\alpha}(\boldsymbol{r}) = \int \mathrm{d}^3\boldsymbol{r} a_\beta \delta^{\perp}_{\alpha\beta}(\boldsymbol{r}) . \tag{2.91}$$

Now we can finally evaluate the new conjugate momentum, (2.83). For its components we obtain

$$p_{i\alpha} = m\dot{x}_{i\alpha} + q_i \sum_\beta \int \mathrm{d}^3\boldsymbol{r} \delta^{\perp}_{\alpha\beta}(\boldsymbol{r}-\boldsymbol{r}_i) A_\beta(\boldsymbol{r}) = m\dot{x}_{i\alpha} + q_i A_\alpha(\boldsymbol{r}_i) . \tag{2.92}$$

Here the fact that $\boldsymbol{A}$ is completely transverse is taken into account. As a result of incorporation of the interaction Lagrangian, the conjugate momentum of the particles becomes dependent on the vector potential $\boldsymbol{A}$.

As we can see, the momentum of a particle is directly affected by the vector potential. We should remind the reader once again that this expression is meaningful only for the Coulomb gauge, and is not applicable to a general gauge. The vector potential is fully defined only in a specific gauge.

2.4.2 Hamiltonian of the Interacting Particle–Field System

In the previous subsections we defined the canonical variables, the Hamiltonian and the Lagrangian of the material system and the radiation field. We also determined the interaction Lagrangian. This allows us now to derive the full Hamiltonian of the interacting material system plus radiation. Combining (2.20), (2.66), (2.67), and (2.81), we can write the Hamiltonian of the interacting system as

$$H = \sum_i \boldsymbol{p}_i \cdot \dot{\boldsymbol{r}}_i + \int \mathrm{d}^3\boldsymbol{r} \boldsymbol{\Pi} \cdot \dot{\boldsymbol{A}} - L_{\text{tot}} , \tag{2.93}$$

where L_{tot} is the total Lagrangian of the material system, the radiation and their interaction, and $\boldsymbol{\Pi}$ is the momentum conjugate to $\boldsymbol{A}$. This leads to

$$\begin{aligned} H = &\sum_i \frac{1}{2m_i}(\boldsymbol{p}_i - q_i\boldsymbol{A}(\boldsymbol{r}_i))^2 + V(\boldsymbol{r}_1,\dots,\boldsymbol{r}_N) \\ &+ \frac{1}{2}\int \mathrm{d}^3\boldsymbol{r}\left[\frac{\Pi^2}{\epsilon_0} + \epsilon_0 c^2(\nabla\times\boldsymbol{A})\right] . \end{aligned} \tag{2.94}$$

Here we introduced the symbol $V(\boldsymbol{r}_1,\dots,\boldsymbol{r}_N)$ instead of ϕ for the electrostatic Coulomb potential, which is now equivalent to the scalar potential of the longitudinal field. Equation (2.94) represents the total classical Hamiltonian of an interacting system of charges and fields.

For our purposes we will group the particles into molecules or supramolecules (such as clusters or aggregates of molecules), and split the potential V into intermolecular and intramolecular parts:

$$V(\xi_1, \ldots, \xi_{N_{\text{mol}}}) = \sum_i V(\xi_i) + \sum_{i<j} V(\xi_i, \xi_j) , \tag{2.95}$$

where ξ_n denotes the particles forming the nth molecule (or supramolecules). This splitting is essential. Some interactions, for example, those inside the molecules or their aggregates, will be treated explicitly (by quantum chemistry, an excitonic model, or a similar theory), and some, for example those occuring between the aggregates or the molecules, can be included in the description of the light–matter interaction.

Our aim is now to write the Hamiltonian in a form suitable for studying the interaction of molecules with light. First, we split the Hamiltonian into three terms, where the first term describes the pure material system, the second term describes the radiation field, and the third term contains the mixed terms:

$$H = H_{\text{mol}} + H_{\text{rad}} + H_{\text{int}} . \tag{2.96}$$

Hamiltonian H_{mol} of the molecules should include only the longitudinal fields, that is,

$$H_{\text{mol}} = \sum_i H_{\text{mol}}(\xi_i) + \sum_{i<j} V(\xi_i, \xi_j) , \tag{2.97}$$

where

$$H_{\text{mol}}(\xi_i) = \sum_{j \in \xi_i} \frac{1}{2m_j} \boldsymbol{p}_j^2 + V(\xi_i) . \tag{2.98}$$

The radiation Hamiltonian is given by (2.68) and thus the rest of (2.94) composes the light–matter interaction Hamiltonian:

$$H_{\text{int}} = \sum_i \frac{q_i}{m_i} \boldsymbol{p}_i \cdot \boldsymbol{A}(\boldsymbol{r}_i) + \sum_i \frac{q_i^2}{2m_i} \boldsymbol{A}(\boldsymbol{r}_i)^2 . \tag{2.99}$$

Equation (2.99) is the so-called *minimal coupling Hamiltonian* or the $\boldsymbol{p}\boldsymbol{A}$ Hamiltonian, which represents a convenient starting point for the discussion of interaction of small molecules with light.

2.4.3
Dipole Approximation

The characteristic dimensions of molecular systems are usually much smaller than the wavelength of light. The radiation field can therefore be assumed to be homogeneous within the extent of the molecule or the molecular aggregate, and the vector potential $\boldsymbol{A}(\boldsymbol{r}_i)$ can be replaced by its value at a chosen reference (e.g., the mass or the charge center) point inside the molecule:

$$A(\boldsymbol{r}_i) \approx A(\boldsymbol{R}_\xi)\,, \quad i \in \xi\,. \tag{2.100}$$

Now we will use the fact that the equations of motion will remain the same if we add a total time derivative of a function of coordinates and time to the Lagrangian (see Section 2.1). Let us for simplicity assume that we have just two supramolecules or aggregates denoted by ξ_1 and ξ_2. We will add the following term:

$$L_{\text{add}} = -\frac{\mathrm{d}}{\mathrm{d}t}\int \mathrm{d}^3\boldsymbol{r}\,\boldsymbol{P}^\perp(\boldsymbol{r})\cdot \boldsymbol{A}(\boldsymbol{r})\,, \tag{2.101}$$

where

$$\begin{aligned} \boldsymbol{P}(\boldsymbol{r}) &= \sum_{n=1}^{2}\sum_{i\in\xi_n} q_i(\boldsymbol{r}_i - R_{\xi_n})\delta(\boldsymbol{r} - \boldsymbol{R}_{\xi_n}) \\ &= \boldsymbol{\mu}_{\xi_1}\delta(\boldsymbol{r} - \boldsymbol{R}_{\xi_1}) + \boldsymbol{\mu}_{\xi_2}\delta(\boldsymbol{r} - \boldsymbol{R}_{\xi_2}) \end{aligned} \tag{2.102}$$

is the polarization of the two molecules, and $\boldsymbol{\mu}_\xi$ is the dipole moment of molecule ξ. In the dipole approximation we can also write

$$\frac{\mathrm{d}}{\mathrm{d}t}\boldsymbol{P}^\perp(\boldsymbol{r}) = \boldsymbol{j}^\perp(\boldsymbol{r})\,, \tag{2.103}$$

and, therefore,

$$L_{\text{add}} = -\int \mathrm{d}^3\boldsymbol{r}\,\boldsymbol{j}^\perp\cdot \boldsymbol{A} - \int \mathrm{d}^3\boldsymbol{r}\;\boldsymbol{P}^\perp\cdot\dot{\boldsymbol{A}}\,. \tag{2.104}$$

Consequently, addition of (2.104) to the total Lagrangian replaces the term containing the product $\boldsymbol{j}^\perp\cdot\boldsymbol{A}$ by a term containing $\boldsymbol{p}^\perp\cdot\dot{\boldsymbol{A}}$. As a result, the conjugate momentum of the particle is again purely kinetic,

$$\boldsymbol{p}_i = m_i\dot{\boldsymbol{r}}_i\,, \quad i\in\xi_1,\xi_2\,, \tag{2.105}$$

and the momentum conjugate to the vector potential reads

$$\boldsymbol{\Pi} = -\epsilon_0\boldsymbol{E}^\perp - \boldsymbol{P}^\perp\,. \tag{2.106}$$

The Hamiltonian that results from the Lagrangian $L' = L + L_{\text{add}}$ reads

$$\begin{aligned} H = H_{\text{mol}} &+ \frac{1}{2}\int \mathrm{d}^3\boldsymbol{r}\left[\frac{\boldsymbol{\Pi}^2}{\epsilon_0} + \epsilon_0 c^2(\nabla\times\boldsymbol{A})^2\right] \\ &+ \frac{1}{\epsilon_0}\int \mathrm{d}^3\boldsymbol{r}\,\boldsymbol{P}^\perp\cdot\boldsymbol{\Pi} + \frac{1}{2\epsilon_0}\int \mathrm{d}^3\boldsymbol{r}|\boldsymbol{P}^\perp|^2\,. \end{aligned} \tag{2.107}$$

The transverse polarization consists of contributions of the two molecules, $\boldsymbol{P}^\perp = \boldsymbol{P}_1^\perp + \boldsymbol{P}_2^\perp$, and thus the last term can be divided into intramolecular and intermolecular parts:

$$\frac{1}{2\epsilon_0}\int \mathrm{d}^3\boldsymbol{r}|\boldsymbol{P}^\perp|^2 = \frac{1}{\epsilon_0}\int \mathrm{d}^3\boldsymbol{r}\,\boldsymbol{P}_1^\perp\cdot\boldsymbol{P}_2^\perp + \frac{1}{2\epsilon_0}\sum_{\xi=1,2}\int \mathrm{d}^3\boldsymbol{r}|\boldsymbol{P}_\xi^\perp|^2\,. \tag{2.108}$$

It can be shown [6] that in the dipole approximation the intermolecular part (the first term on the right-hand side) exactly cancels the interaction between the supramolecules $\sum_{i<j} V(\xi_i, \xi_j)$ provided $V(\xi_i, \xi_j)$ includes only dipole–dipole interaction as well. The Hamiltonian therefore contains only the noninteracting part of $H_{\text{mol}} = \sum_i H_{\text{mol}}(\xi_i)$. The second term on the right-hand side is an intramolecular contribution of the polarization which we will disregard with the provision that it does not play an important role in the radiative processes. The last step of our analysis of the Hamiltonian is the definition of a new field $\boldsymbol{D}(\boldsymbol{r})$, the so-called *displacement vector*, as

$$\boldsymbol{D}(\boldsymbol{r}) = \epsilon_0 \boldsymbol{E}(\boldsymbol{r}) + \boldsymbol{P}(\boldsymbol{r}) \,. \tag{2.109}$$

From (2.105), we can see that $\boldsymbol{\Pi} = -\boldsymbol{D}^{\perp}$. The total Hamiltonian can therefore be written as

$$H = \sum_i H_{\text{mol}}(\xi_i) + \frac{1}{2} \int \mathrm{d}^3 \boldsymbol{r} \left(\frac{\boldsymbol{D}^{\perp 2}}{\epsilon_0} + \epsilon_0 c^2 \boldsymbol{B}^2 \right) - \frac{1}{\epsilon_0} \sum_i \boldsymbol{\mu}_{\xi_i} \cdot \boldsymbol{D}^{\perp}(\boldsymbol{R}_{\xi_i}) \,. \tag{2.110}$$

Hamiltonian (2.110) is a possible starting point for studies of the light–matter interaction for nanoparticles and molecular aggregates. This interaction Hamiltonian (the last term) is in the convenient form of a product of the molecular dipole moment and the transverse field. In practical calculations, we can often assume that the polarization is linearly proportional to the electric field, that is,

$$\boldsymbol{P}(\boldsymbol{r}) = \epsilon_0 \chi \boldsymbol{E}(\boldsymbol{r}) \tag{2.111}$$

and

$$\boldsymbol{D}(\boldsymbol{r}) = \epsilon_0 (1 + \chi) \boldsymbol{E}(\boldsymbol{r}) = \epsilon_0 \epsilon_{\text{r}} \boldsymbol{E}(\boldsymbol{r}) \,, \tag{2.112}$$

where χ and ϵ_{r} are the linear susceptibility and the relative permittivity, respectively.

3
Stochastic Dynamics

Many dynamic processes can be characterized as so-called stochastic. This class of dynamic processes is widely used to describe the time evolution of open systems [8, 9]. In open systems, the degrees of freedom of the system under consideration constitute a small part of the total number of degrees of freedom of the system and its environment. If the environment coordinates are not followed explicitly, we can only observe the system dynamics affected by a large number of *unknown* forces. These forces may drive the system in an unpredictable way and we cannot use simple deterministic differential equations to describe the degrees of freedom of the system. To provide a convenient description of this situation, we can introduce the concept of stochasticity and characterize the stochastic evolution by probabilities that the system is in certain states, and the dynamics between these states is stochastic.

One such stochastic process is the celebrated Brownian motion of a microscopic bead in a liquid, first described by Brown in the 1830s. The bead in a liquid is pushed randomly by fluctuating molecules from the surroundings. As the particle interacts with many degrees of freedom of the liquid simultaneously, the net fluctuating force becomes Gaussian due to the central limit theorem. Not only classical Brownian motion falls under this category, but an arbitrary process in the system coupled to the fluctuating environment can also be described by a stochastic process of some sort. It will be demonstrated later that quantum mechanics describes some intermediate case between the deterministic and probabilistic nature of the dynamics where the wavefunction is defined by the deterministic equation, and the measurement process is purely probabilistic.

3.1
Probability and Random Processes

To introduce random processes we have to turn back to the probability theory. The elementary starting point is the concept of a set of random events. We define certain indivisible elementary events which compose the so-called *probability space*. The probability space can then be divided into various regions covering some groups of elementary events which are associated with observable random events. It is

Molecular Excitation Dynamics and Relaxation, First Edition. L. Valkunas, D. Abramavicius, and T. Mančal.

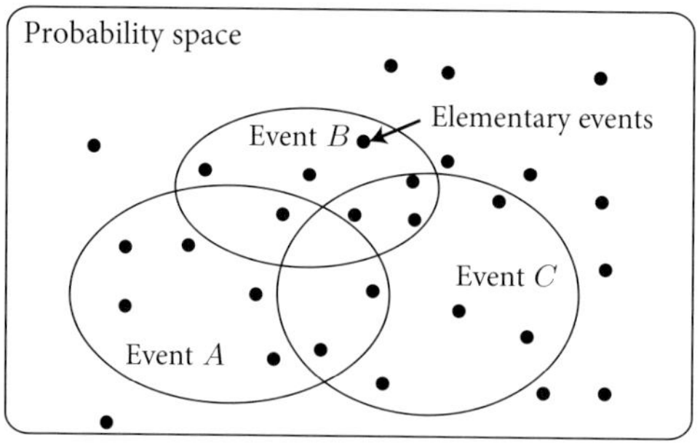

Figure 3.1 Probability space of random events: black dots represent the elementary events, and events A–C cover some elementary events. All elementary events create the probability space.

thus easy to think of a random event as a set of elementary events. This concept is presented graphically in Figure 3.1. Let us now introduce some concepts of set theory which are useful in the description of random events.

Consider two sets A and B. If set A is said to be a part of set B, A is a subset of B, and this dependence is denoted by

$$A \subset B \,. \tag{3.1}$$

Union of sets denoted by

$$C = A \cup B \tag{3.2}$$

creates a new set, the elements of which are given by elements of A and B. Thus, sets A and B become subsets of the resulting set C. Intersection of sets

$$D = A \cap B \tag{3.3}$$

creates set D, which is formed by the elements shared by A and B. For completeness we also introduce the concepts of an empty set $\emptyset$ and the full set Ω. The empty set has no elements and the full set has all elements of the probability space.

A set complement to set A contains all elements of the full space which are not present in A. We denote such a set by A^c. We next introduce the complementarity operation,

$$C = B \setminus A \,, \tag{3.4}$$

which removes all elements from B which are in A, and we can write

$$A^c = \Omega \setminus A \,. \tag{3.5}$$

By definition we thus have

$$A^c \cap A = \emptyset \tag{3.6}$$

and

$$A^c \cup A = \Omega \,. \tag{3.7}$$

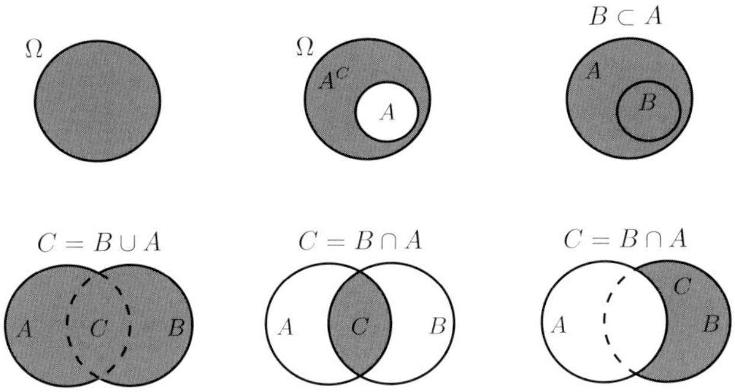

Figure 3.2 Main definitions and operations of sets: Ω denotes the set covering the full space, A is a "finite" event, and A^c is the complement set. The bottom row describes the union, intersection, and subtraction operations as described in the text.

Geometrically these operations are represented in Figure 3.2.

A random event is now understood as a realization of the elementary events belonging to a specific set: event A happens when one of its composing elementary events is realized. To quantify the event we introduce the *probability* of the event $P(A)$. Three axioms of Kolmogorov fully characterize the probability of events. The first axiom states that the probability is defined as a nonnegative number. The second axiom denotes that the full set is characterized by probability 1:

$$P(\Omega) = 1\,. \tag{3.8}$$

The empty set then has

$$P(\emptyset) = 0\,. \tag{3.9}$$

The third axiom states that the probability of the union of *nonintersecting* sets is given by

$$P(\cup_i A_i) = \sum_i P(A_i)\,. \tag{3.10}$$

It follows that a union of an arbitrary two sets has probability

$$P(A \cup B) = P(A) + P(B) - P(A \cap B)\,. \tag{3.11}$$

In practice for all other events (or sets), the probability is given as a limit of the ratio of the realizations of event A, which we denote by m_A, and the number of trials N; thus,

$$P(A) = \lim_{N\to\infty} \frac{m_A}{N}\,. \tag{3.12}$$

This relation is known as the *theorem of large numbers*. It represents the proper recipe to experimentally determine probabilities of various events, and it is behind the idea of the Monte Carlo simulation of stochastic processes. However, this requires a lot of trials, but some events should not be tested experimentally (e.g., the reliability of a nuclear power plant).

Conditional probability is one of the important concepts for discussion of physical processes. Consider an event A. It may happen that in some instances of realization of event A, event B happens at the same time. Such an event should be related to the overlap region of sets A and B. Let event B be the additional necessary condition that we want to include in describing event A. In this case the space of possible events is limited by set B, because realization of B is necessary. We denote this conditional probability as $P(A/B)$. The event that A and B happen at the same time is given by the set $A \cap B$ and the probability is thus proportional to $P(A \cap B)$. However, since event B is a necessary condition with its own probability $P(B)$, the proper normalization requires that the conditional probability satisfies

$$P(A/B) = \frac{P(A \cap B)}{P(B)} . \tag{3.13}$$

Alternatively, we can define the probability of the intersection as

$$P(A \cap B) = P(A/B)P(B) . \tag{3.14}$$

This allows us to define the independent events. If A is independent of B, the conditional probability that A happens with the condition of B is just the probability of A, that is,

$$P(A/B) = \frac{P(A \cap B)}{P(B)} \equiv P(A) . \tag{3.15}$$

This also means that for independent events we have

$$P(A \cap B) = P(A)P(B) . \tag{3.16}$$

Let us now consider the space of events Ω and disjoint events B_j which fill the whole space as shown in Figure 3.3. It holds that

$$\cup_j B_j = \Omega . \tag{3.17}$$

An arbitrary event A can then be given as

$$A = \cup_j (A \cap B_j) , \tag{3.18}$$

which for the probabilities gives

$$P(A) = \sum_j P(A \cap B_j) . \tag{3.19}$$

Now we use the properties of the conditional probability and obtain the following important relation:

$$P(A) = \sum_j P(A/B_j)P(B_j) . \tag{3.20}$$

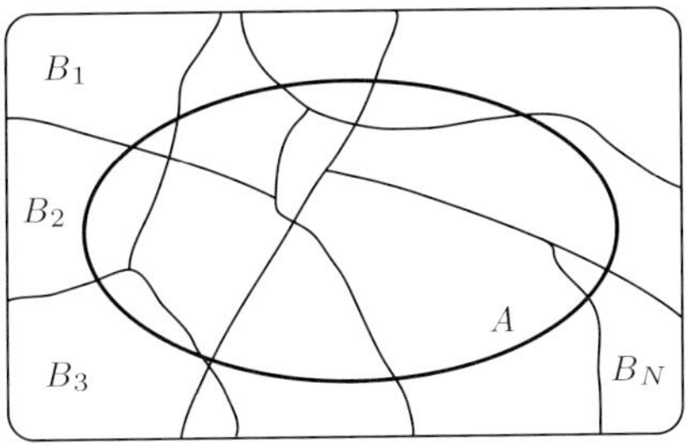

Figure 3.3 Sets B_j are nonintersecting and they span the whole probability space. Set A is given by union of intersections with whole B_j.

The interpretation of (3.20) is as follows. Let us consider the system at a specific time described by a set of states B_j with their probabilities $P(B_j)$. If at a later time the system arrives at state A with probability $P(A)$, the conditional probabilities $P(A/B_j)$ can be regarded as the probabilities of the *transitions* from states B_j to state A. The conditional probability thus becomes a central concept of the stochastic description of dynamics.

Having defined the set of events and the probabilities, we move to another important concept, namely, the one of *random variables*. A random variable is a specific representation or mapping of the elementary event onto real numbers. All elementary events ω are associated with real variables, which we denote by $x(\omega)$. If this mapping is unique, that is, different elementary events are mapped onto different numbers, the random variables can be assigned to probabilities of the corresponding elementary events.

Random variables can be characterized by various means. For instance, the nth moment is defined by the following average:

$$\langle x^n \rangle = \sum_{\omega} x^n(\omega) P(\omega) . \tag{3.21}$$

The most important random variables in the stochastic description are the low moments, the mean value $\langle x^1 \rangle$, and the dispersion

$$D(x) = \langle (x - \langle x \rangle)^2 \rangle = \langle x^2 \rangle - \langle x \rangle^2 . \tag{3.22}$$

The dispersion characterizes the width of the spread of the random numbers. Higher moments are less important in most cases with physical relevance.

When we introduce the time axis, and order the random variables in time, we obtain a set of numbers which describes a so-called *random process*. In this process we have random variable x_1 at time t_1, then x_2 at time t_2, and so on. It is advantageous sometimes to describe a so-called characteristic function of a random process, as described in Appendix A.2.

3.2 Markov Processes

The random process introduced in the previous section describes some time trajectory of variable x. Let us assume that at time t_1 it has value x_1, at time t_2 value x_2, up to value x_n at time t_n. We denote the probability of such a composite event as

$$P(x_n t_n, \ldots, x_2 t_2, x_1 t_1) . \tag{3.23}$$

This probability describes the trajectory as a whole. We may now ask what is the probability of this composite event if we know that before it we had a process $x'_1 t'_1, x'_2 t'_2, \ldots, x'_m t'_m$. As described in the previous section when we introduced conditional probabilities, such a composite event can be characterized by a conditional probability. If we denote the probability of the composite event as

$$P\left(x_n t_n, \ldots, x_2 t_2, x_1 t_1, x'_m t'_m, \ldots, x'_2 t'_2, x'_1 t'_1\right) , \tag{3.24}$$

we can write

$$\begin{aligned} &P\left(x_n t_n, \ldots, x_2 t_2, x_1 t_1 / x'_m t'_m, \ldots, x'_2 t'_2, x'_1 t'_1\right) \\ &= \frac{P\left(x_n t_n, \ldots, x_2 t_2, x_1 t_1, x'_m t'_m, \ldots, x'_2 t'_2, x'_1 t'_1\right)}{P\left(x'_m t'_m, \ldots, x'_2 t'_2, x'_1 t'_1\right)} . \end{aligned} \tag{3.25}$$

We thus imply a relation between all $n + m$ points in time. If we take $t_n > t_{n-1} > \cdots > t_1 > t'_m > \cdots > t'_1$, we imply a long "memory" for the process: the present state depends on a long chain of previous events.

The *Markov process* is a subset of the process described above with the assumption of a short memory. The Markov assumption is that the process or the present state depends only on a single previous time. This condition means that

$$\begin{aligned} &P\left(x_n t_n, \ldots, x_2 t_2, x_1 t_1 / x'_m t'_m, \ldots, x'_2 t'_2, x'_1 t'_1\right) \\ &\equiv P\left(x_n t_n, \ldots, x_2 t_2, x_1 t_1 / x'_m t'_m\right) . \end{aligned} \tag{3.26}$$

Consider now three different time moments. A process with three steps is characterized by the probability $P(x_3 t_3, x_2 t_2, x_1 t_1)$. According to (3.25), we can write

$$P(x_3 t_3, x_2 t_2, x_1 t_1) = P(x_3 t_3 / x_2 t_2, x_1 t_1) P(x_2 t_2, x_1 t_1) . \tag{3.27}$$

From the Markov assumption we have $P(x_3 t_3 / x_2 t_2, x_1 t_1) = P(x_3 t_3 / x_2 t_2)$, and we get

$$P(x_3 t_3, x_2 t_2, x_1 t_1) = P(x_3 t_3 / x_2 t_2) P(x_2 t_2 / x_1 t_1) P(x_1 t_1) . \tag{3.28}$$

Again, the conditional probabilities describe the probability of a transition from state x_1 to state x_3. Also according to (3.16), different transitions in the Markov process are independent. Generalization of (3.28) is straightforward:

$$P(x_n t_n, \ldots, x_2 t_2, x_1 t_1) = P(x_n t_n / x_{n-1} t_{n-1}) \ldots P(x_2 t_2 / x_1 t_1) P(x_1 t_1) . \tag{3.29}$$

The conditional transition probability $P(xt/x't')$ essentially represents a propagator of the process, and it has the following obvious properties. The probability of a transition to an arbitrary state of space is

$$\sum_x P(xt/x't') = 1\,, \tag{3.30}$$

also

$$\sum_{x'} P(xt/x't')P(x't') = P(xt)\,, \tag{3.31}$$

while for a transition at the vanishing time interval

$$\lim_{t\to t'} P(xt/x't') = \delta_{xx'}\,. \tag{3.32}$$

Let us now consider three states in space x again. For the transition from state x_1 to state x_3 we can write

$$P(x_3t_3, x_1t_1) = P(x_3t_3/x_1t_1)P(x_1t_1)\,. \tag{3.33}$$

Also note that

$$P(x_3t_3, x_1t_1) = \sum_{x_2} P(x_3t_3, x_2t_2, x_1t_1)\,, \tag{3.34}$$

and using additionally (3.28), we have

$$P(x_3t_3/x_1t_1) = \sum_{x_2} P(x_3t_3/x_2t_2)P(x_2t_2/x_1t_1)\,. \tag{3.35}$$

This is the so-called *Chapman–Kolmogorov equation* for conditional probabilities. Its interpretation is straightforward. For a system or a particle in state x_1 to reach state x_3, the system must propagate through all possible intermediate points of space, and the resulting probability is additive in all intermediate states. In this description the conditional probability corresponds to the system propagator. The propagator of the Chapman–Kolmogorov equation thus defines the Markov process. It is often not possible to solve the integral Chapman–Kolmogorov equation directly, and its differential form, which is considered next, is often a more convenient alternative.

3.3
Master Equation for Stochastic Processes

As we have already pointed out, the differential form of the Chapman–Kolmogorov equation is often preferential, since the solution of differential equations is easier. It should be noted, however, that differential equations do not cover the whole set of possible Markovian problems. They allow one to describe only continuous-time problems with differentiable propagators. Most of the physical problems considered in this book satisfy this condition, so we will not worry about undifferentiable processes.

Let us assume we have a time-continuous process and let the probability be differentiable with respect to time. In this case we can take the Taylor series for the probability

$$P(x, t + \varDelta) = P(x, t) + a_1\varDelta + a_2\varDelta^2 + \dots , \tag{3.36}$$

where a_j are some constants. We then define the derivative of the probability

$$\frac{\mathrm{d}}{\mathrm{d}t} P(xt) = \lim_{\varDelta \to 0} \frac{1}{\varDelta}[P(x, t + \varDelta) - P(x, t)] . \tag{3.37}$$

Note that when $\varDelta \to 0$, the probability $P(x, t + \varDelta)$ approaches $P(x, t)$ linearly with $\varDelta$, so the difference $P(x, t + \varDelta) - P(x, t) \approx a_1\varDelta$.

We now apply the probability expansion, (3.31), to the probability $P(x, t + \varDelta)$:

$$P(x, t + \varDelta) = \sum_{x_1} P(x, t + \varDelta/x_1 t_1) P(x_1 t_1) . \tag{3.38}$$

For the derivative we thus obtain

$$\frac{\mathrm{d}}{\mathrm{d}t} P(xt) = \lim_{\varDelta \to 0} \frac{1}{\varDelta} \left[\sum_{x_1} P(x, t + \varDelta/x_1 t_1) P(x_1 t_1) - \sum_{x_1} P(x, t/x_1 t_1) P(x_1 t_1) \right] . \tag{3.39}$$

By separating the "diagonal" terms $x_1 = x$, and taking $t_1 = t$, we have

$$\frac{\mathrm{d}}{\mathrm{d}t} P(xt) = \lim_{\varDelta \to 0} \frac{1}{\varDelta} \left[\sum_{x_1}^{x_1 \neq x} P(x, t + \varDelta/x_1 t) P(x_1 t) - (1 - P(x, t + \varDelta/xt)) P(x, t) \right] . \tag{3.40}$$

The conditional probability $P(x, t + \varDelta/x_1 t)$ for $x_1 \neq x$ and $\varDelta = 0$ is actually zero, that is, $P(x, t/x_1 t) = 0$. When $\varDelta \to 0$, the probability should behave at least as $P(x, t + \varDelta/x_1 t) \propto \varDelta$. Similarly, the conditional probability for $x = x_1$ behaves as $P(x, t + \varDelta/xt) \propto 1 - \varDelta$ (the minus sign denotes that the probability is always less than 1).

We can introduce the probability transfer rates:

$$k_{xx_1} = \lim_{\varDelta \to 0} \frac{1}{\varDelta} P(x, t + \varDelta/x_1 t) , \quad x \neq x_1 \tag{3.41}$$

and

$$k_{xx} = \lim_{\Delta \to 0} \frac{1}{\Delta}(1 - P(x, t + \Delta / x t)) . \tag{3.42}$$

The diagonal and off-diagonal rates k_{xy} and k_{xx} are both positive quantities, and we can thus write the differential system of equations:

$$\frac{\mathrm{d}}{\mathrm{d}t} P(x t) = \sum_{x_1}^{x_1 \neq x} k_{x x_1} P(x_1 t) - k_{xx} P(x, t) . \tag{3.43}$$

This equation is known as a *master equation* for probabilities. It can be directly obtained from the Chapman–Kolmogorov equation, (3.35), which means that the master equation, (3.43), is essentially the differential form of the Chapman–Kolmogorov equation.

The rates now have a very clear physical meaning: $k_{x x_1}$ denotes the rate of supply of the probability at state x, originating from state x_1, and k_{xx} denotes the rate of loss of the probability from state x. Note that if we sum up the probabilities over the whole space, we must obtain

$$\sum_x P(x t) = 1 , \tag{3.44}$$

which means that the probability of being found in an arbitrary state or at an arbitrary coordinate is 1. Applying this to the master equation, we have

$$\frac{\mathrm{d}}{\mathrm{d}t} \sum_x P(x t) = 0 \tag{3.45}$$

or

$$\sum_x \left[\sum_{x_1}^{x_1 \neq x} k_{x x_1} P(x_1 t) - k_{xx} P(x, t) \right] = 0 . \tag{3.46}$$

We can now rewrite this as

$$\sum_{x_1} \sum_x \left[(1 - \delta_{x x_1}) k_{x x_1} - \delta_{x_1 x} k_{xx} \right] P(x_1 t) = 0 , \tag{3.47}$$

and the result should be independent of state x_1. We thus get the following requirement for the rates

$$k_{x_1 x_1} = \sum_x^{x \neq x_1} k_{x x_1} , \tag{3.48}$$

which is called the *detailed balance condition*. It signifies that the total rate of the probability loss of a given state is equal to the sum of all possible escape channels from that state. This condition guarantees that the total probability is conserved at all times.

3.3.1 Two-Level System

As an example, let us consider a system with two possible states a and b. The rates of the transitions between these states are k_{ab} and k_{ba}; the system is sketched in Figure 3.4.

The system dynamics is defined by the following master equation:

$$\frac{\mathrm{d}}{\mathrm{d}t} P(a,t) = -k_{ba} P(a,t) + k_{ab} P(b,t) ,$$
$$\frac{\mathrm{d}}{\mathrm{d}t} P(b,t) = +k_{ba} P(a,t) - k_{ab} P(b,t) . \tag{3.49}$$

Since this is a linear first-order equation, its solution must be of the form

$$P(x,t) = A_{x\lambda} \exp(\lambda t) . \tag{3.50}$$

Inserting this form into the master equation, we get its characteristic equation

$$\lambda A_{a\lambda} = -k_{ba} A_{a\lambda} + k_{ab} A_{b\lambda} ,$$
$$\lambda A_{b\lambda} = +k_{ba} A_{a\lambda} - k_{ab} A_{b\lambda} , \tag{3.51}$$

or in the matrix form we get the eigenvalue equation:

$$\begin{pmatrix} -k_{ba} & k_{ab} \\ k_{ba} & -k_{ab} \end{pmatrix} \begin{pmatrix} A_{a\lambda} \\ A_{b\lambda} \end{pmatrix} = \lambda \begin{pmatrix} A_{a\lambda} \\ A_{b\lambda} \end{pmatrix} . \tag{3.52}$$

This allows us to determine the values of $\lambda_1 = 0$ and $\lambda_2 = -(k_{ab} + k_{ba}) \equiv -K$, and correspondingly the probabilities of states a and b are given by

$$P(a,t) = A_{a1} + A_{a2} \exp(-Kt) \tag{3.53}$$

and

$$P(b,t) = A_{b1} + A_{b2} \exp(-Kt) . \tag{3.54}$$

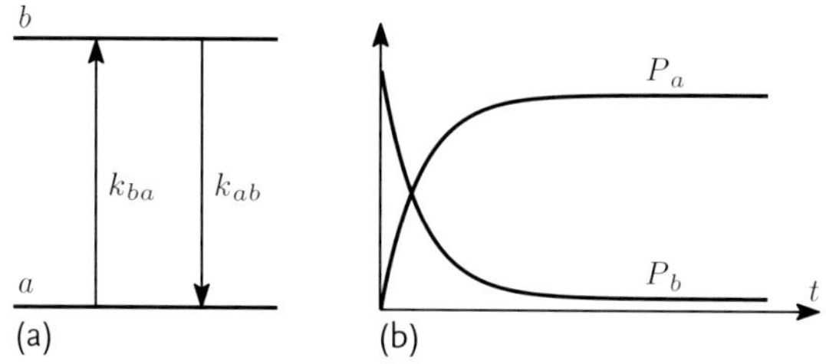

Figure 3.4 The two-level system. It is described by upward k_{ba} and downward k_{ab} rates (a). The exponential relaxation toward equilibrium is shown in (b).

Since for completeness we must have $P(a, t) + P(b, t) = 1$ at all times, this results in

$$A_{a1} + A_{b1} = 1 \tag{3.55}$$

and

$$A_{a2} + A_{b2} = 0 \, . \tag{3.56}$$

We are therefore left with two unknown constants which should be determined from the initial conditions. The exponential dynamics of (3.53) and (3.54) is shown in Figure 3.4b, where the initial occupation $P(b, t = 0)$ of state b is 1 and $P(a, t = 0) = 0$. At long times the probabilities equilibrate exponentially.

We can see that systems of this type should have the following property: one solution of the characteristic equation is $\lambda_1 = 0$. This signifies the existence of an "equilibrium" at $t \to \infty$. In the present case we have $P(a, t = \infty) = A_{a1}$ and $P(b, t = \infty) = 1 - A_{a1}$. The second insight is that all other solutions of the characteristic equation are negative, that is, $\lambda_j \leq 0$. All negative λ signify transient processes.

3.4 Fokker–Planck Equation and Diffusion Processes

So far we have considered a discrete space of events although we allowed arbitrary transitions between system states in continuous time. In this section we consider the continuous space of states, or processes which take place on a continuous axis.

Assume that a process takes place on the real number axis. In the case when the number axis is infinitely dense, the probability $P(x)$, describing the probability of being at coordinate x, has to be replaced by the probability density. We denote the probability of finding the system in the interval from x to $x + \mathrm{d}x$ by $\rho(x)\mathrm{d}x$; $\rho(x)$ thus has the dimension of the inverse coordinate, while the probability remains dimensionless.

The resolution expression for the probability, (3.31), in the case of a continuous space reads

$$\rho(x_3 t_3) = \int p(x_3 t_3 / x_2 t_2) \rho(x_2 t_2) \mathrm{d}x_2 \, . \tag{3.57}$$

The propagator $P(a/b)$ has been replaced by the transition density $p(x_3 t_3 / x_2 t_2)\mathrm{d}x_3$, which denotes the transition from coordinate x_2 into the interval $[x_3, x_3 + \mathrm{d}x_3)$.

Following the derivation of the master equation for probabilities, (3.43), as given in the previous section, we can now obtain the equation for the probability density. The time derivative of (3.57) reads

$$\frac{\mathrm{d}}{\mathrm{d}t} \rho(x t) = \int \mathrm{d}x_2 \frac{\mathrm{d}p(x t / x_2 t_2)}{\mathrm{d}t} \rho(x_2 t_2) \, , \tag{3.58}$$

where

$$\frac{dp(xt/x_2t_2)}{dt} = \lim_{\Delta\to 0} \frac{1}{\Delta} \left(p(x, t+\Delta/x_2t_2) - p(xt/x_2t_2)\right) . \tag{3.59}$$

We now take $t_2 = t$ to get

$$\frac{dp(xt/x_2t)}{dt} = \lim_{\Delta\to 0} \frac{1}{\Delta} \left(p(x, t+\Delta/x_2t) - \delta(x - x_2)\right) . \tag{3.60}$$

Here the expression seems to have a discontinuity at $x = x_2$; however, the first term on the right-hand side $(p(x, t+\Delta/x_2t))$ approaches $\delta(x - x_2)$ as $\Delta \to 0$, and the discontinuity is therefore canceled.

We can now assume the probability of jumping by a distance $y = x - x_2$ to decrease rapidly as the length of the time interval for the jump decreases to zero. This implies that only jumps to nearby positions are possible. To emphasize this behavior, we rewrite the probability of jumping from x_2 to x as follows:

$$\lim_{\Delta\to 0} \frac{1}{\Delta} \left(p(x, t+\Delta/x_2t) - \delta(x - x_2)\right) \equiv f(x_2, x - x_2, t) . \tag{3.61}$$

Now $x - x_2 = y$ can be taken as the small jump increment. Function $f(x, y, t)$ represents the rate of transfer from x to $x + y$ at time t.

Let us now turn back to (3.58). Changing the integration variable to y (while x is kept constant), we get a master equation in continuous space

$$\frac{d}{dt}\rho(xt) = \int dy\, f(x - y, y, t)\rho(x - y, t) . \tag{3.62}$$

As only transitions into regions near x are possible, $f(x, y, t)$ is a smooth function of x, but it is a sharp function of y. It has a maximum at $y = 0$, and it decays as $|y| > 0$. Let us consider now the transition from some original point x_o to the final destination point x_f. Such a transition is characterized by the density increment $f(x_o, y, t)\rho(x_o, t)$. As $f(x_o, y, t)$ is a slowly varying function of the origin x_o, and it is nonzero only for small y, while $\rho(x_o, t)$ is a smooth function of x_0, we can expand the density increment around the final destination point $x_f = x_o + y$ as follows:

$$\begin{aligned} f(x_o, y, t)\rho(x_o, t) \approx \lim_{x_o\to x_f} \Bigg[& \frac{\partial}{\partial x_o} \left(f(x_o, y, t)\rho(x_o, t)\right) \cdot (x_o - x_f) \\ & + \frac{1}{2}\frac{\partial^2}{\partial x_o^2} \left(f(x_o, y, t)\rho(x_o, t)\right) \cdot (x_o - x_f)^2 \Bigg] . \end{aligned} \tag{3.63}$$

Notice that we do not have the zero-derivative term as described in the previous paragraph. We replace $x_o - x_f$ by $-y$ and we obtain

$$\begin{aligned} f(x_o, y, t)\rho(x_o, t) \approx \lim_{x_o\to x_f} \Bigg[& -\frac{\partial}{\partial x_o} \left(f(x_o, y, t)\rho(x_o, t)\right) \cdot y \\ & + \frac{1}{2}\frac{\partial^2}{\partial x_o^2} \left(f(x_o, y, t)\rho(x_o, t)\right) \cdot y^2 \Bigg] . \end{aligned} \tag{3.64}$$

Plugging this expansion in the master equation, (3.62), we get its simplified form where $\lim_{x_0 \to x}$ can now be realized. Denoting

$$\int dy\, f(x, y, t) \cdot y = g(x, t) \tag{3.65}$$

and

$$\int dy\, f(x, y, t) \cdot y^2 = D(x, t)\,, \tag{3.66}$$

we finally obtain the well-known *Fokker–Planck* equation in the form

$$\frac{d}{dt}\rho(xt) = -\frac{\partial}{\partial x}\left[g(x, t)\rho(x, t)\right] + \frac{1}{2}\frac{\partial^2}{\partial x^2}\left[D(x, t)\rho(x, t)\right]\,. \tag{3.67}$$

So far we have obtained the Fokker–Planck equation only formally. The physical interpretation of its parameters will be given below. However, we can now write the equation in the form of the continuity equation:

$$\frac{d}{dt}\rho(x, t) = -\frac{\partial}{\partial x}J(x, t)\,, \tag{3.68}$$

where

$$J(x, t) = g(x, t)\rho(x, t) - \frac{1}{2}\frac{\partial}{\partial x}D(x, t)\rho(x, t) \tag{3.69}$$

is the probability flux in the x space.

Let us consider now as an example the case when $g = 0$ and $D = \text{const}$. In this case we get the famous diffusion equation which describes Brownian motion:

$$\frac{d\rho(xt)}{dt} = \frac{D}{2}\frac{\partial^2\rho(x, t)}{\partial x^2}\,. \tag{3.70}$$

Its solutions are Gaussian functions, and the parameter D is the diffusion coefficient. Thus, one type of process that the Fokker–Planck equation describes is diffusive flow. Other types of processes are described in the next section. A more detailed discussion of various types of stochastic processes can be found in relevant textbooks (e.g., [10]).

3.5 Deterministic Processes

In the previous sections we described the stochastic Markov process. The deterministic processes were described in Chapter 2. It turns out that an arbitrary deterministic process is a Markov process. Both of these types of processes are usually described in different representations, however, they can be unified. Let us consider a process described by a set of dynamic variables x_n. The deterministic evolution of the variables is fully described by a set of differential equations:

$$\frac{\partial x_n}{\partial t} = g_n(x_1 \ldots x_N)\,. \tag{3.71}$$

Instead of specifying all x_n, we will use a vector field $\boldsymbol{x}$. The solution of this equation for the initial value $\boldsymbol{x}_i$ can be denoted by $\phi_t(\boldsymbol{x}_i) = \boldsymbol{x}_f$. Thus, we can also write

$$\phi_{t+s}(\boldsymbol{x}_i) = \phi_t(\phi_s(\boldsymbol{x}_i)) . \tag{3.72}$$

The evolution of the system in the phase space is given by infinitesimally narrow phase space trajectories, and the propagator is therefore given by the following δ distribution of trajectories:

$$p(\boldsymbol{x}_f t_f | \boldsymbol{x}_i t_i) = \delta(\boldsymbol{x}_f - \phi_{t_f - t_i}(\boldsymbol{x}_i)) . \tag{3.73}$$

Using (3.72), we can write

$$\delta(\boldsymbol{x}_f - \phi_{t+s}(\boldsymbol{x}_i)) = \int \mathrm{d}\boldsymbol{x}\, \delta(\boldsymbol{x}_f - \phi_t(\boldsymbol{x}))\delta(\boldsymbol{x} - \phi_s(\boldsymbol{x}_i)) , \tag{3.74}$$

which is essentially the Chapman–Kolmogorov relation for the propagator. So as the deterministic process satisfies the Chapman–Kolmogorov equation, it must be Markov.

Later we will describe statistical ensembles of systems, so we turn to work with probability densities for the deterministic dynamics. If we have the probability density at some initial time $\rho(\boldsymbol{x} t_i)$, its time evolution can also be calculated using (3.62). For the deterministic process we thus have

$$\frac{\mathrm{d}}{\mathrm{d}t}\rho(\boldsymbol{x} t) = \int \mathrm{d}\boldsymbol{x}_2 \frac{\mathrm{d}}{\mathrm{d}t}\delta(\boldsymbol{x} - \phi_{t-t_2}(\boldsymbol{x}_2))\rho(\boldsymbol{x}_2 t_2) . \tag{3.75}$$

Here $\boldsymbol{x}_2$ is a point in phase space different from point $\boldsymbol{x}$, and $\phi_{t-t_2}(\boldsymbol{x}_2) \equiv \boldsymbol{x}'(t)$ is the phase space trajectory as a function of time t. The derivative is calculated by using the usual chain rule

$$\frac{\mathrm{d}}{\mathrm{d}t}\delta(\boldsymbol{x} - \boldsymbol{x}'(t)) = \sum_n \frac{\partial x_n'(t)}{\partial t} \frac{\partial}{\partial x_n'}[\delta(\boldsymbol{x} - \boldsymbol{x}'(t))] , \tag{3.76}$$

and in the integral we can take the limit $t_2 = t$, which also gives $\boldsymbol{x}_2 = \boldsymbol{x}'$. Applying (3.71), we get

$$\frac{\mathrm{d}}{\mathrm{d}t}\rho(\boldsymbol{x} t) = \int \mathrm{d}\boldsymbol{x}' \sum_n g_n(\boldsymbol{x}') \frac{\partial}{\partial x_n'}\left[\delta(\boldsymbol{x} - \boldsymbol{x}')\right] \rho(\boldsymbol{x}' t) . \tag{3.77}$$

Integrating the delta function, we finally obtain

$$\frac{\mathrm{d}}{\mathrm{d}t}\rho(\boldsymbol{x} t) = -\sum_n \frac{\partial}{\partial x_n}\left[g_n(\boldsymbol{x})\rho(\boldsymbol{x} t)\right] . \tag{3.78}$$

Equation (3.78) is the celebrated *Liouville equation* for a deterministic process in the phase space. We can now immediately look back at the Fokker–Planck equation, (3.67), and associate its first term with the deterministic drift of the probability, and the second term describes the diffusive irreversible spread of the probability. Thus, most dynamic processes in physics are Markov processes.

A typical deterministic process can be described by the Hamiltonian dynamics of the classical dynamic system. Given the Hamiltonian H in the space of coordinates q_n and momenta p_n, as described in the previous chapter, the evolution of the system is described by Hamiltonian equations:

$$\frac{\mathrm{d}q_n}{\mathrm{d}t} = \frac{\partial H}{\partial p_n} , \tag{3.79}$$

$$\frac{\mathrm{d}p_n}{\mathrm{d}t} = -\frac{\partial H}{\partial q_n} . \tag{3.80}$$

According to mechanics, the probability density in the phase space follows the classical Liouville equation of the form

$$\frac{\partial \rho(\boldsymbol{q}, \boldsymbol{p})}{\partial t} = \{H, \rho(\boldsymbol{q}, \boldsymbol{p})\} , \tag{3.81}$$

where $\{f, g\}$ denotes the *Poisson brackets*. Denoting the dynamic variables by $q_1 = x_1, \dots q_N = x_N, p_1 = x_{N+1}, \dots p_N = x_{2N}$, and $\partial H/\partial p_n = g_n, \partial H/\partial q_n = -g_{N+n}$, we get the Liouville equation in the form of (3.78). Thus, the classical Hamiltonian dynamics is essentially described by the Fokker–Planck equation without the diffusion part.

3.6
Diffusive Flow on a Parabolic Potential (a Harmonic Oscillator)

Let us now consider a one-dimensional parabolic potential:

$$V(x) = k\frac{x^2}{2} . \tag{3.82}$$

In this case the driving force is defined as

$$F(x) = -kx , \tag{3.83}$$

and the Fokker–Planck equation is given by

$$\frac{\mathrm{d}}{\mathrm{d}t}\rho(x,t) = k\frac{\partial}{\partial x}x\rho(x,t) + \frac{D}{2}\frac{\partial^2}{\partial x^2}\rho(x,t) . \tag{3.84}$$

If we introduce dimensionless time $\tau = kt$, and replace the coordinate by

$$\sqrt{\frac{2k}{D}}x = y , \tag{3.85}$$

we have

$$\frac{\mathrm{d}}{\mathrm{d}\tau}\rho(y,\tau) = \frac{\partial}{\partial y}y\rho(y,t) + \frac{\partial^2}{\partial y^2}\rho(y,t) . \tag{3.86}$$

Consider now a stationary state so that $d\rho/d\tau = 0$. In this case we have the equation

$$\rho'' + y\rho' + \rho = 0 \,. \tag{3.87}$$

To find the solution we make an ansatz $\rho(y) = \exp(g(y))$, which gives

$$1 + g'' = -(yg' + g'^2) \,. \tag{3.88}$$

The solution of this equation is simply

$$g = -\frac{y^2}{2} \,. \tag{3.89}$$

So we find that the steady state or the equilibrium distribution is represented by a Gaussian function:

$$\rho(y) \propto e^{-\frac{y^2}{2}} \,. \tag{3.90}$$

It is centered at zero and decays to the ends. This is tightly connected with the statistical canonical distribution of the probability on the parabolic potential as will be described in Section 7.3.2, which is devoted to problems in statistical physics.

Since the steady-state solution is Gaussian, we look for a Gaussian form of the solution of the nonstationary problem as well. We take an ansatz

$$\rho(y, \tau) = \exp(-A(\tau)y^2 - B(\tau)y - C(\tau)) \,. \tag{3.91}$$

Plugging this into (3.86) and comparing the coefficients at equal powers of y, we get the set of three equations

$$\dot{C} = -1 + 2A - B^2 \,, \tag{3.92}$$

$$\dot{B} = B(1 - 4A) \,, \tag{3.93}$$

$$\dot{A} = 2A(1 - 2A) \,. \tag{3.94}$$

We can solve these equations starting with one for A. The solution is

$$A = -\frac{1}{e^{-2t+a} - 2} \,, \tag{3.95}$$

$$B = \frac{e^{-t+a-b}}{e^{-2t+a} - 2} \,, \tag{3.96}$$

where a and b are still undetermined constants. The function C can be obtained from the normalization of the probability density.

These expressions now have two undetermined constants a and b. We obtain them by assuming that the initial probability distribution is

$$\rho(\tau = 0) = \delta(y - y_0) \,. \tag{3.97}$$

In order to relate this with our form of the solution, we take for the initial condition a Gaussian representation for the delta function, that is, we take a Gaussian centered at y_0 with vanishing dispersion. Around $\tau = 0$ we therefore have

$$\rho(\tau \to 0) = \exp(-S(y - y_0)^2) \tag{3.98}$$

with $S \to \infty$. In this limit we must have

$$-A(\tau)y^2 - B(\tau)y - C(\tau) = -Sy^2 + 2Syy_0 - Sy_0^2 \tag{3.99}$$

or

$$S = A\,, \tag{3.100}$$

$$2Sy_0 = -B\,. \tag{3.101}$$

From the first condition we find that $e^a = 2$ satisfies the requirement that the dispersion approaches infinity as $\tau \to 0$. From the second condition we find $e^{-b} = y_0$. This gives the unnormalized density

$$\rho(y,\tau) = n(\tau)\exp\left(-\frac{1}{2}\frac{y^2 - y_0 y e^{-\tau}}{1 - e^{-2\tau}}\right)\,. \tag{3.102}$$

From the normalization requirement

$$\int dy\rho(y,\tau) = 1 \tag{3.103}$$

at arbitrary time we get the normalization factor

$$n(t) = \frac{1}{\sqrt{2\pi(1 - e^{-2\tau})}}\exp\left(-\frac{1}{2}\frac{(y_0 e^{-\tau})^2}{1 - e^{-2\tau}}\right)\,, \tag{3.104}$$

which finally gives

$$\rho(y,\tau) = \frac{1}{\sqrt{2\pi(1 - e^{-2\tau})}}\exp\left(-\frac{1}{2}\frac{(y - y_0 e^{-\tau})^2}{1 - e^{-2\tau}}\right)\,. \tag{3.105}$$

This expression shows that the distribution remains Gaussian at all times. It approaches the steady-state solution defined by (3.90) exponentially as shown in Figure 3.5. We thus find that in a parabolic potential the relaxation of the displacement is exponential,

$$\langle y \rangle = y_0 e^{-\tau}\,, \tag{3.106}$$

while the dispersion of the distribution is

$$\langle y^2 - \langle y \rangle^2 \rangle = 1 - e^{-2\tau}\,. \tag{3.107}$$

We have described only a simple theoretical model here. A more direct application of stochastic theory to physical problems of this kind can be found in relevant textbooks (e.g., [8]).

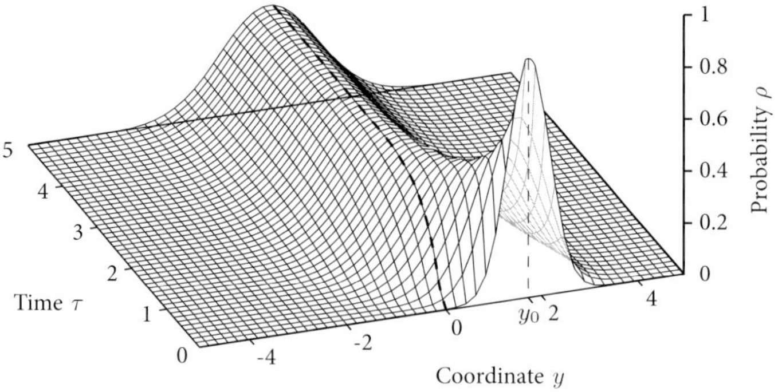

Figure 3.5 Decay of the probability distribution on a parabolic potential into the equilibrium Gaussian according to solution (3.105).

3.7 Partially Deterministic Process and the Monte Carlo Simulation of a Stochastic Process

In practice the system may be often described by the Fokker–Planck equation, which includes a deterministic drift and a diffusive flow. It is, however, not trivial to obtain a solution of the Fokker–Planck equation in many dimensions. This is an important problem especially in the theory of many-body quantum systems. Instead of the Fokker–Planck equation, it is often possible to get specific stochastic trajectories $x(t)$ that describe particular processes occurring in the system. Later the averaging of stochastic trajectories may be used to obtain the dynamics of the probability density.

The *partially deterministic process* (PDP) is one of the processes which can be described by the above scheme [9]. The PDP is considered a deterministic process in certain time intervals. Between these intervals the system makes stochastic jumps. Using the ideas of previous sections about the transition probability, we can write the propagator of the PDP similarly to the description used to characterize the deterministic process. Additionally, we have to introduce the rates of jumping from state x to state x', denoted by $k(x'|x)$. The propagator for a small time interval Δ can be written as

$$p(x, t + \Delta|x', t) = (1 - \Gamma(x')\Delta)\delta(x - \phi_\Delta(x')) + k(x|x')\Delta \,. \tag{3.108}$$

For the deterministic trajectory we can assume a short-time limit $\phi_\Delta(x') \approx x' + g(x')\Delta$, while

$$\Gamma(x') = \int \mathrm{d}x\, k(x|x') \tag{3.109}$$

is the total rate of a jump out of point x'. The quantity $(1 - \Gamma(x')\Delta)$ thus represents the probability that the system does not jump out of state x'. In this case the propagation proceeds according to the deterministic evolution described by the Kronecker delta.

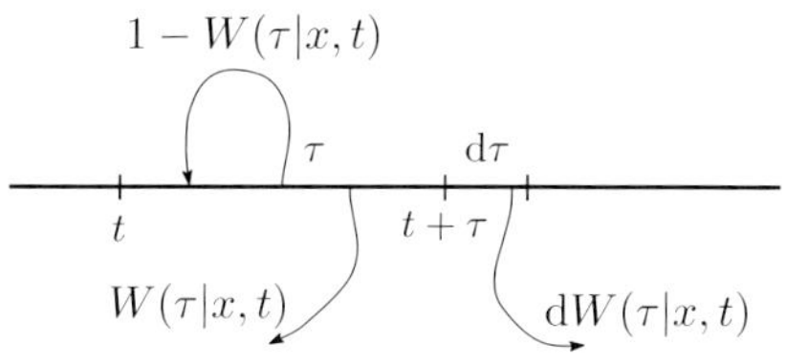

Figure 3.6 Definition of the waiting time distribution in the partially deterministic process.

The Liouville-like equation for the PDP is now given by

$$\frac{\partial}{\partial t}p(x,t) = -\frac{\partial}{\partial x}[g(x)p(x,t)] + \int dx'[k(x|x')p(x',t) - k(x'|x)p(x,t)]\,. \tag{3.110}$$

The first term here describes the deterministic drift, and the second term denotes the stochastic jumps between states. This equation is similar to the Fokker–Planck equation; the jumps, however, do not occur continuously. The equation is also often referred to as the Liouville equation.

Next we introduce the most important quantity in this section – the *waiting time distribution* $W(\tau|x,t)$. It describes the probability of a jump realization during the time interval from t to $t+\tau$ when the system is in state x at time t (see Figure 3.6). The difference

$$W(\tau + d\tau|x,t) - W(\tau|x,t) = d\,W(\tau|x,t) \tag{3.111}$$

is the probability of a jump occurring in the infinitesimal interval $d\tau$ after $t+\tau$. If we now divide $d\,W$ by the probability that no jump occurs in the interval τ, which is $1 - W(\tau|x,t)$, we get the normalized conditional probability of a jump occurring in the interval $d\tau$ after $t+\tau$ when no jumps occurred during τ. If there were no jumps during τ, the system propagated along a deterministic trajectory $\phi_\tau(x)$, and later there was a jump in the interval $d\tau$ at a rate $\Gamma(\phi_\tau(x))d\tau$. This can be summarized in the relation

$$\frac{d\,W(\tau|x,t)}{1 - W(\tau|x,t)} = \Gamma(\phi_\tau(x))d\tau\,. \tag{3.112}$$

Equation (3.112) is a differential equation which can be easily integrated into

$$W(\tau|x,t) = 1 - \exp\left(-\int_0^\tau d\tau'\Gamma(\phi_{\tau'}(x))\right)\,. \tag{3.113}$$

This expression immediately implies an exponential character of the waiting time distribution. For instance, if the process is purely stochastic so that $\phi_{\tau'}(x) = x$, we get the simple exponential distribution

$$W(\tau|x,t) = 1 - \exp(-\Gamma(x)\tau)\,. \tag{3.114}$$

In general the distribution of the PDP can be nonexponential due to deterministic drift, and (3.113) should be integrated numerically for each trajectory. It should also be noted that Γ, being a transition rate, is a nonnegative function, and the integral in (3.113) therefore increases monotonically. When $\tau \to \infty$, the probability W thus grows with time τ until some constant value. If that value is 1, the system will always perform a jump from the original state. However, if the value is smaller than 1, the state may become a trapping state in which the system stays indefinitely.

The waiting time distribution allows us to introduce a continuous time stochastic simulation procedure for the PDP. Let us assume that the initial system state is x at time t. To obtain the next state we need to either propagate the system on the basis of its dynamic set of equations or realize the jump process. The waiting time distribution function $W(\tau|x,t)$ allows us to determine the time required for the deterministic propagation until the jump. For this purpose we draw a random linearly distributed value η from the interval $(0,1]$. We obtain the time τ before the jump from the solution of the equation

$$\eta = W(\tau|x,t) . \tag{3.115}$$

The solution is obtained by numerical simulation of the deterministic propagation by calculating $\phi_{\tau'}(x)$ and at the same time calculating the integral in (3.113) until

$$\int_0^{\tau'} \mathrm{d}s \Gamma(\phi_s(x)) = -\ln(1-\eta) . \tag{3.116}$$

When this equality is reached, the deterministic propagation is stopped, and the jump process is realized. The jump process is realized stochastically according to the set of rates $k(x_\mathrm{f}|x')$, where x' is the final point reached in the deterministic propagation, and x_f is the final point after the jump. The selection of the final state x_f is obtained by taking a linearly distributed random number.

By repeating these steps, we obtain the Monte Carlo algorithm for system dynamics. As we found before in this chapter, the stochasticity leads to relaxation behavior in the master equation for probabilities or the Fokker–Planck equation. The PDP represents a classical or even quantum relaxation algorithm, which can be easily adapted to a particular physical system.

3.8
Langevin Equation and Its Relation to the Fokker–Planck Equation

In the previous section we described a process which is partially defined by the deterministic equations, with their solutions interrupted by stochastic jumps. This type of stochasticity can be embedded in the dynamic equations via external fluctuating forces. Starting with (3.71), we can then postulate the dynamic equation

$$\frac{\mathrm{d}x}{\mathrm{d}t} = \gamma(x) + \xi(t) , \tag{3.117}$$

where the function $\boldsymbol{\gamma}(\boldsymbol{x})$ is the force vector depending on all dynamic variables $\boldsymbol{x} = \{x_1 x_2 \ldots\}$, and $\xi(t)$ is the fluctuating force. Equation (3.117) is usually termed the *Langevin equation* of motion. We will provide a more rigorous backing of the Langevin equation in Section 8.2.

The fluctuating force introduces irreversibility in the dynamics as ξ is assumed to be an irreversible function. Its functional form thus cannot be defined explicitly. Instead, its moments are usually given. We assume that the force has zero mean value

$$\langle \xi_i(t) \rangle = 0 . \tag{3.118}$$

We assume ξ is completely random, and its timescale is much shorter than the system dynamical timescale, so its correlation function is of the form

$$\langle \xi_i(t) \xi_j(0) \rangle = Q \delta_{ij} \delta(t) , \tag{3.119}$$

where Q is the dispersion of the fluctuations. When all odd higher-order moments vanish, and the even moments factorize into products of the second moment, the stochastic force is denoted as the Gaussian noise (it should be noted that only the first two cumulants are nonzero for the Gaussian process as described in Appendix A.2). The external force of this type is denoted as the white noise. Other types of fluctuating forces will be described later.

The Langevin equation for the harmonic potential describes the relaxation of the harmonic oscillator. Let us take, for example, a particle in a parabolic potential (the case of the harmonic oscillator). The force is proportional to the displacement, and we assume the Langevin equation

$$\frac{\mathrm{d}x}{\mathrm{d}t} = -\gamma x + \xi(t) . \tag{3.120}$$

Without the stochastic force the solution is the exponential function $A\mathrm{e}^{-\gamma t}$. Thus, taking $x(t) = y(t)\mathrm{e}^{-\gamma t}$, we simplify the equation and can easily integrate it. For initial condition $x(t = 0) = x_0$, we get

$$x(t) = x_0 \mathrm{e}^{-\gamma t} + \int_0^t \mathrm{d}\tau \mathrm{e}^{-\gamma(t-\tau)} \xi(\tau) . \tag{3.121}$$

This trajectory is not determined as long as $\xi(t)$ is not given. However, using the moments of the stochastic force, we can describe the statistical properties of the x variable. The mean value is given by

$$\langle x(t) \rangle = x_0 \mathrm{e}^{-\gamma t} . \tag{3.122}$$

This result agrees with the solution of the Fokker–Planck equation presented in the previous section which gives the exponential relaxation of the oscillator. The dispersion is

$$\langle (x - \langle x \rangle)^2 \rangle = \int_0^t \mathrm{d}\tau \int_0^t \mathrm{d}\tau' \mathrm{e}^{-\gamma(2t-\tau-\tau')} \langle \xi(\tau) \xi(\tau') \rangle . \tag{3.123}$$

Using (3.119), we get

$$\langle (x - \langle x \rangle)^2 \rangle = \frac{Q}{2\gamma} \left(1 - e^{-2\gamma t}\right) . \tag{3.124}$$

We can see that the dispersion is initially zero, and at later times it grows to a stationary value equal to $Q/2\gamma$. This result is also confirmed by the Fokker–Planck equation (see (3.106) and (3.107)).

Here we found that the two lowest moments of the Langevin equation coincide with the moments obtained by the Fokker–Planck equation. What about higher moments? From (3.121) we observe that the function $x(t)$ is a linear superposition of $\xi(t)$ at various times. Since the linear superposition of Gaussian variables is a Gaussian variable, we conclude that $x(t)$ is a Gaussian process. The Gaussian process is completely described by its first two moments. As the first two moments of the Langevin equation, (3.120), and of the Fokker–Planck equation, (3.84), coincide, and they are both Gaussian processes, these two equations describe the same process. Thus, the description of the process using both equations is equivalent.

The procedure described in this section can be generalized for an arbitrary fluctuating system into the so-called Langevin approach. Here we just discuss the issue, and for a more complete description the reader is referred to other textbooks (e.g., [8]). The first step is to write the deterministic equations describing the drift-related process or forces acting on the system. In the second step, the fluctuating force can be added to describe the thermally induced fluctuations (or fluctuations induced by some other means). In the third step, the strength of the fluctuations has to be adjusted to properly reflect the equilibrium fluctuations on the basis of either the experimental or the physical considerations (see Chapter 7).

This general approach may properly describe the dynamics but its connection to the Fokker–Planck equation is not trivial. Note that in the example of the harmonic potential the dynamic variable x was linear in fluctuating force ξ. For this reason the dynamics of the variable was necessarily Gaussian, and this resulted in an equivalence with the Fokker–Planck equation. In the general case, the forces involved in the system may be nonlinear, and the equivalence relation to the Fokker–Planck equation is not exact. So the linearity becomes the necessary condition. Thus, for equations of the type

$$\dot{x} = F(x) + \xi(x) , \tag{3.125}$$

where $F(x)$ is some nonlinear function, the equation is still linear in the fluctuating force, and the relation to the Fokker–Planck equation

$$\frac{\partial \rho}{\partial t} = -\frac{\partial}{\partial x}(F(x)\rho) + \frac{Q}{2}\frac{\partial^2 \rho}{\partial x^2} \tag{3.126}$$

can be established.

However, equations of the type

$$\dot{x} = F(x) + G(x)\xi(x) \tag{3.127}$$

are intrinsically nonlinear in the fluctuating force, and the general connection with the Fokker–Planck equation cannot be exactly established because the $x(t)$ trajectory is not Gaussian. Further examples in this book are not related to these types of nonlinear equations, so we do not discuss this issue in more depth. The reader interested in this topic may consult specialized textbooks (e.g., [8]).

4
Quantum Mechanics

In this chapter we review the basic concepts of quantum mechanics. We will start with the Schrödinger equation, which defines reversible behavior of a closed quantum system. This fundamental equation introduces eigenstates and the corresponding eigenvalues (energy spectrum) of the system. These quantities play an important role in the representation of various quantum states and quantum dynamics among them. We also introduce the density matrix, which is widely used for the description of system dynamics. Some typical examples of general interest will be presented for demonstration purposes. The perturbative scheme, which is especially important for considering the influence of an external impact on the system, will also be discussed. Finally, special attention will be paid to the effects caused by an external electromagnetic field. However, we will not discuss here all the inconsistencies in observations with classical mechanics and electrodynamics, which itself is very interesting. Some aspects of this issue will be considered in Chapter 6. Over recent decades many textbooks describing quantum mechanics at different levels have been written, and are recommended for further reading [11–16].

4.1
Quantum versus Classical

As follows from Chapter 2, classical physics is based on two fundamental concepts: the concept of a particle, which behaves in accordance with Newton's laws, and the concept of an electromagnetic wave, which describes an extended physical entity. Both of these concepts are inseparable when considering some experimental observations of radiation interacting with matter and some constituents of matter (atoms, molecules, etc.). In order to explain black-body radiation (Max Planck), the photoelectric effect (Albert Einstein), and the Compton effect (see, e.g., [11, 13, 14], for details) a new concept of photons as particles of the electromagnetic field was developed. However, this new concept had to reconcile with the wavelike properties of electromagnetic radiation. The principle of duality was later postulated in an attempt to come to terms with the classically irreconcilable dual character of observations in the microworld. This principle of duality was postulated to be also ap-

Molecular Excitation Dynamics and Relaxation, First Edition. L. Valkunas, D. Abramavicius, and T. Mančal.

plicable to particles of matter (Louis de Broglie) [12–14]. The common relationship between the particle momentum p and its wavelike characteristics, wavelength λ, was defined as

$$\lambda = \frac{h}{p} , \tag{4.1}$$

where $h = 6.626 \times 10^{-34}$ J s is the Planck constant. This wavelength is now usually called the *de Broglie wavelength* of the quantum particle. Many subsequent experiments supported this relationship demonstrating that electrons, protons, neutrons, atoms, and even molecules have wavelike properties. The conceptual implications of these properties are best explored by considering the two-slit interference experiment, which will be described in Chapter 6.

The dual properties of light and matter, that is, the wavelike properties and the particle-like properties, cause the uncertainty relationship between the position and the momentum of the system under consideration (Werner Heisenberg). Indeed, the degree of uncertainty of the position of the particle along a particular direction, Δx, should be connected with the level of uncertainty of the projection of the momentum along the same direction, Δp_x, in accord with the *Heisenberg uncertainty principle*:

$$\Delta x \Delta p_x \approx h . \tag{4.2}$$

This relationship follows directly from (4.1) for a freely moving particle. It asserts that greater accuracy in position is possible only at the expense of greater uncertainty in momentum, and vice versa. Indeed, according to the Heisenberg uncertainty principle, precise determination of the position of a particle, $\Delta x = 0$, is only possible with infinite uncertainty of the momentum Δp_x. Thus, the inferred properties of a quantum particle depend on the experimental conditions. The relevant discussion of this issue is given in Chapter 6.

In order to fulfill the requirement of the duality principle, quantum system behavior does not follow the requirement of classical mechanics, where the position of a particle, $x(t)$, is defined at any given time, t. In this case the position of the particle and its momentum should correspond to the Heisenberg uncertainty principle. In order to take this aspect of duality into account in quantum mechanics, all measurable physical properties are attributed to corresponding mathematical operators according to the so-called correspondence principle, while the system state (and its time evolution) is attributed to a vector from a mathematical vector space, where these operators act. In the so-called coordinate representation, the particle coordinate operator coincides with its coordinate $\boldsymbol{r}$, and the momentum operator is defined as $\hat{\boldsymbol{p}} = -i\hbar\nabla$, where $\hbar = h/2\pi$ is the reduced Planck constant. Thus, the projection of the momentum in a particular x direction is $\hat{p}_x = -i\hbar\partial/\partial x$. In this representation, the state vector coincides with a complex function of coordinates and time, the so-called wavefunction. Similarly to momentum and the coordinate, the Hamiltonian of the quantum system, defined as the sum of kinetic and potential energies, can also be attributed to its corresponding operator. According to the definitions of the kinetic energy given in Chapter 2, it is directly related to the momentum operator,

$$\hat{T} = \frac{\hat{p}^2}{2m} \equiv -\frac{\hbar^2 \nabla^2}{2m} , \tag{4.3}$$

and the potential energy is the corresponding function of the coordinate. The uncertainty between two measurable quantities is reflected in the commutational relationship between the corresponding operators. In the case of coordinate $\hat{x}$ and momentum $\hat{p}_x$ it gives the following result:

$$[\hat{x}, \hat{p}_x] \equiv x\hat{p}_x - \hat{p}_x x = i\hbar . \tag{4.4}$$

4.2 The Schrödinger Equation

Quantum mechanics can be formulated in terms of a set of postulates. For the dynamics, the most important postulate is the one prescribing the time evolution of a quantum system in a state defined by a wavefunction $\Psi(x, t)$. It postulates that the wavefunction should satisfy the Schrödinger equation:

$$i\hbar \frac{\partial}{\partial t} \Psi(x, t) = \hat{H} \Psi(x, t) , \tag{4.5}$$

where $\hat{H}$ is the Hamiltonian of the system under consideration and the variable (a set of variables) x usually represents the coordinates of the system (the wavefunction can, however, also be represented in terms of different variables, such as momentum). An unambiguous solution of (4.5) is defined for fixed initial conditions of the wavefunction $\Psi(x, t_0)$. The Hamiltonian $\hat{H}$ is the operator corresponding to the total energy of the system under consideration. The wavefunction encapsulates the wavelike properties of the quantum system, and it determines the probability amplitude for finding a system in a particular state. Thus, $\Psi(x, t)$ can be a stationary waveform or a localized wavepacket reflecting dual properties of quantum systems. As we show later, the wavefunction is related to the probability of the system being observed in certain states, that is, $|\Psi(x, t)|^2$ determines a probability density in accord with the so-called Born rule. More details on this issue are given in Chapter 6.

The time evolution of the quantum system can be represented as the evolution of the elements (state vectors) in the so-called *Hilbert space*. As we show later, the time evolution of a closed system is the pure phase rotation. The action of an operator on the vector causes the displacement of these elements in the Hilbert space. The wavefunction introduced above is a particular representation of the state vector defined by the eigenstates of the coordinate operator. So (4.5) should be understood as the coordinate representation of the Schrödinger equation and is a linear differential equation. With the following ansatz

$$\Psi(x, t) = \psi(t)\varphi(x) , \tag{4.6}$$

and assuming that the Hamiltonian is independent of time, we can separate it into two equations for the time and space variables. In this way we obtain an eigenvalue equation (the stationary Schrödinger equation),

$$\hat{H}\varphi(x) = E\varphi(x) , \tag{4.7}$$

for the spatial wavefunction $\varphi(x)$. Here E being a number is the eigenvalue and $\varphi(x)$ is the eigenfunction of the Hamiltonian $\hat{H}$. Usually, (4.7) determines a set of eigenvalues E_n together with corresponding eigenfunctions $\varphi_n(x)$; the numbers n become a set of quantum indices or quantum numbers. For $\psi(t)$ we have

$$\mathrm{i}\hbar\frac{\mathrm{d}\psi}{\mathrm{d}t} = E\psi . \tag{4.8}$$

This can be solved for known values of eigenenergies E_n. So, one particular solution of the Schrödinger equation is $\exp(-\mathrm{i}E_n t/\hbar)\varphi_n(x)$, and the complete wavefunction is then defined as a linear superposition of these particular solutions:

$$\Psi_{\mathrm{c}}(x,t) = \sum_n c_n \mathrm{e}^{-\frac{\mathrm{i}}{\hbar}E_n t}\varphi_n(x) . \tag{4.9}$$

It is noteworthy that the eigenfunctions of any operator compose a full basis set. As a result, an arbitrary wavefunction $\phi(x,t)$ of an arbitrary system can be expanded via this (or any other) set of eigenfunctions:

$$\phi(x,t) = \sum_n d_n(t)\varphi_n(x) , \tag{4.10}$$

where the time evolution can be embedded in the coefficient $d_n(t) = d_n \mathrm{e}^{-\mathrm{i}/\hbar\, E_n t}$, and d_n is a corresponding expansion coefficient. This is the outcome of the so-called superposition principle. Evidently, in the case of the continuous spectrum of energy E, (4.10) corresponds to

$$\Psi_{\mathrm{c}}(x,t) = \int c(E)\mathrm{e}^{-\frac{\mathrm{i}}{\hbar}Et}\varphi_E(x)\mathrm{d}E . \tag{4.11}$$

The Schrödinger equation (4.5) defines the "shape" of the wavefunction, but does not define its amplitude. So the wavefunction can be chosen to be additionally normalized, so

$$\int \varphi_n^*(x)\varphi_n(x)\mathrm{d}x = 1 . \tag{4.12}$$

In that case the eigenfunctions of the Hamiltonian form an orthogonal and normalized (orthonormal) set with respect to the scalar product defined by the integral over the variables x:

$$\int \varphi_n^*(x)\varphi_m(x)\mathrm{d}x = \delta_{nm} . \tag{4.13}$$

Here (and further on) the asterisk denotes the complex conjugation. The coefficients of the decomposition

$$\psi(x,t) = \sum_n c_n(t)\varphi_n(x) \tag{4.14}$$

are then

$$c_n(t) = \int \varphi_n^*(x)\psi(x,t)\mathrm{d}x \,, \tag{4.15}$$

and according to the standard interpretation of quantum mechanics, the square of the expansion coefficient $|c_n(t)|^2$ defines the probability of finding the system in the corresponding nth quantum state. As the wavefunction corresponding to a given state is itself a representation of the state in the basis of the eigenstates of the coordinate operator, the values $\varphi_n(x)$ are thus the "coefficients" of this representation. $|\varphi_n(x)|^2$ then determines the probability density for finding the system which is in the nth eigenstate with the value of the coordinate equal to x. More details on the interpretation are given in Chapter 6. This is denoted as the *Born rule* for probability. According to such an interpretation, the wavefunction itself is a *probability amplitude*, and has no direct physical meaning (in the sense of being measurable). While the probability density is real and nonnegative, the wavefunction is complex.

As denoted above, the operators act on wavefunctions and create new wavefunctions:

$$\hat{F}\phi(x) = \psi(x) \,. \tag{4.16}$$

The operators usually represent some physical quantities. The operators are taken as linear. The normalized wavefunction allows one to determine the values of the outcomes of a repeated experimental observation of a system in a particular quantum state. This so-called expectation value $\langle F \rangle$ associated with the operator $\hat{F}$ for a system in a given state $\phi(x)$ is defined as

$$F \equiv \langle \hat{F} \rangle = \int \mathrm{d}x \phi^*(x)\hat{F}\phi(x) \,. \tag{4.17}$$

This is usually termed the *Born rule* as well. We can also introduce the action of an operator *to the left* by $\phi(x)\hat{F} = \hat{F}^{\mathrm{T}}\phi(x)$, where $\hat{F}^{\mathrm{T}}$ denotes the *transposed* operator. For an arbitrary operator the expectation value can be complex. We can thus write

$$\langle \hat{F} \rangle^* = \int \mathrm{d}x \phi(x)(\hat{F}\phi(x))^* \equiv \int \mathrm{d}x \phi^*(x)\hat{F}^\dagger\phi(x) \,. \tag{4.18}$$

Operator $\hat{F}^\dagger$ is denoted as the *conjugate* operator to operator $\hat{F}$. From the equation above we can write

$$(\hat{F}^{\mathrm{T}})^* = \hat{F}^\dagger \,. \tag{4.19}$$

As the expectation value of the physically measurable quantity must be real,

$$\hat{F} = \hat{F}^\dagger \,. \tag{4.20}$$

Such operators always have real expectation values and are denoted as the *Hermitian*.

4.3 Bra-ket Notation

Many quantum mechanical expressions can be greatly simplified by adopting the Dirac bracket notation. In this notion, the fact that the wavefunction $\varphi_n(x)$ of state n in the coordinate representation is a representation of an abstract state vector $|n\rangle$ in the bases of the vectors corresponding to different values of x is represented by

$$\varphi_n(x) = \langle x|n\rangle \,, \tag{4.21}$$

Here bra-ket $\langle x|n\rangle$ represents a scalar product of two vectors, $|x\rangle$ and $|n\rangle$, both defined on an abstract Hilbert space. The part $|n\rangle$ of the bra-ket is the so-called *ket vector*. Similarly, $\langle n|$ is the so-called *bra vector*, denoting the conjugate state in the scalar product. For the wavefunction, the conjugation corresponds to the complex conjugation. More properly $|n\rangle = (\langle n|)^\dagger$, and because the scalar product is a number, we have $\langle x|n\rangle = \langle n|x\rangle^*$. The scalar product $\langle x|n\rangle$ denotes the projection of vector $|n\rangle$ on vector $|x\rangle$ in the Hilbert space.

According to these notations, the integral of two wavefunctions can be expressed as follows:

$$\langle b|a\rangle = \int \varphi_b^*(x)\varphi_a(x)\mathrm{d}x \,. \tag{4.22}$$

And similarly,

$$\langle b|F|a\rangle = \int \varphi_b^*(x)\hat{F}(x)\varphi_a(x)\mathrm{d}x \,. \tag{4.23}$$

On the left-hand side of these expressions we have abstract notation for which the wavefunctions do not have to be expressed in the coordinate representation. For normalized state vectors we have

$$\langle a|a\rangle = 1 \,, \tag{4.24}$$

which means that the elements (vectors) are determined by their direction in the Hilbert space but not by their amplitude.

The eigenfunctions of Hermite operators representing physical observables are orthogonal to each other (unless the eigenvalues are degenerate), which means that the overlap factor is zero. The orthonormal set of eigenfunctions should then satisfy the following conditions:

$$\langle m|n\rangle = \delta_{m,n} \,. \tag{4.25}$$

Since the eigenfunctions of any operator compose the full set, any vector $|a\rangle$ in the Hilbert space can be projected onto this full set, thus giving

$$|a\rangle = \sum_n |n\rangle\langle n|a\rangle \,, \tag{4.26}$$

where $\langle n|a\rangle$ determines the projection of vector $|a\rangle$ onto vector $|n\rangle$. Similarly,

$$\langle x|a\rangle = \sum_n \langle x|n\rangle\langle n|a\rangle \,. \tag{4.27}$$

The latter equation connects wavefunctions in different representations. If the wavefunctions are normalized, that is,

$$\int \langle a|x\rangle\langle x|a\rangle \mathrm{d}x = \int \langle n|x\rangle\langle x|n\rangle \mathrm{d}x = 1 \,, \tag{4.28}$$

we will get

$$\langle a|a\rangle = \sum \langle a|n\rangle\langle n|a\rangle = 1 \,, \tag{4.29}$$

or otherwise

$$\sum_n |n\rangle\langle n| = \hat{I} \,, \tag{4.30}$$

where $\hat{I}$ is the unity operator. Equation (4.30) demonstrates the completeness of the wavefunction set.

Direct generalization of (4.30) can be used for representation of any operator $\hat{F}$:

$$\hat{F} = \sum_{n,m} F_{nm}|n\rangle\langle m| \,, \tag{4.31}$$

where F_{nm} is the matrix element of operator $\hat{F}$ in a particular representation.

4.4 Representations

4.4.1 Schrödinger Representation

The above formulation of quantum mechanics is termed the Schrödinger representation. A formal solution of the Schrödinger equation, (4.5), can be given by

$$\Psi(x,t) = \hat{U}(t,t_0)\Psi(x,t_0) \,, \tag{4.32}$$

where $\hat{U}(t,t_0)$ is the so-called evolution operator, which satisfies the following initial condition:

$$\hat{U}(t_0,t_0) = 1 \,. \tag{4.33}$$

From normalization conditions it follows that the evolution operator has to be a unitary operator since

$$\langle \hat{U}\Psi|\hat{U}\Psi\rangle = \langle \Psi|\hat{U}^\dagger\hat{U}\Psi\rangle = \langle \Psi|\Psi\rangle \,. \tag{4.34}$$

Indeed, inserting (4.32) into the Schrödinger equation, (4.5), we get

$$\mathrm{i}\hbar \frac{\partial \hat{U}(t, t_0)}{\partial t} = \hat{H} \hat{U}(t, t_0) . \tag{4.35}$$

In the case of a time-independent Hamiltonian the solution can be formally expressed by the evolution operator

$$\hat{U}(t, t_0) = \mathrm{e}^{-\frac{\mathrm{i}}{\hbar} \hat{H}(t-t_0)} , \tag{4.36}$$

and the wavefunction is

$$\Psi(x, t) = \mathrm{e}^{-\frac{\mathrm{i}}{\hbar} \hat{H}(t-t_0)} \Psi(x, t_0) . \tag{4.37}$$

In the case of a time-dependent Hamiltonian it follows from (4.35) that

$$\hat{U}(t, t_0) = 1 - \frac{\mathrm{i}}{\hbar} \int_{t_0}^{t} \hat{H}(\tau) \hat{U}(\tau, t_0) \mathrm{d}\tau . \tag{4.38}$$

This integral relation can be solved iteratively. The first iteration of (4.38) gives

$$\begin{aligned} \hat{U}(t, t_0) = 1 &- \frac{\mathrm{i}}{\hbar} \int_{t_0}^{t} \hat{H}(\tau) \mathrm{d}\tau \\ &+ \left(-\frac{\mathrm{i}}{\hbar}\right)^2 \int_{t_0}^{t} \mathrm{d}\tau \int_{t_0}^{\tau} \mathrm{d}\tau_1 \hat{H}(\tau) \hat{H}(\tau_1) \hat{U}(\tau_1, t_0) , \end{aligned} \tag{4.39}$$

and applying further iterations, we get

$$\begin{aligned} \hat{U}(t, t_0) = 1 + \sum_{n=1}^{\infty} \left(-\frac{\mathrm{i}}{\hbar}\right)^n \int_{t_0}^{t} \mathrm{d}\tau_n \int_{t_0}^{\tau_n} \mathrm{d}\tau_{n-1} \dots \int_{t_0}^{\tau_2} \mathrm{d}\tau_1 \\ \times \hat{H}(\tau_n) \hat{H}(\tau_{n-1}) \dots \hat{H}(\tau_1) , \end{aligned} \tag{4.40}$$

with time ordering: $t_0 \leq \tau_1 \leq \dots \leq \tau_{n-1} \leq \tau_n$. By analogy with the definition of an exponential function, it can also be represented symbolically as

$$\hat{U}(t, t_0) = \exp_+ \left(-\frac{\mathrm{i}}{\hbar} \int_{t_0}^{t} \mathrm{d}\tau \hat{H}(\tau) \right) . \tag{4.41}$$

Similarly, the Hermitian conjugated operator can also be defined accordingly:

$$\hat{U}^\dagger(t, t_0) = \exp_- \left(\frac{\mathrm{i}}{\hbar} \int_{t_0}^{t} \mathrm{d}\tau \hat{H}(\tau) \right) . \tag{4.42}$$

Evidently, in the case of the time-independent Hamiltonian the evolution operator and its conjugated operator coincide with the evolution operator given in (4.37) and its conjugated operator.

4.4.2
Heisenberg Representation

In the Schrödinger equation considered so far, the time evolution of the system is reflected in the wavefunction. However, the time evolution can also be relocated from the wavefunction to the operators, resulting in the so-called Heisenberg representation. This can be achieved by means of action of the evolution operator on the wavefunction in the Schrödinger representation:

$$\Psi_{\mathrm{H}}(x) = \hat{U}^{\dagger}(t)\Psi_{\mathrm{S}}(x,t) , \tag{4.43}$$

where $\Psi_{\mathrm{S}}(x,t)$ is the wavefunction in the Schrödinger representation as defined by (4.32). Correspondingly, $\Psi_{\mathrm{H}}(x)$ is the wavefunction in the Heisenberg representation and it is time-independent. Evidently, any operator should also be transformed by means of the evolution operator accordingly, giving

$$\hat{F}_{\mathrm{H}}(t) = \hat{U}^{\dagger}(t)\hat{F}\hat{U}(t) . \tag{4.44}$$

The operator $\hat{F}_{\mathrm{H}}(t)$ is thus the operator $\hat{F}$ in the Heisenberg representation, and it is now time-dependent. Since $\hat{U}^{\dagger}(0) = \hat{U}(0) = 1$, we get

$$\hat{F}_{\mathrm{H}}(0) = \hat{F} , \tag{4.45}$$

which means that operators in both representations coincide at some chosen initial time. From (4.44) it follows that the time evolution of the operator $\hat{F}_{\mathrm{H}}(t)$ over a short period of time Δt can be given as

$$\hat{F}_{\mathrm{H}}(t+\Delta t) = \hat{U}^{\dagger}(\Delta t)\hat{F}_{\mathrm{H}}(t)\hat{U}(\Delta t) . \tag{4.46}$$

Assuming that the Hamiltonian of the system is time-independent, and expanding the exponent in (4.36), we get

$$\hat{F}_{\mathrm{H}}(t+\Delta t) = \hat{F}_{\mathrm{H}}(t) + \frac{\mathrm{i}}{\hbar}[\hat{H}, \hat{F}_{\mathrm{H}}(t)]\Delta t + \ldots \tag{4.47}$$

From this equation it follows that

$$\frac{\mathrm{d}\hat{F}_{\mathrm{H}}}{\mathrm{d}t} = \frac{1}{\mathrm{i}\hbar}[\hat{F}_{\mathrm{H}}, \hat{H}] . \tag{4.48}$$

The equation obtained allows us to conclude that operators which commute with the Hamiltonian are time-independent in the Heisenberg representation. Since operators in the Schrödinger and Heisenberg representations coincide at $t = 0$, it is evident that the operators commuting with the Hamiltonian are the same in both representations. As a result, the Hamiltonian of the system is the same in both representations.

4.4.3
Interaction Representation

Let us consider a situation in which the Hamiltonian of a system can be represented as a sum of two terms:

$$\hat{H} = \hat{H}_0(t) + \hat{H}^1(t) . \tag{4.49}$$

If we invoke the definition of the evolution operator given in (4.35), the evolution operator for the system defined by Hamiltonian $\hat{H}_0(t)$ should satisfy the following equation of motion:

$$\frac{\partial}{\partial t}\hat{U}_0(t, t_0) = -\frac{\mathrm{i}}{\hbar}\hat{H}_0(t)\,\hat{U}_0(t, t_0) . \tag{4.50}$$

The solution of this equation can be given in terms of a corresponding evolution operator, (4.40):

$$\hat{U}_0(t, t_0) = \exp_+\left(-\frac{\mathrm{i}}{\hbar}\int_{t_0}^{t} \mathrm{d}\tau\,\hat{H}_0(\tau)\right) . \tag{4.51}$$

Let us define the wavefunction in the interaction representation, $\Psi_\mathrm{I}(t)$, via the following relation with the wavefunction in the Schrödinger representation, $\Psi_\mathrm{S}(t)$:

$$\Psi_\mathrm{S}(t) = \hat{U}_0(t, t_0)\,\Psi_\mathrm{I}(t) . \tag{4.52}$$

By inserting this relation into the Schrödinger equation, (4.5), we get the equation of motion for $\Psi_\mathrm{I}(t)$ in the form

$$\frac{\partial}{\partial t}\Psi_\mathrm{I}(t) = -\frac{\mathrm{i}}{\hbar}\hat{H}_\mathrm{I}^1(t)\,\Psi_\mathrm{I}(t) , \tag{4.53}$$

where

$$\hat{H}_\mathrm{I}^1(t) = \hat{U}_0^\dagger(t, t_0)\,\hat{H}^1(t)\,\hat{U}_0(t, t_0) . \tag{4.54}$$

Similarly to (4.32), let us now define the evolution operator $\hat{U}_\mathrm{I}(t, t_0)$ describing the solution of (4.53) as

$$\Psi_\mathrm{I}(t) = \hat{U}_\mathrm{I}(t, t_0)\,\Psi_\mathrm{I}(t_0) . \tag{4.55}$$

Evidently, the evolution operator $\hat{U}_\mathrm{I}(t, t_0)$ should satisfy the equation of motion with Hamiltonian,(4.54):

$$\frac{\partial}{\partial t}\hat{U}_\mathrm{I}(t, t_0) = -\frac{\mathrm{i}}{\hbar}\hat{H}_\mathrm{I}^1(t)\,\hat{U}_\mathrm{I}(t, t_0) . \tag{4.56}$$

By analogy with (4.35), it follows that

$$\hat{U}_\mathrm{I}(t, t_0) = \exp_+\left(-\frac{\mathrm{i}}{\hbar}\int_{t_0}^{t} \mathrm{d}\tau\,\hat{H}_\mathrm{I}^1(\tau)\right) . \tag{4.57}$$

Taking into account definition (4.52), we get

$$\Psi_S(t) = \hat{U}_0(t, t_0)\Psi_I(t) = \hat{U}_0(t, t_0)\hat{U}_I(t, t_0)\Psi_I(t_0)$$
$$= \hat{U}_0(t, t_0)\hat{U}_I(t, t_0)\Psi_S(t_0)\,, \tag{4.58}$$

because $\Psi_I(t_0) = \Psi_S(t_0)$. From comparison with (4.32), it follows that

$$\hat{U}(t, t_0) = \hat{U}_0(t, t_0)\hat{U}_I(t, t_0) = \hat{U}_0(t, t_0)\exp_+\left(-\frac{i}{\hbar}\int_{t_0}^{t} d\tau\,\hat{H}_I^1(\tau)\right)\,, \tag{4.59}$$

or more explicitly

$$\hat{U}(t, t_0) = \hat{U}_0(t, t_0) + \sum_{n=1}^{\infty}\left(-\frac{i}{\hbar}\right)^n \int_{t_0}^{t} d\tau_n \int_{t_0}^{\tau_n} d\tau_{n-1}\ldots\int_{t_0}^{\tau_2} d\tau_1$$
$$\times\,\hat{U}_0(t, \tau_n)\hat{H}^1(\tau_n)\hat{U}_0(\tau_n, \tau_{n-1})\hat{H}^1(\tau_{n-1})\ldots$$
$$\times\,\hat{U}_0(\tau_2, \tau_1)\hat{H}^1(\tau_1)\hat{U}_0(\tau_1, t_0)\,. \tag{4.60}$$

Thus, both wavefunctions and operators are time-dependent in the interaction representation. The time dependence of the latter is defined by analogy with (4.54) for any operator $\hat{F}$ in the Schrödinger representation by the following relation:

$$\hat{F}_I = \hat{U}_0^+(t, t_0)\hat{F}\hat{U}_0(t, t_0)\,. \tag{4.61}$$

Similarly to (4.48) derived in the Heisenberg representation, operator $\hat{F}_I$ in the interaction representation should satisfy the following equation:

$$\frac{d\hat{F}_I(t)}{dt} = \frac{1}{i\hbar}\left[\hat{F}_I, \hat{H}_0\right]\,. \tag{4.62}$$

All representations described in this section are equivalent to each other. They should be used depending on the convenience they provide in the search for the solution of the problem under consideration.

4.5 Density Matrix

4.5.1 Definition

The wavefunction introduced above is not an experimentally observable physical entity. Its square determines the probability density for finding a quantum mechanical system in some associated state. It is possible to construct a quantity, the so-called *density matrix*, which operates directly with the notion of the "square" of the wavefunction and has therefore a more direct interpretation. The density matrix provides an alternative description of a system state. Formally, the density matrix represents the so-called *density operator*. The main advantage of such a representation of the quantum system is due to the possibility to encompass statistical ensembles beyond those of identical molecules assumed in the probabilistic interpretation of the wavefunction.

Let us assume that the system is characterized by a wavefunction Ψ defined in the basis set of wavefunctions φ_n:

$$\Psi(x) = \sum_n c_n \varphi_n(x) . \tag{4.63}$$

The probability of finding the system in the state characterized by the wavefunction φ_n is given by $c_n^* c_n$. The expectation value $\langle A \rangle$ of an operator $\hat{A}$ for a system in state Ψ is given by (4.17) or in the chosen basis of φ_n functions,

$$\langle A \rangle = \sum_{mn} A_{mn} c_n c_m^* , \tag{4.64}$$

where

$$A_{mn} = \int dx \varphi_m^*(x) \hat{A}(x) \varphi_n(x) . \tag{4.65}$$

It follows from (4.64) that the expansion coefficients c_n of the system state are the only thing necessary to calculate the expectation value of a physical quantity, once the operator is represented in a given basis, that is, A_{mn} is known. Thus, a quantity which holds these coefficients can be defined,

$$\rho_{mn} = c_m c_n^* , \tag{4.66}$$

and the expectation value $\langle A \rangle$ can be expressed as

$$\langle A \rangle = \sum_{mn} A_{mn} \rho_{nm} . \tag{4.67}$$

The quantity ρ_{nm} is termed the *density matrix*.

Equation (4.67) has a structure of a trace over a matrix obtained by multiplication of the density matrix and the matrix representation of the operator $\hat{A}$. We can thus alternatively present (4.67) as

$$\langle A \rangle = \mathrm{tr}\{\hat{A}\rho\} . \tag{4.68}$$

The trace operation on an arbitrary operator can be defined in an abstract way for operators acting on any Hilbert space. Equation (4.68) suggests that ρ is an operator, which could also be represented in an abstract Hilbert space. The density matrix is thus just a representation of the abstract density operator in a specific basis. The basis can be discrete or continuous.

The interpretation of the density matrix elements in the discrete representation is rather straightforward. According to the Born rule, the diagonal elements $\rho_{nn} =$

$|c_n|^2$ represent the probabilities (or probability densities) of finding the system in a state characterized by the wavefunction φ_n. The off-diagonal elements ρ_{mn}, often denoted as coherences, represent the coherence between states φ_m and φ_n, that is, the degree to which the state Ψ of the system is a linear combination of φ_m and φ_n. In some sense this is the degree of interference of the system between these states. Because the diagonal elements represent probabilities, their sum has to be normalized to 1. We can thus introduce the general normalization condition for the density matrix:

$$\sum_n \rho_{nn} = \mathrm{tr}\{\rho\} = 1\,. \tag{4.69}$$

The density matrix in a continuous coordinate representation has a form analogous to (4.66), that is, $\rho(x', x) = \Psi(x')\Psi^*(x)$. The diagonal "elements" of this continuous density matrix are $\rho(x, x)\mathrm{d}x = |\Psi(x)|^2\mathrm{d}x$, and they therefore represent the probability density for finding the system in the interval $\mathrm{d}x$ around coordinate x. It is normalized accordingly as $\int \mathrm{d}x\rho(x, x) = 1$, and the expectation value of the operator is

$$\langle A\rangle = \iint \mathrm{d}x\mathrm{d}x' A_{x'x}\rho(x, x')\,. \tag{4.70}$$

Here it is important to note that in the coordinate representation the operator has to have the form $A_{x'x} = \delta(x-x')A(x)$. The Dirac delta function is required for (4.70) to be equivalent with (4.17).

The density matrix can be equivalently presented in bra-ket notation. Consider the system described by a wavefunction $\psi(x)$, which can also be denoted by $|\psi\rangle$. The density matrix is then given by

$$\rho = |\psi\rangle\langle\psi|\,. \tag{4.71}$$

The state can be expanded in another basis $|a\rangle$ as $|\psi\rangle = \sum_a c_a|a\rangle$. Now the product with the conjugate wavefunction according to (4.71) gives the density operator:

$$\rho = \sum_{ab} c_a c_b^*|a\rangle\langle b|\,. \tag{4.72}$$

Thus, in this basis set $\rho_{ab} = c_a c_b^*$, which is equivalent to (4.66).

The main properties of the density matrix follow from its definition: $\rho_{aa} = |c_a|^2$ is the probability with normalization $\sum_a \rho_{aa} = 1$, $\rho_{ab} = \rho_{ba}^*$, and $|\rho_{ab}|^2 = |c_a|^2|c_b|^2 = \rho_{aa}\rho_{bb}$. The density operator clearly does not carry more information than the wavefunction as described above. Instead, it often disregards the auxiliary unobservable information from the wavefunction. Its advantage becomes apparent when considering statistical ensembles.

4.5.2 Pure versus Mixed States

In the usual interpretation of quantum mechanics, the wavefunction, (4.63), represents a system in state Ψ and all properties of the system are thus defined. However, when preparing an individual system for measurement, there might be some uncertainty in the states in which the system is actually prepared. One might prepare states ψ_a and ψ_b with probabilities p_a and p_b. The state of the system is no longer a linear superposition of states ψ_a and ψ_b, but is rather "in state ψ_a or state ψ_b with certain probabilities," which is denoted as a statistical mixture. The difference between these two situations is a lack of coherence between states ψ_a and ψ_b in the latter case.

It turns out that the density operator in such a scenario is completely determined. Adding two wavefunctions automatically introduces the coherence, but adding two density operators (weighted by the probabilities of finding systems with the corresponding states) does not. Thus,

$$\rho = \sum_{n=a,b} p_n |\psi_n\rangle\langle\psi_n| \tag{4.73}$$

correctly describes a statistical ensemble of system states prepared in either state ψ_a or state ψ_b. We can verify that the expectation value for operator $\hat{A}$ is defined consistently as a sum of expectation values for states ψ_a and ψ_b:

$$\langle A\rangle = \text{tr}\{\rho A\} = \sum_{n=a,b} p_n \langle\psi_n|\hat{A}|\psi_n\rangle \; . \tag{4.74}$$

The physical quantity in this case is well defined, and the system is described by a statistical ensemble of wavefunctions. We will describe statistical ensembles more extensively in Chapter 7. States that are represented by a single wavefunction are called *pure* states, and states that cannot be written as a single wavefunction and require a density operator for their characterization are called *mixed* states. The mixed states represent ensembles. This does not imply the existence of physical replicas of the system under consideration, but instead describes the limited knowledge of the system state.

To distinguish the pure and mixed states from each other formally, we can use the following inequality:

$$\text{tr}\{\rho^2\} < 1 \; , \tag{4.75}$$

which holds for a mixed state. It is easy to show that the pure state $\rho = |\psi\rangle\langle\psi|$ has

$$\text{tr}\{\rho\} = \sum_n \langle\varphi_n|\psi\rangle\langle\psi|\varphi_n\rangle = \langle\psi|\psi\rangle = 1 \tag{4.76}$$

as a result of the normalization conditions for the wavefunction. In a mixed state $\rho = \sum_n p_n \rho^{(n)}$, where $\rho^{(n)} = |\psi_n\rangle\langle\psi_n|$ are some density operators representing (different but not necessarily orthogonal) pure states, the probabilities have to satisfy $\sum_n p_n = 1$. From this it follows that

$$\text{tr}\{\rho^2\} = \sum_{nm} p_n p_m \text{tr}\{\rho^{(n)}\rho^{(m)}\} < \sum_{nm} p_n p_m = 1 \; , \tag{4.77}$$

because

$$\text{tr}\{\rho^{(n)}\rho^{(m)}\} = |\langle\psi_n|\psi_m\rangle|^2 < 1 \; . \tag{4.78}$$

4.5.3
Dynamics in the Liouville Space

The Schrödinger equation, (4.5), describes quantum dynamics of a system wavefunction:

$$i\hbar\frac{d|\Psi\rangle}{dt} = \hat{H}|\Psi\rangle \tag{4.79}$$

where $\hat{H}$ is the total system Hamiltonian, given by the sum of kinetic and potential energies. For the time-independent Hamiltonian the formal solution, (4.37), reads

$$|\Psi(t)\rangle = \exp\left(-\frac{i}{\hbar}\hat{H}t\right)|\Psi(0)\rangle\ . \tag{4.80}$$

The Hermitian conjugate equation for the wavefunction can also be defined, and it reads

$$i\hbar\frac{d\langle\Psi|}{dt} = -\langle\Psi|\hat{H}\ . \tag{4.81}$$

With use of these definitions it is straightforward to write the equation for the density matrix of the pure state:

$$i\hbar\frac{d}{dt}\rho = i\hbar\frac{d}{dt}|\Psi\rangle\langle\Psi| = \hat{H}\rho - \rho\hat{H} = [\hat{H},\rho]\ . \tag{4.82}$$

Expanding the equation in some basis set, we get

$$i\hbar\frac{d}{dt}\rho_{ab} = \sum_c (H_{ac}\rho_{cb} - \rho_{ac}H_{cb})\ . \tag{4.83}$$

Let us assume that the system eigenstates are $|\varphi_a\rangle$ with energies ε_a. We then have for the Hamiltonian and the density matrix in this basis set $H_{ab} \equiv \delta_{ab}\varepsilon_a$ and $\rho \equiv \rho_{ab}$, and from (4.83) we get

$$i\hbar\frac{d\rho_{ab}}{dt} = (\varepsilon_a - \varepsilon_b)\rho_{ab}\ , \tag{4.84}$$

and the solution

$$\rho_{ab}(t) = e^{-i\omega_{ab}t}\rho_{ab}(0)\ , \tag{4.85}$$

where $\omega_{ab} = (\varepsilon_a - \varepsilon_b)/\hbar$ describes the unitary evolution of the density matrix, which is simply the phase rotation of coherences. As the state is stationary, the probabilities are time-independent, $\rho_{aa} = \text{const}$.

It is sometimes convenient to introduce a different notation with respect to the density matrix dynamics. In (4.83) both the density matrix and the Hamiltonian are square matrices. We can reorganize elements of the density matrix into a single row and assume it to be a vector. We can write (4.83) in the form

$$\frac{d}{dt}\rho_{ab} = -i\sum_{cd}\mathcal{L}_{ab,cd}\rho_{cd}\ , \tag{4.86}$$

or without specifying the basis set simply as $d\rho/dt = -i\mathcal{L}\rho$, where the *superoperator* $\mathcal{L}$ acting on *operator* ρ is introduced. In this so-called *Liouville representation*, the possible density matrices are vectors in the Liouville space, that is, we treat each regular operator, in our particular case the density matrix, as a vector, and the superoperators, which are the tetradic matrices in a regular case, operate on these square-matrix operators.

Two types of superoperators are the most important. The Liouville operator is a construct out of the system Hamiltonian. For the arbitrary orthogonal basis set we have $\mathcal{L}_{ab,cd} = \hbar^{-1}(\delta_{bd}H_{ac} - \delta_{ac}H_{db})$. For the expansion in Hamiltonian eigenstates, the Liouville operator becomes *diagonal*, that is, $\mathcal{L}_{ab,cd} = \omega_{ab}\delta_{ac}\delta_{bd}$. The second important superoperator is the propagator, or the Green's function for the density matrix. It is the time-domain forward propagator. The time evolution of the wavefunction can be expressed as $|\Psi(t)\rangle = G(t)|\Psi(0)\rangle$, where

$$G(t) = \theta(t)\exp\left(-\frac{i}{\hbar}\hat{H}t\right) \equiv \theta(t)\hat{U}(t) , \tag{4.87}$$

and $\theta(t)$ is the Heaviside step function, which is equal to 1 for positive times and 0 for negative times. The operator $G(t)$ is often called a Green's function, and it coincides with the evolution operator defined by (4.37), when considering the forward propagation in the time domain. In the Liouville space according to (4.86), the formal solution of the Liouville equation for the time-independent Liouville operator is $\rho(t) = \exp(-i\mathcal{L}t)\rho(0)$. Similarly to the Hilbert space, the Liouville space Green's function is defined by forward propagation of the density operator $\rho(t) = \mathcal{G}(t)\rho(0)$, where

$$\mathcal{G}(t) = \theta(t)\exp(-i\mathcal{L}t) \tag{4.88}$$

is the formal expression of the Green's function. These types of superoperators will be used later in this book.

4.6 Model Systems

4.6.1 Harmonic Oscillator

The classical harmonic oscillator is the model describing the motion of a particle in a harmonic potential (Figure 4.1):

$$V(x) = \frac{k}{2}x^2 , \tag{4.89}$$

where k determines the strength of the restoring force and x is the coordinate of the particle. The classical dynamics of this oscillator was described in Section 2.1.3. We now consider the quantum case.

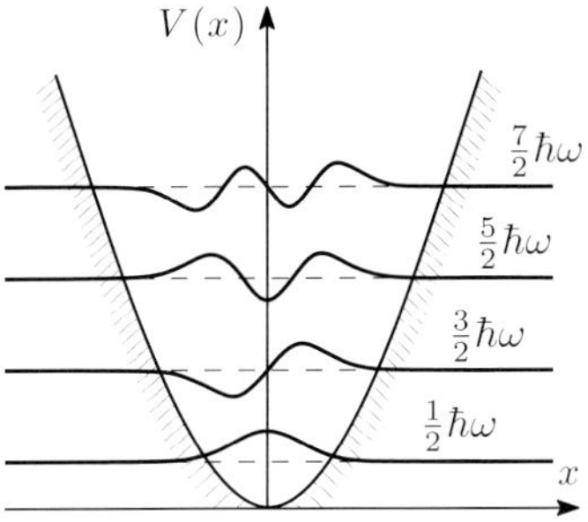

Figure 4.1 Low-lying energy levels of a particle in a one-dimensional harmonic potential $V(x)$. Wavy curves represent the eigenfunctions, the dashed lines – the eigenenergies.

In the case of a one-dimensional harmonic oscillator, the coordinate describes the motion along this particular coordinate. The Hamiltonian of such a system reads as follows:

$$\hat{H} = -\frac{\hbar^2}{2m}\frac{d^2}{dx^2} + V(x) . \tag{4.90}$$

By introducing the notation $k = m\omega^2$, we can rewrite the Schrödinger equation, (4.7), as

$$-\frac{\hbar^2}{2m}\frac{d^2\varphi(x)}{dx^2} + \frac{m\omega^2}{2}x^2\varphi(x) = E\varphi(x) . \tag{4.91}$$

Introducing also the dimensionless coordinate $\xi = \sqrt{m\omega/\hbar}x$ and energy $\varepsilon = E/(\hbar\omega)$, we get

$$\left(\frac{d^2}{d\xi^2} - \xi^2\right)\varphi(\xi) = -2\varepsilon\varphi(\xi) . \tag{4.92}$$

This is now a second-order differential equation with $\varphi(\xi)$ representing the desired wavefunction, and ε representing the corresponding energy of the oscillator.

We will solve (4.92) by first noting that

$$\left(\xi + \frac{d}{d\xi}\right)\left(\xi - \frac{d}{d\xi}\right)\varphi(\xi) = \left(\xi^2 - \frac{d^2}{d\xi^2} + 1\right)\varphi(\xi) , \tag{4.93}$$

which can be used to rewrite (4.92) as

$$\left(\xi + \frac{d}{d\xi}\right)\left(\xi - \frac{d}{d\xi}\right)\varphi(\xi) = (2\varepsilon + 1)\varphi(\xi) . \tag{4.94}$$

Similarly we obtain

$$\left(\xi - \frac{d}{d\xi}\right)\left(\xi + \frac{d}{d\xi}\right)\varphi(\xi) = (2\varepsilon - 1)\varphi(\xi) . \tag{4.95}$$

Thus, the wavefunction satisfies both (4.94) and (4.95). If we now take (4.94) and act on the left with operator $\xi - \mathrm{d}/\mathrm{d}\xi$, we have

$$\left(\xi - \frac{\mathrm{d}}{\mathrm{d}\xi}\right)\left(\xi + \frac{\mathrm{d}}{\mathrm{d}\xi}\right)\left(\xi - \frac{\mathrm{d}}{\mathrm{d}\xi}\right)\varphi(\xi)$$
$$= [2(\varepsilon + 1) - 1]\left(\xi - \frac{\mathrm{d}}{\mathrm{d}\xi}\right)\varphi(\xi)\,, \tag{4.96}$$

that is, the function

$$\phi_{+}(\xi) = \left(\xi - \frac{\mathrm{d}}{\mathrm{d}\xi}\right)\varphi(\xi)$$

is the wavefunction with energy $\varepsilon + 1$. Using (4.95), we can show that

$$\phi_{-}(\xi) = \left(\xi + \frac{\mathrm{d}}{\mathrm{d}\xi}\right)\varphi(\xi)$$

is the wavefunction with energy $\varepsilon - 1$. Thus, any wavefunction corresponding to eigenvalue ε can be constructed when the "first" wavefunction is found.

The first wavefunction may be obtained from (4.94). Taking $\varepsilon = -1/2$, we get the requirement

$$\left(\xi - \frac{\mathrm{d}}{\mathrm{d}\xi}\right)\varphi(\xi) = 0\,. \tag{4.97}$$

It is easy to verify that

$$\varphi(\xi) = A\mathrm{e}^{\xi^2/2}\,, \tag{4.98}$$

where A is the normalization constant, is the solution of (4.97). However, this solution is rejected because it does not satisfy the boundary conditions, $\varphi(\xi) \to 0$ when $\xi \to \pm\infty$, needed for the normalization requirement of the wavefunction.

Equation (4.95) can be satisfied if $\varepsilon = 1/2$ and

$$\left(\xi + \frac{\mathrm{d}}{\mathrm{d}\xi}\right)\varphi(\xi) = 0 \tag{4.99}$$

at the same time. In this case the differential equation has the solution

$$\varphi(\xi) = A\mathrm{e}^{-\xi^2/2}\,, \tag{4.100}$$

which represents an acceptable eigenfunction because it satisfies the boundary conditions.

Let us try to find a wavefunction which corresponds to an energy lower than $\varepsilon = 1/2$. We will try

$$\phi_{-}(\xi) = \left(\xi + \frac{\mathrm{d}}{\mathrm{d}\xi}\right)\mathrm{e}^{-\xi^2/2} = 0\,,$$

which means that states with lower energy are not possible. Therefore, the eigenstate defined by (4.100) and the corresponding eigenvalue determine the oscillator ground state, that is, we can denote them by $\varphi_0(\xi)$ and ε_0.

To determine the wavefunction $\varphi_n(\xi)$ of the excited states characterized by quantum number n we will use (4.94) and apply operator $(\xi - \mathrm{d}/\mathrm{d}\xi)$ as described above. The discussion above shows that operator $(\xi - \mathrm{d}/\mathrm{d}\xi)$ is an energy-raising operator. When this operator acts on eigenfunction $\varphi_n(\xi)$ corresponding to eigenvalue ε_n, eigenfunction $\varphi_{n+1}(\xi)$ with eigenvalue ε_{n+1} is created. Thus, starting with (4.100), we can generate the rest of the states of the harmonic oscillator. The first excited state is given by

$$\varphi_1(\xi) = A_1\left(\xi - \frac{\mathrm{d}}{\mathrm{d}\xi}\right)\varphi_0(\xi) \tag{4.101}$$

and $\varepsilon_1 = 3/2$, and the second excited state is defined accordingly:

$$\varphi_2(\xi) = A_2\left(\xi - \frac{\mathrm{d}}{\mathrm{d}\xi}\right)^2\varphi_0(\xi) \tag{4.102}$$

and $\varepsilon_2 = 5/2$. Similarly,

$$\varphi_n(\xi) = A_n\left(\xi - \frac{\mathrm{d}}{\mathrm{d}\xi}\right)^n\varphi_0(\xi) \tag{4.103}$$

and $\varepsilon_n = n + 1/2$. Coefficients A_n are normalization factors.

Wavefunctions for an arbitrary n can be explicitly defined via Hermite polynomials,

$$H_n(\xi) = (-1)^n \mathrm{e}^{\xi^2}\frac{\mathrm{d}^n}{\mathrm{d}\xi^n}\mathrm{e}^{-\xi^2}\,, \tag{4.104}$$

as

$$\varphi_n(\xi) = N_n H_n(\xi)\mathrm{e}^{-\xi^2/2}\,, \tag{4.105}$$

where

$$N_n = \frac{1}{\sqrt{\sqrt{\pi}\,n!2^n}} \tag{4.106}$$

accounts for the normalization.

Let us now apply operator $(\xi + \mathrm{d}/\mathrm{d}\xi)$ to both sides of (4.95) to give

$$\left(\xi + \frac{\mathrm{d}}{\mathrm{d}\xi}\right)\left(\xi - \frac{\mathrm{d}}{\mathrm{d}\xi}\right)\left(\xi + \frac{\mathrm{d}}{\mathrm{d}\xi}\right)\varphi_n(\xi) = (2\varepsilon_n - 1)\left(\xi + \frac{\mathrm{d}}{\mathrm{d}\xi}\right)\varphi_n(\xi)\,. \tag{4.107}$$

By comparison with (4.94), we can conclude that $(\xi + \mathrm{d}/\mathrm{d}\xi)$ is the energy-lowering operator.

We thus find that the energy states of the harmonic oscillator form a ladder with equal energy gaps. This is one of the most important results, allowing one to introduce the particles – bosons – and their number operators. This so-called second-quantization procedure is introduced in Section 4.9.

4.6.2
Quantum Well

Let us consider the one-dimensional quantum well shown in Figure 4.2. The potential energy for the quantum particle in this well is defined as

$$V(x) = \begin{cases} 0\,, & \text{if} \quad 0 \le x \le a\,, \\ \infty\,, & \text{otherwise}\,. \end{cases} \tag{4.108}$$

The quantum particle in the quantum well is completely free, except at the two walls, which confine the particle. The Hamiltonian of a free particle is equal to the kinetic energy, $\hat{T} = \hat{p}^2/(2m)$, defined in Chapter 2. The momentum operator in quantum mechanics is $\hat{\boldsymbol{p}} = -\mathrm{i}\hbar\nabla$ [2, 15]. Thus, the Schrödinger equation, (4.7), inside the one-dimensional well reads

$$-\frac{\hbar^2}{2m}\frac{\mathrm{d}^2\varphi(x)}{\mathrm{d}x^2} = E\varphi(x)\,, \tag{4.109}$$

where m is the mass of the quantum particle. A general solution of this equation can be found in the form

$$\varphi(x) = A\sin kx + B\cos kx\,, \tag{4.110}$$

where $k = \sqrt{2mE/\hbar^2}$, and A and B are arbitrary constants, which should satisfy the boundary conditions of the well and the normalization of the wavefunction. Since the wavefunction at the walls should be zero, that is, $\varphi(0) = \varphi(a) = 0$, it follows that $B = 0$ and the wavefunction, (4.110), is given by

$$\varphi(x) = A\sin kx\,. \tag{4.111}$$

From the boundary conditions it follows that $\sin ka = 0$ and thus

$$ka = 0, \pm\pi, \pm 2\pi, \pm 3\pi, \ldots \tag{4.112}$$

This allows us to determine k as

$$k_n = \frac{\pi n}{a} \tag{4.113}$$

with $n = 1, 2, 3, \ldots$ and hence the possible eigenvalues of E_n:

$$E_n = \frac{\pi^2\hbar^2 n^2}{2ma^2}\,. \tag{4.114}$$

By taking into account the normalization conditions for the wavefunction, we can define the amplitude A accordingly:

$$\varphi_n(x) = \sqrt{\frac{2}{a}}\sin\left(\frac{\pi n x}{a}\right)\,. \tag{4.115}$$

The wavefunctions obtained determine quantum states which satisfy the conditions of standing waves in this well. The state with $n = 1$ has the lowest energy and it represents the ground state. States with higher energies proportional to n^2 are excited states. The eigenfunctions corresponding to different values of n are orthogonal to each other.

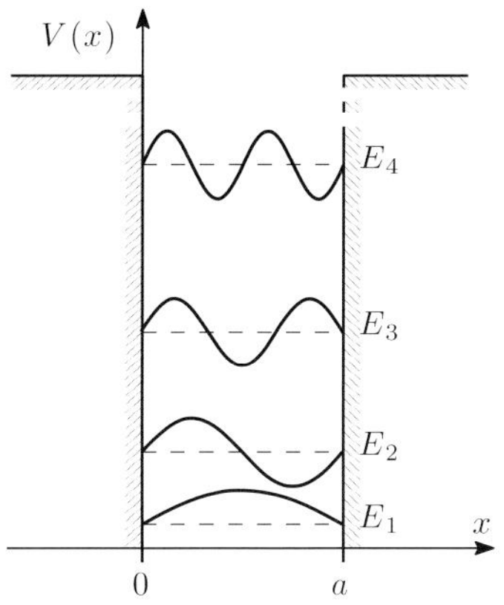

Figure 4.2 Low-lying energy levels of a particle of mass m in an infinite square-well potential $V(x)$ with width a. Wavy curves represent the eigenfunctions, the dashed lines – the eigenenergies.

4.6.3
Tunneling

The ability to penetrate the barrier, which is classically forbidden, is one of the exceptional properties of a quantum particle. To demonstrate this, let us consider a free particle moving in a one-dimensional space and interacting with a potential barrier, which determines the following potential energy (see Figure 4.3):

$$V(x) = \begin{cases} 0\,, & -\infty < x \le 0\,, \\ V_0\,, & 0 < x < a\,, \\ 0\,, & a \le x < \infty\,. \end{cases} \tag{4.116}$$

If a classical particle is approaching this barrier from the side of negative x values, it is reflected if its kinetic energy is below V_0, and it is transmitted in the opposite

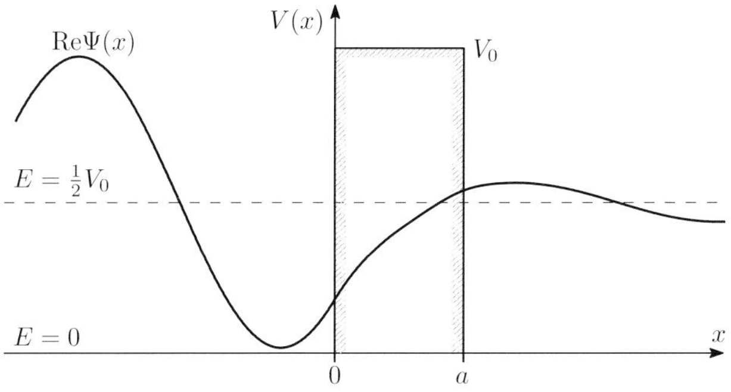

Figure 4.3 Penetration of a quantum particle through a potential barrier of height V_0 and width a. The spatial distribution of the real part of the wavefunction of the particle initially positioned on the left side of the potential barrier with energy $E = 1/2V_0$ is also shown.

case. In general, the behavior of a quantum particle of mass m in the presence of potential $V(x)$ should satisfy the Schrödinger equation:

$$-\frac{\hbar^2}{2m}\frac{\partial^2 \Psi}{\partial x^2} + V(x)\Psi = E\Psi \,. \tag{4.117}$$

In order to demonstrate the tunneling effect, let us consider the possibility of reflection and transmission of an incoming particle by the barrier in the case when the particle energy E is lower than V_0. The stationary Schrödinger equation in regions $-\infty < x \leq 0$ and $a \leq x < \infty$ corresponds to the free particle case:

$$\frac{d^2 \Psi_k}{dx^2} = -k^2 \Psi_k \,, \tag{4.118}$$

where k determines the eigenenergy, $E = \hbar^2 k^2/(2m)$. The general solution of this equation can be represented by free particle states:

$$\Psi_k = A e^{ikx} + B e^{-ikx} \,, \tag{4.119}$$

where A and B are arbitrary constants.

In the region $0 \leq x \leq a$ the energy of the particle is classically forbidden since $E < V_0$. In this case the stationary Schrödinger equation can be given by

$$\frac{d^2 \Psi_\chi}{dx^2} = \chi^2 \Psi_\chi \,, \tag{4.120}$$

where $V_0 - E = \hbar^2\chi^2/(2m)$. The general solution in this case is as follows:

$$\Psi_\chi = C e^{-\chi x} + D e^{\chi x} \,, \tag{4.121}$$

with arbitrary constants C and D.

In order for the wavefunction to be smooth, we have to determine the arbitrary constants. The wavefunctions and their derivatives have to be smoothly joined at points $x = 0$ and $x = a$ by taking into account the normalization condition.

To consider the transparency of the barrier we have to analyze the relation between the particle moving toward the barrier in the region $-\infty < x \leq 0$ and away from the barrier in the region $a \leq x < \infty$ (terms proportional to A in (4.119)). The ratio of these coefficients will determine the probability of transmission, T, of the particle through the barrier, $T = |A_R|^2/|A_L|^2$, where indices R and L indicate the right and left side of the barrier, respectively. Assuming in addition that the barrier is wide enough so that $e^{-2\chi a} \ll 1$, we get

$$T \approx \frac{16k^2\chi^2}{(k^2+\chi^2)^2} e^{-2\chi a} \,, \tag{4.122}$$

or using the energy,

$$T \approx \frac{16E(V_0 - E)}{V_0^2} e^{-2\chi a} \,. \tag{4.123}$$

Since this approximate expression is valid when $e^{-2\chi a} \ll 1$, the wavefunction inside the barrier is mainly defined by the first term of (4.121), and the tunneling probability is proportional to $e^{-2\chi a}$. Such an exponential dependence of the tunneling probability on the barrier penetration value χ and the width a of the barrier is the main factor determining the quantum character of the particle (see Figure 4.3).

4.6.4
Two-Level System

When considering atoms or molecules resonantly interacting with an electromagnetic field, one can disregard all states for which the transition frequencies are sufficiently different from the frequency of the light. One can often disregard all states except the ground state and one excited state resonant with the exciting light. Let us therefore consider a two-level system interacting with an external field. The Hamiltonian of such a system is given by

$$\hat{H} = \hat{H}_0 + \hat{H}^1(t) \,, \tag{4.124}$$

where $\hat{H}_0$ is the Hamiltonian of the two-level system and $\hat{H}^1(t)$ describes the external field. Since it is assumed to be time-dependent, we have to consider the nonstationary Schrödinger equation.

Let us assume that the eigenstates corresponding to the two-level system (Hamiltonian $\hat{H}_0$) have eigenvalues E_1^0 and E_2^0. We assume the external field is an oscillating field with frequency ω, that is,

$$\hat{H}^1(t) = 2\hat{H}^1 \cos \omega t \equiv \hat{H}^1(e^{i\omega t} + e^{-i\omega t}) \,. \tag{4.125}$$

Since the two-level system is characterized by two states, they evidently constitute the complete set of functions. By denoting eigenfunction φ_1 as corresponding to eigenvalue E_1^0 and eigenfunction φ_2 as corresponding to eigenvalue E_2^0, we can represent any wavefunction accordingly:

$$\Psi(t) = c_1(t)\varphi_1 + c_2(t)\varphi_2 \,. \tag{4.126}$$

Substituting this wavefunction into the Schrödinger equation, (4.5), we get

$$i\hbar \frac{\partial}{\partial t}\left[c_1(t)\varphi_1 + c_2(t)\varphi_2\right] = \hat{H}\left[c_1(t)\varphi_1 + c_2(t)\varphi_2\right] \,, \tag{4.127}$$

and now multiplying (4.127) either by φ_1^* or by φ_2^* and integrating over all variables of the system, we get

$$i\hbar \frac{\partial}{\partial t} c_1(t) = E_1^0 c_1(t) + H_{11}^1(t) c_1(t) + H_{12}^1(t) c_2(t) \tag{4.128}$$

and

$$i\hbar \frac{\partial}{\partial t} c_2(t) = E_2^0 c_2(t) + H_{22}^1(t) c_2(t) + H_{21}^1(t) c_1(t) \,, \tag{4.129}$$

where $H^1_{ij}(t) = \langle i|\hat{H}^1(t)|j\rangle$.

In the absence of the external field, the solutions of (4.127) and (4.129) give

$$\Psi(t) = a_1(0)e^{-i\omega_1 t}\varphi_1 + a_2(0)e^{-i\omega_2 t}\varphi_2 , \tag{4.130}$$

where $\omega_i = E_i^0/\hbar$ and coefficients $a_i(0)$ allow one to satisfy the normalization conditions. Evidently, all coefficients $c_j(t)$ should contain the phase factor $e^{-i\omega_j t}$ due to the first terms on the right-hand side of (4.127) and (4.129) even in the presence of the external field. By inserting

$$c_j(t) = a_j(t)e^{-i\omega_j t} \tag{4.131}$$

into (4.127) and (4.129), we get

$$i\hbar\frac{\partial}{\partial t}a_1(t) = H^1_{11}(t)a_1(t) + H^1_{12}(t)a_2(t)e^{-i\omega_{21}t} \tag{4.132}$$

and

$$i\hbar\frac{\partial}{\partial t}a_2(t) = H^1_{22}(t)a_2(t) + H^1_{21}(t)a_1(t)e^{-i\omega_{12}t} , \tag{4.133}$$

where $\omega_{ij} = \omega_i - \omega_j$. By considering the resonance interaction with the external field, we can assume that diagonal elements are absent, that is, $H^1_{ii}(t) = 0$. Equations (4.132) and (4.133) then reduce to

$$\frac{\partial}{\partial t}a_1(t) = \frac{1}{i\hbar}H^1_{12}(t)e^{-i\omega_{21}t}a_2(t) \tag{4.134}$$

and

$$\frac{\partial}{\partial t}a_2(t) = \frac{1}{i\hbar}H^1_{21}(t)e^{-i\omega_{12}t}a_1(t) , \tag{4.135}$$

and then substituting the time dependence of the field term defined by (4.125), we get explicitly

$$\frac{\partial}{\partial t}a_1(t) = \frac{1}{i\hbar}H^1_{12}\left[e^{i(\omega-\omega_{21})t} + e^{-i(\omega+\omega_{21})t}\right]a_2(t) \tag{4.136}$$

and

$$\frac{\partial}{\partial t}a_2(t) = \frac{1}{i\hbar}H^1_{21}\left[e^{i(\omega+\omega_{21})t} + e^{-i(\omega-\omega_{21})t}\right]a(t) , \tag{4.137}$$

where $H^1_{21} = \langle 2|H^1|1\rangle$.

Let us assume now that $\omega = 0$, that is, the external field is constant. By differentiating (4.135), with subsequent substitution of (4.134), we get

$$\frac{\partial^2}{\partial t^2}a_2 = -\frac{|H^1_{21}|^2}{\hbar^2}a_2 + i\omega_{21}\frac{\partial}{\partial t}a_2 . \tag{4.138}$$

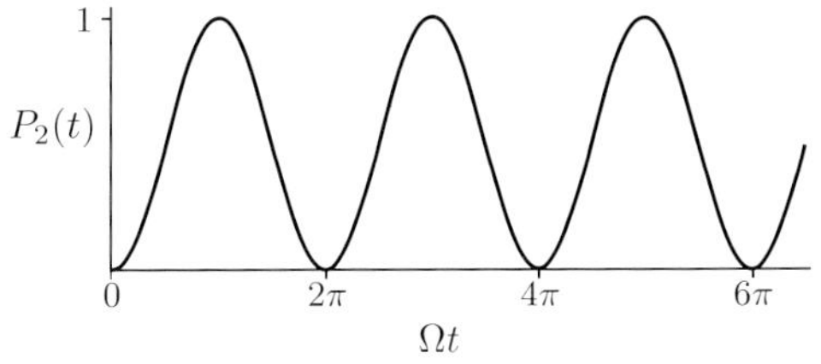

Figure 4.4 The time-dependent probability of a two-level system being found in an initially unoccupied state.

Similarly, we can also derive a separate equation for a_1.

The general solution of (4.138) is given by

$$a_2(t) = \left(A e^{i\Omega t} + B e^{-i\Omega t}\right) e^{i\omega_{21} t/2} , \tag{4.139}$$

where $\Omega = 1/2\sqrt{\omega_{21}^2 + 4|H_{21}^1|^2/\hbar^2}$, and A and B are constants determined by the initial conditions.

Let us assume that $a_1(0) = 1$ and $a_2(0) = 0$ at $t = 0$, that is, the system is in state 1. Then we get

$$a_2(t) = -i\frac{|H_{21}^1|}{\Omega} e^{i\omega_{21} t} \sin \Omega t \tag{4.140}$$

and

$$a_1(t) = \left(\cos \Omega t + i\frac{\omega_{21}}{2\Omega} \sin \Omega t\right) e^{-i\omega_{21} t/2} . \tag{4.141}$$

Now we can define the probability of finding the system in any of these two states. $P_1(t) = |a_1(t)|^2$ is the probability of finding the system in state 1, and $P_2(t) = |a_2(t)|^2$ is the probability of finding the system in state 2. The latter is

$$P_2(t) = \frac{4|H_{21}^1|^2}{\hbar^2\omega_{21}^2 + 4|H_{21}^1|^2} \sin^2(\Omega t) , \tag{4.142}$$

and

$$\Omega = \frac{1}{2}\sqrt{\omega_{21}^2 + 4\frac{|H_{21}^1|^2}{\hbar^2}} , \tag{4.143}$$

which is the so-called Rabi formula [11]. The time evolution of this probability is shown in Figure 4.4. It is evident that $P_1(t) = 1 - P_2(t)$. According to (4.142), the system oscillates between the two states with frequency Ω as long as the external perturbation is acting.

In the case when the external field is oscillating in time according to (4.125) and the oscillation frequency ω is close to resonance with the transition frequency ω_{21} of the system, the two exponents in (4.136) and (4.137) are very different. Indeed, $e^{\pm i(\omega+\omega_{21})t}$ is strongly oscillating with frequency close to $2\omega_{21}$ and $e^{\pm i(\omega-\omega_{21})t}$ is very slowly oscillating. Considering the evolution on a timescale much longer than $1/2\omega_{21}$, we might disregard the highly oscillating terms in (4.136) and (4.137), and thus we will come to the following approximation:

$$\frac{\partial}{\partial t}a_1(t) = \frac{1}{\mathrm{i}\hbar}H^1_{12}\mathrm{e}^{-\mathrm{i}(\omega_{21}-\omega)t}a_2(t) , \tag{4.144}$$

$$\frac{\partial}{\partial t}a_2(t) = \frac{1}{\mathrm{i}\hbar}H^1_{21}\mathrm{e}^{-\mathrm{i}(\omega_{12}-\omega)t}a_1(t) . \tag{4.145}$$

Evidently, the solution of these equations is the same as the solution of (4.136) and (4.137) given by (4.139) with substitution of $\omega_{21} - \omega$ instead of ω_{21} and $\Omega = 1/2\sqrt{(\omega_{21}-\omega)^2 + 4|H^1_{21}|^2/\hbar^2}$. In this case the Rabi formula is given by

$$P_2(t) = \frac{4|H^1_{21}|^2}{\hbar^2(\omega_{21}-\omega)^2 + 4|H^1_{21}|^2}\sin^2\left[\frac{1}{2}\sqrt{(\omega_{21}-\omega)^2 + 4\frac{|H^1_{21}|^2}{\hbar^2}}t\right] . \tag{4.146}$$

In the limiting case when $\omega = \omega_{21}$, we get

$$P_2(t) = \sin^2\frac{H^1_{21}}{\hbar}t , \tag{4.147}$$

that is, the oscillations are determined by the strength of the external field.

4.6.5
Periodic Structures and the Kronig–Penney Model

In this subsection we consider yet another important model problem – the periodic potential for a particle. Let us consider potential $V(x)$ with periodicity c, so

$$V(x + c) = V(x) . \tag{4.148}$$

Since the wavefunction is defined up to a constant, the wavefunction of a particle in this potential should satisfy the same demands of periodicity:

$$\psi(x + c) = C\psi(x) , \tag{4.149}$$

where

$$|C| = 1 . \tag{4.150}$$

If we continue shifting the wavefunction, we should have

$$\psi(x + nc) = C^n\psi(x) . \tag{4.151}$$

To take into account an infinitely long periodic system, we assume the periodic boundary condition at the length of N sites:

$$\psi(x + Nc) = \psi(x) , \tag{4.152}$$

or

$$C^N = 1 \,. \tag{4.153}$$

This can be considered as the polynomial equation of Nth power, whose solution is

$$C = \exp(\mathrm{i}\boldsymbol{k}x) \,, \tag{4.154}$$

where

$$\boldsymbol{k} = \frac{2\pi n}{cN} \tag{4.155}$$

is the lattice wavevector and $n = 0, \ldots, N-1$.

As the zero coordinate for an infinite lattice is arbitrary, we can choose

$$\psi(x) = \mathrm{e}^{\mathrm{i}kx} u_k(x) \,, \tag{4.156}$$

where $u_k(x)$ is the periodic function with period c. The final expression is known as the Bloch theorem and $u_k(x)$ is known as the Bloch wavefunction.

Let us consider the case where the potential V is defined as the infinite periodic set of barriers as shown in Figure 4.5:

$$V(x) = \begin{cases} 0 \,, & 0 < x < a \,, \\ V_0 \,, & a < x < a + b \,. \end{cases} \tag{4.157}$$

So the total length determining the periodicity is equal to $a + b$. This constitutes the unit cell. For larger or smaller values of x coordinates, the cell is translated horizontally by a distance $a + b$. The Schrödinger equation of the problem now reads

$$\left(-\frac{\hbar^2}{2m}\frac{\mathrm{d}^2}{\mathrm{d}x^2} + V(x)\right)\psi(x) = E\psi(x) \tag{4.158}$$

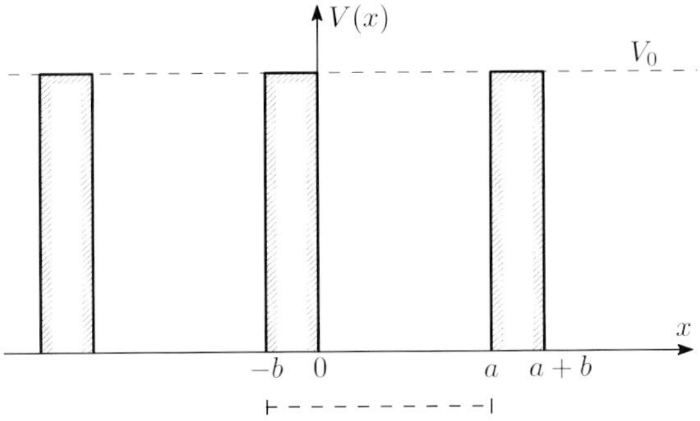

Figure 4.5 Potential surface of an infinite periodic system constituting the Kronig–Penney model. The unit cell is marked by a dashed line.

or

$$\frac{d^2\psi(x)}{dx^2} + (W - U(x))\psi(x) = 0 , \tag{4.159}$$

where we denoted $2m/\hbar^2 E = W$, and similarly $2m/\hbar^2 V(x) = U(x)$. Inserting the Bloch wavefunction, (4.156), into the wave equation, (4.159), we get

$$u_k''(x) + 2ik u_k'(x) + (W - U(x) - k^2)u_k(x) = 0 . \tag{4.160}$$

We will look for its solution in the form of the wave

$$u_k(x) = A\exp(iwx) \tag{4.161}$$

in the region $0 < x < a$, while for $-b < x < 0$ we will look for the solution in the form

$$u_k(x) = B\exp(ivx) . \tag{4.162}$$

Here A and B are constants to be determined. Inserting these definitions into (4.160), we get

$$w^2 + 2wk - (W - k^2) = 0 \tag{4.163}$$

and

$$v^2 + 2vk - (W - U_0 - k^2) = 0 . \tag{4.164}$$

The solutions of (4.163) and (4.164) are

$$w = -k \pm \sqrt{W} \tag{4.165}$$

and

$$v = -k \pm i\sqrt{U_0 - W} . \tag{4.166}$$

The value W is proportional to the energy of the particle, and it is considered to lie in the interval $0 < W < U_0$. Thus, w is a real number, and $v + k$ must be imaginary. Keeping this in mind and taking $\alpha^2 = W$ and $\beta^2 = U_0 - W$ (both α and β are taken as positive numbers), we get

$$u_k(x) = a e^{-ikx} e^{i\alpha x} + b e^{-ikx} e^{-i\alpha x} \tag{4.167}$$

in the interval $0 < x < a$, while for $a < x < b$, we get

$$u_k(x) = c e^{-ikx} e^{\beta x} + d e^{-ikx} e^{-\beta x} . \tag{4.168}$$

Coefficients a, b, c, and d are to be determined from the boundary conditions requiring that the function $u_k(x)$ and its first derivative are smooth at boundaries $x = 0$ and $x = a \equiv x = -b$. For them we get

$$a + b = c + d , \tag{4.169}$$

$$a(\alpha - k) - b(\alpha + k) = c(-\mathrm{i}\beta - k) + d(\mathrm{i}\beta - k)\,, \tag{4.170}$$

$$a\mathrm{e}^{\mathrm{i}(\alpha-k)a} + b\mathrm{e}^{-\mathrm{i}(\alpha+k)a} = c\mathrm{e}^{(\mathrm{i}k-\beta)b} + d\mathrm{e}^{(\mathrm{i}k+\beta)b}\,, \tag{4.171}$$

and

$$\begin{aligned} &a(\alpha - k)\mathrm{e}^{\mathrm{i}(\alpha-k)a} - b(\alpha + k)\mathrm{e}^{-\mathrm{i}(\alpha+k)a} \\ &= c(-\mathrm{i}\beta - k)\mathrm{e}^{(\mathrm{i}k-\beta)b} + d(\mathrm{i}\beta - k)\mathrm{e}^{(\mathrm{i}k+\beta)b}\,. \end{aligned}$$

These four equations make up a homogeneous set of equations which give a non-trivial solution when the determinant

$$\begin{vmatrix} 1 & 1 & 1 & 1 \\ k-\alpha & k+\alpha & k+\mathrm{i}\beta & k-\mathrm{i}\beta \\ \mathrm{e}^{\mathrm{i}\alpha a} & \mathrm{e}^{-\mathrm{i}\alpha a} & \mathrm{e}^{\mathrm{i}k(a+b)-\beta b} & \mathrm{e}^{\mathrm{i}k(a+b)+\beta b} \\ (k-\alpha)\mathrm{e}^{\mathrm{i}\alpha a} & (k+\alpha)\mathrm{e}^{-\mathrm{i}\alpha a} & (k+\mathrm{i}\beta)\mathrm{e}^{\mathrm{i}k(a+b)-\beta b} & (k-\mathrm{i}\beta)\mathrm{e}^{\mathrm{i}k(a+b)+\beta b} \end{vmatrix} \tag{4.172}$$

is zero. This leads to

$$\frac{\beta^2 - \alpha^2}{2\alpha\beta}\sin(a\alpha)\sinh(b\beta) + \cos(a\alpha)\cosh(b\beta) = \cos((a+b)k)\,. \tag{4.173}$$

Equation (4.173) provides the possible values of energy $W = \alpha^2$ determined by the height of the barrier U_0, the lengths of the regions a and b, and the lattice wavevector k. The parameter $\beta = \sqrt{U_0 - \alpha^2}$ is not independent. The solution can only be obtained numerically.

We can, however, obtain a qualitative picture by considering the case when the width of the barrier is small and the height is increasing. In this case we have $b \to 0$, $\cosh(b\beta) \to 1$, $\sinh(b\beta) \to b\beta$, and $\beta^2 \gg \alpha^2$, while $\beta^2 b = U_0 b$ is constant. This gives

$$\frac{U_0 b a}{2}\frac{\sin(a\alpha)}{a\alpha} + \cos(a\alpha) = \cos(ak)\,. \tag{4.174}$$

In Figure 4.6 we have plotted the left-hand side of (4.174), that is, function $\pi\sin(x)/x + \cos(x)$, with $x = a\alpha$, and with the choice $U_0 ba = 2\pi$. According to (4.174) it should be equal to $\cos(y)$, with $y = ak$. The $\cos(y)$ function has values between -1 and 1, and we have drawn this interval of values in Figure 4.6. This allows us to obtain several shaded areas, where we can have solutions of

$$\pi\frac{\sin(x)}{x} + \cos(x) = \cos(y) \tag{4.175}$$

for a given y. The energies corresponding to these shaded areas represent bands of allowed states. These are shown in Figure 4.7. There we show the dependence of the particle energy $x = a\sqrt{W}$ on the momentum $y = ak$ in the interval $(-\pi, \pi)$. As the cosine function is periodic, this is the smallest interval required to describe

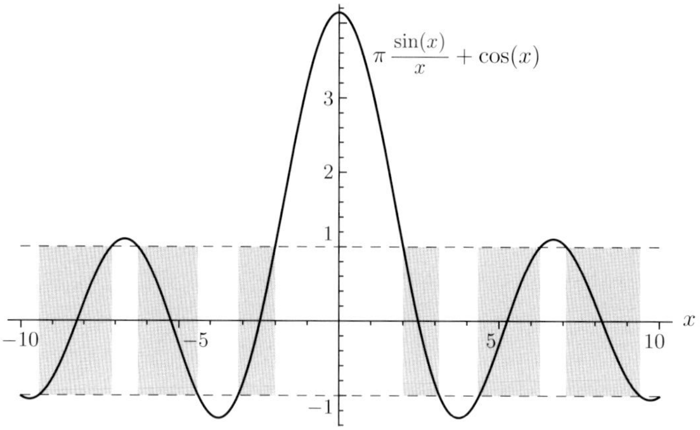

Figure 4.6 Plot of function $\pi \sin(x)/x + \cos(x)$. Shaded areas indicate the solution of (4.175).

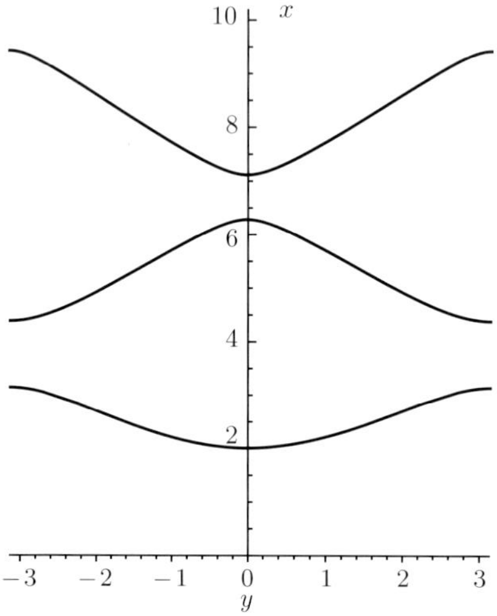

Figure 4.7 Energy bands of the Kronig–Penney model ($x^2 \propto E$) as a function of momentum $y = ka$.

the particle energies. It can thus be denoted as the unit cell in the momentum space.

The solution obtained (the band structure) obviously depends on the barrier height. As the height gets smaller, the bands start to overlap and all energy values become possible. At the other extreme, as the barrier becomes infinitely tall, we can disregard cosines in (4.174) and then we have

$$\frac{\sin(x)}{x} \approx 0 \,. \tag{4.176}$$

This is satisfied when $x = \pi n$, $n = 1, 2, \ldots$, and the energy becomes $E \propto x^2 \propto n^2$. This reminds us of the result for a particle in a box, which was described in the previous section.

In the case of crystals, each site is occupied by an atom and can donate an electron. Therefore, many electrons fill the electronic bands. Metals are obtained when one band is half filled by electrons. In this case electrons can easily change momentum and respond to the applied electric field. In the case of semiconductors and dielectrics, one band is completely filled (the valence band) and the higher-energy band is empty (the conduction band). In this case the properties of the system are determined by the electrons around the minimum of the conduction band and by the holes around the maximum of the valence band. One of the main parameters is the so-called effective mass of the electron or hole, which is defined by the curvature of the band around the minimum or maximum of the band. It is defined by analogy with the free electron, whose momentum is $\hbar k$. For a free electron we have

$$\frac{d^2 E_{\text{free}}}{dk^2} = \frac{\hbar}{m_{\text{free}}} . \tag{4.177}$$

Similarly, the effective mass inside the crystal is given by

$$m^* = \hbar \left(\frac{d^2 E(k)}{dk^2} \right)^{-1} . \tag{4.178}$$

All properties of infinite periodic structures are characterized by a unit cell in the real space. Equivalently, we can use the momentum representation, which is essentially the spatial Fourier transform of the problem. In the momentum interval from $-\pi/a$ to π/a, in every dimension the system is then fully characterized.

4.7 Perturbation Theory

In the previous examples we described exactly solvable models. However, real applications of quantum mechanics usually face the problem that the dynamic equations cannot be solved exactly. Often even numerical solutions are not feasible. Even if a numerical solution is possible, approximative methods provide invaluable insight into the physics of the problem.

4.7.1 Time-Independent Perturbation Theory

Let us consider the situation when the Hamiltonian is represented by the sum of two constituents: Hamiltonian $\hat{H}_0$ characterizing an unperturbed part of the system and a time-independent perturbation $\hat{H}^1$. Our aim is to generate the wavefunctions and energies of the perturbed system using the characteristics of the unperturbed system. The eigenvalues $E_n^{(0)}$ and eigenfunctions $\varphi_n \equiv |n\rangle$ of the unperturbed system are defined as solutions of the following Schrödinger equation and are assumed to be known:

$$\hat{H}_0 \varphi_n = E_n^{(0)} \varphi_n \,. \tag{4.179}$$

Let us assume further that $\hat{H}^1 = \lambda \hat{W}$, where λ is a small parameter. We will look for the solution of the Schrödinger equation of the perturbed system,

$$(\hat{H}_0 + \lambda \hat{W})\Psi = E\Psi \,, \tag{4.180}$$

in the form

$$\Psi = \sum_n a_n \varphi_n \,. \tag{4.181}$$

By inserting (4.181) into the stationary Schrödinger equation, (4.7), we get

$$(E - \hat{H}_0) \sum_n a_n \varphi_n = \lambda \hat{W} \sum_n a_n \varphi_n \,. \tag{4.182}$$

Multiplying this by φ_m^* and subsequently integrating over the space of the variables of Hamiltonian $\hat{H}_0$, we obtain the following equation:

$$E a_m - E_m^{(0)} a_m = \lambda \sum_n W_{mn} a_n \,, \tag{4.183}$$

where $W_{mn} = \langle m | \hat{W} | n \rangle$.

Now let us consider a particular state l. The corresponding energy term E_l and the expansion coefficient a_m of the eigenfunction defined by (4.181) then also contain correction terms of various orders:

$$E_l = E_l^{(0)} + \lambda E_l^{(1)} + \lambda^2 E_l^{(2)} + \ldots \tag{4.184}$$

and

$$a_m = \delta_{ml} + \lambda a_m^{(1)} + \lambda^2 a_m^{(2)} + \ldots \tag{4.185}$$

Inserting these expressions into (4.183) and collecting terms of the same power of λ, we get the relationships

$$E_l^{(1)} = W_{ll} \tag{4.186}$$

and

$$E_l^{(2)} + E_l^{(1)} a_l^{(1)} = \sum_n W_{ln} a_n^{(1)} \tag{4.187}$$

in the case when $m = l$, also

$$\left(E_l^{(0)} - E_m^{(0)} \right) a_m^{(1)} = W_{ml} \tag{4.188}$$

and

$$E_l^{(1)} a_m^{(1)} + \left(E_l^{(0)} - E_m^{(0)} \right) a_m^{(2)} = \sum_n W_{mn} a_n^{(1)} \tag{4.189}$$

when $m \neq l$.

From these relationships we obtain the first-order corrections for the energy,

$$E_l = E_l^{(0)} + \lambda E_l^{(1)} = E_l^{(0)} + \lambda W_{ll} \equiv E_l^{(0)} + V_{ll} , \tag{4.190}$$

where $V_{mn} = \langle m | \hat{H}^1 | n \rangle$, and for the wavefunction,

$$\Psi_l = \varphi_l + \lambda a_l^{(1)} \varphi_l + \sum_{m \neq l} \frac{V_{ml}}{E_l^{(0)} - E_m^{(0)}} \varphi_m . \tag{4.191}$$

Coefficient $\lambda a_l^{(1)}$ corresponding to the first-order correction can be determined from the normalization conditions. Since eigenfunctions φ_l of the unperturbed Hamiltonian are normalized, it follows that $1 + \lambda (a_l^{(1)} + a_l^{(1)*}) = 1$, where we disregarded terms proportional to λ^2. Thus, we must have $a_l^{(1)} + a_l^{(1)*} = 0$, that is, $\mathrm{Re}\, a_l^{(1)} = 0$. Choosing an appropriate phase for the wavefunction, we can set $a_l^{(1)} = 0$, and we obtain

$$\Psi_l = \varphi_l + \sum_{m \neq l} \frac{V_{ml}}{E_l^{(0)} - E_m^{(0)}} \varphi_m . \tag{4.192}$$

Evidently, if the spectrum of the unperturbed Hamiltonian contains both discrete and continuous parts, the additional term containing the integral over the continuous spectrum should also be taken into account:

$$\Psi_l = \varphi_l + \sum_{m \neq l} \frac{V_{ml}}{E_l^{(0)} - E_m^{(0)}} \varphi_m + \int \frac{V_{\nu l}}{E_l^{(0)} - E_\nu^{(0)}} \varphi_\nu \mathrm{d}\nu , \tag{4.193}$$

where ν enumerates the continuum part of the spectrum.

Similarly, the energy corrections in the second-order approach are given by

$$E_l^{(2)} = \sum_{n \neq l} \frac{W_{ln} W_{nl}}{E_l^{(0)} - E_n^{(0)}} , \tag{4.194}$$

and the desired energy up to the second order is therefore

$$E_l = E_l^{(0)} + V_{ll} + \sum_{n \neq l} \frac{V_{ln} V_{nl}}{E_l^{(0)} - E_n^{(0)}} . \tag{4.195}$$

Thus, it follows that the second-order correction to the ground-state energy is always negative since $E_l^{(0)} < E_n^{(0)}$ for any n, when $l = 0$. In practice it is often enough to take into account the first-order correction for the eigenfunctions and the second-order corrections to the eigenenergies.

Alternatively we can solve the problem in the original unperturbed basis set. Let us now return to the full Hamiltonian, $\hat{H} = \hat{H}_0 + \hat{H}^1$. The Schrödinger equation for the system characterized by such a Hamiltonian can also be given in the basis of the unperturbed Hamiltonian (see (4.179)):

$$\sum_n (H_{mn} - E\delta_{mn})a_n = 0 , \tag{4.196}$$

where $H_{mn} = \langle m|\hat{H}|n\rangle$. The matrix elements H_{nm} are equal to $E_n^{(0)} + V_{nn}$ when $n = m$, and V_{mn} when $m \neq n$. In order to obtain a nontrivial solution of (4.196), the following requirement should be fulfilled:

$$\det[H_{mn} - E\delta_{mn}] = 0 . \tag{4.197}$$

This provides the possibility to determine the eigenenergies of the system in the representation of the unperturbed Hamiltonian. Equation (4.197) is called the secular equation, and can be used to consider degenerate as well as nondegenerate spectra of the unperturbed Hamiltonian. In the latter case and for weak nondiagonal terms, that is, when $|H_{lm}| \ll |E_l^{(0)} - E_m^{(0)}|$, (4.197) can be simplified significantly. For instance, by disregarding all nondiagonal terms, we obtain $E = H_{ll} = E_l^{(0)} + V_{ll}$, that is, the result coincides with the result which follows from the first-order corrections with respect to the perturbation. In order to obtain the second-order perturbation corrections, let us consider a particular state $l = 1$ defined according to the unperturbed Hamiltonian. If we take into account only the nondiagonal terms, which mix all other states with the chosen state $l = 1$, it follows from (4.197) that the perturbed eigenvalue can be determined by the following transcendent relation:

$$E = H_{11} - \sum_{m=2}^{\infty} \frac{|H_{1m}|^2}{H_{mm} - E} . \tag{4.198}$$

This equation can be solved approximately. In the first-order approximation we get

$$E = E_1^{(0)} + V_{11} + \sum_{m=2}^{\infty} \frac{|V_{1m}|^2}{E_1^{(0)} + V_{11} - \left(E_m^{(0)} + V_{mm}\right)} . \tag{4.199}$$

As an example let us consider the model containing two coupled energy states characterized by φ_1 and φ_2 as the eigenfunctions of the unperturbed Hamiltonian H_0. A secular equation, (4.197), then corresponds to the second-order determinant

$$\begin{vmatrix} H_{11} - E & H_{12} \\ H_{21} & H_{22} - E \end{vmatrix} = 0 , \tag{4.200}$$

giving the solution

$$E_{1,2} = \frac{1}{2}\left[(H_{11} + H_{22}) \pm \sqrt{(H_{11} - H_{22})^2 + 4|H_{12}|^2}\right] . \tag{4.201}$$

In the case when the perturbation conditions are satisfied, that is,

$$|H_{11} - H_{22}| \gg |H_{12}| , \tag{4.202}$$

the solution presented in (4.201) for state $i = 1$ coincides with the result which follows from (4.199):

$$E_1 = E_1^{(0)} + V_{11} + \frac{|V_{12}|^2}{E_1^{(0)} + V_{11} - \left(E_2^{(0)} + V_{22}\right)} , \tag{4.203}$$

where $H = H_0 + V$ is chosen as the sum of the unperturbed Hamiltonian and the perturbation, respectively. Similarly,

$$E_2 = E_2^{(0)} + V_{22} + \frac{|V_{21}|^2}{E_2^0 + V_{22} - \left(E_1^{(0)} + V_{11}\right)} . \tag{4.204}$$

In the opposite case, when

$$|H_{11} - H_{22}| \ll |H_{12}| , \tag{4.205}$$

from (4.201) it follows that

$$E_{1,2} = \frac{H_{11} + H_{22}}{2} \pm \left[|H_{12}| + \frac{(H_{11} - H_{22})^2}{8|H_{12}|}\right] . \tag{4.206}$$

The dependences of the values of energies E_1 and E_2 on the difference of the diagonal matrix elements $H_{11} - H_{22}$ at some fixed value of H_{12} are presented in Figure 4.8. The dotted lines indicate the linear dependences of the H_{11} and H_{22} values tracing the asymptotic values of energies E_1 and E_2. It is worthwhile mentioning that the second-order corrections always contain the difference between the energy positions corresponding to the H_{11} and H_{22} values.

Coefficients determining the strengths of perturbation can also be defined by using (4.196), giving in this case

$$\frac{a_1}{a_2} = \frac{H_{12}}{E - H_{11}} . \tag{4.207}$$

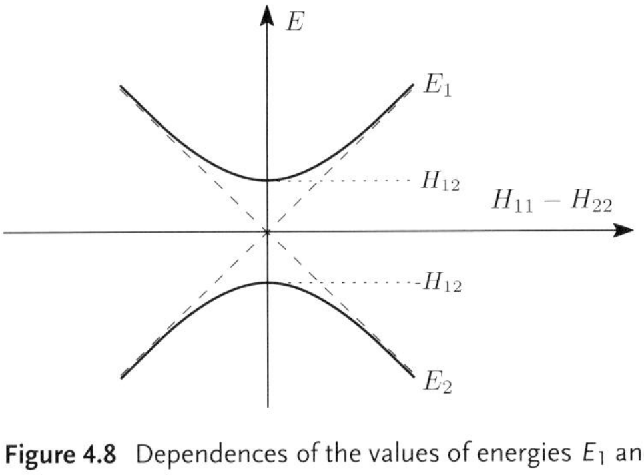

Figure 4.8 Dependences of the values of energies E_1 and E_2 on the difference of the diagonal matrix elements $H_{11} - H_{22}$ at some fixed value of H_{12}.

This allows us to determine the ratio between coefficients. The normalization condition is the additional requirement when determining both coefficients.

4.7.2
Time-Dependent Perturbation Theory

Let us now consider the system described by Hamiltonian $\hat{H}_0$ in the presence of time-dependent perturbation $\hat{H}^1(t)$. In this case the time-dependent Schrödinger equation, (4.5), has to be solved. We use the interaction representation as described earlier in this chapter.

We start with the time-dependent Schrödinger equation for the unperturbed Hamiltonian:

$$\mathrm{i}\hbar\frac{\partial}{\partial t}\varphi_n(t) = \hat{H}_0\varphi_n(t) \,. \tag{4.208}$$

The eigenfunctions are

$$\varphi_n(t) = \mathrm{e}^{-\frac{\mathrm{i}}{\hbar}E_n^0 t}\varphi_n \,, \tag{4.209}$$

where E_n^0 are the eigenvalues of the unperturbed Hamiltonian. With use of the expansion

$$\Psi(t) = \sum_n c_n(t)\varphi_n \,, \tag{4.210}$$

the time-dependent Schrödinger equation with the full Hamiltonian then reads

$$\mathrm{i}\hbar\frac{\partial}{\partial t}\sum_n c_n(t)\varphi_n(t) = \hat{H}(t)\sum_n c_n(t)\varphi_n(t) \,, \tag{4.211}$$

where

$$\hat{H}(t) = \hat{H}_0 + \hat{H}^1(t) \,. \tag{4.212}$$

For coefficients c slowly varying in time, we can use the following expansion:

$$c_n(t) = c_n^{(0)}(t) + c_n^{(1)}(t) + c_n^{(2)}(t) + \ldots \tag{4.213}$$

We assume that at the initial time $\hat{H}^1(0) = 0$ (so that the system is unperturbed at zero time) and consider the system starting with its initial state φ_i. In this case it is evident that $c_n(0) = c_n^{(0)}(0) = \delta_{ni}$ and $c_n^{(1)}(0) = c_n^{(2)}(0) = \ldots = 0$. If we take into into account these initial conditions, it follows from (4.211) that

$$\mathrm{i}\hbar\frac{\partial}{\partial t}\left(c_j^{(1)}(t) + c_j^{(2)}(t) + \ldots\right) = \sum_k V_{jk}(t)\left(\delta_{ki} + c_k^{(1)}(t) + c_k^{(2)}(t) + \ldots\right) , \tag{4.214}$$

where

$$V_{jk}(t) = \langle j(t)|\hat{H}^1(t)|k(t)\rangle \,. \tag{4.215}$$

In the first order of the expansion, it follows that

$$\mathrm{i}\hbar \frac{\partial c_j^{(1)}(t)}{\partial t} = V_{ji} \,, \tag{4.216}$$

and the solution is given by

$$c_j^{(1)}(t) = \frac{1}{\mathrm{i}\hbar} \int_0^t V_{ji}(t')\mathrm{d}t' \,. \tag{4.217}$$

Subsequently, in the second order of the expansion we get

$$\mathrm{i}\hbar \frac{\partial c_j^{(2)}(t)}{\partial t} = \sum_k V_{jk}(t) c_k^{(1)}(t) \,, \tag{4.218}$$

and the solution is given by

$$c_j^{(2)}(t) = \left(\frac{1}{\mathrm{i}\hbar}\right)^2 \sum_k \int_0^t V_{jk}(t')\mathrm{d}t' \int_0^{t'} V_{ki}(t'')\mathrm{d}t'' \,. \tag{4.219}$$

Coefficients corresponding to higher orders of the expansion can be defined in a similar way. However, let us restrict ourselves to the first order of the expansion and consider the quantity $W_{ji}(t) = |c_j^{(1)}(t)|^2$. It determines the probability of the quantum system emerging in state j after time t when it was initially in state i. In other words, $W_{ji}(t)$ corresponds to the probability of transition from state i to state j after time t:

$$W_{ji}(t) = \frac{1}{\hbar^2} \left| \int_0^t V_{ji}(t')\mathrm{d}t' \right|^2 \,. \tag{4.220}$$

Let us now assume that perturbation $\hat{H}^1$ remains constant during the time interval from $t = 0$ to $t = t_1$. Then the time dependence of the matrix element defined by (4.215) persists in the time dependence of the wavefunctions only in the time interval defined by the action of the perturbation:

$$V_{ji}(t) = \mathrm{e}^{\mathrm{i}\omega_{ji} t} V_{ji} \,, \tag{4.221}$$

where $\omega_{ji} = (E_j^0 - E_i^0)/\hbar$, and $V_{jk} = \langle j|\hat{H}^1|k\rangle$, $|j\rangle = \varphi_j$. In this case (4.220) gives

$$W_{ji}(t_1) = \frac{|V_{ji}|^2}{\hbar^2} \left| \int_0^{t_1} \mathrm{e}^{\mathrm{i}\omega_{ji} t}\mathrm{d}t \right|^2 = \frac{|V_{ji}|^2}{\hbar^2} \frac{\sin^2 \frac{\omega_{ji} t_1}{2}}{\left(\frac{\omega_{ji}}{2}\right)^2} \,. \tag{4.222}$$

If the final state j belongs to the continuum of states, we have to consider the probability of a transition to a particular energy interval $E_j^0 \div E_j^0 + \mathrm{d}E_j^0$ containing some number of states, which is usually defined as $\rho(E_j^0)\mathrm{d}E_j^0$, where $\rho(E_j^0)$ is a density of states. The total probability of a transition from the ith state is then given by

$$W_i(t_1) = \int W_{ji}(t_1)\rho\left(E_j^0\right)\mathrm{d}E_j^0 \,. \qquad (4.223)$$

For long time t_1, we can use the following approximate relationship

$$\frac{\sin^2\frac{\left(E_j^0 - E_i^0\right)t_1}{2\hbar}}{\left(\frac{E_j^0 - E_i^0}{2\hbar}\right)^2} \approx \pi t_1 \delta\left(\frac{E_j^0 - E_i^0}{2\hbar}\right) = 2\pi\hbar t_1 \delta\left(E_j^0 - E_i^0\right) \qquad (4.224)$$

since

$$\frac{1}{\pi}\lim_{t\to\infty}\frac{\sin^2 \alpha t}{\alpha^2 t} = \delta(\alpha) \,. \qquad (4.225)$$

Taking into account approximate relation (4.224) and by integrating (4.223) over the full energy spectrum of the final state (see Figure 4.9), we get

$$W_i(t_1) = \frac{2\pi}{\hbar} t_1 |V_{ji}|^2 \rho\left(E_j^0\right) \,. \qquad (4.226)$$

The transition probability obtained is proportional to the time t_1 for which the perturbation acts. Thus, we can define the transition probability per unit time as

$$w = \frac{W_i(t_1)}{t_1} = \frac{2\pi}{\hbar}|V_{Ei}|^2 \rho(E) \,, \qquad (4.227)$$

where E indicates the energy region in resonance with E_i^0. Returning to the transition between two states, we can also define formally the transition probability per unit time as

$$w = \frac{W_{ji}(t_1)}{t_1} = \frac{2\pi}{\hbar}|V_{ji}|^2 \delta\left(E_j^0 - E_i^0\right) \,. \qquad (4.228)$$

The relation obtained is usually called the Fermi golden rule.

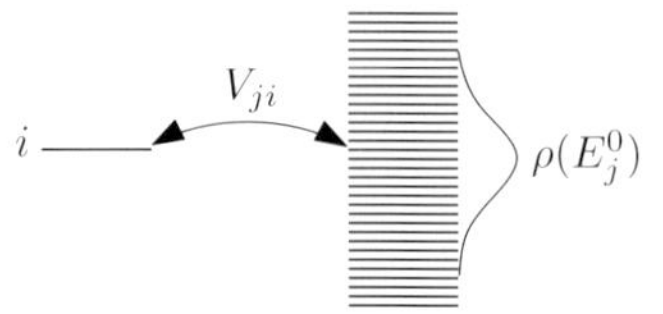

Figure 4.9 Level coupling scheme with the continuum spectrum characterized by the density of states.

In the case of a periodic perturbation, that is, when $H^1(t) = V e^{\pm i\omega t}$, all derivations presented above can be easily generalized, giving

$$w = \frac{2\pi}{\hbar} |V_{ji}|^2 \delta\left(E_j^0 - E_i^0 \mp \hbar\omega\right) \tag{4.229}$$

for two separate levels, and

$$w = \frac{2\pi}{\hbar} |V_{Ei}|^2 \rho(E) \tag{4.230}$$

in the case of transitions into the continuum states. Here $E = E_i^0 \mp \hbar\omega$.

The above expressions determining the probability of transition per unit time (transition rate) point to the notion of the lifetime of the initial state. The relationship between the lifetime, τ, and the transition probability per unit time w is $1/\tau = w$.

For the description of the time-dependent perturbation, the interaction representation is used with advantage, because it allows us to consider higher-order terms of the time-dependent perturbation conveniently.

This level of theory can be efficiently used for the description of the optical properties of a quantum systems. It is, however, more convenient to use the Liouville space description as in Section 13.1.

4.8 Einstein Coefficients

Interaction of a quantum system with an external electromagnetic field can be considered a typical time-dependent perturbation. Indeed, the plane electromagnetic field is usually described by

$$\boldsymbol{\mathcal{E}} = \boldsymbol{\mathcal{E}}_0 \cos(\boldsymbol{k} \cdot \boldsymbol{r} - \omega t) , \tag{4.231}$$

where the polarization direction and amplitude of the wave are given by $\boldsymbol{\mathcal{E}}_0$, $\boldsymbol{k}$ is the wave vector, and ω is the frequency. Equation (4.231) represents a monochromatic electromagnetic field. In the situation when the system is exposed to an incoherent electromagnetic field, the latter is characterized by a continuous spectrum of frequencies. The dominating characteristic of the field is then the density of states of the electromagnetic field at the resonance frequency, and the transition rates can be defined by using (4.230). For a dipole-allowed transition, the amplitude of the interaction with the electromagnetic field is given by $(-\hat{\boldsymbol{\mu}}\boldsymbol{\mathcal{E}})$, where $\hat{\boldsymbol{\mu}}$ is the electric dipole moment operator of the system (see Chapter 2). In the dipole (or the long-wavelength) approximation ($\boldsymbol{kr} \ll 1$, which means that the wavelength is much larger than the size of the system) the matrix element describing the transition between eigenstates j and i of the system induced by the electromagnetic field can be defined as

$$V_{ji} = -\mathcal{E}_0 \mu_{ji} \cos\left(\boldsymbol{e}^0 \cdot \boldsymbol{\mu}_{ji}^0\right) , \tag{4.232}$$

where $\mu_{ji} = |\langle j|\boldsymbol{\mu}|i\rangle|$ is the corresponding transition dipole moment amplitude, and $\boldsymbol{e}^0$ and $\boldsymbol{\mu}^0_{ji}$ are corresponding unit vectors of the field and the dipole moment. Using the Fermi golden rule (4.230), we find the rate of the transition from the inital state i to the excited state f is

$$w_{fi} = \frac{2\pi}{\hbar}|\mu_{ji}|^2|\mathcal{E}_0|^2\rho_{\text{rad}}(\hbar\omega) , \tag{4.233}$$

where $\rho_{\text{rad}}(\hbar\omega)$ gives the density of states of the electromagnetic field at the resonance frequency. By taking into account the definitions of the energy of the electromagnetic field given in Section 2.2, we can give the energy of the electromagnetic field in volume V by $E_{\text{field}} = 2\varepsilon_0|\mathcal{E}_0|^2 V$, and thus $|\mathcal{E}_0|^2 = E_{\text{field}}/(2\varepsilon_0 V)$. For an isotropic sample, when all orientations of the transition dipole moment μ_{ji} are likely to be equal, the averaging over all orientations leads to

$$\left\langle \cos^2\left(\boldsymbol{e}^0 \cdot \boldsymbol{\mu}^0_{ji}\right)\right\rangle = \frac{1}{3} . \tag{4.234}$$

Thus,

$$w_{fi} = B_{fi}\rho(\hbar\omega) , \tag{4.235}$$

where $\rho(\hbar\omega) = E_{\text{field}}\rho_{\text{rad}}(\hbar\omega)/V$ is the energy density of radiation states, and B_{j0} is the Einstein coefficient of stimulated absorption (see [11] for more details):

$$B_{fi} = \frac{|\mu_{fi}|^2}{3\epsilon_0\hbar^2} , \tag{4.236}$$

where ϵ_0 is the vacuum permittivity.

Similar calculations for the rate of the transition from an excited state f to the ground state lead to the same result, that is, to proportionality to the intensity of radiation. The Einstein coefficient of stimulated emission B_{if} equals the coefficient of stimulated absorption, that is, $B_{fi} = B_{if}$.

Let us now consider the ensemble of the equivalent two-level system (e.g., atoms) interacting with an electromagnetic field with frequency which is in resonance with the $i \to f$ transition of each atom. The master equation for the occupation of states (or probabilities) describing the radiative transitions between these two states can then be written as described in Chapter 3. As the transition rates are known, we can immediately write

$$\frac{dN_f}{dt} = N_i B_{fi}\rho(\hbar\omega) - N_f B_{if}\rho(\hbar\omega) - N_f A_{if} , \tag{4.237}$$

where N_i and N_f denote the populations of the initial and excited states, respectively, and A_{if} is the Einstein coefficient of spontaneous emission. The latter has to be introduced to guarantee the Boltzmann relationship between the equilibrium populations. Indeed, in the case of thermal equilibrium between the system and the radiation, all transitions should be compensated, that is,

$$N_i B_{fi}\rho(\hbar\omega) = N_f B_{if}\rho(\hbar\omega) + N_f A_{if} . \tag{4.238}$$

Since both transition rates are equal, that is, $w_{fi} = w_{if}$ and $B_{fi}\rho(\hbar\omega) = B_{if}\rho(\hbar\omega)$, it follows that the population rates should be equal in the stationary conditions if the spontaneous emission is not taken into account (when assuming $A_{if} = 0$). However, that is in conflict with the Boltzmann distribution of the populations given by

$$\frac{N_f}{N_i} = e^{-(E_f^0 - E_i^0)/(k_B T)} , \quad \text{i.e.,} \quad B_{fi}\rho(\hbar\omega) = B_{if}\rho(\hbar\omega) , \tag{4.239}$$

which should be reached in stationary conditions. To avoid this conflict, Einstein proposed the spontaneous emission term. If the Boltzmann ratio for the populations, (4.239), is taken into account, it follows that

$$\rho(\hbar\omega) = \frac{A_{fi}}{B_{fi}e^{(E_f^0 - E_i^0)/(k_B T)} - B_{if}} , \tag{4.240}$$

where $\hbar\omega = E_f^0 - E_i^0$.

In thermal equilibrium, the density of states of the thermal electromagnetic field is given by the Planck distribution [11]:

$$\rho(\hbar\omega) = \frac{\hbar\omega^3/c^3}{e^{\hbar\omega/(k_B T)} - 1} , \tag{4.241}$$

where $\omega = 2\pi\nu$. Comparison of the last two expressions confirms that $B_{if} = B_{fi}$ and the Einstein coefficient of spontaneous emission is

$$A_{if} = \frac{\omega^3|\mu_{fi}|^2}{3\varepsilon_0\hbar\pi c^3} . \tag{4.242}$$

It is noteworthy that the spontaneous emission coefficient increases as ω^3 and thus it is potentially of great importance at very high frequencies. The spontaneous emission process can be viewed as the outcome of the interaction of an excited state with zero-point fluctuations of the electromagnetic field as described in Section 14.7.

4.9 Second Quantization

The description of quantum systems as discussed above constitutes the so-called first quantization. In the first quantization we described the system by operators corresponding to the physical observable quantities. In this section we briefly review different types of operations over the system states.

Let us once again consider the harmonic oscillator problem as described in Section 4.6.1. We found that the operator $\xi - d/d\xi$ acting on a wavefunction $\varphi_n(\xi)$ with energy ε_n creates a new wavefunction $\varphi_{n+1}(\xi)$ with energy $\varepsilon_n + 1$. Similarly, the operator $\xi + d/d\xi$ creates a new wavefunction $\varphi_{n-1}(\xi)$ with energy $\varepsilon_n - 1$.

The operator $\xi - \mathrm{d}/\mathrm{d}\xi$ is thus the energy-raising operator, and $(\xi + \mathrm{d}/\mathrm{d}\xi)$ is the energy-lowering operator.

With proper normalization we define the energy-raising and energy-lowering operators as

$$\hat{a}^\dagger = \frac{1}{\sqrt{2}}\left(\xi - \frac{\mathrm{d}}{\mathrm{d}\xi}\right) \tag{4.243}$$

and

$$\hat{a} = \frac{1}{\sqrt{2}}\left(\xi + \frac{\mathrm{d}}{\mathrm{d}\xi}\right) . \tag{4.244}$$

The following rules of action on the wavefunctions then apply:

$$\hat{a}^\dagger \varphi_n(\xi) = \sqrt{n+1}\varphi_{n+1}(\xi) \tag{4.245}$$

and

$$\hat{a}\varphi_n(\xi) = \sqrt{n}\varphi_{n-1}(\xi) . \tag{4.246}$$

Let us now switch to a bra-ket notation. Let states $|n\rangle$ and $|n+1\rangle$ have energies ε_n and $\varepsilon_n + 1$, respectively. We can introduce the notion of an energy "quantum," a particle, and associate it with the state of the oscillator. We denote state $|n\rangle$ as the state with n quanta. The energy of this state of n quanta is $\varepsilon_n = n + 1/2$. The zero-quantum state is to be taken as the ground state of the oscillator. The energy of the zero-quantum state, also known as the vacuum state, is $\varepsilon_0 = 1/2$.

In this language the operators $\hat{a}^\dagger$ and $\hat{a}$ become the operators which manipulate the number of quanta. They are now defined in the space of these quantum number states, and they represent the creation or annihilation of the energy quanta. In bra-ket notation, (4.245) and (4.246) become

$$\hat{a}|n\rangle = \sqrt{n}|n-1\rangle \tag{4.247}$$

and

$$\hat{a}^\dagger|n\rangle = \sqrt{n+1}|n+1\rangle . \tag{4.248}$$

We also properly get that $\hat{a}|0\rangle = 0$, so the vacuum state is the state with the lowest possible energy.

Let us now check the action of $\hat{a}^\dagger\hat{a}$:

$$\hat{a}^\dagger\hat{a}|n\rangle = \hat{a}^\dagger\sqrt{n}|n-1\rangle = n|n\rangle . \tag{4.249}$$

We find that the operator $\hat{n} = \hat{a}^\dagger\hat{a}$ has $|n\rangle$ as its eigenstate, with the eigenvalue corresponding to the number of energy quanta represented by state $|n\rangle$. It therefore corresponds to a particle or quantum number operator.

The creation and annihilation operators satisfy the following commutational relation:

$$[\hat{a}, \hat{a}^\dagger] = 1 . \tag{4.250}$$

All operators needed for the description of the harmonic operators can be expressed in terms of the creation and annihilation operators. Their commutation relations allow us to represent and evaluate conveniently all observable properties of this quantum system. For instance, the Hamiltonian of the harmonic oscillator, (4.90), in terms of the creation and annihilation operators reads

$$\hat{H} = \hbar\omega \left(\hat{a}^\dagger \hat{a} + \frac{1}{2} \right) . \tag{4.251}$$

It is worthwhile mentioning that the excitations of the harmonic oscillator satisfy Bose–Einstein statistics. This will be described thoroughly in Chapter 7. Later we assign this property to all particles with even spin (whole number); such particles are called bosons. Fermions, another type of quantum particle, have fractional spin, and we can also define creation and annihilation operators for them, only with a different commutation relation.

4.9.1 Bosons and Fermions

The creation and annihilation operators introduced in the previous section are convenient to describe systems containing many particles of the same origin. Due to the uncertainty principle, quantum particles are indistinguishable, and the wavefunction of the system has to be either symmetric or antisymmetric with respect to permutation of indistinguishable particles. The system containing many particles should follow statistical rules [15, 17]. The photons, which characterize the electromagnetic field, as well as the vibrational quanta introduced in Section 4.9, satisfy the so-called Bose–Einstein statistics. From the point of view of permutation, the wavefunction of the system containing many bosons is symmetric.

Fermions are another type of particle, and are characterized by an antisymmetric wavefunction with respect to particle permutation. They satisfy the so-called Fermi–Dirac statistics. Due to this permutation requirement, the wavefunction of a many-fermion system cannot contain two particles in the same quantum state. The particles of this type satisfy the so-called Pauli exclusion principle, which leads to Fermi–Dirac statistics. It is noteworthy that the type of statistics the particles satisfy correlates with their spins. The particles characterized by half-integer spins ($1/2, 3/2, \ldots$) are fermions, while the particles with whole-number spins are bosons. These properties are described in more detail in Section 7.3.3.

In the system containing many particles, the quantum states can be defined in the space of so-called "occupation numbers" of particular states. As demonstrated for bosons in the previous sections, these states are directly defined by the creation operators; see (4.278). The bosonic character of a particle is determined by the commutation rules

$$[\hat{a}, \hat{a}^\dagger] = 1 \tag{4.252}$$

and

$$[\hat{a}, \hat{a}] = [\hat{a}^\dagger, \hat{a}^\dagger] = 0 , \tag{4.253}$$

which are satisfied by the creation and annihilation operators. In the case of fermions, the commutation (anticommutation) rules are as follows:

$$\hat{a}\hat{a}^\dagger + \hat{a}^\dagger\hat{a} \equiv \{\hat{a}, \hat{a}^\dagger\} = 1 \tag{4.254}$$

and

$$\hat{a}\hat{a} = \hat{a}^\dagger\hat{a}^\dagger = 0\,. \tag{4.255}$$

In the case of a many-particle system containing different types of particles, the commutation rules have to be formulated for all of them. In the case of bosons, the creation/annihilation operators for different particles commute, and for fermions they satisfy the anticommutation rules.

4.9.2
Photons

In this subsection we consider the quantization of the electromagnetic field. In Section 2.3.2, we showed that the electromagnetic field can be represented by an ensemble of (infinitely many) harmonic oscillators, so-called field modes. Here we will denote the quanta of these modes as photons.

Equations (2.76) and (2.77) link the Fourier coefficients $A_{\lambda k}$ of vector potential $\boldsymbol{A}$ with momenta $\boldsymbol{p}_{\lambda k}$ and coordinates $\boldsymbol{q}_{\lambda k}$ of the mode oscillators. The modes are considered independent so that each of them can be treated and quantized as an isolated system. Quantization itself is performed simply by promoting the momenta and coordinates to operators $\hat{\boldsymbol{p}}_{\lambda k}$ and $\hat{\boldsymbol{q}}_{\lambda k}$ and postulating commutation relations according to (4.4), that is,

$$[\hat{p}_{\lambda k}, \hat{q}_{\lambda' k'}] = -\mathrm{i}\hbar\delta_{\lambda\lambda'}\delta_{kk'}\,. \tag{4.256}$$

As we have already demonstrated, it is very useful to express all quantities in terms of creation and annihilation operators introduced for the harmonic oscillator by (4.243) and (4.244). Let us introduce operators

$$\hat{a}_{\lambda k} = \alpha\hat{q}_{\lambda k} + \beta\hat{p}_{\lambda k} \tag{4.257}$$

and

$$\hat{a}^\dagger_{\lambda k} = \alpha^*\hat{q}_{\lambda k} + \beta^*\hat{p}_{\lambda k}\,, \tag{4.258}$$

for each mode characterized by λ and $\boldsymbol{k}$. We require that the commutation relation, (4.252), holds, and the Hamiltonian of a single mode has the form of (4.251). The latter requirement gives the following relations for coefficients:

$$\mathrm{Re}\,[\beta^*\alpha] = 0\,, \quad |\alpha|^2 = \frac{\omega}{2\hbar}\,, \quad |\beta|^2 = \frac{1}{2\hbar\omega}\,, \quad \beta^*\alpha = -\frac{\mathrm{i}}{2\hbar}\,. \tag{4.259}$$

These are consistent with the commutation relation for a particular mode:

$$[\hat{a}^\dagger, \hat{a}] = 2\hbar\,\mathrm{Im}\,[\beta^*\alpha] = -1\,. \tag{4.260}$$

Conditions (4.259) are satisfied if $\alpha = \sqrt{\omega/(2\hbar)}$ and $\beta = \mathrm{i}(1/\sqrt{2\hbar\omega})$, and the creation and annihilation operators of the field are then

$$\hat{a}_{\lambda k} = \sqrt{\frac{\omega_k}{2\hbar}}\left(\hat{q}_{\lambda k} + \frac{\mathrm{i}}{\omega_k}\hat{p}_{\lambda k}\right) \tag{4.261}$$

and

$$\hat{a}^{\dagger}_{\lambda k} = \sqrt{\frac{\omega_k}{2\hbar}}\left(\hat{q}_{\lambda k} - \frac{\mathrm{i}}{\omega_k}\hat{p}_{\lambda k}\right) . \tag{4.262}$$

The inverse relations

$$\hat{p}_{\lambda k} = -\mathrm{i}\omega_k\sqrt{\frac{\hbar}{2\omega_k}}\left(\hat{a}_{\lambda k} - \hat{a}^{\dagger}_{\lambda k}\right) \tag{4.263}$$

and

$$\hat{q}_{\lambda k} = \sqrt{\frac{\hbar}{2\omega_k}}\left(\hat{a}_{\lambda k} + \hat{a}^{\dagger}_{\lambda k}\right) \tag{4.264}$$

can be used to express the Fourier components $A_{\lambda k}$ (now promoted to operators $\hat{A}_{\lambda k}$) of the vector potential $\hat{A}_{\lambda k}$ in terms of the creation and annihilation operators. Comparing (2.76) and (2.77) with (4.263) and (4.264), we get

$$A_{\lambda k} \rightarrow \hat{A}_{\lambda k} = \sqrt{\frac{\hbar}{2\Omega\epsilon_0\omega_k}}\hat{a}_{\lambda k} \tag{4.265}$$

and

$$A^{*}_{\lambda k} \rightarrow \hat{A}^{\dagger}_{\lambda k} = \sqrt{\frac{\hbar}{2\Omega\epsilon_0\omega_k}}\hat{a}^{\dagger}_{\lambda k} . \tag{4.266}$$

Now we will express the operator of the vector potential in terms of $\hat{a}$ and $\hat{a}^{\dagger}$. We obtain [7]

$$\hat{\boldsymbol{A}}(\boldsymbol{r}) = \sum_{k\lambda}\left(\boldsymbol{f}_{\lambda k}(\boldsymbol{r})\hat{a}_{\lambda k} + \boldsymbol{f}_{\lambda -k}(\boldsymbol{r})\hat{a}^{\dagger}_{\lambda k}\right) . \tag{4.267}$$

Here we introduced a spatial vector function

$$\boldsymbol{f}_{\lambda k}(\boldsymbol{r}) = \boldsymbol{e}_{\lambda k}\sqrt{\frac{\hbar}{2\epsilon_0\omega_k\Omega}}\mathrm{e}^{\mathrm{i}\boldsymbol{k}\cdot\boldsymbol{r}} \tag{4.268}$$

containing the polarization of the mode. The electric field vector is related to the vector potential by a time derivative (see Section 2.2). In the Heisenberg representation, the time derivative corresponds to the commutator with the Hamiltonian (see (4.48)). The operator $\hat{\boldsymbol{E}}(\boldsymbol{r})$ thus reads

$$\hat{\boldsymbol{E}}(\boldsymbol{r}) = -\frac{\mathrm{i}}{\hbar}\left[\hat{H}_{\mathrm{T}}, \hat{\boldsymbol{A}}(\boldsymbol{r})\right] = \sum_{k\lambda}\left(\mathrm{i}\omega_k\boldsymbol{f}_{\lambda k}(\boldsymbol{r})\hat{a}_{\lambda k} - \mathrm{i}\omega_k\boldsymbol{f}_{\lambda -k}(\boldsymbol{r})\hat{a}^{\dagger}_{\lambda k}\right) , \tag{4.269}$$

where

$$\hat{H}_{\mathrm{T}} = \sum_{\lambda k} \hbar \omega_k \left(a^{\dagger}_{\lambda k} a_{\lambda k} + \frac{1}{2} \right) \tag{4.270}$$

is the Hamiltonian operator of radiation in the empty space. In a similar way one can obtain operators for all relevant electrodynamic quantities. The states of light can be constructed from the eigenstates of the Hamiltonian. In each mode these eigenstates follow (4.247) and (4.248). It is worthwhile mentioning that the quanta of the electromagnetic field – photons – behave according to Bose–Einstein statistics, that is, they are bosons.

4.9.3 Coherent States

Any state of the electromagnetic field can be described in terms of the eigenstates of the field Hamiltonian. However, if we interpret states of the electromagnetic field in the same way as we understand the states of matter, we may get confusing results.

Let us consider, for example, a state with n photons at some specific mode:

$$|\psi\rangle = |n\rangle \equiv \frac{1}{\sqrt{n!}} \hat{a}^{\dagger n} |0\rangle \,. \tag{4.271}$$

According to (2.70) and promoting the field amplitudes to the operators along (4.265) and (4.266), we can obtain the time domain vector potential as

$$\hat{\mathbf{A}}(\mathbf{r}, t) = \sum_k \sqrt{\frac{\hbar}{2\Omega \epsilon_0 \omega_k}} \left(\hat{a}_{\lambda k} \mathrm{e}^{-\mathrm{i}\omega_k t + \mathrm{i}\mathbf{k}\cdot\mathbf{r}} + \hat{a}^{\dagger}_{\lambda k} \mathrm{e}^{\mathrm{i}\omega_k t - \mathrm{i}\mathbf{k}\cdot\mathbf{r}} \right) , \tag{4.272}$$

while according to (2.60) the electric field reads

$$\hat{\mathbf{E}}(\mathbf{r}, t) = \mathrm{i} \sum_k \sqrt{\frac{\hbar \omega_k}{2\Omega \epsilon_0}} \left(-\hat{a}_{\lambda k} \mathrm{e}^{-\mathrm{i}\omega_k t + \mathrm{i}\mathbf{k}\cdot\mathbf{r}} + \hat{a}^{\dagger}_{\lambda k} \mathrm{e}^{\mathrm{i}\omega_k t - \mathrm{i}\mathbf{k}\cdot\mathbf{r}} \right) . \tag{4.273}$$

Let us consider the expectation value of the electric field in the state $|\psi\rangle = |n\rangle$. For that we need to calculate expectation values

$$\langle \psi | \hat{a} | \psi \rangle \tag{4.274}$$

and

$$\langle \psi | \hat{a}^{\dagger} | \psi \rangle \,. \tag{4.275}$$

However, both of these quantities are zero, and the expectation value of the field is therefore

$$\mathbf{E}(\mathbf{r}, t) \equiv \langle \psi | \hat{\mathbf{E}}(\mathbf{r}, t) | \psi \rangle = 0 \,. \tag{4.276}$$

This is a surprising result since the number-state of the field does not display an electric (or magnetic) field. The field state when the observable electric field is "on" is therefore not the number-state. We thus next consider different types of states of the field.

A different type of state enables us to describe states which exhibit electric fields more similar to those known from classical intuition. Let us consider the bosonic annihilation operator of the harmonic oscillator $\hat{a}$. The annihilation operator together with the creation operator $\hat{a}^\dagger$ forms the number operator $\hat{n} = \hat{a}^\dagger \hat{a}$, whose eigenstates are the same as those of the Hamiltonian, but the eigenvalues are the number of excitations:

$$\hat{n}|n\rangle = n|n\rangle \,. \tag{4.277}$$

Here

$$|n\rangle = \frac{\hat{a}^{\dagger n}}{\sqrt{n!}}|0\rangle \tag{4.278}$$

are the number states generated from the vacuum.

Equivalently to the stationary Schrödinger equation, we can write the eigenequation for the annihilation operator [18, 19]:

$$\hat{a}|\alpha\rangle = \alpha|\alpha\rangle \,. \tag{4.279}$$

To find the expression for $|\alpha\rangle$, we expand it in the number states:

$$|\alpha\rangle = \sum_n c_n|n\rangle \,. \tag{4.280}$$

Inserting this expansion into (4.279), we find

$$\sum_n c_n \hat{a}\frac{\hat{a}^{\dagger n}}{\sqrt{n!}}|0\rangle = \alpha \sum_n c_n \frac{\hat{a}^{\dagger n}}{\sqrt{n!}}|0\rangle \,. \tag{4.281}$$

Now on the left-hand side we use the commutation relation (4.260), that is, $\hat{a}\hat{a}^\dagger = 1 + \hat{a}^\dagger \hat{a}$, to change

$$\begin{aligned} \hat{a}\hat{a}^{\dagger n} &= (1 + \hat{a}^\dagger \hat{a})\hat{a}^{\dagger(n-1)} \\ &= \hat{a}^{\dagger(n-1)} + \hat{a}^\dagger(1 + \hat{a}^\dagger \hat{a})\hat{a}^{\dagger(n-2)} = \ldots = n\hat{a}^{\dagger(n-1)} + \hat{a}^{\dagger n}\hat{a} \end{aligned} \tag{4.282}$$

and we get an iterative expression:

$$\alpha c_n = \sqrt{n+1}c_{n+1} \tag{4.283}$$

or

$$c_n = \frac{\alpha}{\sqrt{n}}c_{n-1} \,. \tag{4.284}$$

This allow us to get the solution for the coefficients in terms of c_0:

$$c_n = \frac{\alpha}{\sqrt{n}} c_{n-1} = \frac{\alpha^2}{\sqrt{n(n-1)}} c_{n-1} = \frac{\alpha^n}{\sqrt{n!}} c_0 \,. \tag{4.285}$$

And we get

$$|\alpha\rangle = c_0 \sum_n \frac{\alpha^n}{\sqrt{n!}} |n\rangle \,. \tag{4.286}$$

The normalization requirement

$$\langle\alpha|\alpha\rangle = c_0^2 \sum_m \frac{|\alpha|^2}{m!} = c_0^2 \exp(|\alpha|^2) = 1 \tag{4.287}$$

allows us to determine the last unknown parameter:

$$c_0 = \exp\left(-\frac{|\alpha|^2}{2}\right) \,. \tag{4.288}$$

Thus, the eigenstate $|\alpha\rangle$ is given by the expansion in the number-states as

$$|\alpha\rangle = \exp\left(-\frac{|\alpha|^2}{2}\right) \sum_n \frac{\alpha^n}{\sqrt{n!}} |n\rangle \,. \tag{4.289}$$

The state $|\alpha\rangle$ can be considered as a new type of state described by the "quantum number" α. The $|\alpha\rangle$ states obtained are not orthogonal since

$$\langle\alpha|\beta\rangle = \exp\left(\alpha^*\beta - \frac{|\alpha|^2}{2} - \frac{|\beta|^2}{2}\right) , \tag{4.290}$$

and also

$$|\langle\alpha|\beta\rangle|^2 = \exp(-|\alpha - \beta|^2) \,. \tag{4.291}$$

The states $|\alpha\rangle$ are called *coherent states*. Additional properties of coherent states are that the expectation value for the coordinate operator $\hat{q}$ is

$$\langle\alpha|\hat{q}|\alpha\rangle = \mathrm{Re}\,(\alpha) \,, \tag{4.292}$$

while that of the momentum operator $\hat{p}$ is

$$\langle\alpha|\hat{p}|\alpha\rangle = \mathrm{Im}\,(\alpha) \,. \tag{4.293}$$

Now suppose that the state of the field $|\psi\rangle$ is a coherent state $|\alpha\rangle$. This coherent state will now generate the electric field since

$$\langle\alpha|\hat{a}|\alpha\rangle = \alpha \tag{4.294}$$

and

$$\langle \alpha | \hat{a}^\dagger | \alpha \rangle = \alpha^* \,. \tag{4.295}$$

Thus, the value of the electric field becomes completely defined. The correct state of the "emitting" field is the coherent state.

Let us now express the coherent state in terms of the field ground state (the vacuum). From the definition of the number-state we have $|n\rangle = (n!)^{-1/2} \hat{a}^{\dagger n} |0\rangle$. Equation (4.286) can be rewritten in terms of an exponential operator as

$$|\alpha\rangle = \exp\left(-\frac{|\alpha|^2}{2} + \alpha \hat{a}^\dagger\right) |0\rangle \,. \tag{4.296}$$

From the operator identity $\hat{a}|0\rangle = 0$ it follows that

$$\exp(\alpha \hat{a}^\dagger)|0\rangle = \exp(\alpha \hat{a}^\dagger) \exp(x \hat{a})|0\rangle \tag{4.297}$$

for an arbitrary number x. Applying the Weyl identity (see Appendix A.3), we obtain

$$\exp(\alpha \hat{a}^\dagger) \exp(x \hat{a}) = \exp(\alpha \hat{a}^\dagger + x \hat{a}) \exp\left(-\frac{\alpha x}{2}\right) \,. \tag{4.298}$$

Thus, taking $x = -\alpha^*$, we obtain

$$|\alpha\rangle = \exp(\alpha \hat{a}^\dagger - \alpha^* \hat{a})|0\rangle \,. \tag{4.299}$$

The operator

$$\hat{D}(\alpha) = \exp(\alpha \hat{a}^\dagger - \alpha^* \hat{a}) \tag{4.300}$$

is known as the displacement operator. It can easily be checked that $\hat{D}(0) = \hat{I}$. The commutator of an operator $\hat{S}(\alpha) = \alpha \hat{a}^\dagger - \alpha^* \hat{a}$,

$$[\hat{S}(\alpha), \hat{S}(\beta)] = \alpha\beta^* - \beta\alpha^* \equiv 2\mathrm{i}s \,, \tag{4.301}$$

where $s = \operatorname{Im} \alpha\beta^*$, leads us to an important property of the displacement operator, namely,

$$\hat{D}(\alpha)\hat{D}(\beta) = \exp(\mathrm{i}s)\hat{D}(\alpha + \beta) \,. \tag{4.302}$$

It follows now that $\hat{D}^n(\alpha) = \hat{D}(n\alpha)$ and $\hat{D}(-\alpha) = \hat{D}^\dagger(\alpha)$. Thus, the operator $\hat{D}$ indeed performs a displacement of the coherent state.

It can be shown that the coordinate and momentum variances in an arbitrary coherent state are the same as those of the ground state of the harmonic oscillator. Thus, the coherent state is equivalent to the ground state of the harmonic oscillator, which is a Gaussian wavepacket shifted by α in the phase space of the oscillator. The coherent state is the state with the closest relationship to classical physics [19].

One more relation is important for coherent states. Consider the outer product of the coherent states $|\alpha\rangle\langle\alpha|$. Using the expansion in the number-states, (4.278), we have

$$|\alpha\rangle\langle\alpha| = \exp\left(-|\alpha|^2\right)\sum_n\sum_m \frac{\alpha^n}{\sqrt{n!}}\frac{\alpha^{*m}}{\sqrt{m!}}|n\rangle\langle m| \,. \tag{4.303}$$

We now take $\alpha = r\exp(\mathrm{i}\varphi)$. By integrating the outer product in the full complex plane (we denote $\mathrm{d}^2\alpha \equiv \mathrm{d}\alpha'\mathrm{d}\alpha''$, where $\alpha = \alpha' + \mathrm{i}\alpha''$) and changing the integration to polar coordinates, we have

$$\int \mathrm{d}^2\alpha|\alpha\rangle\langle\alpha| = \sum_{mn}\frac{|n\rangle\langle m|}{\sqrt{n!m!}}\int_0^\infty r^{n+m+1}\exp(-r^2)\mathrm{d}r\int_0^{2\pi}\mathrm{d}\varphi\exp(\mathrm{i}\varphi(n-m)) \,. \tag{4.304}$$

The integration over angles is

$$\int_0^{2\pi}\mathrm{d}\varphi\exp(\mathrm{i}\varphi(n-m)) = 2\pi\delta_{mn} \,, \tag{4.305}$$

and the integration over the radius is

$$\int_0^\infty r^{2n+1}\mathrm{e}^{-r^2}\mathrm{d}r = \frac{1}{2}\int_0^\infty x^n\mathrm{e}^{-x}\mathrm{d}x = \frac{1}{2}n! \,. \tag{4.306}$$

We thus find that

$$\int \mathrm{d}^2\alpha|\alpha\rangle\langle\alpha| = \pi\sum_n |n\rangle\langle n| = \pi\hat{I} \,. \tag{4.307}$$

We can see that the coherent states form a complete (overcomplete) set of states and they resolve unity up to a constant. These states can therefore be used for a complete expansion of arbitrary quantum states in an arbitrary quantum problem. We use the coherent states to describe relaxation phenomena for an arbitrary second-quantized Hamiltonian in Section 10.5.

5
Quantum States of Molecules and Aggregates

As described in Section 4.7, pure transitions between quantum states should correspond to infinitely sharp lines (stick spectra) reflecting their transition frequencies according to (4.228). In reality, electronic molecular transitions are not infinitely sharp. Vibrational degrees of freedom of molecules and/or the molecular surroundings are coupled to the electronic excitations, thereby leading to significant broadening of the transition bands in the spectra. Different broadening mechanisms will be introduced later and some consequences for spectroscopy will be discussed in Part Two. Here the molecular exciton approach, which describes the coherent superposition of the excited states of molecules in molecular aggregates, will be considered. A coherent relationship of this type between excitations of the molecules should evidently manifest itself in the stationary and time-resolved absorption and fluorescence spectra. Exciton models corresponding to the so-called Frenkel, Wannier–Mott, and charge-transfer (CT) excitons will be described. The concept of exciton self-trapping will also be introduced.

5.1
Potential Energy Surfaces, Adiabatic Approximation

A molecule is essentially a collection of positively charged nuclei and negative electrons, which all interact through electrostatic interactions. Their states and dynamics are usually described using quantum mechanics. We denote coordinates of the nuclei by $\boldsymbol{R}$ and coordinates of the electrons by $\boldsymbol{r}$. The starting point of the system's theoretical description is its Hamiltonian. It consists of the kinetic and potential energy of all particles:

$$\hat{H}_0 = \sum_i \frac{\hat{P}_i^2}{2M_i} + \sum_j \frac{\hat{p}_j^2}{2m_j} + k_e \sum_{mn}^{m>n} \frac{q_m q_n}{|\hat{\boldsymbol{R}}_m - \hat{\boldsymbol{R}}_n|} + k_e \sum_{ij}^{i>j} \frac{e^2}{|\hat{\boldsymbol{r}}_i - \hat{\boldsymbol{r}}_j|} - k_e \sum_{mj} \frac{q_m e}{|\hat{\boldsymbol{R}}_m - \hat{\boldsymbol{r}}_j|}\,, \quad (5.1)$$

Molecular Excitation Dynamics and Relaxation, First Edition. L. Valkunas, D. Abramavicius, and T. Mančal.

where $k_e = (4\pi\epsilon\epsilon_0)^{-1}$, $\hat{P}_i$ and $\hat{p}_j$ are the operators of the momenta of nuclei and electrons, respectively, q_n is the charge of the nth nucleus, and e is the charge of an electron (the elementary charge).

The notion of electronic transitions of a molecule implies a major simplification for the molecular description. This is justified because of the difference in the mass of an electron (m_e) and the mass of a nucleus (M_i). Since the mass of a nucleus is much larger than that of an electron, the motion of nuclei is much slower than that of electrons, and therefore it can be assumed in the first approximation that the nuclei are fixed at their equilibrium positions denoting the structure or configuration of the molecule. The so-called adiabatic approximation (or Born–Oppenheimer approximation) is then used to describe the quantum properties of molecules. The basic parameter is then the ratio m_e/M_i, which is the small parameter of the approximation [11, 20]. In the adiabatic approximation, the Hamiltonian of any molecule or of a molecular system can be formally given as

$$\hat{H} = \hat{H}_e(\boldsymbol{r}) + \hat{V}(\boldsymbol{r}, \boldsymbol{R}) , \tag{5.2}$$

where $\boldsymbol{r}$ and $\boldsymbol{R}$ are the coordinates of electrons and nuclei of the system, respectively. $\boldsymbol{R}$ is taken as a number (nonoperator). $\hat{H}_e(\boldsymbol{r})$ is the electronic Hamiltonian including the kinetic energy of electrons, and the electron–electron interaction energy, $V(\boldsymbol{r}, \boldsymbol{R})$, is the electron–nuclear interaction energy taken at fixed positions of the nuclei. The kinetic energy of nuclei is disregarded. Since the nuclei are at fixed positions, their Coulombic interactions only add an offset to the potential energy. By diagonalizing this Hamiltonian, we obtain the eigenvalues $\epsilon^e(\boldsymbol{R})$ and eigenfunctions $\varphi^e(\boldsymbol{r}, \boldsymbol{R})$ for the electronic subsystem from the following Schrödinger equation:

$$\hat{H}\varphi^e(\boldsymbol{r}, \boldsymbol{R}) = \epsilon^e(\boldsymbol{R})\varphi^e(\boldsymbol{r}, \boldsymbol{R}) . \tag{5.3}$$

Note that this eigenvalue equation is only for electronic degrees of freedom, and all values are parametrically dependent on $\boldsymbol{R}$, that is, the equation is solved for a fixed configuration of nuclei represented by fixed $\boldsymbol{R}$ values. This is essentially a variational problem as well, since the nuclear coordinates should be varied to minimize the electronic energy. This optimization guarantees the proper equilibrium configuration of the molecule.

In the next step the nuclear kinetic energy $\hat{T}(\boldsymbol{R})$ may be included to describe vibrational degrees of freedom of the nuclei. In the adiabatic approach, the nuclear kinetic energy is treated as a perturbation.[1] The electronic wavefunction $\varphi^e(\boldsymbol{r}, \boldsymbol{R})$ obtained from (5.3) can be used as a basis set, that is, we will use the following expansion for the total function to account for the nuclear motion:

$$\Psi(\boldsymbol{r}, \boldsymbol{R}) = \sum_e \Phi_e(\boldsymbol{R})\varphi^e(\boldsymbol{r}, \boldsymbol{R}) . \tag{5.4}$$

1) It is worthwhile mentioning that there is a different approach called the diabatic (or nonadiabatic) approach, which considers $V(\boldsymbol{r}, \boldsymbol{R})$ as a perturbation. Both approaches lead to similar conclusions.

The Schrödinger equation, which has to be solved to obtain the whole (including nuclear motion) spectrum, is as follows:

$$(\hat{T}(\boldsymbol{R}) + \hat{H})\Psi(\boldsymbol{r}, \boldsymbol{R}) = E\Psi(\boldsymbol{r}, \boldsymbol{R}) \,. \tag{5.5}$$

By inserting (5.4) into (5.2), we obtain the following equation for the expansion coefficients $\Phi_e(\boldsymbol{R})$:

$$(\hat{T}(\boldsymbol{R}) + U^e(\boldsymbol{R}) - E)\Phi_e(\boldsymbol{R}) + \sum_{e'} \Lambda_{ee'}(\boldsymbol{R})\Phi_{e'}(\boldsymbol{R}) = 0 \,, \tag{5.6}$$

where

$$U^e(\boldsymbol{R}) = \epsilon^e(\boldsymbol{R}) + \Lambda_{ee}(\boldsymbol{R}) \,. \tag{5.7}$$

$\epsilon^e(\boldsymbol{R})$ is the so-called adiabatic potential energy in the eth electronic state, and

$$\Lambda_{ee'}(\boldsymbol{R}) = -\sum_i \frac{\hbar^2}{M_i} \int \left[\varphi^{e*}(\boldsymbol{r}, \boldsymbol{R}) \frac{\partial}{\partial R_i} \varphi^{e'}(\boldsymbol{r}, \boldsymbol{R}) + \frac{1}{2} \varphi^{e*}(\boldsymbol{r}, \boldsymbol{R}) \frac{\partial^2}{\partial R_i^2} \varphi^{e'}(\boldsymbol{r}, \boldsymbol{R}) \right] \mathrm{d}\boldsymbol{r} \tag{5.8}$$

defines the so-called nonadiabaticity operator. This operator appears because the nuclear kinetic energy operator acts on the electronic wavefunctions, which parametrically depend on the nuclear coordinates.

By disregarding off-diagonal terms in (5.6) ($e \neq e'$), we arrive at the "real" adiabatic results, where the electronic quantum number e determines the energy levels of the molecule characterized by the eigenfunctions

$$\Psi_e(\boldsymbol{r}, \boldsymbol{R}) = \Phi_e(\boldsymbol{R})\varphi^e(\boldsymbol{r}, \boldsymbol{R}) \tag{5.9}$$

for the following diagonal Hamiltonian:

$$\hat{H}_e = \hat{T}(\boldsymbol{R}) + \hat{U}^e(\boldsymbol{R}) \,. \tag{5.10}$$

The electronic energies $U^e(\boldsymbol{R})$ have become well-defined potential energies and determine the spectrum of the molecule. Figure 5.1 shows a projection of the ground-state and excited-state potential energy surface on a particular nuclear coordinate, demonstrating the change of the nuclear equilibrium position between two states. We will be mainly interested in the behavior of the system close to the minima of the corresponding potential energy surfaces. In that case the harmonic approximation can be used to define the potential energy surfaces, which yields

$$\hat{U}^e(\boldsymbol{R}) = U^e(\boldsymbol{R}_0) + \frac{1}{2} \sum_{i,j} \frac{\partial^2}{\partial R_i \partial R_j} U^e(\boldsymbol{R}_0)(R_i - R_{i0})(R_j - R_{j0}) \,. \tag{5.11}$$

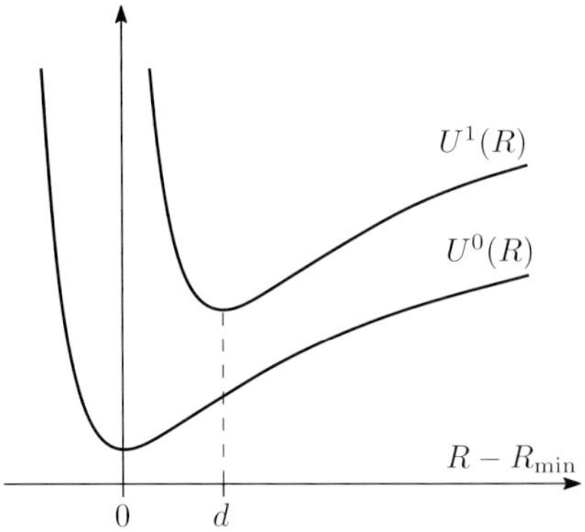

Figure 5.1 Projections of the potential energy surfaces of the ground state and the first excited state on a particular nuclear coordinate, where d is the relative shift of the minima of both curves.

The potential energy can be written in terms of noninteracting degrees of freedom, so-called normal modes. In the normal coordinate representation, we arrive at some new nuclear coordinates Q, and the oscillator potential is obtained in quadratic form [21], that is,

$$\hat{U}^e(Q) = U^e(0) + \sum_i \frac{\hbar\omega_i^e}{2} Q_i^2 , \tag{5.12}$$

where ω_i^e is the vibration frequency of the ith mode in the eth electronic state. Then (5.10) for the Hamiltonian $\hat{H}_e$ can be rewritten as

$$\hat{H}_e = U^e(0) + \sum_i \frac{\hbar\omega_i^e}{2}\left(Q_i^2 - \frac{\partial^2}{\partial Q_i^2}\right) \tag{5.13}$$

and the corresponding Schrödinger equation for the harmonic oscillator,

$$(\hat{H}_e - E_e)\Phi_e(Q) = 0 , \tag{5.14}$$

can be easily solved as shown in Section 4.6.1 [11, 20]. It is noteworthy that the Hamiltonian of the harmonic oscillator can be represented in the second-quantization representation by means of creation and annihilation operators (see Section 4.9). The solution of (5.5) gives quantum energy levels of the harmonic oscillator with internal vibrational quantum numbers ν_i. The total energy of the system has eigenvalues

$$E_{e\nu} = U^e(0) + \sum_i \left(\nu_i + \frac{1}{2}\right)\hbar\omega_i^e , \tag{5.15}$$

with corresponding eigenfunctions

$$\Phi_{e\nu}(Q) = \frac{\exp\left(-\frac{Q^2}{2}\right)}{\sqrt{2^\nu \nu! \sqrt{\pi}}} H_\nu(Q) , \tag{5.16}$$

where

$$H_\nu(Q) = (-1)^\nu \exp(Q^2) \frac{\mathrm{d}^\nu}{\mathrm{d}Q^\nu} \exp(-Q^2) \tag{5.17}$$

is the νth-order Hermitian polynomial.

Considering optical energy gaps, the lowest-energy state is the state in which the molecule is usually observed under room temperature conditions. When the molecule is electronically excited into an excited state e, the Hamiltonian corresponding to the nuclear degrees of freedom can be written as

$$\hat{H}_e(Q) = \hat{H}_g(Q) + \Delta \hat{H}_e(Q) , \tag{5.18}$$

where $\hat{H}_g(Q)$ is the Hamiltonian describing the system in the electronic ground state, and $\Delta \hat{H}_e(Q)$ is the change of the system energy due to electronic excitation. Such a representation allows us to define terms responsible for the interaction of the molecular excitation with molecular vibrations. The difference term corresponds to the difference between the ground-state and the excited-state nuclear potential. This can be verified by making the decomposition

$$\hat{H}_e(Q) = \hat{T}(Q) + \hat{U}^g(Q) + (\hat{U}^e(Q) - \hat{U}^g(Q)) , \tag{5.19}$$

and comparing it with (5.18).

Let us now consider the case of a single vibrational mode, since the generalization to the case of many vibrational modes is straightforward. For convenience in this case we will skip the mode numbering index i and will use the following definition: $\omega_i^e = \omega_e$. So the potential energy surfaces of the excited and ground electronic states are

$$U^e(Q) = U^e(0) + \frac{\hbar\omega_e}{2}(Q - d)^2 \tag{5.20}$$

and

$$U^g(Q) = U^g(0) + \frac{\hbar\omega_g}{2} Q^2 , \tag{5.21}$$

respectively. Here d is the displacement of the minimum of the excited-state potential from the minimum of the ground-state potential; see (5.12). Thus, the energy difference between both potentials is given by

$$\hat{U}^e(Q) - \hat{U}^g(Q) = E_e + \lambda - d\hbar\omega_e Q + \delta Q^2 , \tag{5.22}$$

where

$$E_e = U^e(0) - U^g(0) \tag{5.23}$$

is the electronic excitation energy,

$$\lambda = \hbar\omega_e \frac{d^2}{2} \tag{5.24}$$

is the so-called Franck–Condon energy [20, 22], which is analogous to the reorganization energy defined in the theory of CT in polar solvents, and

$$\delta = \frac{\hbar\omega_e - \hbar\omega_g}{2} \tag{5.25}$$

is the change of the vibrational energy in the excited electronic state in comparison with the vibrational energy in the ground state. The last two terms in (5.22) determine the vibronic interaction (or the electron–phonon interaction).

5.2
Interaction between Molecules

Intermolecular interactions may significantly change the electronic excitation spectrum of molecular aggregates with respect to the spectrum of an isolated molecule. When intermolecular interactions are strong, this may result in exciton formation [22–24]. They also determine van der Waals forces (or so-called dispersion forces), describing the repulsion of molecules at the closest distances and their attraction at larger distances as shown schematically in Figure 5.1. The intermolecular interaction can also be presented as a sum of terms inversely proportional to growing powers of the intermolecular distance using the multipole expansion formula. Indeed, when the charge distribution of two molecules is $\rho_1(\boldsymbol{r})$ and $\rho_2(\boldsymbol{r})$, the Coulomb interaction between them

$$V_{\text{Coulomb}} = k_e \iint \frac{\rho_1(\boldsymbol{r})\rho_2(\boldsymbol{r}')}{|\boldsymbol{r} - \boldsymbol{r}'|} \mathrm{d}\boldsymbol{r}\mathrm{d}\boldsymbol{r}' \tag{5.26}$$

may be expressed in terms of the multiple moments of the respective charge distributions. Taking the center $\boldsymbol{R}_i$ of the ith molecule as a reference point, for instance, its center of mass, one defines the total charge

$$q_i = \int \rho_i(\boldsymbol{r})\mathrm{d}\boldsymbol{r} , \tag{5.27}$$

the dipole moment

$$\boldsymbol{D}_i = \int \rho_i(\boldsymbol{r})(\boldsymbol{r} - \boldsymbol{R}_i)\mathrm{d}\boldsymbol{r} , \tag{5.28}$$

the quadrupole tensor

$$Q_{i,\alpha\beta} = \int \rho_i(\boldsymbol{r}) \left[3(\boldsymbol{r} - \boldsymbol{R}_i)_\alpha(\boldsymbol{r} - \boldsymbol{R}_i)_\beta - \delta_{\alpha\beta}(\boldsymbol{r} - \boldsymbol{R}_i)^2\right] \mathrm{d}\boldsymbol{r} , \tag{5.29}$$

and higher multiple moments of its charge distribution $\rho_i(\boldsymbol{r})$ relative to this center. The energy of the interaction between two molecules with the relative position vector $\boldsymbol{R} = \boldsymbol{R}_j - \boldsymbol{R}_i$ and the corresponding unit vector $\boldsymbol{R}_0$ is then expressed as

$$\begin{aligned} V(\boldsymbol{R}) = k_e \Bigg[& \frac{q_i q_j}{R} + \frac{q_i(\boldsymbol{D}_j \boldsymbol{R}_0) - q_j(\boldsymbol{D}_i \boldsymbol{R}_0)}{R^2} \\ & + \frac{\boldsymbol{D}_i \boldsymbol{D}_j - 3(\boldsymbol{D}_i \boldsymbol{R}_0)(\boldsymbol{D}_j \boldsymbol{R}_0)}{R^3} + \ldots \Bigg] . \end{aligned} \tag{5.30}$$

For such an expansion to be valid and meaningful it is essential that the distance R between the molecules is considerably larger than the size of the charge distribution in individual molecules to guarantee a proper convergence of the series. The first two terms are important for charged molecules and in the case of coupling between CT states. However, in many situations molecules can be considered uncharged, and the third term of the expansion is dominant. It reflects the dipole–dipole interaction between two molecules.

It is noteworthy that the dipole–dipole approximation is reasonable not only if the intermolecular distance R is larger than the size of the molecule but also when R is larger than the dipole radius a, where $a = \mu/e$ (μ is the transition dipole moment). Since the transition dipole moment is at most on the order of a few debyes ($1\,\text{D} = 3.336 \times 10^{30}\,\text{C m}$; e.g., an electron separated from a hole by 1 Å corresponds to 4.8 D), the dipole radius is on the order of a few angstroms. As a consequence, in many cases the dipole–dipole coupling term to a large extent determines the observable spectroscopic properties of the interacting molecules as well as the excitation energy transfer properties. Higher-order terms fall off more rapidly with distance, and their importance is limited.

For the case when the intermolecular distances become comparable to the molecular characteristic dimensions, the multipole series expansion loses meaning and one must resort to the full expression for the Coulomb energy, (5.26). The molecules are then no longer strictly distinct entities. However, even then one invokes the dipole–dipole approximation as an effective model that may give at least a qualitative description.

Molecular aggregation leads to substantial changes in observed energy spectra due to the emergence of nonvanishing intermolecular interactions. These spectral changes are usually related to exciton formation [6, 24, 25]. The excitons are also widely used to explain energy transfer through the molecular system.

5.3 Excitonically Coupled Dimer

The simplest system for which exciton effects can be demonstrated is a pair of interacting molecules. We call such a pair an excitonically coupled dimer, where it should be noted that the word "dimer" does not imply that the molecules are in van der Waals contact; in fact they can be spatially separated. We speak of a physical dimer to distinguish it from a chemical dimer, where chemical bonds are present between the two monomers. In a physical dimer only electrostatic intermolecular interactions through space are relevant and exchange of electrons between molecules is negligible.

First we consider two similar molecules in a vacuum at a fixed distance R_{12} having a fixed orientation and each having only two energy levels. For a given Hamiltonian $\hat{H}$ the isolated molecules have their two eigenstates φ^i determined by

$$\hat{H}_n \varphi_n^{(i)} = \varepsilon_n^{(i)} \varphi_n^{(i)} , \tag{5.31}$$

where the subscript n identifies the pigment (either pigment 1 or pigment 2) and the superscript i refers to the ground state (g) and the excited state (e). We will describe different and identical eigenenergies of both molecules. Later we take the ground-state energy to be zero, that is, $\varepsilon_1^{(g)} = \varepsilon_2^{(g)} = 0$.

When the molecules interact, the total Hamiltonian besides $\hat{H}_1$ and $\hat{H}_2$ also includes interaction $\hat{V}$ between them. In that case φ_1 and φ_2 are no longer the correct eigenstates of the dimer and the eigenenergies will also be different. Thus, intermolecular interaction gives rise to perturbation of the energy spectrum. Due to the weak intermolecular interactions, the perturbation theory for degenerate states can be used with the Heitler–London approximation, which means that the eigenfunctions of the dimer are equal to superpositions of the product of the molecular eigenfunctions.

We describe the electronic ground state as $\Psi_g = \varphi_1^{(g)}\varphi_2^{(g)}$, which corresponds to the condition of noninteracting molecules, and the corresponding Hilbert space of the dimer is determined as the product of the Hilbert spaces corresponding to monomers. Note that we do not properly antisymmetrize Ψ_g, thereby implicitly assuming that exchange of electrons between participating molecules 1 and 2 does not occur. The corresponding ground-state energy of the dimer is now expressed by

$$\begin{aligned} E_g &= \left\langle \varphi_1^{(g)}\varphi_2^{(g)} \right| \hat{H}_1 + \hat{H}_2 + \hat{V} \left| \varphi_1^{(g)}\varphi_2^{(g)} \right\rangle \\ &= \varepsilon_1^{(g)} + \varepsilon_2^{(g)} + \left\langle \varphi_1^{(g)}\varphi_2^{(g)} \right| \hat{V} \left| \varphi_1^{(g)}\varphi_2^{(g)} \right\rangle = \varepsilon_1^{(g)} + \varepsilon_2^{(g)} + V_{gg} \, . \end{aligned} \tag{5.32}$$

This indicates that coupling between molecules leads to a displacement of the ground-state energy by V_{gg}. The excited states can be formally written as

$$\Psi_e = c_{e1}\varphi_1^{(e)}\varphi_2^{(g)} + c_{e2}\varphi_1^{(g)}\varphi_2^{(e)} \, . \tag{5.33}$$

The excited states are normalized and orthogonal, and the coefficients c_{e1} and c_{e2} thus fulfill the following equation:

$$|c_{e1}|^2 + |c_{e2}|^2 = 1 \, . \tag{5.34}$$

Consequently, the excited state e of the coupled system is a linear combination of two terms in which one or the other molecule is excited. The relative contributions of these two terms are determined by coefficients $c_{e1,2}$. The new eigenstates are required to be stationary solutions of the Schrödinger equation of the dimer; thus,

$$(\hat{H}_1 + \hat{H}_2 + \hat{V})\Psi_e = E_e \Psi_e \, . \tag{5.35}$$

Multiplying both sides from the left with either $\varphi_1^{(e)}\varphi_2^{(g)}$ or $\varphi_1^{(g)}\varphi_2^{(e)}$ and integrating over the entire space gives the two equations

$$c_{e1}\left(\varepsilon_1^{(e)} + \left\langle \varphi_1^{(e)}\varphi_2^{(g)} \right| \hat{V} \left| \varphi_1^{(e)}\varphi_2^{(g)} \right\rangle\right) + c_{e2}\left\langle \varphi_1^{(e)}\varphi_2^{(g)} \right| \hat{V} \left| \varphi_1^{(g)}\varphi_2^{(e)} \right\rangle = c_{e1} E_e \tag{5.36}$$

and

$$c_{e1}\left\langle \varphi_1^{(g)}\varphi_2^{(e)} \right| \hat{V} \left| \varphi_1^{(e)}\varphi_2^{(g)} \right\rangle + c_{e2}\left(\varepsilon_2^{(e)} + \left\langle \varphi_1^{(g)}\varphi_2^{(e)} \right| \hat{V} \left| \varphi_1^{(g)}\varphi_2^{(e)} \right\rangle\right) = c_{e2} E_e \, , \tag{5.37}$$

which can be written in shorthand notation as

$$c_{e1}\left(\varepsilon_1^{(e)} + V_{11} - E_e\right) + c_{e2} V_{12} = 0 \tag{5.38}$$

and

$$c_{e1} V_{21} + c_{e2}\left(\varepsilon_2^{(e)} + V_{22} - E_e\right) = 0\,. \tag{5.39}$$

Here we get energy shifts

$$V_{11} = \left\langle \varphi_1^{(e)} \varphi_2^{(g)} \right| \hat{V} \left| \varphi_1^{(e)} \varphi_2^{(g)} \right\rangle , \tag{5.40}$$

$$V_{22} = \left\langle \varphi_1^{(g)} \varphi_2^{(e)} \right| \hat{V} \left| \varphi_1^{(g)} \varphi_2^{(e)} \right\rangle \tag{5.41}$$

and

$$V_{12} = \left\langle \varphi_1^{(e)} \varphi_2^{(g)} \right| \hat{V} \left| \varphi_1^{(g)} \varphi_2^{(e)} \right\rangle , \tag{5.42}$$

$$V_{21} = \left\langle \varphi_1^{(g)} \varphi_2^{(e)} \right| \hat{V} \left| \varphi_1^{(g)} \varphi_2^{(e)} \right\rangle \tag{5.43}$$

are resonance interaction terms which are also important for energy-transfer processes. Evidently, a nontrivial solution is obtained only if

$$\begin{vmatrix} \varepsilon_1^{(e)} + V_{11} - E_e & V_{12} \\ V_{21} & \varepsilon_2^{(e)} + V_{22} - E_e \end{vmatrix} = 0\,. \tag{5.44}$$

Let us first consider a homodimer, where both molecules are identical. For this case $\varepsilon_1^{(e)} = \varepsilon_2^{(e)} = \varepsilon^{(e)}$, $V_{11} = V_{22}$, and $V_{12} = V_{21}$, and from (5.44) it follows that

$$\left(\varepsilon^{(e)} + V_{11} - E_e\right)^2 = V_{12}^2\,, \tag{5.45}$$

leading to two eigenenergies:

$$E_1 = \varepsilon^{(e)} + V_{11} + V_{12} \tag{5.46}$$

and

$$E_2 = \varepsilon^{(e)} + V_{11} - V_{12}\,. \tag{5.47}$$

Thus, the transition energy of the dimer has changed as compared with the transition energy of a constituent single molecule. The energy levels have split by the amount $2V_{12}$ (so-called Davydov splitting) and the average energy of these two levels has been shifted with respect to the ground state by the amount $V_{11} - V_{gg}$, which is the so-called displacement energy Δ (see Figure 5.2). In fact, Δ is comparable to the spectral change which a molecule experiences when it goes from the gas phase into solution, and using the definition given by (5.26), we get

$$\Delta = k_e \int \mathrm{d}\boldsymbol{r}_1 \int \mathrm{d}\boldsymbol{r}_2 \frac{\left(\rho_i^{(ee)}(\boldsymbol{r}_1) - \rho_i^{(gg)}(\boldsymbol{r}_1)\right) \rho_j^{(gg)}(\boldsymbol{r}_2)}{|\boldsymbol{r}_1 - \boldsymbol{r}_2|}\,, \tag{5.48}$$

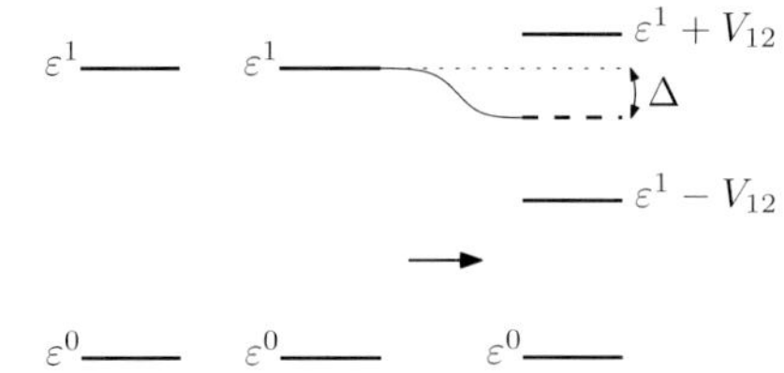

Figure 5.2 The energy scheme for the homodimer, where ε^0 and ε^1 are the energies of the ground and excited states of the monomers, Δ is the displacement energy, and V_{12} is the resonance interaction.

where $i = 1$ and $j = 2$ in this case. Here $\rho_i^{(gg)}(\boldsymbol{r})$ is the total charge density of the ith molecule in its ground state and $\rho_i^{(ee)}(\boldsymbol{r}_1)$ is the charge density of the electronic excited state of the ith molecule. For neutral molecules we have

$$\int \mathrm{d}\boldsymbol{r}\rho_i^{(ss)}(\boldsymbol{r}) = 0\,, \tag{5.49}$$

where (ss) is either (gg) or (ee).

The value of the resonance interaction V_{12} can also then be given in terms of transition charge densities accordingly as

$$V_{12} = k_\mathrm{e} \int \mathrm{d}\boldsymbol{r}_1 \int \mathrm{d}\boldsymbol{r}_2 \frac{\rho_1^{(ge)}(\boldsymbol{r}_1)\rho_2^{(eg)}(\boldsymbol{r}_2)}{|\boldsymbol{r}_1 - \boldsymbol{r}_2|}\,, \tag{5.50}$$

where $\rho_i^{(eg)}(\boldsymbol{r})$ and $\rho_i^{(ge)}(\boldsymbol{r})$ represent the transition charge densities of molecule i. When the intermolecular distances are larger than the molecular dimensions, a dipole approximation for charge densities is often assumed and the resonance interaction can be calculated by using an expression for the intermolecular interaction defined by (5.30), thus giving

$$V_{12} = k_\mathrm{e}\left[\frac{(\boldsymbol{\mu}_1 \cdot \boldsymbol{\mu}_2)}{|R_{12}|^3} - 3\frac{(\boldsymbol{R}_{12} \cdot \boldsymbol{\mu}_1)(\boldsymbol{R}_{12} \cdot \boldsymbol{\mu}_2)}{|R_{12}|^5}\right]\,, \tag{5.51}$$

where $\boldsymbol{\mu}_j = \langle \varphi_j^{(g)} | \hat{\boldsymbol{D}}_j | \varphi_j^{(e)} \rangle$ is the transition dipole moment in the jth molecule.

In the case of nonidentical transition energies of the monomers (let us assume that the difference equals δ), the calculation of the excitonic eigenenergies and eigenstates can also be obtained by solving (5.44). It is then convenient to define a new zero energy as the average of the transition energies of the two molecules. Then one of the molecules has the excited-state "site energy" equal to $\delta/2$, and the site energy of the other molecule is $-\delta/2$ (see Figure 5.3). To find new eigenenergies we have to solve the following pair of equations:

$$c_{e1}\left(\frac{\delta}{2} - E_\mathrm{e}\right) + c_{e2}V_{12} = 0\,, \tag{5.52}$$

$$c_{e1}V_{21} + c_{e2}\left(-\frac{\delta}{2} - E_\mathrm{e}\right) = 0\,. \tag{5.53}$$

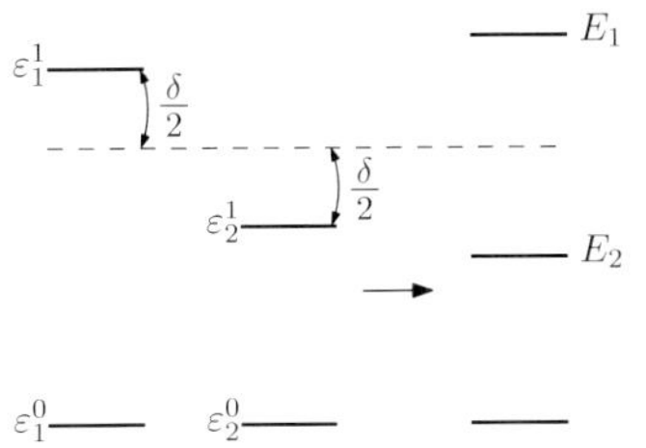

Figure 5.3 The energy scheme for the heterodimer, where ε_i^0 and ε_i^1 are the energies of the ground and excited states of the ith monomer, and δ is the difference between the excitation energies of the monomers.

The solutions can be written as

$$E_1 = V_{12}\sqrt{1+\Gamma^2}\,, \quad E_2 = -V_{12}\sqrt{1+\Gamma^2}\,, \tag{5.54}$$

where

$$\Gamma = \frac{\delta}{2V_{12}}\,. \tag{5.55}$$

Since $|c_{e1}|^2 + |c_{e2}|^2 = 1$, we can replace c_{e1} and c_{e2} by $\cos\alpha_e$ and $\sin\alpha_e$, respectively, where α_e still has to be determined. After insertion of (5.54) into (5.53), it directly follows that α_e is given by

$$\tan\alpha_1 = -\Gamma + \sqrt{1+\Gamma^2}\,,$$

$$\tan\alpha_2 = -\Gamma - \sqrt{1+\Gamma^2}\,, \tag{5.56}$$

and in general

$$\tan 2\alpha_e = \frac{1}{\Gamma}\,. \tag{5.57}$$

The angle α_e is sometimes denoted as the mixing angle of the dimer. Explicit expressions for the coefficients are now

$$c_{11} = \frac{1}{\sqrt{1+\left(-\Gamma+\sqrt{1+\Gamma^2}\right)^2}}\,, \tag{5.58}$$

$$c_{12} = c_{11}\left(-\Gamma+\sqrt{1+\Gamma^2}\right)\,, \tag{5.59}$$

$$c_{21} = \frac{1}{\sqrt{1+\left(-\Gamma-\sqrt{1+\Gamma^2}\right)^2}}\,, \tag{5.60}$$

$$c_{22} = c_{21}\left(-\Gamma-\sqrt{1+\Gamma^2}\right)\,. \tag{5.61}$$

In the case of identical molecules, $\delta = 0$, and the coefficients are $c_{11} = 1/\sqrt{2}$, $c_{12} = 1/\sqrt{2}$, $c_{21} = 1/\sqrt{2}$, and $c_{22} = -1/\sqrt{2}$, indicating that exciton states are completely coherently delocalized over the whole dimer. The two excited-state energy levels are split by two times the "resonance" interaction energy V_{12}. In the opposite case, when $V_{12} \ll \delta$, we approach the case of independent molecules, both for the energy splitting and for the degree of delocalization, and the excitations on the different molecules maintain their identity:

$$E_{1,2} = \pm\frac{\delta}{2}\left(1 + \frac{2V_{12}^2}{\delta^2}\right) . \tag{5.62}$$

The probability of finding the excitation on either of the molecules is $(1/(2\Gamma))^2$ and $1 - (1/(2\Gamma))^2$, respectively, that is, with large Γ the excitation is almost entirely localized on the individual molecules.

Another valid state of a dimer is one where both monomers are excited, the so-called doubly excited state. In the Heitler–London approximation the wavefunction of the double excited state is given by

$$\Psi_f = \varphi_1^{(e)}\varphi_2^{(e)} , \tag{5.63}$$

and the excitation energy is

$$E_f = \varepsilon_1^{(e)} + \varepsilon_2^{(e)} + K . \tag{5.64}$$

Here

$$K = k_e \int \mathrm{d}\boldsymbol{r}_1 \int \mathrm{d}\boldsymbol{r}_2 \frac{\left[\rho_1^{(ee)}(\boldsymbol{r}_1) - \rho_1^{(gg)}(\boldsymbol{r}_1)\right]\left[\rho_2^{(ee)}(\boldsymbol{r}_2) - \rho_2^{(gg)}(\boldsymbol{r}_2)\right]}{|\boldsymbol{r}_1 - \boldsymbol{r}_2|} \tag{5.65}$$

describes the excitation energy shift due to the presence of another excitation. Often this coupling is denoted as K-type coupling, and the resonance interaction is denoted as J-type coupling. Since there is only one doubly excited state in a dimer, it is an eigenstate of the system.

5.4 Frenkel Excitons of Molecular Aggregates

Now let us consider a molecular aggregate of a fixed number N of molecules organized in a specific geometrical configuration. Each molecule consists of a large number of electrons and nuclei; however, for the present purpose, only valence electrons need be considered. If the electrons (and nuclei) cannot interchange between different molecules, chemical reactions and electron exchange can be disregarded. We denote by $\boldsymbol{r}_j^{(m)}$ the coordinate vectors of all relevant electrons belonging to molecule m. $\boldsymbol{R}_n^{(m)}$ are the corresponding coordinates of the nuclei. The Hamiltonian of such a complex can be written as

$$\hat{H}_M = \sum_m \hat{H}_m + \sum_{mm'}^{m>m'} \hat{V}_{mm'} , \tag{5.66}$$

where $\hat{H}_m$ is the Hamiltonian of a single isolated molecule and $\hat{V}_{mm'}$ is the intermolecular Coulomb interaction.

If the molecules were noninteracting, the state of the whole aggregate would be a direct product of local molecular states. Similarly to the case of the dimer, let us assume that each molecule can be in two quantum states: the ground state and an excited state. The ground-state wavefunction of the complex can then be written as

$$|g\rangle = \prod_n^N \varphi_n^{(g)} , \tag{5.67}$$

corresponding to the state where all molecules are in their ground states. The excited state of the complex is obtained by promoting one molecule to its excited state:

$$|e_j\rangle = \varphi_j^{(e)} \prod_n^{N,n\neq j} \varphi_n^{(g)} . \tag{5.68}$$

By changing the place of excitation in the aggregate, one can obtain N possible *singly excited* states (single-exciton states). The *doubly excited* states (double-exciton states) are obtained by promoting two molecules in the aggregate to their excited states:

$$|f_{(kl)}\rangle = \varphi_k^{(e)} \varphi_l^{(e)} \prod_n^{N,n\neq k,l} \varphi_n^{(g)} . \tag{5.69}$$

By counting all possible pairs of molecules, we obtain $N(N-1)/2$ double excitations. We can continue this procedure until we arrive at the state where all molecules are excited. This is the single highest possible excited state of the aggregate:

$$|h\rangle = \prod_n^N \varphi_n^{(e)} . \tag{5.70}$$

In aggregates of the same type or similar types of molecules, the states with the same number of excited molecules form a band (manifold) of states with similar energies. In the following we consider only ground-state, single-exciton, and double-exciton manifolds. These three bands are the only bands directly accessible in a resonant third-order nonlinear optical laser experiment. Higher-lying states can be disregarded.

The basis set defined above is easily translated into an excitation creation/annihilation operator picture (see Section 4.9). As already described, the vacuum state $|g\rangle$ is the ground state of the aggregate having no excitations. This can be considered as the vacuum of particles. The state where the mth molecule is excited (a single-exciton state) is obtained by $|e_m\rangle \equiv \hat{B}_m^\dagger|g\rangle$, and for a pair of excited molecules (a double-exciton state) $|f_{(mn)}\rangle \equiv \hat{B}_m^\dagger \hat{B}_n^\dagger|g\rangle$. Since a molecule cannot be excited twice, the particles and excitations behave as fermions on the same molecule; thus, the particle operators $\hat{B}$ satisfy the Pauli commutation relation [25, 26]:

$$\left[\hat{B}_n, \hat{B}_m^\dagger\right] = \delta_{mn}\left(1 - 2\hat{B}_m^\dagger \hat{B}_m\right) . \tag{5.71}$$

In the space of single and double excitations, the Hamiltonian of the aggregate can be represented using these operators:

$$\hat{H} = \sum_i^N \varepsilon_i \hat{B}_i^\dagger \hat{B}_i + \sum_{i \neq j}^N J_{ij} \hat{B}_i^\dagger \hat{B}_j + \sum_{i \neq j}^N K_{ij} \hat{B}_i^\dagger \hat{B}_j^\dagger \hat{B}_i \hat{B}_j , \tag{5.72}$$

where

$$\varepsilon_i = \varepsilon_i^{(0)} + k_e \sum_{j \neq i}^N \int \mathrm{d}\boldsymbol{r}_1 \int \mathrm{d}\boldsymbol{r}_2 \frac{\left(\rho_i^{(ee)}(\boldsymbol{r}_1) - \rho_i^{(gg)}(\boldsymbol{r}_1)\right) \rho_j^{(gg)}(\boldsymbol{r}_2)}{|\boldsymbol{r}_1 - \boldsymbol{r}_2|} \tag{5.73}$$

is the transition energy of the molecular excitation in the presence of other molecules in their ground states. The parameters are as follows: $\varepsilon_i^{(0)}$ is the transition energy of the isolated molecule, $\rho_i^{(gg)}(\boldsymbol{r})$ is the total charge density of the ith molecule in its ground state, and $\rho_i^{(ee)}(\boldsymbol{r}_1)$ is the charge density of the electronic excited state of the ith molecule. Similarly to the case of the dimer described in the previous section, there are two types of intermolecular couplings:

$$J_{ij} = k_e \int \mathrm{d}\boldsymbol{r}_1 \int \mathrm{d}\boldsymbol{r}_2 \frac{\rho_i^{(ge)}(\boldsymbol{r}_1)\rho_j^{(eg)}(\boldsymbol{r}_2)}{|\boldsymbol{r}_1 - \boldsymbol{r}_2|} \tag{5.74}$$

is the resonant Coulomb interaction between transition charge densities, and

$$K_{ij} = k_e \int \mathrm{d}\boldsymbol{r}_1 \int \mathrm{d}\boldsymbol{r}_2 \frac{\left(\rho_i^{(ee)}(\boldsymbol{r}_1) - \rho_i^{(gg)}(\boldsymbol{r}_1)\right)\left(\rho_j^{(ee)}(\boldsymbol{r}_2) - \rho_j^{(gg)}(\boldsymbol{r}_2)\right)}{|\boldsymbol{r}_1 - \boldsymbol{r}_2|} \tag{5.75}$$

describes the excitation energy shift due to the presence of another excitation. K_{ij} may be understood as the exciton–exciton binding parameter. When the intermolecular distances are larger than the molecular dimensions, the dipole approximation for charge densities is often assumed [24]. It approximates all charge densities by simple dipole vectors: transition dipoles $\boldsymbol{\mu}_i$ represent the transition charge densities $\rho_i^{(eg)}(\boldsymbol{r})$ and permanent dipoles $\boldsymbol{d}_i$ represent the difference densities $(\rho_i^{(ee)}(\boldsymbol{r}) - \rho_i^{(gg)}(\boldsymbol{r}))$. In that case we obtain the dipole–dipole coupling expressions:

$$J_{ij} = k_e \left[\frac{(\boldsymbol{\mu}_i \cdot \boldsymbol{\mu}_j)}{|\boldsymbol{R}_{ij}|^3} - 3\frac{(\boldsymbol{R}_{ij} \cdot \boldsymbol{\mu}_i)(\boldsymbol{R}_{ij} \cdot \boldsymbol{\mu}_j)}{|\boldsymbol{R}_{ij}|^5}\right] \tag{5.76}$$

and similarly

$$K_{ij} = k_e \left[\frac{(\boldsymbol{d}_i \cdot \boldsymbol{d}_j)}{|\boldsymbol{R}_{ij}|^3} - 3\frac{(\boldsymbol{R}_{ij} \cdot \boldsymbol{d}_i)(\boldsymbol{d}_{ij} \cdot \boldsymbol{\mu}_j)}{|\boldsymbol{R}_{ij}|^5}\right] . \tag{5.77}$$

The dipole vectors can be obtained from charge densities by calculating their first moments. Alternatively, they are given by wavefunction-based dipole operator expectation values: $\boldsymbol{\mu}_j = \langle g|\hat{\boldsymbol{D}}_j|e_j\rangle$ and $\boldsymbol{d}_j = \langle e_j|\hat{\boldsymbol{D}}_j|e_j\rangle - \langle g|\hat{\boldsymbol{D}}_j|g\rangle$.

Due to the equivalence of the molecules in the aggregate (assuming that $\varepsilon_i^0 = \varepsilon$), the corresponding excited-state wavefunction has to account for the probability of each of the molecules being excited (see the perturbation procedure for the degenerate states described in Section 4.7). Thus, we define a single excited state of the aggregate as a superposition of molecular single excitations (in the same way as we did for the dimer; see (5.33)):

$$\Psi_e = \sum_m c_{em} \hat{B}_m^\dagger |g\rangle \,, \tag{5.78}$$

where the expansion coefficients are indicated by c_{em}. The doubly excited state is similarly

$$\Phi_f = \sum_{kl}^{k>l} C_{f,(kl)} \hat{B}_k^\dagger \hat{B}_l^\dagger |g\rangle \,. \tag{5.79}$$

The coefficients c_{em} and $C_{f,(kl)}$ are determined from the Schrödinger equation for the aggregate.

Let us consider as an example the translationally invariant molecular aggregate. These types of systems are relevant for so-called J aggregates and molecular crystals [23, 25, 27]. The translational invariance implies that all molecules are identical. As we described in Section 4.6.5, the wavefunction of a one-dimensional aggregate of N sites has to be an eigenfunction of the translational operator $\hat{T}_n$, where $n = 0, 1, \ldots, N-1$ is an integer number of translations,

$$\hat{T}_n \Psi_k = \mathrm{e}^{-ikn} \Psi_k \,, \tag{5.80}$$

where k is the quantum number of the translational operator. In the case of a single equivalent molecule per unit cell, it follows directly from (5.80) that

$$c_{en}(k) = \frac{1}{\sqrt{N}} \mathrm{e}^{ikn} \,, \tag{5.81}$$

which implies that all molecules in the crystal are equally likely to be excited. For the eigenenergies we get

$$E_k = \langle \Psi_k | \hat{H}_M | \Psi_k \rangle - E_g = \varepsilon + \Delta + L(k) \,, \tag{5.82}$$

where ε is the molecular excitation energy, Δ is the renormalization due to other molecules (see Section 5.3), and

$$L(k) = \sum_n J_{0,m} \mathrm{e}^{ikm} \,, \tag{5.83}$$

where

$$J_{n,m} = \left\langle \varphi_n^{(e)} \varphi_m^{(g)} \right| V_{nm} \left| \varphi_n^{(g)} \varphi_m^{(e)} \right\rangle \tag{5.84}$$

is the resonance interaction between excitations on the nth and mth molecules, respectively. Again due to translational symmetry, the value of $L(k)$ is independent of the molecular position (taken as zero in (5.83)).

Equation (5.81) demonstrates that the excitation is delocalized with equal probability over the whole aggregate of equivalent molecules, and the coefficients $c_{en}(k)$ differ only by their phase factors. The excitations corresponding to eigenstates are defined in the k-space (or the reciprocal lattice space) and they are usually termed excitons. The exciton energies E_k determine the exciton energy band or spatial dispersion of the exciton energy. The number of energy levels E_k is determined by the size of the system. Taking periodic boundary conditions, we assume that the system contains N_s molecules along the s-axis, and it is periodically repeated along this axis. Correspondingly, translation of the system as a whole does not change the wavefunction defined by (5.78). The corresponding N_s values for k_s are determined as follows [21, 23, 25]:

$$k_s = \frac{2\pi}{N_s} j_s , \tag{5.85}$$

where s enumerates the crystallographic axes (in the case of the crystal). k_s is the projection of the wave vector $\boldsymbol{k}$ on the s-axis. Here the intermolecular distances are expressed in terms of the distance between two nearest neighbors along the corresponding axis. The distance between the nearest neighbors is assumed to be equal to unity. Thus, j_s is an integer with possible range of values

$$-\frac{N_s}{2} < j_s \le \frac{N_s}{2} . \tag{5.86}$$

For example, if only the nearest-neighbor interaction is important in a linear periodic system with N sites, we have

$$L(k) = 2 J_{01} \cos(k) , \tag{5.87}$$

and $-\pi < k < \pi$.

For small absolute values of $\boldsymbol{k}$, the exponent in (5.83) can be expanded in a series which gives the following approximation of (5.82):

$$E_k = E_s + \frac{\hbar^2 k_s^2}{2 m_s} . \tag{5.88}$$

Again s reflects the crystallographic direction, m_s is the effective mass corresponding to the exciton motion along this direction,

$$m_s = \frac{-\hbar^2}{\sum_n n_s^2 J_{0,n}} , \tag{5.89}$$

and

$$E_s = \epsilon + \Delta + \sum_n J_{0,n} . \tag{5.90}$$

Due to the dependence on the wavevector direction s, the expression for the eigenenergy, (5.88), is a nonanalytical function for very small $\boldsymbol{k}$ values. According to definition (5.86), the difference between the nearest $\boldsymbol{k}$ values along the s-axis is very small for very large values of N_s. For small k values, the exciton can be considered a quasiparticle characterized by the phase and group velocity. A large resonance interaction, which corresponds to the small mass, gives rise to excitons with high mobility.

Let us now consider a more complex periodic structure which contains σ molecules per unit cell. In this case we will obtain a number (σ) of different subbands (so-called Davydov subbands). We have to choose the following excitation amplitudes (wavefunction expansion coefficients)

$$c_{\alpha\nu} = \frac{1}{\sqrt{N}} u_{\alpha\nu}(\boldsymbol{k}) \mathrm{e}^{\mathrm{i}\boldsymbol{k}\boldsymbol{r}_{n\alpha}} \,, \tag{5.91}$$

which diagonalize the Hamiltonian, (5.66). The matrix $u_{\alpha\nu}(\boldsymbol{k})$ is unitary, and the index α enumerates molecules in the unit cell; $\boldsymbol{r} = \boldsymbol{n} + \boldsymbol{\rho}_\alpha$, where $\boldsymbol{\rho}_\alpha$ are the coordinates of the molecules within the unit cell, and ν accounts for splitting of the corresponding degenerate molecular states. This gives rise to a number σ of molecular subbands in the exciton band. The transformation coefficients have to be normalized:

$$\sum_\alpha |c_{\alpha\nu}(\boldsymbol{k})|^2 = 1 \,. \tag{5.92}$$

Due to the presence of σ molecules per unit cell, the numbering by n has to be changed into $n\alpha$ everywhere. Now the exciton energies are determined by solving the Schrödinger equation given in (5.82), which gives the following set of equations for coefficients $u_{\alpha\nu}(\boldsymbol{k})$:

$$\sum_{\beta=1}^{\sigma} [(\epsilon + \Delta_\alpha)\delta_{\alpha\beta} + L_{\alpha\beta}(\boldsymbol{k})] u_{\beta\nu}(\boldsymbol{k}) = E_\nu(\boldsymbol{k}) u_{\alpha\nu}(\boldsymbol{k}) \,, \tag{5.93}$$

where

$$\Delta_\alpha = \sum_{m\beta} \left[\left\langle \varphi_{n\alpha}^{(e)} \varphi_{m\beta}^{(g)} \right| V_{n\alpha m\beta} \left| \varphi_{n\alpha}^{(e)} \varphi_{m\beta}^{(g)} \right\rangle - \left\langle \varphi_{n\alpha}^{(g)} \varphi_{m\beta}^{(g)} \right| V_{n\alpha m\beta} \left| \varphi_{n\alpha}^{(g)} \varphi_{m\beta}^{(g)} \right\rangle \right] \tag{5.94}$$

is the displacement energy and

$$L_{\alpha\beta}(\boldsymbol{k}) = \sum_m \left\langle \varphi_{n\alpha}^{(e)} \varphi_{m\beta}^{(g)} \right| V_{n\alpha m\beta} \left| \varphi_{n\alpha}^{(g)} \varphi_{m\beta}^{(e)} \right\rangle \mathrm{e}^{\mathrm{i}\boldsymbol{k}(\boldsymbol{r}_{m\beta} - \boldsymbol{r}_{n\alpha})} \tag{5.95}$$

is the resonance interaction matrix for an arbitrary $\boldsymbol{k}$ value. The exciton spectrum is determined via the corresponding characteristic equation

$$\det(L_{\alpha\beta}(\boldsymbol{k}) + \delta_{\alpha\beta}[\varepsilon + \Delta_\alpha - E_\nu(\boldsymbol{k})]) = 0 \,. \tag{5.96}$$

The matrix given by (5.95) is Hermitian, and all σ values of $E_\nu(\boldsymbol{k})$ are correspondingly real. They define exciton subbands, and the phenomenon of the organization of the band into subbands is called *Davydov splitting* [23, 25]. It should be emphasized that the procedure we performed for periodic molecular aggregates in three dimensions represents an extension of the simple case of a dimer, which we described in Section 5.3.

If the system contains two molecules per unit cell, the unitary matrix $u_{\alpha\nu}(k)$ can be written in the following form:

$$u_{\alpha\nu}(k) = \begin{pmatrix} \cos\varphi(k) & \sin\varphi(k) \\ -\sin\varphi(k) & \cos\varphi(k) \end{pmatrix} . \tag{5.97}$$

The corresponding Hamiltonian matrix which determines the left-hand side of the set of (5.93) in this case becomes

$$\begin{pmatrix} \epsilon + \Delta_1 + L_{11}(k) & L_{12}(k) \\ L_{21}(k) & \epsilon + \Delta_2 + L_{22}(k) \end{pmatrix} . \tag{5.98}$$

Diagonalization of this matrix leads to the following analytical solution:

$$\begin{aligned} E_\nu(k) = \epsilon + \frac{\Delta_1 + \Delta_2}{2} \\ - (-1)^\nu \sqrt{\left(\frac{L_{11}(k) - L_{22}(k) + \Delta_1 - \Delta_2}{2}\right)^2 + |L_{12}(k)|^2} , \end{aligned} \tag{5.99}$$

where $\nu = 1, 2$. For each energy band $E_\nu(k)$ the transformation function $\varphi_\nu(k)$ in (5.97) is determined by the following relation:

$$\tan\varphi_\nu(k) = \frac{\epsilon + \Delta_1 + L_{11}(k) - E_\nu(k)}{L_{12}(k)} . \tag{5.100}$$

If $\Delta_1 = \Delta_2$ and $L_{11}(k) = L_{22}(k)$, then the value of the Davydov splitting according to (5.99) is

$$\Delta E(k) = 2L_{12}(k) . \tag{5.101}$$

Summarizing, the resonance interaction between molecules within the unit cell leads to Davydov splitting, which for every k gives rise to two new states. The amount of splitting between these states leads to two subbands. In general, the number of subbands is equal to the number of molecules within a unit cell.

5.5 Wannier–Mott Excitons

Let us now consider an atomic crystal or a crystal with covalent/ionic bonds. In such a crystal, interactions between crystal constituents are not weak, and the

Heitler–London approximation is not applicable. Due to strong interactions the electrons can be considered as belonging to the whole crystal and are characterized by a periodic potential, instead of a single lattice site [21, 23]. To understand the dissipative electron dynamics in a solid of this type, it is sufficient to consider an electron in a periodic potential as we did in Section 4.6.5. For simplicity, let us assume there is a single electron in such a crystal. In three dimensions the Hamiltonian can then be written as

$$\hat{H} = -\frac{\nabla^2}{2m_e} + V(\boldsymbol{r}) , \qquad (5.102)$$

where $V(\boldsymbol{r})$ is a periodic function with the property $V(\boldsymbol{r}) = V(\boldsymbol{r} + \boldsymbol{a})$. Here $\boldsymbol{a}$ is the translational vector of the system – the smallest possible vector – which translates the system into itself. According to the Bloch theorem (Section 4.6.5) the wavefunction of this Hamiltonian must have the same translational symmetry apart from a different phase, so the wavefunction is of the form

$$\psi(\boldsymbol{r}) = \varphi(\boldsymbol{r}) \exp(i\boldsymbol{k}\boldsymbol{a}) . \qquad (5.103)$$

Here $\varphi(\boldsymbol{r})$ is a periodic function with translational invariance $\varphi(\boldsymbol{r}) = \varphi(\boldsymbol{r} + \boldsymbol{a})$. This allows us to consider free electrons distributed in the conduction band of the crystal. The electron is characterized by a specific effective mass in the vicinity of edges of the band [21].

The exciton spectrum should be considered as defined by two particles: an electron in the conduction band and a hole in the valence band. They interact with each other by Coulomb coupling (see Figure 5.4). The Hamiltonian of such an electron–hole system (the so-called Hamiltonian of the *Wannier–Mott exciton*) can be written as

$$\hat{H}_{WM} = \frac{\hat{\boldsymbol{p}}_e^2}{2m_e} + \frac{\hat{\boldsymbol{p}}_h^2}{2m_h} - \frac{k_e e^2}{|\boldsymbol{r}_e - \boldsymbol{r}_h|} , \qquad (5.104)$$

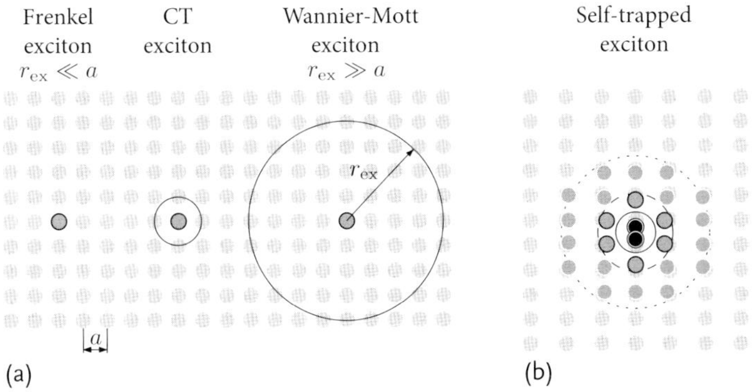

Figure 5.4 Possible models of excitons in crystals with the translational constant a. The zero-radius (Frenkel) exciton, the intermediate-radius (charge-transfer, CT) exciton, and the large-radius (Wannier–Mott) exciton are presented in (a), and the possible lattice polarization in the case of the self-trapped excitons is shown in (b).

where $\hat{\boldsymbol{p}}_e$ and $\hat{\boldsymbol{p}}_h$, m_e and m_h, and $\boldsymbol{r}_e$ and $\boldsymbol{r}_h$ are momenta, masses, and radius vectors of the electron and hole, respectively. This is a two-body problem, conveniently solved by transforming it into a problem of the center of mass and the relative motions. After the transformation we obtain the following equivalent representation of the Hamiltonian:

$$\hat{H}_{\mathrm{WM}} = \frac{\hat{\boldsymbol{P}}^2}{2(m_e + m_h)} + \frac{\hat{\boldsymbol{p}}^2}{2\mu} - \frac{k_e e^2}{r} , \tag{5.105}$$

where $\hat{\boldsymbol{P}}$ and $\hat{\boldsymbol{p}}$ are the momenta of the center mass and the relative movements, respectively. The quantity $r = |\boldsymbol{r}_e - \boldsymbol{r}_h|$ is the distance between the electron and the hole, and μ is the reduced mass:

$$\frac{1}{\mu} = \frac{1}{m_e} + \frac{1}{m_h} . \tag{5.106}$$

The new coordinates describing the movement of the center of mass (as a free particle) and the relative movement of the electron and the hole are independent. The wavefunction of the problem can then be presented as

$$\Psi(\boldsymbol{R}, \boldsymbol{r}) = \frac{1}{\sqrt{L^d}} e^{i\boldsymbol{K}\boldsymbol{R}} \varphi(\boldsymbol{r}) , \tag{5.107}$$

where the initial factor is due to the normalization of the wavefunction. Here L is the linear size of the crystal, d is its dimension, and $\boldsymbol{K}$ is the wavevector corresponding to the translational symmetry of the center of mass of the electron and the hole. The relative movement of the electron and the hole is described by the wavefunction $\varphi(\boldsymbol{r})$. The problem is equivalent to that of one particle in a Coulomb potential, and it correspondingly exhibits a hydrogen-type energy spectrum [28]:

$$E_n = E_g - \frac{E_e}{\left(n + \frac{d-3}{2}\right)^2} . \tag{5.108}$$

Here E_g is the energy gap for the transition of the electron from the valence band to the conduction band (the reference point of free electrons and holes), and $E_e = \mathrm{Ry} \cdot k_e \mu / m_e$, with Ry representing the Rydberg energy of the hydrogen atom. Similarly, the exciton binding energy and the exciton radius corresponding to the lowest excited state ($n = 1$) can be determined, resulting in

$$E_b = E_g - E_1 = \frac{2}{d-1} E_e \tag{5.109}$$

and

$$r_{ex} = \frac{d-1}{2} a_e , \tag{5.110}$$

where $a_e = a_B m_e / (k_e \mu)$ is the effective Bohr radius, and a_B is the Bohr radius defined by considering the spectrum of the hydrogen atom.

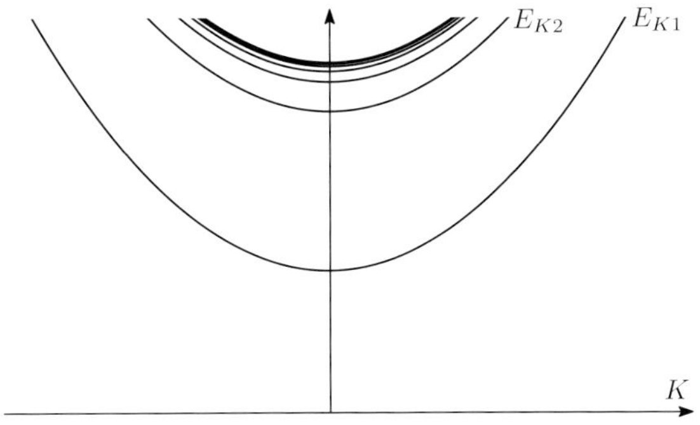

Figure 5.5 The energy levels of the Wannier–Mott exciton.

For typical parameters of a semiconducting crystal, E_e is on the order of tens or hundreds of millielectronvolts. In the case of $d = 3$, the exciton binding energy is E_e and $r_{ex} = a_e$. In the case of $d = 2$, the exciton binding energy increases four times and the exciton radius decreases four times, while for $d = 1$, $E_b = \infty$ and $r_{ex} = 0$. The divergence of the exciton binding energy and the δ-type wavefunction in the case of $d = 1$ can be understood as follows: in the case $d = 2$ or $d = 3$, a charged particle can freely move around the origin of the Coulomb potential, while in the case $d = 1$, it would have to move through the origin because of the spatial restriction. In the origin the coupling has a pole. A typical example of the quasi-one-dimensional system, where the Wannier–Mott excitons are expected, is carbon nanotubes [29].

By taking into account the energy values corresponding to the center of mass, we can express the total eigenvalues of the exciton energy spectrum as

$$E_{Kn} = E_n + \frac{\hbar^2 K^2}{2(m_e + m_h)} . \tag{5.111}$$

Thus, the exciton spectrum shown in Figure 5.5 defines the bound electron and a hole freely moving together through the crystal. Their binding states are quantized in a hydrogen-atom fashion.

5.6 Charge-Transfer Excitons

In addition to neutral, small-radius (Frenkel) excitons, where the excited electron remains correlated with the hole and both are located on the same molecular site, excited states where the excited electron is transferred to the nearest or next-nearest neighboring molecules are also expected to occur when exchange of electrons between molecules is allowed. Such states, which are intermediate between Frenkel excitons and Wannier–Mott excitons, are called charge-transfer (CT) states (see Figure 5.4) [23]. Similar to Wannier–Mott excitons, these CT states are positioned below the lowest conduction band, because of the Coulomb interaction between the electron and the hole (or between molecular ions) freely moving together through the crystal. Evidently, the CT state is determined by the ability of the molecules to accept/donate electrons. This ability corresponds to the molecular characteristics, such as the ionization potential (I_g) and the electron affinity (A_g). Due to the presence of charges in the system, the polarization energy ($P_{eh}(r)$) of an electron–hole pair separated by some distance r also has to be taken into consideration. The polarization energy reflects the adaptation of the system to the presence of a CT state. Thus, the energies of the CT excitons are given by [23, 27]

$$E_{CT} = I_g - A_g - P_{eh}(r) - C(r) , \tag{5.112}$$

where $C(r)$ denotes the Coulomb energy of the charged pairs separated by distance r [30].

Evidently, the existence of CT excitons is expected in heteromolecular crystals constructed by electron-donating and electron-accepting molecular pairs. The lowest CT state corresponds to the transfer of an electron from the highest occupied molecular orbital (HOMO) of the donor molecule to the lowest unoccupied molecular orbital (LUMO) of the acceptor molecule. From the theoretical point of view, the ion-pair states must be included in the complete basis system of wavefunctions describing excited states of the crystal. Correspondingly, the CT states interact with the Frenkel excitons and the two types of states intermix, resulting in a change of the resonance interaction values for the Frenkel excitons.

CT excitons are usually precursors to photoinduced generation of charges. Indeed, the optically excited Frenkel exciton states can be in resonance with the CT states. A transition from the initial Frenkel-type exciton state to a CT state is possible due to such resonance, and a separation of charges in the presence of an external electric field is expected. In homomolecular complexes and crystals, optical transitions corresponding to CT excitons are usually extremely weak, and they are not expected to be resolved in the absorption spectra. However, this is not the case for heteromolecular systems arranged in sequential stacks of electron–donor and electron–acceptor molecular configurations. In such systems, the CT excitons are well resolved in both absorption and fluorescence spectra. Due to the strong polar character of CT excitons, the effect of the exciton self-trapping is significant in such systems, resulting in large values of the Stokes shift.

A combined representation of all excitations can be obtained by using the so-called tight-binding description [31]. In this model, each molecule is represented by two electronic orbitals: the HOMO and the LUMO. Let us define operator $\hat{e}_m^\dagger$ ($\hat{e}_m$) which creates (annihilates) an electron in the LUMO of site m and operator $\hat{h}_m^\dagger$ ($\hat{h}_m$) which creates (annihilates) a hole in the HOMO. When the electron in the LUMO and the hole in the HOMO reside at the same site m, we have molecular excitation (the Frenkel excitonic state $m^* \Leftrightarrow \hat{e}_m^\dagger \hat{h}_m^\dagger |g\rangle$, where $|g\rangle$ denotes the ground-vacuum state). On the other hand, when the electron and the hole are excited at different sites m and n, we obtain the CT state $n^+ m^- \Leftrightarrow \hat{e}_m^\dagger \hat{h}_n^\dagger |g\rangle$. These states are illustrated in Figure 5.4.

The Hamiltonian of the system in the tight-binding model is [31]:

$$\begin{aligned}\hat{H}_{\mathrm{S}} = &\sum_{m,n} t^{\mathrm{e}}_{mn}\hat{e}^{\dagger}_{m}\hat{e}_{n} + \sum_{m,n} t^{\mathrm{h}}_{mn}\hat{h}^{\dagger}_{m}\hat{h}_{n} \\ &+ \sum_{m,n}^{m\neq n} W^{\mathrm{d}}_{mn}\hat{e}^{\dagger}_{m}\hat{h}^{\dagger}_{m}\hat{h}_{n}\hat{e}_{n} - \sum_{m,n} V^{\mathrm{eh}}_{mn}\hat{e}^{\dagger}_{m}\hat{h}^{\dagger}_{n}\hat{h}_{n}\hat{e}_{m} \\ &+ \frac{1}{2}\sum_{m,n}^{m\neq n} V^{\mathrm{e}}_{mn}\hat{e}^{\dagger}_{m}\hat{e}^{\dagger}_{n}\hat{e}_{n}\hat{e}_{m} + \frac{1}{2}\sum_{m,n}^{m\neq n} V^{\mathrm{h}}_{mn}\hat{h}^{\dagger}_{m}\hat{h}^{\dagger}_{n}\hat{h}_{n}\hat{h}_{m} \\ &+ \frac{1}{4}\sum_{k,m}^{k\neq m}\sum_{l,n}^{l\neq n} K_{kl,mn}\hat{e}^{\dagger}_{k}\hat{h}^{\dagger}_{l}\hat{e}^{\dagger}_{m}\hat{h}^{\dagger}_{n}\hat{h}_{n}\hat{e}_{m}\hat{h}_{l}\hat{e}_{k}\,. \end{aligned} \tag{5.113}$$

Here t^{e}_{mn} (t^{h}_{mn}) is the electron (hole) hopping rate between LUMOs (HOMOs). W^{d}_{mn} is the dipole–dipole-type resonance interaction between excitons at sites m and n. $V^{\mathrm{e}}_{mn} = V(\boldsymbol{r}_m - \boldsymbol{r}_n)$ is the electron–electron Coulomb repulsion between sites m and n, $V^{\mathrm{h}}_{mn} = V(\boldsymbol{r}_m - \boldsymbol{r}_n)$ is the hole–hole Coulomb repulsion between sites m and n, and $V^{\mathrm{eh}}_{mn} = V(\boldsymbol{r}_m - \boldsymbol{r}_n)$ is the electron–hole Coulomb attraction between sites m and n. Note that in the simplest case, $V^{\mathrm{e}}_{mn} = V^{\mathrm{h}}_{mn} = V^{\mathrm{eh}}_{mn}$; however, at short distances these values may be not equal since the delocalization of electrons and holes at the single site is usually different. Additionally, terms representing coupling K responsible for shifting the energies of the doubly excited states are also added to the Hamiltonian. They are, however, of higher order in terms of the number of creation and annihilation operators.

For the single-excitation manifold, where one electron and one hole are created, we have states $|e_k h_l\rangle = \hat{e}^{\dagger}_{k}\hat{h}^{\dagger}_{l}|\mathrm{g}\rangle$. The matrix elements of the Hamiltonian for the singly excited states are as follows [30]:

$$\begin{aligned}\langle e_k h_l|\hat{H}_{\mathrm{S}}|e_m h_n\rangle = &\; t^{\mathrm{e}}_{km}\delta_{ln} + t^{\mathrm{h}}_{ln}\delta_{km} - V^{\mathrm{eh}}_{kl}\delta_{ln}\delta_{km} \\ &+ W^{\mathrm{d}}_{km}(1-\delta_{km})\delta_{lk}\delta_{mn}\,. \end{aligned} \tag{5.114}$$

These thus give the following matrix elements:

$$\varepsilon_{a^*} = t^{\mathrm{e}}_{aa} + t^{\mathrm{h}}_{aa} - V^{\mathrm{eh}}_{aa}\,, \tag{5.115}$$

$$\varepsilon_{a^+b^-} = t^{\mathrm{e}}_{bb} + t^{\mathrm{h}}_{aa} - V^{\mathrm{eh}}_{ba}\,, \tag{5.116}$$

$$J_{a^*,b^*} = W^{\mathrm{d}}_{ab}\,, \tag{5.117}$$

$$J_{a^*,b^+c^-} = t^{\mathrm{e}}_{ac}\delta_{ab} + t^{\mathrm{h}}_{ab}\delta_{ac}\,, \tag{5.118}$$

and

$$J_{a^+b^-,c^+d^-} = t^{\mathrm{e}}_{bd}\delta_{ac} + t^{\mathrm{h}}_{ac}\delta_{bd}\,. \tag{5.119}$$

Here ε_{a^*} is the excitation energy of the Frenkel exciton a^*, $\varepsilon_{a^+b^-}$ is the excitation energy of the CT state a^+b^-, $J_{a^*b^*}$ defines the mixing between Frenkel exciton states a^* and b^*, J_{a^*,b^+c^-} is the coupling between the Frenkel exciton state a^* and the CT state b^+c^-, and $J_{a^+b^-,c^+d^-}$ is the coupling between CT states a^+b^- and c^+d^-. This information is sufficient to describe the single-excitation properties.

In the double-excitation space, two electrons and two holes are excited; thus, we have states $|e_k e_l h_m h_n\rangle = \hat{e}_k^\dagger \hat{e}_l^\dagger \hat{h}_m^\dagger \hat{h}_n^\dagger |g\rangle$, with $k > l$ and $m > n$. The Hamiltonian matrix elements for doubly excited states are given by elements of type $\langle e_k e_l h_m h_n | \hat{H}_S | e_{k'} e_{l'} h_{m'} h_{n'} \rangle$. Three kinds of doubly excited states are obtained: (1) Frenkel excition–Frenkel exciton states $a^* b^*$, (2) Frenkel exciton–CT states $a^* b^+ c^-$, and (3) CT–CT states with two electrons and two holes $a^+ b^- c^+ d^-$.

The transformations to the single-excitation eigenstates

$$|e\rangle = \sum_{k,l} \psi_{e,kl} |e_k h_l\rangle$$

and double-excitation eigenstates

$$|f\rangle = \sum_{k,l}^{k>l} \sum_{m,n}^{m>n} \Psi_{f,klmn} |e_k e_l h_m h_n\rangle$$

are given in terms of transformation matrices $\psi_{e,kl}$ and $\Psi_{f,klmn}$, which are calculated by solving the eigenvalue problems of the Hamiltonian.

It is noteworthy that a presentation of this kind is convenient for describing nonlinear spectroscopy data of the system where both types of excitons occur [31].

5.7
Vibronic Interaction and Exciton Self-Trapping

The exciton energy states discussed so far are defined at fixed positions of the nuclei of the molecules and their surroundings. The nuclear motion modulates the energy levels and this has an influence on the exciton spectrum. The effect can be obtained by adding the vibrational Hamiltonian defined for every molecule. In the adiabatic approach we consider (5.22) and the exciton Hamiltonian, (5.72). We thus take

$$\begin{aligned}\hat{H} = &\sum_{i}^{N} \varepsilon_i \hat{B}_i^\dagger \hat{B}_i + \sum_{i \neq j}^{N} J_{ij} \hat{B}_i^\dagger \hat{B}_j + \sum_{i \neq j}^{N} K_{ij} \hat{B}_i^\dagger \hat{B}_j^\dagger \hat{B}_i \hat{B}_j \\ &+ \sum_{i,s} \frac{\hbar\omega_s^i}{2} \left(Q_{i,s}^2 - \frac{\partial^2}{\partial Q_{i,s}^2} \right) + \sum_{i,s} \hat{B}_i^\dagger \hat{B}_i \left(a_s \hbar\omega_s^i Q_{i,s} + \delta_{i,s} Q_{i,s}^2 \right) ,\end{aligned} \tag{5.120}$$

where ω_s^i is the s vibration in the electronic ground state of the ith molecule along the $Q_{i,s}$ coordinate, $a_{i,s}$ is the displacement of the excited-state potential of the ith molecule along the s coordinate of the vibrations, and $\delta_{i,s}$ is the change of the vibrational energies in the electronic excited states, that is, $\delta_{i,s} = (\hbar\omega_s^{e,i} - \hbar\omega_s^i)/2$.

A similar, however, more general Hamiltonian determining the exciton–phonon interaction is defined by assuming that all parameters of the exciton Hamiltonian depend on the nuclear coordinates. Apart from the kinetic energy terms, the Hamiltonian reads

$$\hat{H} = \sum_{i}^{N} \varepsilon_i(\boldsymbol{R}) \hat{B}_i^\dagger \hat{B}_i + \sum_{i \neq j}^{N} J_{ij}(\boldsymbol{R}) \hat{B}_i^\dagger \hat{B}_j + \sum_{i \neq j}^{N} K_{ij}(\boldsymbol{R}) \hat{B}_i^\dagger \hat{B}_j^\dagger \hat{B}_i \hat{B}_j \,. \quad (5.121)$$

The first term reflects the $\boldsymbol{R}$ dependence of the displacement energy. The second, resonance interaction, term reflects the lattice point displacements and thus the lattice deformation. The third term, which is the exciton–exciton interaction, necessarily depends on the lattice deformation.

Let us consider the properties of a single excitation, so the third term in the Hamiltonian can be disregarded. In a molecular crystal the excitations and vibrations are considered as delocalized excitons and phonons characterized by their wavevectors $\boldsymbol{k}$ and $\boldsymbol{q}$, respectively. The exciton–phonon Hamiltonian is then given by [21, 25, 27]

$$\hat{H} = \hat{H}_{\text{ex}} + \hat{H}_{\text{ph}} + \hat{H}_{\text{int}}^{(1)} + \hat{H}_{\text{int}}^{(2)} \,, \quad (5.122)$$

where $\hat{H}_{\text{ex}}$ is the $\boldsymbol{R}$-independent exciton Hamiltonian, and

$$\hat{H}_{\text{ph}} = \sum_{s,\boldsymbol{q}} \hbar\omega_s(\boldsymbol{q}) \left(\hat{b}_{s,\boldsymbol{q}}^\dagger \hat{b}_{s,\boldsymbol{q}} + \frac{1}{2} \right) \quad (5.123)$$

is the phonon Hamiltonian with phonon frequency $\omega_s(\boldsymbol{q})$. The index s enumerates the phonon branches, $\boldsymbol{q}$ is the phonon wavevector, and $\hat{b}_{s,\boldsymbol{q}}^\dagger$ and $\hat{b}_{s,\boldsymbol{q}}$ are the phonon creation and annihilation operators, respectively. The Hamiltonian

$$\hat{H}_{\text{int}}^{(1)} = \frac{1}{\sqrt{N}} \sum_{s,\boldsymbol{k},\boldsymbol{q}} \hat{F}_s(\boldsymbol{k}, \boldsymbol{q}) \hat{B}^\dagger(\boldsymbol{k} + \boldsymbol{q}) \hat{B}(\boldsymbol{k}) \hat{\varphi}_s(\boldsymbol{q}) \quad (5.124)$$

represents the momentum transfer from the phonon to the exciton. Here $\hat{B}(\boldsymbol{k})$ and $\hat{B}^\dagger(\boldsymbol{k})$ are exciton annihilation and creation operators for wavevector $\boldsymbol{k}$,

$$\hat{B}(\boldsymbol{k}) = \frac{1}{\sqrt{N}} \sum_{n} \hat{B}_n \mathrm{e}^{-\mathrm{i}\boldsymbol{k}\boldsymbol{n}} \quad (5.125)$$

(a similar canonical transformation is used for the creation operator), and

$$\hat{\varphi}_s(\boldsymbol{q}) = \hat{b}_{s,\boldsymbol{q}} + \hat{b}_{s,-\boldsymbol{q}}^\dagger \quad (5.126)$$

is the phonon coordinate in the second-quantization representation (see Chapter 4). The term

$$\hat{H}_{\text{int}}^{(2)} = \frac{1}{\sqrt{N}} \sum_{s,\boldsymbol{k},\boldsymbol{q}} \hat{\chi}_s(\boldsymbol{q}) \hat{B}^\dagger(\boldsymbol{k}) \hat{B}(\boldsymbol{k}) \hat{\varphi}_s(\boldsymbol{q}) \quad (5.127)$$

is caused by the expansion of the displacement energy. Here $\hat{F}_s(\boldsymbol{k}, \boldsymbol{q})$ and $\hat{\chi}_s(\boldsymbol{q})$ are given by the following equations:

$$\hat{F}_s(\boldsymbol{k}, \boldsymbol{q}) = \sum_{l,m\neq 0} \epsilon_s^l(\boldsymbol{q}) \sqrt{\frac{\hbar}{2M_i\omega_s(\boldsymbol{q})}} \left[\left(\frac{\partial}{\partial R_0^l} + e^{iqm} \frac{\partial}{\partial R_m^l} \right) J_{0m} \right]_0 e^{ikm} \tag{5.128}$$

and

$$\hat{\chi}_s(\boldsymbol{q}) = \sum_{l,m\neq 0} \epsilon_s^l(\boldsymbol{q}) \sqrt{\frac{\hbar}{2M_i\omega_s(\boldsymbol{q})}} \left[\left(\frac{\partial}{\partial R_0^l} + e^{iqm} \frac{\partial}{\partial R_m^l} \right) \Delta_m \right]_0 , \tag{5.129}$$

where $\epsilon_s^l(\boldsymbol{q})$ are the components of the unit vector of polarization of the s component of the phonons, Δ_m is the displacement energy of the mth molecule, N is the number of unit cells in the system, and M_i is the corresponding nuclear mass.

By comparing these terms of the exciton–phonon interaction, we can consider two limiting cases. When

$$\hat{H}_{\text{int}}^{(1)} \gg \hat{H}_{\text{int}}^{(2)} , \tag{5.130}$$

we are dealing with a weak exciton–phonon interaction which leads to exciton scattering by phonons. The excitonic properties are retained and the phonons cause shifts of energy values and induce decoherence. In the opposite case, when

$$\hat{H}_{\text{int}}^{(1)} \ll \hat{H}_{\text{int}}^{(2)} , \tag{5.131}$$

strong exciton–phonon interaction prevails and the Hamiltonian $H_{\text{int}}^{(2)}$ determines the lattice deformation in the region where the exciton is present. In the strong-coupling limit, the exciton interaction with the phonon field can become so strong that it can result in exciton self-trapping due to lattice deformation in the vicinity of the excited molecule (see Figure 5.4).

Let us briefly describe the outcome of exciton self-trapping. There are two possible pathways for evolution of the exciton: (1) free coherent exciton formation in a rigid lattice corresponding to delocalization and (2) self-trapping of the excitation caused by lattice deformation with a subsequent localization. The loss of energy caused by the lattice deformation is called the energy of the lattice relaxation E_{LR}. On the other hand, the measure of the loss of energy resulting from free exciton relaxation is half the exciton bandwidth, $B = E_{\text{band}}/2$. The ratio $g = E_{\text{LR}}/B$ is used as a measure of the strength of the electron–phonon coupling and it is called the constant of the exciton–phonon coupling. Thus, when $g < 1$, exciton delocalization is most favorable, and when $g > 1$, exciton self-trapping is more probable. One can define the energy of exciton self-trapping as $E_{\text{ST}} = B - E_{\text{LR}}$.

Due to the possibility of experiencing exciton self-trapping (or exciton–polaron formation), the dynamics of the excitation defined by (5.122) is very complex. Below we will present an approach which can be used to describe the exciton dynamics in the presence of interaction with molecular vibrations/phonons. Thus, let us consider the Hamiltonian defined by (5.122) by introducing the simplified parametrization of the exciton–phonon coupling:

$$\hat{H} = \sum_{n,m} J_{mn} \hat{B}_m^\dagger \hat{B}_n + \sum_q \omega_q \hat{b}_q^\dagger \hat{b}_q + \sum_{n,q} g_{qn} \hat{B}_n^\dagger \hat{B}_n \left(\hat{b}_q^\dagger + \hat{b}_q\right) , \tag{5.132}$$

where g_{qn} is the linear coupling strength between an exciton state at site n and a phonon mode q.

The dynamics of the system can be derived using the Dirac–Frenkel variational principle [30]. For this purpose let us assume there is a trial wavefunction for the system to be characterized by a set of parameters $\{x_n(t)\}$; thus, $|\Psi\rangle = |\{x_n(t)\}\rangle$. If we use the Hamiltonian and the trial wavefunction, the Lagrangian L is given by

$$L = \langle \Psi(t)| \frac{\mathrm{i}}{2} \frac{\overleftrightarrow{\partial}}{\partial t} - \hat{H} |\Psi(t)\rangle , \tag{5.133}$$

where for simplicity $\hbar$ is assumed to be 1 everywhere in the following of this section and the double-overhead time derivative $\langle \Psi(t)|\mathrm{i}/2 \overleftrightarrow{\partial} /\partial t|\Psi(t)\rangle$ is defined as $\mathrm{i}/2(\langle \Psi(t)|\dot{\Psi}(t)\rangle - \langle \dot{\Psi}(t)|\Psi(t)\rangle)$.

If the trial wavefunction does not obey the time-dependent Schrödinger equation, then

$$\left(\mathrm{i}\frac{\mathrm{d}}{\mathrm{d}t} - \hat{H}\right) |\Psi(t)\rangle = |\delta(t)\rangle , \tag{5.134}$$

where $|\delta(t)\rangle$ is the deviation vector. Our aim is to minimize this value, resulting in a set of equations for the parameters $\{x_n(t)\}$ that make our trial wavefunction as close as possible to the exact wavefunction of the system. In order to do this, a variation of the Lagrangian with respect to the parameters is performed to obtain a set of Euler–Lagrange differential equations.

For minimization it is usually convenient to adopt some specific trial functions. Let us consider now the so-called Davydov D1 ansatz [32] as the trial wavefunction:

$$\Psi_{\mathrm{D1}}(t) = \sum_n \left\{ \alpha_n(t) \hat{B}_n^\dagger \exp\left[\sum_q \lambda_{qn}(t) \hat{b}_q^\dagger - \mathrm{h.c.}\right]\right\} |0\rangle , \tag{5.135}$$

with the set of time-dependent variational parameters α_n denoting the exciton amplitudes and λ_{qn} characterizing the vibrational displacements. This form denotes the vibrational displacement of a coherent vibrational wavepacket due to the electronic excitation (notice the form of the displacement operator (4.300)).

To obtain equations of motion for the wavefunction parameters we should first construct the Lagrangian L (5.133) as a function of α_n and λ_{qn}, and the equations of motion will be given by a set of Euler–Lagrange equations:

$$\frac{\mathrm{d}}{\mathrm{d}t}\left(\frac{\partial L}{\partial \dot{\alpha}_n}\right) - \frac{\partial L}{\partial \alpha_n} = 0 , \tag{5.136}$$

$$\frac{\mathrm{d}}{\mathrm{d}t}\left(\frac{\partial L}{\partial \dot{\lambda}_{qn}}\right) - \frac{\partial L}{\partial \lambda_{qn}} = 0\,. \tag{5.137}$$

First, we calculate the time derivative of the $\Psi_{\mathrm{D1}}(t)$ trial function

$$\dot{\Psi}_{\mathrm{D1}}(t) = \sum_n \hat{B}_n^\dagger \left\{ \dot{\alpha}_n + \alpha_n \sum_q \left[\dot{\lambda}_{qn} \hat{b}_q^\dagger - \frac{1}{2}\left(\dot{\lambda}_{qn}\lambda_{qn}^* + \text{c.c.}\right)\right]\right\} \times \exp\left(\sum_q \lambda_{qn} \hat{b}_q^\dagger - \text{h.c.}\right) |0\rangle\,. \tag{5.138}$$

Using (4.294)–(4.299) we get

$$\langle 0| \exp\left(\lambda_{qn}^* \hat{b}_q - \text{h.c.}\right)\left(\dot{\lambda}_{qn}\hat{b}_q^\dagger - \text{h.c.}\right) \exp\left(\lambda_{qn}\hat{b}_q^\dagger - \text{h.c.}\right)|0\rangle = \dot{\lambda}_{qn}\lambda_{qn}^* - \lambda_{qn}\dot{\lambda}_{qn}^* \tag{5.139}$$

which leads to the overlap of the wavefunction and its derivative:

$$\langle \Psi_{\mathrm{D1}}(t)|\dot{\Psi}_{\mathrm{D1}}(t)\rangle = \sum_n \left[\dot{\alpha}_n \alpha_n^* + \frac{1}{2}|\alpha_n|^2 \sum_q \left(\dot{\lambda}_{qn}\lambda_{qn}^* - \text{c.c.}\right)\right]. \tag{5.140}$$

Substracting the complex conjugate of this expression from itself and simplifying the result, we obtain:

$$\langle \Psi_{\mathrm{D1}}(t)|\frac{\overleftrightarrow{\partial}}{\partial t}|\Psi_{\mathrm{D1}}(t)\rangle = \sum_n \left[\dot{\alpha}_n\alpha_n^* - \dot{\alpha}_n^*\alpha_n + |\alpha_n|^2 \sum_q \left(\dot{\lambda}_{qn}\lambda_{qn}^* - \dot{\lambda}_{qn}^*\lambda_{qn}\right)\right]. \tag{5.141}$$

Second, calculating the expression for $\langle \Psi_{\mathrm{D1}}(t)|\hat{H}|\Psi_{\mathrm{D1}}(t)\rangle$, we break the result into three energy terms:

$$E_{\mathrm{ex}}(t) = \sum_{m,n} J_{mn}\alpha_m^*\alpha_n \cdot S_{mn}\,, \tag{5.142}$$

$$E_{\mathrm{ph}}(t) = \sum_n |\alpha_n|^2 \sum_q \omega_q |\lambda_{qn}|^2\,, \tag{5.143}$$

$$E_{\mathrm{int}}(t) = \sum_n |\alpha_n|^2 \sum_q g_{qn}\left(\lambda_{qn}^* + \lambda_{qn}\right)\,, \tag{5.144}$$

where

$$S_{nm} = \exp\left\{ \sum_q \left[\lambda_{qn}^*\lambda_{qm} - \frac{1}{2}\left(|\lambda_{qn}|^2 + |\lambda_{qm}|^2\right)\right]\right\} \tag{5.145}$$

is the so-called Debye–Waller factor. $E_{\text{ex}}(t)$ describes the electronic Hamiltonian part, $E_{\text{ph}}(t)$ denotes the phonon energy and $E_{\text{int}}(t)$ is the phonon–electron interaction part. Combining these expressions we find the resulting Lagrangian as

$$\begin{aligned} L = \frac{\mathrm{i}}{2}\sum_n &\left\{\dot{\alpha}_n\alpha_n^* - \dot{\alpha}_n^*\alpha_n + |\alpha_n|^2\sum_q\left(\dot{\lambda}_{qn}\lambda_{qn}^* - \text{c.c.}\right)\right\} \\ &- \sum_{m,n} J_{mn}S_{mn}\alpha_m^*\alpha_n - \sum_n|\alpha_n|^2\sum_q\omega_q|\lambda_{qn}|^2 \\ &- \sum_n|\alpha_n|^2\sum_q g_{qn}\left(\lambda_{qn}^* + \lambda_{qn}\right) . \end{aligned} \tag{5.146}$$

This Lagrangian describes the system of coupled electronic and vibrational modes with respect to the trial wavefunction. The Euler–Lagrange equations give the following terms

$$\frac{\mathrm{d}}{\mathrm{d}t}\left(\frac{\partial L}{\partial\dot{\alpha}_n}\right) = \frac{\mathrm{i}}{2}\dot{\alpha}_n^*, \tag{5.147}$$

$$\begin{aligned} \frac{\partial L}{\partial\alpha_n} &= -\frac{\mathrm{i}}{2}\dot{\alpha}_n^* + \frac{\mathrm{i}}{2}\alpha_n^*\sum_q\left(\dot{\lambda}_{qn}\lambda_{qn}^* - \text{c.c.}\right) \\ &\quad - \sum_m J_{mn}S_{mn}\alpha_m^* - \alpha_n^*\sum_q\omega_q|\lambda_{qn}|^2 \\ &\quad - \alpha_n^*\sum_q g_{qn}\left(\lambda_{qn} + \lambda_{qn}^*\right) , \end{aligned} \tag{5.148}$$

$$\frac{\mathrm{d}}{\mathrm{d}t}\left(\frac{\partial L}{\partial\dot{\lambda}_{qn}}\right) = \frac{\mathrm{i}}{2}\left(\dot{\alpha}_n\alpha_n^*\lambda_{qn}^* + \alpha_n\dot{\alpha}_n^*\lambda_{qn}^* + |\alpha_n|^2\dot{\lambda}_{qn}^*\right) \tag{5.149}$$

$$\begin{aligned} \frac{\partial L}{\partial\lambda_{qn}} &= -\frac{\mathrm{i}}{2}|\alpha_n|^2\dot{\lambda}_{qn}^* - \sum_m J_{mn}S_{mn}\alpha_m^*\alpha_n\lambda_{qm}^* \\ &\quad + \frac{1}{2}\sum_m J_{nm}S_{nm}\alpha_n^*\alpha_m\lambda_{qn}^* + \frac{1}{2}\sum_m J_{mn}S_{mn}\alpha_m^*\alpha_n\lambda_{qn}^* \\ &\quad - |\alpha_n|^2\omega_q\lambda_{qn}^* - |\alpha_n|^2 g_{qn} . \end{aligned} \tag{5.150}$$

Rearranging the terms, we obtain the differential equations for the time evolution of parameters:

$$\begin{aligned} \dot{\alpha}_n = \frac{i}{2}\sum_m &\left[J_{nm}S_{nm}\frac{\alpha_n^*\alpha_m}{|\alpha_n|^2}\sum_q\left(\lambda_{qn}^*\lambda_{qm} - |\lambda_{qn}|^2\right) + \text{c.c.}\right]\alpha_n \\ &- i\sum_m J_{nm}S_{nm}\alpha_m - \frac{i}{2}\sum_q g_{qn}\left(\lambda_{qn} + \lambda_{qn}^*\right)\alpha_n , \end{aligned} \tag{5.151}$$

$$\begin{aligned} \dot{\lambda}_{qn} &= -i\omega_q\lambda_{qn} - i\sum_m J_{nm}S_{nm}\frac{\alpha_n^*\alpha_m}{|\alpha_n|^2}(\lambda_{qm} - \lambda_{qn}) \\ &\quad - ig_{qn} . \end{aligned} \tag{5.152}$$

This set of equations can be used to simulate the coupled dynamics. However, the equations are highly nonlinear and special care should be taken when numerically propagating these equations due to terms like $\alpha_n^* \alpha_m / |\alpha_n|^2$.

A simplified D2 ansatz is sometimes useful for the description of phonon-like modes, which are delocalized along the linear molecular chain. In this case we reduce the variational parameter space by assuming $\lambda_{qn} = \lambda_q$. The trial wavefunction then assumes the form

$$\Psi_{D2}(t) = \sum_n \left\{ \alpha_n(t) \hat{B}_n^\dagger \exp\left[\sum_q \lambda_q(t) \hat{b}_q^\dagger - \text{h.c.} \right] \right\} |0\rangle , \tag{5.153}$$

which assumes that the phonon wavelength is much longer than the size of the aggregate. The resulting equations of motion are

$$\dot{\alpha}_n = -\mathrm{i} \sum_m J_{nm} \alpha_m - \frac{i}{2} \sum_q g_q \left(\lambda_q + \lambda_q^* \right) \alpha_n , \tag{5.154}$$

$$\dot{\lambda}_q = -i\omega_q \lambda_q - i g_q . \tag{5.155}$$

5.8 Trapped Excitons

Defects caused by chemical impurities or vacancies perturb the crystal lattice. If the impurity does not fit well into the lattice, a ring of distorted host molecules will surround the impurity. A similar distortion may surround faults or vacancies in the host lattice. These misaligned molecules give rise to the so-called X-traps (for Frenkel excitons) [23]. Evidently, the wavefunctions corresponding to the trapped exciton states are not delocalized anymore and describe the localization characteristics of the excitation in the vicinity of the trap. The localization radius corresponding to this state is directly related to the trapping energy of the exciton, that is, the shallower the trapped exciton, the larger the localization radius. As a result of the increase of the exciton localization radius for the shallow trapped states, the optical transition into these states dramatically increases, resulting in the so-called Rashba effect [22, 24]. Detailed studies of the absorption and fluorescence spectra of trapped excitons allow one to determine the parameters of the exciton band.

The problem of localization was formulated for the first time by Anderson in 1958 [23, 24]. It was demonstrated that due to sufficiently large randomly distributed disorder, the delocalization of the wavefunctions, describing the band states, changes essentially: above a certain value of the disorder, a transition from the delocalized to the fully localized state takes place. This transition is called the Anderson transition and the localization is called the Anderson localization. Among the multiple models used to describe the energy spectrum of disordered systems [33], the most widely used ones are based on the assumption of a random distribution of impurities or other kind of imperfections within the crystal structure.

The models which are based on random disorder of the excitation energies of the molecules in the lattice (the diagonal disorder) assume variations only in the transition energies, while the lattice itself remains unperturbed. Models taking into account various aspects of the randomness of the molecular orientation and/or their position have also been developed [33]. These models are mainly related to amorphous systems, glasses, liquids, and gases which are called topologically disordered systems.

The surface of the crystal also disturbs the translational symmetry of the lattice. Thus, the surface exciton states might also be present in the energy spectra of the crystal. Their presence in the spectrum depends on the interplay of the exciton parameters. They can be localized either above the exciton band or below it.

6
The Concept of Decoherence

Quantum mechanics has changed physics not just by introducing us to new observable effects not predicted in the framework of classical mechanics (such as tunneling, nonlocality, and discrete energy spectrum; see Chapter 5), but also by fundamentally limiting the scope of practically attainable knowledge of the systems studied. One aspect of these limitations is expressed by various uncertainty relations, (4.2). These are consequences of the wavelike character of the state in quantum theory. Another limitation is embodied in the apparent probabilistic nature of macroscopic measurements on quantum systems. It appears that for a measurement of a particular quantity of a quantum system, the outcomes can be predicted only in a probabilistic manner. We reviewed classical stochasticity in Chapter 3, but the fundamental stochasticity embedded in quantum mechanics appears to be of a more general kind. A single-value measurement of a quantum mechanical quantity provides only very little information about the system before it was measured. Only repeated measurements on identically prepared systems can provide us with some idea about the distribution of the probability of finding the system with a particular value of the measured observable. This, on the other hand, can be calculated from quantum theory, and compared with experiment, yielding exceptionally good agreement. Thus, quantum theory seems to be limited to probabilistic predictions about ensembles.

6.1
Determinism in Quantum Evolution

To conclude, however, that the time evolution of a quantum system proceeds in some stochastic way would be a gross mistake. The Schrödinger equation is a completely deterministic equation for the quantum mechanical state, the time evolution of which is thus deterministic. This feature of quantum mechanics can be demonstrated in the following experiment with quantum coin tossing.

Let us imagine a set of deep potential traps created by some switchable trapping field and a particle prepared with probability equal to 1 in one of the traps (Figure 6.1a). We label this initial trap by index 0. The trapped particle can be in two internal states, $|R\rangle$ and $|L\rangle$, *left* and *right*, respectively, which determine the direc-

Molecular Excitation Dynamics and Relaxation, First Edition. L. Valkunas, D. Abramavicius, and T. Mančal.

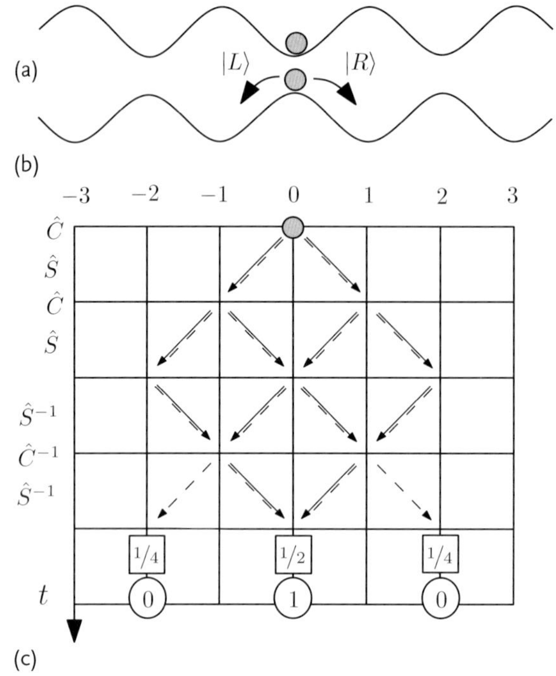

Figure 6.1 Quantum random walk scheme: a particle is prepared in a periodic trapping field in a trap with index 0 (a); the field is inverted and the particle moves to the right (left) when its internal state is $|R\rangle$ ($|L\rangle$) (b). Four steps of the random walk with an inversion in the middle (c). The boxed numbers give the outcome of a classical random walk; the numbers in circles give the corresponding quantum result for traps −2, 0, and 2.

tion in which the particle will move when we invert the trapping field (Figure 6.1b) and apply some external field. A laboratory realization of such an experiment can be found in [34].

A complete description of the particle's state can be achieved by determining its position and its internal state. For the description of the position we introduce state vectors $|P_n\rangle$, where n denotes the trap. Let us initially have the particle in an internal state $|R\rangle$ and a position state $|P_0\rangle$:

$$|\psi_0\rangle = |P_0\rangle|R\rangle\ . \tag{6.1}$$

The whole process of inverting the trapping field and applying an external field which drags the particle to a new position can be encapsulated in a "shift" operator $\hat{S}$. It shifts the particle to the right if the internal state is $|R\rangle$, and to the left if the internal state is $|L\rangle$:

$$\hat{S}|P_k\rangle|R\rangle = |P_{k+1}\rangle|R\rangle\ , \quad \hat{S}|P_k\rangle|L\rangle = |P_{k-1}\rangle|L\rangle\ . \tag{6.2}$$

Our aim here is to simulate the quantum random walk among the traps. In the classical version we would toss a coin, which would decide with probability of 1/2 for one of the states $|R\rangle$ or $|L\rangle$ of the particle. In quantum mechanics, we can prepare the state in a superposition of states, say, $(|R\rangle + |L\rangle)/\sqrt{2}$. This state has the property of providing the outcome of $|R\rangle$ or $|L\rangle$ with the same (1/2) probability as in the classical case. To be in the linear combination or coherent superposition of the possible outcomes is something radically different from being with a certain probability in one or the other state. The quantum random walk demonstrates this nicely. Tossing the coin will be realized in our quantum case by an operator $\hat{C}$ which acts only in the Hilbert space of the internal states of the particle. We prescribe the following operations:

$$\hat{C}|R\rangle = \frac{1}{\sqrt{2}}(|R\rangle - |L\rangle) \tag{6.3}$$

and

$$\hat{C}|L\rangle = \frac{1}{\sqrt{2}}(|R\rangle + |L\rangle) . \tag{6.4}$$

The random walk can now be realized by successive application of $\hat{C}$ and $\hat{S}$ (see Figure 6.1). Using (6.1), (6.3), and (6.4), we find that the first step in the random walk will result in

$$|\psi_0\rangle \stackrel{\hat{C}}{\rightarrow} \frac{1}{\sqrt{2}}|P_0\rangle(|R\rangle - |L\rangle) \stackrel{\hat{S}}{\rightarrow} |\psi_1\rangle = \frac{1}{\sqrt{2}}(|P_1\rangle|R\rangle - |P_{-1}\rangle|L\rangle) . \tag{6.5}$$

The second step yields

$$\hat{S}\hat{C}|\psi_0\rangle \stackrel{\hat{S}\hat{C}}{\rightarrow} |\psi_2\rangle = \frac{1}{2}(|P_2\rangle|R\rangle - |P_0\rangle|L\rangle - |P_0\rangle|R\rangle - |P_{-2}\rangle|L\rangle) . \tag{6.6}$$

After the two steps we can measure the position of the particle. Quantum mechanics prescribes that the probability p_n of finding the particle in the trap denoted by n is given by the expectation value of the projection operator $|P_n\rangle\langle P_n|$, so

$$p_n = \langle\psi|P_n\rangle\langle P_n|\psi\rangle . \tag{6.7}$$

We obtain the probabilities $p_{-2} = 1/4$, $p_0 = 1/2$, and $p_2 = 1/4$, that is, the same probabilities as for the classical random walk after two tosses.

We have chosen prescriptions given by (6.3) and (6.4) in order to be able to define an inverse operator $\hat{C}^{-1}$. It is clear that with the inverse shift operator $\hat{S}^{-1}$, that is, with the opposite meaning of the internal states $|R\rangle$ and $|L\rangle$ with respect to the shift, and with the operator $\hat{C}^{-1}$, we can also realize a random walk. By continuing the walk with the inverse operator $(\hat{S}\hat{C})^{-1} = \hat{C}^{-1}\hat{S}^{-1}$, we arrive at the initial state after the next two tosses:

$$(\hat{S}\hat{C})^{-1}(\hat{S}\hat{C})^{-1}|\psi_2\rangle = (\hat{S}\hat{C})^{-1}(\hat{S}\hat{C})^{-1}(\hat{S}\hat{C})(\hat{S}\hat{C})|\psi_0\rangle = |\psi_0\rangle . \tag{6.8}$$

In this state the probability $p_0 = 1$. This is in stark contrast with the classical case, where the probabilities after the next two steps with inverted dragging are again $p_{-2} = 1/4$, $p_0 = 1/2$, and $p_2 = 1/4$ (see Figure 6.1c).

Despite the fact that after each application of the operators $\hat{C}$ and $\hat{S}$ (or $\hat{C}^{-1}$ and $\hat{S}^{-1}$) we could verify that the probabilities of shifting to the right and left are 1/2, we found that the system is (with certainty) in trap 0 after four steps of the quantum random walk. Thus, we can conclude that the quantum system did not behave in a stochastic manner during the walk, and it was rather governed by the deterministic wavefunction evolution.

6.2
Entanglement

To handle the quantum and classical random walks in a unified manner, that is, by a single (quantum) theory, we have to introduce another important quantum mechanical concept, the concept of entanglement. For the description of a two-particle state, quantum mechanics prescribes a Hilbert space composed of the Hilbert spaces of individual particles by direct product, $\mathcal{H}_{1+2} = \mathcal{H}_1 \otimes \mathcal{H}_2$ (see the description of the Heitler–London approximation in Chapter 5). Practically, this means that we can construct an orthonormal basis $\{|m_{1+2}\rangle\}$ of the Hilbert space $\mathcal{H}_{1+2}$ from the orthonormal bases $\{|n_1\rangle\}$ and $\{|n_2\rangle\}$ of the Hilbert spaces $\mathcal{H}_1$ and $\mathcal{H}_2$, respectively, as

$$|1_{1+2}\rangle = |1_1\rangle|1_2\rangle\,, \quad |2_{1+2}\rangle = |1_1\rangle|2_2\rangle, \ldots\,, \tag{6.9}$$

where we list all $N_1 \times N_2$ states of the N_1-dimensional Hilbert space $\mathcal{H}_1$ and the N_2-dimensional Hilbert space $\mathcal{H}_2$. Because of the superposition principle, a state constructed as a linear combination of any of the states, defined by (6.9), is also a valid state. Let us take, for example, the linear combination

$$|\psi_{1+2}\rangle = \frac{1}{\sqrt{2}}(|1_1\rangle|3_2\rangle + |3_1\rangle|1_2\rangle)\,. \tag{6.10}$$

This type of wavefunction is similar to the exciton wavefunction defined by (5.33). The particles prepared in this state are found with equal probability of 1/2 in states $|1\rangle$ and $|3\rangle$. However, if we decide to measure the state of particle 1 and find it in state $|1\rangle$, we have to find the other particle in state $|3\rangle$ with certainty. This does not seem to depend in any way on how far apart the particles have traveled, and it constitutes one of the most surprising and nonintuitive consequences of quantum mechanics. The phenomenon we describe here is usually termed *entanglement* and it demonstrates the *nonlocality* of quantum mechanics. To be more precise, we will say that two particles are entangled if their state vector cannot be factorized into a product of their respective state vectors. Quantum mechanics is said to be nonlocal, because under certain conditions the measurement of one part of the composite system influences the other part of the same system in a way which cannot be explained by any local interaction, that is, the transfer of information from one part of the system to the other part is not limited by the speed of light.

Another consequence of the entanglement is that no wavefunction completely describing one part of a composite system can in general be defined. We have to use the total wavefunction, (6.10), and no simpler wavefunction assigned to, say, just particle 1 can be found that would contain all the information necessary to completely describe its state. This is not surprising, because its state clearly depends on the state of the other particle. The way out of the problem to describe the interacting subsystems is the concept of the density operator, which was introduced in Chapter 5.

Entanglement is not some exotic property that has to be carefully prepared in the laboratory and kept protected from interaction with the outside world as it might seem from some modern experiment. Entanglement is created spontaneously by interacting systems, and consequently the pair of particles described by state $|\psi_{1+2}\rangle$, (6.10), will entangle with all quantum systems with which it interacts. The difficulty we meet when studying entanglement in the laboratory is to keep it exclusive for the two particles we have under control. It is this exclusiveness of entanglement that has to be protected.

6.3
Creating Entanglement by Interaction

Let us now demonstrate how entanglement can be created by the interaction of two particles. Let us consider two particles a and b with internal states $|1_n\rangle$ and $|2_n\rangle$, for $n = a, b$ (see Figure 6.2). The particles move freely in space, prepared in states

$$|\phi_a^0\rangle = \frac{1}{\sqrt{2}}(|1_a\rangle + |2_a\rangle) \tag{6.11}$$

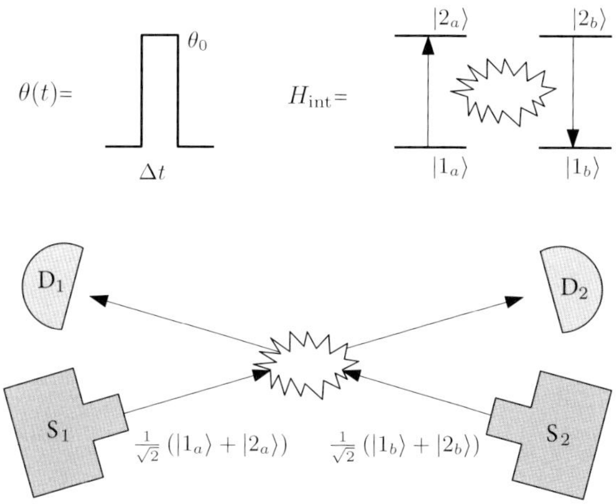

Figure 6.2 Formation of an entangled state by an elastic collision. Two sources S_1 and S_2 prepare two identical particles in superposition states of their internal states. During the collision, which takes time Δt, their interaction is described by the Hamiltonian H_{int} defined in (6.13) and depicted here.

and

$$|\phi_b^0\rangle = \frac{1}{\sqrt{2}}(|1_b\rangle + |2_b\rangle) , \tag{6.12}$$

and they are approaching each other. When the particles meet, they interact for a short interval of time Δt. Let us assume their interaction is defined as an elastic collision, that is, when one particle is excited, it can transfer its energy to the other one and vice versa, but no energy is lost in the collision. This can be described by an interaction Hamiltonian (see Figure 6.2):

$$\hat{H}_{\text{int}} = \theta(t)(|1_a\rangle|2_b\rangle\langle 2_a|\langle 1_b| + |2_a\rangle|1_b\rangle\langle 1_a|\langle 2_b|) . \tag{6.13}$$

The function $\theta(t)$ is equal to zero except for in a short interval $\{t_0, t_0 + \Delta t\}$ (where t_0 is the time of the collision), and $\theta(t) = \theta_0$ for $t \in \{t_0, t_0 + \Delta t\}$. The Schrödinger equation in this case reads

$$\frac{\partial}{\partial t}|\psi(t)\rangle = -\frac{\mathrm{i}}{\hbar}(\hat{H}_{\text{S}} + \hat{H}_{\text{int}})|\psi(t)\rangle , \tag{6.14}$$

with the initial condition $|\psi(t_0)\rangle = |\phi_a^0\rangle|\phi_b^0\rangle$ and the particle Hamiltonian

$$H_{\text{S}} = \epsilon_a^{(1)}|1_a\rangle\langle 1_a| + \epsilon_a^{(2)}|2_a\rangle\langle 2_a| + \epsilon_b^{(1)}|1_b\rangle\langle 1_b| + \epsilon_b^{(2)}|2_b\rangle\langle 2_b| . \tag{6.15}$$

Without loss of generality, we can set the energies of all states to zero, and therefore disregard Hamiltonian $\hat{H}_{\text{S}}$. The solution of (6.14) to first order reads

$$|\psi(t_0 + \Delta t)\rangle = |\psi(t_0)\rangle - \frac{\mathrm{i}}{\hbar}\int_{t_0}^{t_0+\Delta t} \mathrm{d}\tau\, \hat{H}_{\text{int}}(\tau)|\psi(t_0)\rangle + \ldots \tag{6.16}$$

and if Δt is sufficiently short, we get

$$|\psi(t_0 + \Delta t)\rangle \approx |\psi(t_0)\rangle - \frac{\mathrm{i}}{\hbar}\Delta t\, \hat{H}_{\text{int}}(t_0)|\psi(t_0)\rangle , \tag{6.17}$$

while higher orders in $\hat{H}_{\text{int}}$ are also higher orders in Δt, which is small. The initial state $|\psi(t_0)\rangle$ is not entangled, because it can be factorized into parts belonging to particles a and b, respectively. However, after the interaction the particles are entangled,

$$|\psi(t_0 + \Delta t)\rangle \approx |\psi(t_0)\rangle - \frac{\mathrm{i}}{2\hbar}\theta_0\Delta t(|1_a\rangle|2_b\rangle + |2_a\rangle|1_b\rangle) , \tag{6.18}$$

because the new wavefunction cannot be factorized anymore.

According to what we said above, we can conclude that almost all interactions of a quantum system with other (neighboring) systems lead to entanglement and an increased loss of the ability to assign a wavefunction to the individual components of a composite system. Most of interactions that occur in nature have a continuous character. It might be extremely difficult to solve equations of motion in some particular cases, but it is often possible to guess how the final entangled state will look like. Instead of presenting the whole time evolution, we will denote the transition from the initial state to the final state by an arrow:

$$|\psi_{\text{initial}}\rangle \rightarrow |\psi_{\text{final}}\rangle \; . \tag{6.19}$$

We can always describe the evolution of the total system on the combined Hilbert spaces of the subsystems, that is, we can use the eigenvectors of subsystems corresponding to their Hamiltonians to represent the initial and final states. Since the system evolves from an initially unentangled state $|\psi_{\text{initial}}\rangle = |\phi_a^0\rangle|\phi_b^0\rangle$ to an entangled state, this transition can be expressed as follows:

$$\begin{aligned} |\psi_{\text{initial}}\rangle = |\phi_a^0\rangle\,|\phi_b^0\rangle &= \left(\sum_k \alpha_k^a|k_a\rangle\right)\left(\sum_l \alpha_l^b|l_b\rangle\right) \\ &\rightarrow \sum_{kl} \alpha_{kl}^{(ab)}|k_a\rangle|l_b\rangle = |\psi(t_0+\Delta t)\rangle \; , \end{aligned} \tag{6.20}$$

where $|k_a\rangle$ and $|l_b\rangle$ are the eigenvectors of separate (noninteracting) particles a and b. Choosing to observe just subsystem a, we can also write equivalently

$$|\phi_a^0\rangle\,|\phi_b^0\rangle = \left(\sum_k \alpha_k^a|k_a\rangle\right)|\phi_b^0\rangle \rightarrow \sum_k n_a^{(k)}|k_a\rangle\left|\phi_b^{(k,a)}\right\rangle \; , \tag{6.21}$$

where $|\phi_b^{(k,a)}\rangle = 1/n_b^{(k)} \sum_l \alpha_{kl}^{(ab)}|l_b\rangle$ is the state of system b relative to the state $|k_a\rangle$ of system a, and $n_b^{(k)}$ is a normalization factor of the wavefunction $|\phi_b^{(k,a)}\rangle$. In (6.20) and (6.21), the brackets stress separable parts of the wavefunctions. On the right-hand side, we cannot separate the two systems from each other unless all vectors $|\phi_b^{(k,a)}\rangle$ are the same.

6.4 Decoherence

Entanglement is a ubiquitous phenomenon in nature. It is created spontaneously by interaction of quantum systems, and it cannot be undone in a simple way. In this section we will discuss the most famous demonstration of quantum behavior, the double-slit experiment, where entanglement plays a crucial role. This consideration allows us to introduce the important concept of *decoherence*.

Let us assume that we have a source of identically prepared particles which move toward a wall with two slits. Behind the slit there is a flat array of detectors (a CCD chip or a photographic plate) which records the particle's impact position after it has passed through the double slit (see Figure 6.3). The particle is expected to pass with equal probability through one or the other slit. We can test this by covering first slit 1 and then slit 2, and measuring the number of particles that reach the

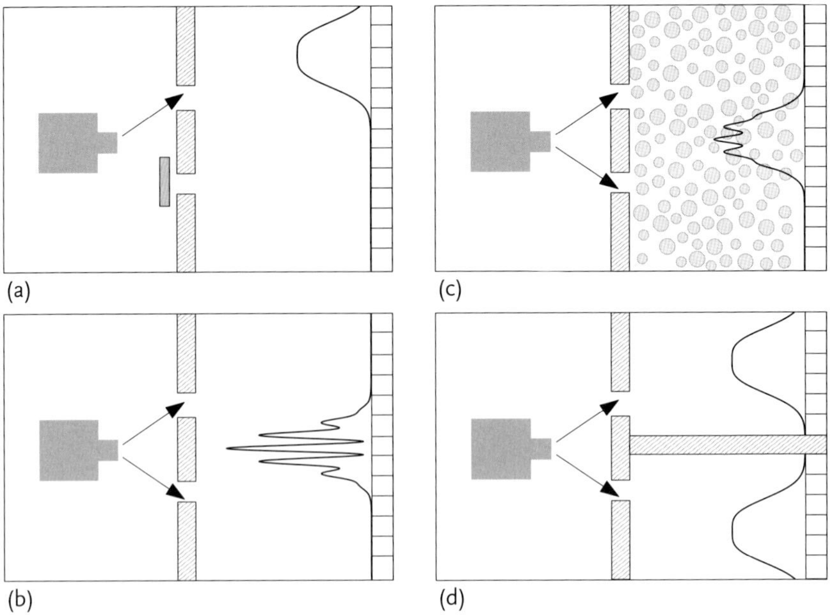

Figure 6.3 Double-slit experiment and the distribution of the detection on the array of detectors behind the double slit under different conditions: (a) one slit covered; (b) double-slit experiment in a vacuum; (c) double-slit experiment with the environment; (d) double-slit arrangement converted into a set of two detectors.

detector. It should always be half the number of particles that pass through when both slits are open.

When slit 1 is open (see Figure 6.3a), no matter how the particles were prepared in the source, those that appear behind the slit had to have passed through the slit, and we label their state as $|1\rangle$. The slit acts as a filter of particles, allowing only those in an expected state to pass through. The distribution of the landing positions of the particle when only slit 1 is open will be denoted $P_1(x)$. When slit 2 is open, we similarly describe this state as $|2\rangle$. In this case the particles land with probability distribution $P_2(x)$ at position x of the detector. The classical theory of probability dictates that when both slits are open, the probability distribution is simply $P_1(x) + P_2(x)$ (see Chapter 3). Standard quantum mechanics states that when both slits are open the probability distribution is not simply $P_1(x) + P_2(x)$. The experimentally observed probability $P_{1+2}(x)$ shows an interference term, like the one we would obtain from light passing through the double slit.

Quantum mechanics postulates that the state of each particle after passing through the double slit is a superposition of states $|1\rangle$ and $|2\rangle$:

$$|\psi_{1+2}\rangle = \alpha_1|1\rangle + \alpha_2|2\rangle \,, \tag{6.22}$$

where constants α_1 and α_2 are complex probability amplitudes for finding the particle in particular states. Normalization requires that $|\alpha_1|^2 = |\alpha_2|^2 = 1/2$. In braket notation the probability distributions are the expectation values of operators

$|x\rangle\langle x|$ projecting the state on the spatial coordinate x. For probability $P_{1+2}(x)$ we have

$$P_{1+2}(x) = \langle \psi_{1+2}|x\rangle\langle x|\psi_{1+2}\rangle$$
$$= \frac{1}{2}\left[P_1(x) + P_2(x) + 2\mathrm{Re}\left(\langle 1|x\rangle\langle x|2\rangle\right)\right] . \tag{6.23}$$

The interference pattern seen experimentally (Figure 6.3b) is explained by the wave properties of the quantum particles.

Let us now set up the experiment in such a way that particles can collide with particles of a gas on their way to the detector (see Figure 6.3c). The gas, which forms an environment of our quantum system, is itself a quantum system described by Hamiltonian $\hat{H}_{\mathrm{env}}$, and the interaction of the particles with the gas molecules (environment) is defined by Hamiltonian $\hat{H}_{\mathrm{int}}$. The total Hamiltonian then reads

$$\hat{H} = \hat{H}_{\mathrm{env}} + \hat{H}_{\mathrm{S}} + \hat{H}_{\mathrm{int}} , \tag{6.24}$$

where $\hat{H}_{\mathrm{S}}$ is responsible for the "flight" of the particle in the absence of the environment. After passing though the slits, the particle can only be in two states $|1\rangle$ and $|2\rangle$ (and their superpositions), so we can use the expression $|1\rangle\langle 1| + |2\rangle\langle 2|$ as a definition of the unity operator in the Hilbert space of a particle. Let us choose the particle–environment interaction as

$$\hat{H}_{\mathrm{int}} = \Xi_1|1\rangle\langle 1| + \Xi_2|2\rangle\langle 2| , \quad \Xi_1 \neq \Xi_2 . \tag{6.25}$$

This form expresses the fact that the gas only performs detection of the particle's state which is conserved by the interaction. Additionally, the gas interacts differently with a particle in state $|1\rangle$ and a particle in state $|2\rangle$. Note that if $\Xi \equiv \Xi_1 = \Xi_2$, the interaction Hamiltonian would have the form of a product of an environmental operator and the unity operator in the Hilbert space of the particle. Operator Ξ could then be added to the Hamiltonian of the environment, $\hat{H}_{\mathrm{env}}$, and the total Hamiltonian, (6.24), would be the sum of two Hamiltonians. Consequently, the wavefunction of the system and the environment would never entangle, the two wavefunctions would never influence each other, and by definition there would be no interaction between the particle and its environment. In other words, the environment would not be able to distinguish different states of the particle. Thus, operators Ξ_1 and Ξ_2 have to be different for any interaction to take place.

For convenience, we can now rewrite the interaction Hamiltonian as

$$\hat{H}_{\mathrm{int}} = \Xi_1 + \Delta\Xi|2\rangle\langle 2| , \tag{6.26}$$

where $\Delta\Xi = \Xi_2 - \Xi_1$, and where we used the definition of unity on the Hilbert space of the particle ($1 = |1\rangle\langle 1| + |2\rangle\langle 2|$). Taking into account such a definition of the interaction Hamiltonian, we now define new environment and interaction Hamiltonian operators:

$$\hat{H}'_{\mathrm{env}} = \hat{H}_{\mathrm{env}} + \Xi_1 , \tag{6.27}$$

and

$$\hat{H}'_{\text{int}} = \hat{H}_{\text{int}} - \Xi_1 \equiv \Delta\Xi|2\rangle\langle 2| \,. \tag{6.28}$$

After this redefinition, the interaction Hamiltonian is zero if Ξ_1 and Ξ_2 are the same, that is, in the situation when there is no interaction, no entanglement between the particles and the environment.

The initial state of the system including the state of the environment, $|\eta_0\rangle$, contains evidently no entanglement:

$$|\psi_0\rangle = |\psi_{1+2}\rangle|\eta_0\rangle \,. \tag{6.29}$$

From its initial state, the system propagates for a time t until it reaches the detector. At the detector the system will be in state $|\psi(t)\rangle = \hat{U}(t)|\psi_0\rangle$, where $\hat{U}(t)$ is the evolution operator of the total Hamiltonian. Because all Hamiltonians are diagonal in the basis of states $|1\rangle$ and $|2\rangle$ of the particle, we can write $\hat{U}(t)$ using just two matrix elements $\hat{U}_1(t) = \langle 1|\hat{U}(t)|1\rangle$ and $\hat{U}_2(t) = \langle 2|\hat{U}(t)|2\rangle$. Consequently, using the definition of $|\psi_{1+2}\rangle$, (6.22), we get

$$|\psi(t)\rangle = \alpha_1\hat{U}_1(t)|\eta_0\rangle|1\rangle + \alpha_2\hat{U}_2(t)|\eta_0\rangle|2\rangle = \alpha_1|\eta_1\rangle|1\rangle + \alpha_2|\eta_2\rangle|2\rangle \,. \tag{6.30}$$

Both $\hat{U}_1(t)$ and $\hat{U}_2(t)$ are operators in the environmental Hilbert space and their action on the initial state vector of the environment $|\eta_0\rangle$ can be expressed as

$$\hat{U}_1(t)|\eta_0\rangle = \exp\left(-\frac{\mathrm{i}}{\hbar}\hat{H}'_{\text{env}}t\right)|\eta_0\rangle \equiv |\eta_1\rangle \,, \tag{6.31}$$

$$\hat{U}_2(t)|\eta_0\rangle = \exp\left(-\frac{\mathrm{i}}{\hbar}(\hat{H}'_{\text{env}} + \Delta\Xi)t\right)|\eta_0\rangle \equiv |\eta_2\rangle \,. \tag{6.32}$$

The probability distribution for finding the particle with coordinate x now also includes the overlap of the environmental wavefunction:

$$\begin{aligned} P(x) &= \langle\psi(t)|x\rangle\langle x|\psi(t)\rangle \\ &= \frac{1}{2}\left[P_1(x) + P_2(x) + 2\text{Re}\left(\langle 1|x\rangle\langle x|2\rangle\langle\eta_1|\eta_2\rangle\right)\right] \,. \end{aligned} \tag{6.33}$$

Thus, the interference term is influenced only by the presence of the interaction with the environment.

The term that changes the interference pattern due to the presence of the gas reads

$$\langle\eta_1|\eta_2\rangle = \langle\eta_0|\hat{U}_1^\dagger(t)\hat{U}_2(t)|\eta_0\rangle \,. \tag{6.34}$$

To express the operator $\hat{U}_1^\dagger(t)\hat{U}_2(t)$ in a suitable form, we will use (6.31) and (6.32). First, we find that we can derive a simple differential equation by taking a time derivative of $\hat{U}_1^\dagger(t)\hat{U}_2(t)$

$$\frac{\partial}{\partial t}\left(\hat{U}_1^\dagger(t)\hat{U}_2(t)\right) = -\frac{\mathrm{i}}{\hbar}\hat{U}_1^\dagger(t)\Delta\Xi\,\hat{U}_1(t)\left(\hat{U}_1^\dagger(t)\hat{U}_2(t)\right) \,. \tag{6.35}$$

Equation (6.35) has a formal solution in terms of a time-ordered exponential (see (4.41) for the definition):

$$\hat{U}_1^\dagger(t)\,\hat{U}_2(t) = \exp_+\left(-\frac{i}{\hbar}\int_0^t d\tau\,\hat{U}_1^\dagger(\tau)\Delta\Xi\,\hat{U}_1(\tau)\right), \tag{6.36}$$

where we assumed $\hat{U}_1(0) = \hat{U}_2(0) = 1$. The desired overlap is therefore just a matrix element of a certain time-ordered exponential and it can be approximately evaluated for some models of the environment. Details on an approximate evaluation of (6.34) are presented in Appendix A.8. Here it is enough to know that the particle–environment interaction can often be described by some two-point correlation function $C(t)$. If the interaction can be regarded as stochastic, we can try to model it using some simple assumptions about the corresponding stochastic process. The overlap, (6.34), then decays exponentially

$$\langle\eta_1|\eta_2\rangle = e^{-\xi(t)}, \tag{6.37}$$

where $\xi(t)$ is proportional to the double integral of the correlation function $C(t)$ (A109).

For instance, when consecutive interactions between the particle and its environment are not correlated at all,

$$C(t) \approx \delta(t), \tag{6.38}$$

and (A109) yields

$$\xi(t) \approx t. \tag{6.39}$$

If the correlation itself decays as an exponential function, for example,

$$C(t) \approx e^{-\Gamma t}, \tag{6.40}$$

we get

$$\xi(t) \approx \frac{1}{\Gamma}(e^{-\Gamma t} - \Gamma t + 1) \approx \frac{1}{2}t^2. \tag{6.41}$$

Thus, the overlap $\langle\eta_1|\eta_2\rangle$ decays with time t according to the two model cases described above as

$$\langle\eta_1|\eta_2\rangle = e^{-\gamma t}, \quad \langle\eta_1|\eta_2\rangle = e^{-\frac{\Delta^2}{2}t^2}. \tag{6.42}$$

Constants γ and Δ are related to the strength of the interaction of the particle with its environment because they are proportional to the square of the interaction term (see Appendix A.8):

$$\gamma \approx \Delta^2 \approx \langle\eta_0|\Delta\Xi\Delta\Xi|\eta_0\rangle. \tag{6.43}$$

It is possible to show that $\xi(t)$ is always a growing function at long times ($t \gg \tau$) for the case of an arbitrary correlation function which decays to zero at a certain correlation time τ. We can thus conclude that the overlap $\langle \eta_1 | \eta_2 \rangle$ decays for the chaotic gas and the interference term therefore disappears in the course of time.

As follows from the analysis presented above, the state of the particle in the presence of the gas becomes entangled with the gas and thus cannot be given in the form of a wavefunction associated only with the particles. However, the corresponding reduced density matrix of particles can be given as follows:

$$\rho(t) = \mathrm{tr}_{\mathrm{env}}\{|\psi(t)\rangle\langle\psi(t)|\} = \frac{1}{2}\begin{pmatrix} 1 & \mathrm{e}^{-\xi(t)} \\ \mathrm{e}^{-\xi^*(t)} & 1 \end{pmatrix} . \tag{6.44}$$

While at $t = 0$ this reduced density matrix represents a pure state, at long times $t \gg 1/\gamma$ it represents a statistical mixture of the system in states $|1\rangle$ and $|2\rangle$:

$$\rho\left(t \gg \frac{1}{\gamma}\right) = \frac{1}{2}\begin{pmatrix} 1 & 0 \\ 0 & 1 \end{pmatrix} . \tag{6.45}$$

Thus, we have arrived at a diagonal density matrix, so the particle never seems to be in a state represented by a superposition of states $|1\rangle$ and $|2\rangle$. This is true, however, only after the averaging over the states of the environment, that is, after disregarding a part of the entangled composite system.

In our description, the diagonal form of the density matrix does not mean that we have an ensemble of particles found in state $|1\rangle$ and particles found in state $|2\rangle$. However, in the measurement, the system will behave as such, because the interference is not present. The effect that we have identified here is usually called *decoherence*, and the approach in which we take seriously the superposition principle and the rapidly spreading entanglement between interacting quantum systems is termed *decoherence theory*.

6.5
Preferred States

Similar to the transition from the pure to the mixed state described in Chapter 4, the decoherence approach does not require any notion of the collapse of the wavefunction. Sticking to simple principles, one can explain many phenomena in a unified and elegant manner without invoking the collapse at all. This idea will be discussed later.

In terms of decoherence theory we can answer the question why only a small subset of states are in principle allowed by the superposition principle. Indeed, some states destructively interfere due to the superposition and their signatures decay, while other states are stable with respect to the decoherence. Apparently, some states are selected by the decoherence process over other states.

By describing the double-slit experiment, we have shown that the corresponding reduced density operator becomes diagonal (in some basis) due to the interaction of the system with its environment. Thus, some basis of states was selected from all possible basis by the decoherence process. These selected states are therefore called the *preferred states* and they compose the *preferred basis*, and the process in which this basis is selected (by the environment or the measuring apparatus) is called *environment-induced superselection* [35]. In the absence of the environment, the superposition principle lets us describe the quantum system in an arbitrary basis, and all superpositions of states are equivalent. Decoherence changes this freedom.

The selection of the preferred basis is, however, not so straightforward as the example in Section 6.4 might suggest. It seems that the loss of coherence as defined by (6.44) leads directly to states $|1\rangle$ and $|2\rangle$ as the preferred states. However, the final density matrix, (6.45), looks exactly the same in all bases. What is even worse is that the complete state vector, (6.30), apparently does not help us to distinguish the preferred basis either. We can, for instance, introduce new basis states $|\alpha\rangle$ and $|\beta\rangle$ so that

$$|1\rangle = \langle\alpha|1\rangle|\alpha\rangle + \langle\beta|1\rangle|\beta\rangle\,, \quad |2\rangle = \langle\alpha|2\rangle|\alpha\rangle + \langle\beta|2\rangle|\beta\rangle\,. \tag{6.46}$$

The time evolution will have the form

$$\begin{aligned}\frac{1}{\sqrt{2}}(|1\rangle + |2\rangle)|\eta_0\rangle \to \frac{1}{\sqrt{2}}\big[(\langle\alpha|1\rangle|\alpha\rangle + \langle\beta|1\rangle|\beta\rangle + |2\rangle)|\eta_1\rangle \\ +(\langle\alpha|2\rangle|\alpha\rangle - \langle\beta|2\rangle|\beta\rangle)|\eta_2\rangle\big]\,.\end{aligned} \tag{6.47}$$

The right-hand side can be rewritten as

$$\frac{1}{\sqrt{2}}\left(|\alpha\rangle|\eta_\alpha\rangle + |\beta\rangle|\eta_\beta\rangle\right)\,, \tag{6.48}$$

where

$$|\eta_\alpha\rangle = \langle\alpha|1\rangle|\eta_1\rangle + \langle\alpha|2\rangle|\eta_2\rangle\,, \quad |\eta_\beta\rangle = \langle\beta|1\rangle|\eta_1\rangle + \langle\beta|2\rangle|\eta_2\rangle\,. \tag{6.49}$$

States $|\eta_\alpha\rangle$ and $|\eta_\beta\rangle$ are bath states and they are orthogonal. Not surprisingly, they prevent any coherence between states $|\alpha\rangle$ and $|\beta\rangle$ from surviving.

We have to step back to the whole notion of the system–environment interaction again. We have stated that the interaction between the system and the environment is possible only when the environment can distinguish between different states of the system. We have postulated that operators Ξ_1 and Ξ_2 are different. If they are not, there would be no interaction between the particles and the environment in the sense that it would not be possible to distinguish from the state of the *environment* whether the particle is in state $|1\rangle$ or state $|2\rangle$. Therefore, we can conclude that in our case states $|1\rangle$ and $|2\rangle$ can be distinguished; the particles *encode* their state (or the information about their state) in the environment: the environment acts as a detector.

We can show by an extreme example that not all states are distinguishable from the point of view of the environmental states. Let us include an observable, which measures the bath state, in the whole system under consideration. The observable should assign different eigenvalues $\mathcal{E}_n$ to different states of the bath $|\mathcal{E}_n\rangle$. The corresponding operator reads

$$\hat{E} = \sum_n \mathcal{E}_n |\mathcal{E}_n\rangle\langle\mathcal{E}_n| \ . \tag{6.50}$$

Let us measure this observable for the final environmental states corresponding to the following linear combinations of the particle states:

$$|+\rangle = \frac{1}{\sqrt{2}}(|1\rangle + |2\rangle) \ , \quad |-\rangle = \frac{1}{\sqrt{2}}(|1\rangle - |2\rangle) \ . \tag{6.51}$$

The corresponding bath states read, according to (6.49), $|\eta_+\rangle = (|\eta_1\rangle + |\eta_2\rangle)/\sqrt{2}$, and $|\eta_-\rangle = (|\eta_1\rangle - |\eta_2\rangle)/\sqrt{2}$. It is easy to verify that we get

$$\langle +|\hat{E}|+\rangle = \langle -|\hat{E}|-\rangle = \sum_n \mathcal{E}_n \frac{\langle\eta_1|\mathcal{E}_n\rangle\langle\mathcal{E}_n|\eta_1\rangle + \langle\eta_2|\mathcal{E}_n\rangle\langle\mathcal{E}_n|\eta_2\rangle}{2} \ . \tag{6.52}$$

This result does not depend on the way in which we measured the information contained in the two bath states; the results are the same for all conditions. This suggests that the environment cannot distinguish states $|+\rangle$ and $|-\rangle$ from each other, and consequently it would not be able to destroy coherence of their linear combinations of this specific type. For instance, the linear combination $(|+\rangle + |-\rangle)/\sqrt{2}$ should not dephase at all. This is easy to see, because $(|+\rangle + |-\rangle)/\sqrt{2} = |1\rangle$, and starting with this state, we get

$$|1\rangle|\eta_0\rangle \rightarrow |1\rangle|\eta_1\rangle \ . \tag{6.53}$$

Thus, this time evolution does not lead to any entanglement.

It is important to note here that we had to look at the environment to be able to distinguish the preferred basis in the Hilbert space of the particle. This basis is determined by the environment and its interaction with the particles. We have dealt with a particularly simple system–environment interaction Hamiltonian, which was diagonal in the basis built from the eigenstate of the system Hamiltonian.

The role of the environment can often be played by a macroscopic measuring device. The nonoverlapping scalar product of the environment states is the condition for the superselection of the system states. Because the construction of detectors is in our hands, and with it, at least to some extent, also the properties of the corresponding interaction Hamiltonian, one can construct detectors detecting the system in different states. A simple example is presented in Figure 6.3d, where by adding a wall creating separated chambers we have made sure that the time evolution of the environment, in this case the particles of the chamber walls, will enable us to distinguish whether the particle has passed through one or the other hole. Were the wall not present, the final state of the environment would not allow us to determine the hole through which the particle passed.

6.6 Decoherence in Quantum Random Walk

The concept of decoherence helps us to reconcile the deterministic evolution of the wavefunction and the superposition principle, on the one hand, with the classical prediction of the random walk, on the other hand. What we have so far completely ignored is the fact that the particle in a trap interacts with some environment, which is sensitive to its position on the grid. At least, the particle has to relax to the bottom of the new potential well after the field has been inverted. Thus, the fluctuations of the environment to which the particle is exposed in different traps will be different, or (expressed from the opposite point of view) the particle interacts with the bath differently when in different traps. Because the environment causes neither the transitions of the particle between the traps nor the transitions between internal states of the particle, it must be the environment which is driven in a different way by the presence of the particle in one or the other trap. Again, information about the particle gets encoded in the environment by this interaction. This is exactly the situation found in the double-slit experiment described in Section 6.4.

The state of the environment for a particle prepared in the same initial state (including the environment) but traveling through different positions on the grid will be different. We will denote the initial state of the particle–environment system by a state vector

$$|\psi_0\rangle = |P_0\rangle|R\rangle|0\rangle \,, \tag{6.54}$$

where $|0\rangle$ describes the environment evolving with the particle occupying trap 0. Let us assume the particle was transferred to position -1, and then back to 0 after time Δt. We denote the state of the environment as $|0, -1_{\Delta t}, 0\rangle$, reflecting this movement. If the particle were instead moved to position $+1$, the state of its environment would be $|0, +1_{\Delta t}, 0\rangle$. As any state vectors, also those of the environment have to be normalized to 1, so

$$\langle 0, -1_{\Delta t}, 0|0, -1_{\Delta t}, 0\rangle = \langle 0, +1_{\Delta t}, 0|0, +1_{\Delta t}, 0\rangle = 1 \,. \tag{6.55}$$

The state vectors for different bath histories might not overlap perfectly, that is, $\langle 0, +1_{\Delta t}, 0|0, -1_{\Delta t}, 0\rangle < 1$. Because the bath is assumed to consist of a large number of degrees of freedom, the overlap of different histories can be assumed to decay in the same way as described in Section 6.4. This happens for the intervals when the bra and ket baths evolve under the influence of the particles in different traps. If the history differs in the particle position for time t, the overlap will decay by the factor $e^{-\gamma t}$, where γ is the decoherence rate introduced in Section 6.4. We will therefore have

$$\langle 0, +1_{\Delta t}, 0|0, -1_{\Delta t}, 0\rangle = e^{-\gamma \Delta t} \,. \tag{6.56}$$

This simple model of decoherence will enable us to treat the transition from the coherent to the incoherent (quantum to classical) random walk in a consistent manner. We will rewrite (6.5) and (6.6) so that they also include a consistent description of the environment. Equation (6.5) becomes

$$
\begin{aligned}
|\psi_0\rangle &\overset{\hat{C}}{\rightarrow} \frac{1}{\sqrt{2}}|P_0\rangle|0\rangle(|R\rangle - |L\rangle) \\
&\overset{\hat{S}}{\rightarrow} \frac{1}{\sqrt{2}}(|P_1\rangle|0,1\rangle|R\rangle - |P_{-1}\rangle|0,-1\rangle|L\rangle) = |\psi_1\rangle\,,
\end{aligned}
\tag{6.57}
$$

where we dropped the subindex Δt, and assumed that each step of the walk took time Δt. The second step yields

$$
\begin{aligned}
\hat{S}\hat{C}|\psi_0\rangle \overset{\hat{S}\hat{C}}{\rightarrow} \frac{1}{2}(&|P_2\rangle|0,1,2\rangle|R\rangle - |P_0\rangle|0,1,0\rangle|L\rangle \\
&- |P_0\rangle|0,-1,0\rangle|R\rangle - |P_{-2}\rangle|0,-1,-2\rangle|L\rangle) = |\psi_2\rangle\,.
\end{aligned}
\tag{6.58}
$$

We can check that nothing changed in the probabilities of finding the particle at position n after two steps. Using the condition defined by (6.55), we obtain $p_{-2} = 1/4$, $p_0 = 1/2$, $p_2 = 1/4$.

Since the operator $\hat{C}^{-1}\hat{S}^{-1}$ acts only on the particle and not on the environment, its action does not lead to a complete inversion of the dynamics. We see that the bath still keeps the memory of the previous evolution. The next step in the random walk is therefore

$$
\begin{aligned}
\hat{S}^{-1}|\psi_2\rangle = \frac{1}{2}\Big(&|P_1\rangle|0,1,2,1\rangle|R\rangle - |P_1\rangle|0,1,0,1\rangle|L\rangle \\
&- |P_{-1}\rangle|0,-1,0,-1\rangle|R\rangle - |P_{-1}\rangle|0,-1,-2,-1\rangle|L\rangle\Big) = |\psi_3\rangle\,,
\end{aligned}
\tag{6.59}
$$

and

$$
\begin{aligned}
\hat{S}^{-1}\hat{C}^{-1}|\psi_3\rangle = |\psi_4\rangle \equiv \left(\frac{1}{\sqrt{2}}\right)^3 (&|P_0\rangle|0,1,2,1,0\rangle|R\rangle + |P_2\rangle|0,1,2,1,2\rangle|L\rangle \\
&+ |P_0\rangle|0,1,0,1,0\rangle|R\rangle - |P_2\rangle|0,1,0,1,2\rangle|L\rangle \\
&- |P_{-2}\rangle|0,-1,0,-1,-2\rangle|R\rangle - |P_0\rangle|0,-1,0,-1,0\rangle|L\rangle \\
&+ |P_{-2}\rangle|0,-1,-2,-1,-2\rangle|R\rangle - |P_0\rangle|0,-1,-2,-1,0\rangle|L\rangle)\,.
\end{aligned}
\tag{6.60}
$$

Now the evaluation of the probabilities of finding the particle at position n also includes the overlaps of the environmental wavefunctions. The probabilities read

$$
\begin{aligned}
p_2 &= \frac{1}{8}(\langle 0,1,2,1,2| - \langle 0,1,0,1,2|)(|0,1,2,1,2\rangle - |0,1,0,1,2\rangle) \\
&= \frac{1}{4}(1 - \langle 0,1,0,1,2|0,1,2,1,2\rangle) = \frac{1}{4}(1 - e^{-\gamma\Delta t})\,,
\end{aligned}
\tag{6.61}
$$

$$
p_{-2} = \frac{1}{4}(1 - \langle 0,-1,0,-1,-2|0,-1,-2,-1,-2\rangle) = \frac{1}{4}(1 - e^{-\gamma\Delta t})\,, \tag{6.62}
$$

and

$$p_0 = \frac{1}{2}\left(1 + e^{-\gamma \Delta t}\right) . \tag{6.63}$$

We can distinguish two important limits: The limit of the very fast or strong decoherence is when $\gamma > \Delta t^{-1}$, and the limit of weak decoherence is when $\gamma \ll \Delta t^{-1}$. In the former case the probabilities converge to those of the incoherent classical random walk. In the weak decoherence regime, all probabilities tend to be zero except for the trap where the particle started.

We have thus demonstrated how we can describe the quantum and classical random walk within one (quantum mechanical) description. We will now discuss in two steps the origin of (6.7) for the probability.

6.7 Quantum Mechanical Measurement

In the previous sections we assumed that we know how to measure the properties of quantum mechanical systems. We have taken it for granted that there are macroscopic measurement devices that will yield a definite result for every measurement. The usual assumption is that when quantity a, to which we can assign an operator $\hat{A}$, is measured, each of its measurements yields one of its eigenvalues a_i, and the system is found in eigenstate $|a_i\rangle$ of operator $\hat{A}$ after this measurement. The probability of obtaining the value a_i is given by the scalar product of the system state $|\psi\rangle$ and the corresponding eigenvector as

$$p_i = |\langle a_i|\psi\rangle|^2 . \tag{6.64}$$

This is referred to as the Born rule. Since the system is found in state $|a_i\rangle$ after the measurement, this process could be naturally attributed to the collapse of the wavefunction $|\psi\rangle \to |a_i\rangle$. The discussion of the meaning of this apparent collapse and how it happens is continuing today. Decoherence theory cannot explain the apparent collapse, but it can clarify some of the statements above.

First, let us think about what the measurement actually means. We will consider a nondestructive type of measurement discussed first by von Neumann. He was the first to consistently discuss the measuring apparatus as a part of the quantum mechanical problem. He suggested that during the measurement the state of the apparatus evolves together with the state of the system. If the system is in some state $|s_i\rangle$ measurable by the apparatus (e.g., it is a certain position state for a detector being able to measure the position), then the apparatus evolves from its initial "ready" state $|d_r\rangle$ to state $|d_i\rangle$ corresponding to the measured value of the physical quantity:

$$|s_i\rangle|d_r\rangle \to |s_i\rangle|d_i\rangle . \tag{6.65}$$

An ideal detector would not distort the state of the system, and thus due to linearity of the time evolution, we should have

$$|\psi\rangle = \sum_i c_i|s_i\rangle|d_r\rangle \rightarrow \sum_i c_i|s_i\rangle|d_i\rangle \,. \tag{6.66}$$

In other words, the apparatus entangles with the quantum system.

Let us now construct the detectors in such a way that the measurement can be performed in a chosen basis. This choice is sometimes simple to achieve (we can, e.g., choose to measure a spin along any axis); on other occasions it might be more difficult to achieve experimentally. One should not forget that the environment might also play the role of the preferred basis selector, and certain types of states (such as linear combinations of very distant position states of massive objects) might be destroyed before even reaching our detectors. An idealized measurement process can be analyzed and formalized with the help of *filters*. Let us put a filter between our apparatus and an approaching quantum system. The filter does not allow the specific system states to reach our detector unless they are in a certain desired state (see the discussion of the double-slit experiment in Section 6.4). This selection can be easily represented formally by a projection operator:

$$\hat{f}_i = |a_i\rangle\langle a_i| \,. \tag{6.67}$$

This operator sorts out all states and prepares state $|a_i\rangle$ (up to a complex prefactor) out of them:

$$\hat{f}_i|\psi\rangle = \langle a_i|\psi\rangle|a_i\rangle \,. \tag{6.68}$$

An example of such a filter is a monochromator, which selects a given frequency of the spectrum of light. A slit in the double-slit experiment is also a good example of a filter.

One nice thing about filters is that if we set a detector behind them and if the detector registers the presence of a particle, we can claim that we have measured the particle in a particular state filtered out with the filter, that is, with a particular value a_i. The filter formalism nicely reflects the only type of access we have to quantum mechanical entities, namely, access through the readings of detectors.

We may now describe the state of the particle $|a_i\rangle$ after it has passed through filter f_i, and assign the quantity measured by filter f_i by the operator

$$\hat{A}_i \equiv a_i\hat{f}_i = a_i|a_i\rangle\langle a_i| \,. \tag{6.69}$$

The measured value can then be obtained as a regular expectation value of state $|a_i\rangle$:

$$a_i = \langle a_i|\hat{A}_i|a_i\rangle \,. \tag{6.70}$$

This is all by construction, and the only requirement is that states $|a_i\rangle$ are normalized.

The filter and the detector that we have constructed so far are very simple. The detector can only point out that the system was found in state $|a_i\rangle$ or not found at all. If it happens to be in some state that is orthogonal to $|a_i\rangle$, we have no information about it. However, nothing can prevent us from constructing (at least virtually) other filters that account for other states of the particle, just as a CCD chip is used as an array of detectors in a spectrometer. If we completely cover the "spectrum" of the quantity we measure, we expect with certainty that one of our detectors fires if a particle was present in our instrument. If two detectors fired due to one particle, the detector would be useless for determining the state of the particle.

A proper detector can be constructed by requiring that it interacts with particles in different states in a sufficiently different manner so that such interaction brings the detectors into orthogonal internal states. We can remind ourselves of the situation in Section 6.4 where we created a detector which can distinguish the path of the particles by dividing the space behind the slits into isolated compartments.

Now that we have many filters, we can add them together into one large filter:

$$\hat{f} = \sum_i \hat{f}_i = \sum_i |a_i\rangle\langle a_i| \, . \tag{6.71}$$

If all possible states are covered, the presence of the complete filter should not actually matter at all to the state of the particle. The particle will pass through the filter untouched, because all of its possible states are allowed to pass through, which means that $\hat{f}$, (6.71), represents the unity operator. To make a measuring device out of this filter, we have to equip each individual filter $|a_i\rangle\langle a_i|$ with a detector. The array of detectors will be denoted by the state $|d_1, d_2, \dots, d_N\rangle$, where $|d_n\rangle$ is the state of the nth detector. The detector has at least two states $|0_n\rangle$ (nothing detected) and $|1_n\rangle$ (particle detected) which are orthogonal, that is, $\langle 0_n|1_n\rangle = 0$. The initial state of the combined system of the particle and the detectors reads

$$|\Psi_0\rangle = |\psi_0\rangle|0_1, \dots, 0_N\rangle \, . \tag{6.72}$$

After the particle has passed through the filter, it finds itself in state

$$|\Psi_0\rangle = \sum_i \langle a_i|\psi_0\rangle|a_i\rangle|0_1, \dots, 0_N\rangle \, , \tag{6.73}$$

which is the same state (the filter does not change the state of the system) as before it passed through. Now the particle interacts with the detectors whose internal states will change according to the state of the particle:

$$|\Psi_{\rm d}\rangle = \sum_i \langle a_i|\psi_0\rangle|a_i\rangle|0_1, \dots, 1_i, \dots, 0_N\rangle \, . \tag{6.74}$$

Apparently, the particle has entangled with the array of detectors, and cannot now be assigned a state without reference to the state of the detectors. If we discarded the information from the detectors (which we have not looked at yet), we could at best assign a reduced density matrix to the particle. This reduced density matrix would be diagonal in the basis of states $|a_i\rangle$, and it would therefore represent a statistical mixture. It is not surprising that we will find the particle in one of the states $|a_i\rangle$ and not in a superposition, because that is exactly what the reduced density matrix, corresponding to (6.74), would tell us.

The last step in the chain of processes is the observer, who comes to register the state of the detectors. We have to take into account the interaction of the observer with the detectors, and by analogy with previous cases, we expect that the observer has to interact with them in such a way that his/her internal states corresponding to observing different detector outcomes are orthogonal, so that he/she is able to distinguish the states of the detectors.

Let us denote the state of the observer who registers the firing of the ith detector as $|M_i\rangle$. Now the complete state vector reads

$$|\Psi_{\mathrm{d+o}}\rangle = \sum_i \langle a_i|\psi_0\rangle|a_i\rangle|0_1,\ldots,1_i,\ldots,0_N\rangle|M_i\rangle\ , \qquad (6.75)$$

and it contains the observer who registers all the different states of the detectors. Now imagine you are the observer, and you have just seen a certain detector go off, say, the one with index k. You are certain that the particle has been detected with the value of the quantity $\hat{A}$ equal to a_i. You can hardly register yourself in a different state of the mind (in $|M_j\rangle$ where $j \neq k$) simultaneously, that is, you cannot see the other outcomes of the experiment. One thing that we know for certain is that our mind does not enable detection of such superposition states. We postpone to Section 6.9 the question of what happened to the other states of the mind and why you ended up registering the kth detector. Already we can now reveal that no definite answer exists to these questions. For now let us note that for the observer subjectively, the state of the system plus the detector reads

$$|\Psi_M\rangle = |a_k\rangle|0_1,\ldots,1_k,\ldots,0_N\rangle\ . \qquad (6.76)$$

We normalized the state formally because all other *branches* of the wavefunction are effectively locked away from us by decoherence, and all the probabilities registered by the observer will be relative to his/her branch.

The above analysis explains why we get the system in a single state out of the whole superposition. The process of detection entangles the measuring instrument and the observer with the state of the particle. Both the detector and the observer are "constructed" in such a way as to distinguish between some states by means of decoherence. We as observers rely on the device to be a broker between us and the quantum systems. When it comes to the question of which of the outcomes an observer will see, to the best of our knowledge, particular results appear to occur randomly. In order to predict quantum mechanical averages of the measured quantities (which we know from comparison with experiments that we can) we would need to know at least the frequency $\nu_i = \lim_{N\to\infty} N_i/N$ of these occurrences. Here N_i is the number of occurrences out of N measurements that the ith detector goes off. Then the expected average value of the measured quantity would be given by

$$\langle A \rangle = \sum_i a_i \nu_i \,. \tag{6.77}$$

Interestingly, the frequency ν_i can be calculated without resorting to any new postulates about quantum mechanics.

6.8 Born Rule

In previous sections we calculated the probability of finding the system at a certain trap in the quantum random walk experiment. Now we will try to do it based on our discussion of measurement. As before, the central idea is that if a quantum system is in state $|a_i\rangle$ which corresponds to a filter $\hat{f}_i = |a_i\rangle\langle a_i|$, it passes through this filter with certainty.

The measurable quantity a can be assigned an operator constructed from all possible filters $\hat{f}_i$, and to each filter we assign the corresponding value a_i of the quantity a:

$$\hat{A} = \sum_i a_i \hat{f}_i \,. \tag{6.78}$$

Every time we find the system in eigenstate $|a_i\rangle$, we can obtain the corresponding value as

$$a_i = \langle a_i|\hat{A}|a_i\rangle \,. \tag{6.79}$$

The rule for calculating the expected value of the quantity a from its operator $\hat{A}$ works only for the eigenstates of the operator $\hat{A}$. It would work for an arbitrary state vector $|\psi\rangle$ if we could show that

$$\langle\psi|\hat{A}|\psi\rangle = \sum_i a_i |\langle\psi|a_i\rangle|^2 \tag{6.80}$$

coincides with (6.77), that is, if we could show that the frequency ν_i of the occurrence of the value a_i is $|\langle\psi|a_i\rangle|^2$.

Let us assume a related measurement of the quantity a on many systems that are prepared in identical states $|\psi\rangle$. It will be more convenient to consider the whole ensemble of N such systems described by a total wavefunction:

$$|\Psi^{(N)}\rangle = |\psi\rangle_1|\psi\rangle_2 \ldots |\psi\rangle_N \,. \tag{6.81}$$

The subscript indices number the systems in the ensemble, all of which are in state $|\psi\rangle$. We will now construct an operator $\hat{\mathcal{F}}_i^{(N)}$ which would give us the frequency $\nu_i^{(N)}$ of the occurrence of state $|a_i\rangle$ among the states constructed out of the eigenstates $|a_i\rangle$ of an ensemble of N identical systems. For instance, for state

$$|a_1\rangle_1|a_5\rangle_2|a_{99}\rangle_3 \tag{6.82}$$

we want the operator $\hat{\mathcal{F}}_5^{(3)}$ to return to $1/3|a_1\rangle_1|a_5\rangle_2|a_{99}\rangle_3$ so that we will find $\nu_5^{(3)} = 1/3$. In general form, the operator reads

$$\hat{\mathcal{F}}_i^{(N)} = \sum_{k_1,k_2,\dots,k_N} \nu_i^{(N)} |a_{k_1}\rangle_1 \dots |a_{k_N}\rangle_N \langle a_{k_N}|_N \dots \langle a_{k_1}|_1 \, . \tag{6.83}$$

The eigenvalue $\nu_i^{(N)}$ of the operator reads

$$\nu_i^{(N)} = \frac{1}{N} \sum_{\alpha=1} \delta_{ik_\alpha} \, . \tag{6.84}$$

This operator represents the quantity corresponding to the frequency of the occurrence of state $|a_i\rangle$ in the ensemble of N systems. We know what the eigenstates of this operator look like, because we have constructed it from them. For all these eigenstates $|s\rangle$

$$\mathcal{F}_i^{(N)}|s\rangle = \nu_i^{(N)}|s\rangle \tag{6.85}$$

holds.

From the above we know that the rule, (6.80), for calculating the expected values works for the eigenstate of the "measured" operator. Interestingly, we can verify that if we increase the number N to infinity, any state $|\Psi\rangle = \lim_{N\to\infty} |\Psi^{(N)}\rangle$ constructed out of N copies of an arbitrary state $|\psi\rangle$ is an eigenstate of the corresponding frequency operator $\hat{\mathcal{F}}_i = \lim_{N\to\infty} \hat{\mathcal{F}}_i^{(N)}$. By analogy with filters, all vectors pass through the filters corresponding to the operator $\hat{\mathcal{F}}_i$, and for all of them we will with certainty measure one of the eigenvalues. Moreover, we can show that this eigenvalue is $|\langle\psi|a_i\rangle|^2$ as was suggested by (6.80). The derivation of this result is presented in Appendix A.6. We can therefore conclude that the frequency of the occurrence of state $|a_i\rangle$ in state $|\Psi\rangle$ is

$$\nu_i = |\langle\psi|a_i\rangle|^2 \, , \tag{6.86}$$

which is the Born rule. This confirms at the same time that

$$\langle A\rangle = \langle\psi|A|\psi\rangle \, . \tag{6.87}$$

We have thus calculated the statistical probability that we obtain a certain eigenvalue from a single measurement of quantity a.

6.9
Everett or Relative State Interpretation of Quantum Mechanics

The previous discussion of quantum mechanics proposed solutions to most of the problems with the transition from the quantum to the classical world. The problem that remains is to find some way to decide which of the possible outcomes, that is, the eigenvalues of the preferred states, will be measured. Unfortunately, decoherence theory has no answer to this question. To the best of our knowledge, there is no theory supported by experiment which would be able to provide a definite answer.

We will discuss one direction in the interpretation of quantum mechanics consistently. This interpretation is connected with the name of Hugh Everett [36]. Everett realized that the mechanism of entanglement as we discussed above in (6.75) can also apply to the observer, and he suggested that this content of information is sufficient, and nothing more is needed, to describe the world in which we live consistently. Indeed, if the whole universe is assigned by the wavefunction $|\Psi\rangle$, and if the possible states $|M_i\rangle$ corresponding to the observer form a complete set, this total state vector can be written in a form of a sum over the states that are relative to the state of the observer:

$$|\Psi\rangle = \sum_i p_i |M_i\rangle |U_i\rangle \,. \tag{6.88}$$

Here the state vectors $|U_i\rangle$ describe the rest of the universe, which is not reached by the observer. We can leave it to philosophers and future generations of physicists to decide how deep in us or in our brains the observer state $|M_i\rangle$ resides. The most important observation here is that the observer (as we have discussed already) can find himself subjectively only in one of the states $|M_i\rangle$. The observer then perceives only one of the many possible states of the universe, namely, the one which corresponds to the subjective branch. Because the observer perceives himself as classical, and as such stands outside the quantum wavefunction, he can assign the rest of the universe to the wavefunction $|U_i\rangle$, which is normalized to 1. This wavefunction is obviously relative, even "subjective," to the observer, but the observer (as far as we know today) has no influence on the choice of the possible "Everett branch" of the wavefunction, (6.88), he will find himself in. Individual branches of the total wavefunction cannot communicate with each other because of decoherence, and the observer is not aware of the probability $|p_i|^2$ with which he has arrived in his particular branch.

As observers, we can construct detectors so that we can classically observe only certain outcomes, in other words, devices such that our state $|M_i\rangle$ is determined by a given state of the detector $|D_i\rangle$ (which is only a part of the state vector of the whole branch). We can predict the probability that we will end up with a particular outcome of the measurement using the Born rule, but of course, we still have no way to determine beforehand which outcome will be realized in our particular measurement. It is important to note that other branches of the total wavefunction effectively cease to exist for the observer, and the relative state interpretation is thus equivalent to the interpretations that assume some "collapse of the wavefunction" during the measurement process. It so far remains a matter of taste whether leaving the unobservable branches of the wavefunction to live their own lives is a more gruesome offense to *Occam's razor* than the supposed existence of an unobservable mechanism which deletes them. From the point of view of quantum theory, whose basic tenets, including the superposition principle and nonlocality, seem today to

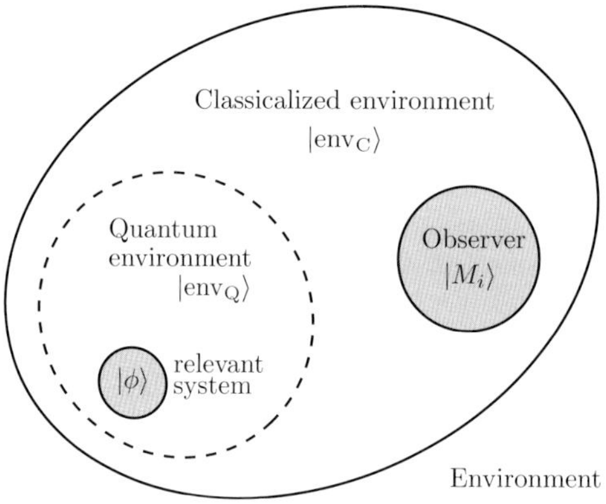

Figure 6.4 Hierarchy of environments of a relevant quantum system (state vector $|\phi\rangle$). The immediate quantum environment is described by the state vector $|\mathrm{env_Q}\rangle$, the degrees of freedom of the environment determined by the observer are described by $|\mathrm{env_C}\rangle$, and the classical state corresponding to the observer is described by the state vector $|M_i\rangle$.

be extremely well tested, it is the "collapse" of the wavefunction which might seem to be an unjustified extension of the theory into the realm of the unobservable. As we have already pointed out however, the objective collapse of the wavefunction cannot be excluded, and the mechanism of the wavefunction branching or collapsing is not known. We do not know what determines the branch of the wavefunction we live in, or how this determination is done.

6.10
Consequences of Decoherence for Transfer and Relaxation Phenomena

Let us now discuss the consequences of the picture that we have just created about the quantum to classical transition for our understanding of energy transfer and relaxation phenomena in molecular aggregates and small molecules. It is an undeniable fact that observers behave in the most profound sense as classical objects and they can find themselves only in classical states. If the state vector $|\Psi\rangle$ of the universe delimited in Figure 6.4 by the outer full line is given, then in the branch $|M_i\rangle|U_i\rangle$, relative to a particular observer in state $|M_i\rangle$, there are also other classical degrees of freedom distinct from the observer. All macroscopic degrees of freedom that we know and observe are classical. On the other hand, we have at least some evidence that at the microscopic level systems behave quantum mechanically, unless we entangle them with some macroscopic device to determine their state. The situation is depicted in Figure 6.4, and it can be formalized as follows.

Let us start with the observer in a certain state $|M\rangle$, and a microscopic quantum system in state $|\phi\rangle = \alpha|1\rangle + \beta|2\rangle$. A large number of degrees of freedom correspond to the observer and to the environment of the quantum system: $|\mathrm{env}\rangle = |\psi_i\rangle \ldots |\psi_N\rangle$. We assume an initially nonentangled state, which could be a result of some careful preparation, in which states of all degrees of freedom of the environment were determined, and the quantum mechanical system prepared in state $|\phi\rangle$. From now on, let us switch on the interaction among all parts of the system. To discuss this interaction we will take the liberty to describe the time evolution in a sequential fashion, bearing in mind, however, that all processes occur simultaneously.

Let us first discuss the interaction of the observer with his environment. The observer interacts strongly with some degrees of freedom of the environment and these degrees of freedom become entangled with the possible Everett branches. The observer follows a single branch $|M_i\rangle$ out of many branches, which determines the state of some degrees of freedom that are observed as classical. This step of the interaction is described as

$$|M\rangle|\psi_1\rangle\dots|\psi_N\rangle|\phi\rangle \to |M_i\rangle|\text{env}_\text{C}\rangle|\psi_{k+1}\rangle\dots|\psi_N\rangle|\phi\rangle\,, \tag{6.89}$$

where the first k degrees of freedom of the environment were determined by the observer in their state $|\text{env}_\text{C}\rangle = |\psi_1\rangle\dots|\psi_k\rangle$. At the same time, however, the degrees of freedom undetermined by the interaction with the observer interact and entangle with the quantum mechanical system, so

$$\begin{aligned}&|M_i\rangle|\text{env}_\text{C}\rangle|\psi_{k+1}\rangle\dots|\psi_N\rangle|\phi\rangle\\&\to |M_i\rangle|\text{env}_\text{C}\rangle\left(\alpha\left|\text{env}_\text{Q}^{(1)}\right\rangle|1\rangle + \beta\left|\text{env}_\text{Q}^{(2)}\right\rangle|2\rangle\right)\,.\end{aligned} \tag{6.90}$$

The state of the quantum subsystem can be assigned the wavefunction $|\psi_\text{Q}\rangle = \alpha|\text{env}_\text{Q}^{(1)}\rangle|1\rangle + \beta|\text{env}_\text{Q}^{(2)}\rangle|2\rangle$ relative to the observer and the classical environment. However, the relevant system cannot be assigned a wavefunction. There exist no wavefunctions $|\phi_\text{r}\rangle$ that would describe the relevant system without having to refer to the state of its quantum environment. We can write a density operator for the whole quantum part,

$$W_\text{Q} = |\psi_\text{Q}\rangle\langle\psi_\text{Q}|\,, \tag{6.91}$$

and this density operator can be reduced to provide a density operator for the relevant system $\rho_\text{r} = \text{tr}_\text{env}\{W_Q\}$:

$$\begin{aligned}\rho_\text{r} &= |\alpha|^2|1\rangle\langle 1| + |\beta|^2|2\rangle\langle 2|\\&\quad + \alpha\beta^*\left\langle\text{env}_\text{Q}^{(2)}\middle|\,\text{env}_\text{Q}^{(1)}\right\rangle|1\rangle\langle 2| + \alpha^*\beta\left\langle\text{env}_\text{Q}^{(1)}\middle|\,\text{env}_\text{Q}^{(2)}\right\rangle|2\rangle\langle 1|\,.\end{aligned} \tag{6.92}$$

Because the scalar products $\langle\text{env}_\text{Q}^{(2)}|\text{env}_\text{Q}^{(1)}\rangle < 1$, the relevant system is, in general, in a mixed state. It is important to note that this mixed state can still be partially coherent depending on the evolution of its quantum environment.

When the observer decides to measure the relevant state, a direct strong interaction link to it has to be established so that different states of the relevant system entangle with different branches of the observer. This can be done by appropriately constructing the detector.

The observer could also attempt to measure the degrees of freedom of the quantum environment by some detector corresponding to a set of filters,

$$1 = \sum_k |\mathcal{E}_k\rangle\langle\mathcal{E}_k| \,, \tag{6.93}$$

as described in Section 6.5. The measurement would result in choosing one of the possible outcome states $|\mathcal{E}_k\rangle$ of the environment, and the state of the universe would then read

$$|M_i\rangle|\mathrm{env_C}\rangle|\mathcal{E}_k\rangle \left(\alpha \left\langle\mathcal{E}_k\middle|\, \mathrm{env}_\mathrm{Q}^{(1)}\right\rangle |1\rangle + \beta \left\langle\mathcal{E}_k\middle|\, \mathrm{env}_\mathrm{Q}^{(2)}\right\rangle |2\rangle\right) \,. \tag{6.94}$$

Thus, the state of the sole undetermined part of the whole system is a pure state. It can correspond to a superposition of states $|1\rangle$ and $|2\rangle$ even if the overlap $\langle \mathrm{env}_\mathrm{Q}^{(1)}|\mathrm{env}_\mathrm{Q}^{(2)}\rangle = 0$, because the states of the quantum environment can still have nonzero overlap with $|\mathcal{E}_k\rangle$ both at the same time. If the state of the environment is fully determined, the border between the classical and quantum degrees of freedom is shifted all the way to the microscopic quantum system. The quantum system will then act as if it was subject to a classical stochastic environment.

We do not know where exactly to put the cut between the classical and the quantum description in real systems. The only question is whether the cut is somewhere close to the relevant system so that we would be justified in using a classical stochastic description for its environment, and a wavefunction description for the relevant system, or whether there is only a substantially large environment which requires a quantum description. This question is not easy to answer; molecular aggregates, which were introduced in Section 5.4, are often constructed by nature from a surrounding (protein) scaffold and selected chromophores. We know that for many functions of proteins, a classical molecular dynamics description is sufficient. Does this mean that we can indeed describe the energy transfer and relaxation phenomena in a semiclassical way where the electronic degrees of freedom are quantum and protein and intramolecular vibrations are classical?

In general, the answer is no. It will be demonstrated in Chapter 14 that disregarding the entanglement between the environment and the electronic degrees of freedom leads to a clear deviation from the quantum mechanical description already at the level of simple optical absorption experiments. The resulting effects are in principle observable and could be experimentally verified. More importantly, we will discuss in Chapter 8 that the quantum nature of the environment is intimately linked with the detailed balance condition which has to be satisfied for transitions between quantum mechanical levels. Classical environmental fluctuations are not able to drive a quantum system into canonical equilibrium, and the quantum and classical relaxations meet only at the infinite-temperature limit. This is a problem which limits seriously the application of classical stochastic approaches to the simulation of a thermodynamic bath, and it will be discussed in Chapter 7. The fact that real-world molecular systems follow thermodynamic relaxation and the detailed balance quite well lets us conclude that the border between classical and quantum degrees of freedom is enough to include a thermodynamically significant number of degrees of freedom in the quantum part. The optical transitions of molecules are therefore embedded in a quantum thermodynamic bath.

This leads us to several important conclusions. First, *both the relevant systems and the bath have to be described quantum mechanically*. All other approaches are approximations, and they have to be carefully checked for unwanted consequences. Second, *electronic degrees of freedom of molecules embedded in a quantum bath cannot be assigned by a wavefunction*. They have to be described by a density operator. Third, this *density operator*, however, *cannot be interpreted statistically* as a sum over individual molecules,

$$\rho = \frac{1}{N}\sum_{n=1}^{N} |\psi_n\rangle\langle\psi_n| \,, \tag{6.95}$$

where individual molecules would be assigned by a state vector $|\psi_n\rangle$ (this is due to the second conclusion). There are a few more conclusions that are consequences of the three conclusions discussed here, and we will mention them in Chapter 8.

Decoherence in quantum mechanics is a powerful instrument for qualitative discussion of the delicate issues of interaction between microscopic and macroscopic systems. However, in this chapter we have rarely used the Schrödinger equation to calculate some quantitative results, and when discussing the consequences of decoherence, we often resorted to the density operator. This is a consequence of the fact that the density operator technique is extremely convenient for representing the state of a quantum system embedded in an environment. It provides much more advanced tools for studying the dynamics of microsystems coupled to a possibly macroscopic bath.

7
Statistical Physics

The concepts of statistical physics are necessary to properly describe naturally observed phenomena of relaxation toward *equilibrium*. This topic is extensively described in various textbooks. For theoretical concepts the reader is referred to [37, 38], while the book of Cengel [39] covers application aspects of thermodynamics. It is worthwhile mentioning that the processes described by both classical and quantum mechanics are reversible. It was explained in Chapter 6 how the transition of the quantum system behavior into classical stochastic dynamics evolves and for that the "observer" as a part of the environment has to be taken into account. The realistic irreversible dynamics is thus intimately related to the concept of the environment. As the environment is macroscopic, it can be described using statistical arguments for an infinite number of particles. Thermodynamics and statistical physics describe such types of systems.

Thermodynamics is a purely phenomenological theory based on a few laws. A more theoretical background for thermodynamics is given by statistical physics. It forms the microscopic foundation of thermodynamics; however, statistical physics itself is based on a postulate of equal probabilities of microscopic states. This principle has no proof; it arises from heuristic arguments. Statistical physics and thermodynamics are thus basic concepts for the analysis of systems interacting with their environment. Before immersing ourselves in the realm of statistical physics, we will first review concepts of thermodynamics.

7.1
Concepts of Classical Thermodynamics

Classical thermodynamics considers macroscopic systems, which are phenomenologically described by a set of observable parameters. The parameters are characterized as intensive and extensive. The intensive parameters do not depend on the system size, that is, if the system is divided into two parts, the intensive parameters have the same values for all parts. In contrast, the extensive parameters are proportional to the system size.

The well-known intensive parameters are temperature T, pressure p, and density ρ, and the extensive parameters are mass m, volume V, energy E, enthalpy H,

Molecular Excitation Dynamics and Relaxation, First Edition. L. Valkunas, D. Abramavicius, and T. Mančal.

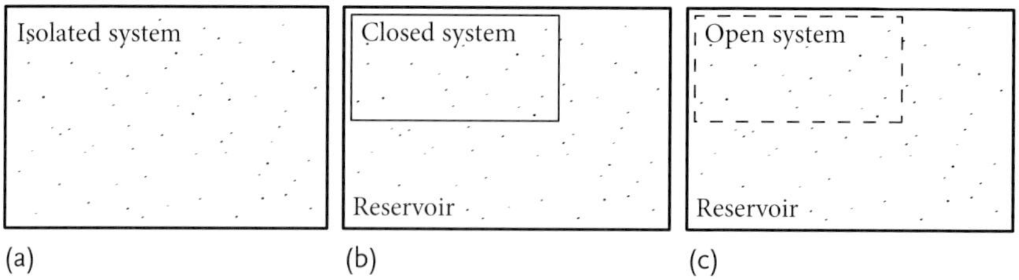

Figure 7.1 Three different types of systems identified by thermodynamics: isolated system (a), closed system embedded in a thermal reservoir (b), and open system embedded in a thermal reservoir (c).

and so on. We can get an intensive parameter from two extensive parameters. For instance, we can introduce the specific volume $v = V/m$, which is the volume taken by a unit mass. In the same way we can introduce the specific energy, specific enthalpy, and so on.

The thermodynamic system under consideration is denoted as *isolated* when the system has no interaction with the surroundings. This includes energy exchange and matter exchange. The laws of conservation of energy and mass thus apply. The *closed* system is one which may have exchange of energy with the surroundings, but the matter of the system is isolated; thus, the mass conservation law is implied. The third type of system is the *open* system, which has exchange of both energy and matter with the environment. These three types of systems are shown in Figure 7.1.

One of the most important primary concepts in thermodynamics is the concept of the *state* of the system. The state is attributed to the whole system under consideration and is defined by the quasi-equilibrium condition. This is the condition when all intensive quantities are the same throughout the whole system volume; the system is thus considered as homogeneous. Heterogeneous systems are characterized by different values of the intensive parameters at different parts of the system; that is, there is no equilibrium state and the system has some transient fluxes which lead toward a homogeneous state of the system. Alternatively, constant heterogeneity can be maintained by specific external conditions. Such a state is denoted as the steady state.

Thermodynamics is based on several laws. These are based on specific experimental observations, which cannot be derived from other theoretical arguments. We briefly review the laws below.

The *zeroth law* of thermodynamics is defined by an empirical fact that there exists a specific intensive state parameter called the temperature. If two systems characterized by different temperatures are brought into contact, while no matter is exchanged, the states of the systems change until the whole combined system comes into a new equilibrium. When there is no exchange of matter, the contact is called a thermal contact and the new equilibrium is the thermal equilibrium. The final state is characterized by a new temperature. The *change* of temperature is thus defined by this law. This law also defines the temperature-measuring device. Since the volume or pressure of a gas or length of some pencil-like solid material changes with temperature, these geometrical parameters can be measured by creating thermal contact.

In thermodynamics, concepts of work and heat are among the main quantities determining the main energy exchange parameters. The work characterizes the content of *mechanical* energy that can be brought into the system or out of the system through the surface. The heat also determines the flow of energy, however it is not of the mechanical form; it corresponds to the flow of the *thermal* energy. The work and heat are thus the properties of interaction and are not system parameters.

Once the system changes its state, the system undergoes a thermodynamic process. Reversible and irreversible processes are distinguished. The *internally reversible* process proceeds over quasi-equilibrium system states until a new equilibrium state is achieved. Such conditions can, in principle, be maintained by the *irreversible external process*. When the process is maintained by reversible external conditions, the process is completely reversible; however, such a process is never possible. The *irreversible* process involves irreversible changes of the system. Such processes involve a nonequilibrium system state when the system is heterogeneous and its thermodynamic parameters are poorly defined. In practice all processes are irreversible.

The *first law* of thermodynamics is the thermodynamic expression of the conservation of energy. We define the system parameter, which is the internal energy U. Its change is due to interaction with the surroundings by accomplishment of work or by heat exchange:

$$\mathrm{d}U = \delta Q - \delta W \,. \tag{7.1}$$

Here the direction of heat transfer is assumed to be from the environment to the system, while the work interaction is the opposite: thus, adding more heat to the system increases its internal energy ($\delta Q > 0 \rightarrow \mathrm{d}U > 0$), while the work performed by the system on the environment reduces the system energy ($\delta W > 0 \rightarrow \mathrm{d}U < 0$). Note that the internal energy is the system parameter. However, the heat and the work are not characteristics of the system; thus, their changes are not directly related to the changes of system states, which are denoted by different letters.

The *second law* of thermodynamics defines the direction of the irreversible process. To define the direction we introduce the system property, the entropy. For a closed system, the entropy S defines the change of the system state when the system *reversibly* interchanges the heat with the environment at fixed temperature, that is,

$$\mathrm{d}S = \frac{1}{T}\delta Q_{\mathrm{rev}} \,. \tag{7.2}$$

Similar to temperature, this definition describes only the change of the entropy.

Now let us consider two systems: a small system A and its environment as a large heat reservoir denoted as system B. System A performs the process and at the end of the process it returns to the same state. In such a cyclic process, all parameters of system A (including its entropy) return to the same value (since system A returns to the same state). If the process is realistic, from our experience, the heat is generated due to friction and the generated portion of heat in system A must to be transferred from system A to reservoir B during the cycle. Thus, if we consider the entropy of system B, $\mathrm{d}S_\mathrm{B} = 1/T\delta Q_\mathrm{gen} > 0$. Note that the entropy change of system A is zero, since it returns to the same state. We thus get that the total entropy of both systems $\mathrm{d}S = \mathrm{d}S_\mathrm{A} + \mathrm{d}S_\mathrm{B} > 0$.

The second law thus determines the direction of the process: the realistic irreversible process in the isolated system is always characterized by increasing entropy

$$\mathrm{d}S > 0 \,, \tag{7.3}$$

which becomes largest in the equilibrium state. The reversible cyclic process has zero entropy change,

$$\mathrm{d}S_\mathrm{rev} = 0 \,. \tag{7.4}$$

The two laws of thermodynamics describe the possibility and the direction of processes. The *third law* of thermodynamics defines the entropy of a system at zero temperature. The entropy at this point is minimal and can be assigned a zero value as it is a finite constant. Analysis of thermal processes allows us to conclude that zero absolute temperature and the lowest value of the entropy are unreachable in a finite number of processes.

Thermodynamics considers quasi-equilibrium processes. The change of the internal energy is the central quantity of interest. It changes due to thermal interaction

$$\delta Q_\mathrm{rev} = T\mathrm{d}S \,, \tag{7.5}$$

mechanical interaction

$$\delta W = p\mathrm{d}V \,, \tag{7.6}$$

and chemical interaction in open systems, which changes the number of constituent particles of the system N, and any other types of interaction. All the interactions can be described in the form

$$\sum_i \mathrm{force}_i \cdot \mathrm{displacement}_i \,. \tag{7.7}$$

For thermal, mechanical, and chemical interactions we have

$$\mathrm{d}U = T\mathrm{d}S - p\mathrm{d}V + \mu\mathrm{d}N \,, \tag{7.8}$$

where μ is the chemical potential. It is the energy which is being added to the system when the number of particles is increased by 1.

Thermodynamic potentials, such as the internal energy, U, the free energy

$$F = U - TS\ , \tag{7.9}$$

and the enthalpy

$$H = U + pV \tag{7.10}$$

are the most useful for description of thermodynamic systems. Additional thermodynamic potentials and relations between them are briefly described in Appendix A.4. The equilibrium of all constituent forces determines the thermal equilibrium between different parts of the system.

7.2 Microstates, Statistics, and Entropy

In statistical physics the concept of a microstate is defined and it is related to the system entropy. Let us consider a classical system consisting of N particles. In the classical description the state of such a system is completely defined when the coordinates q_i and the momenta p_i of all particles are known. In the $6N$-dimensional space consisting of all coordinates and momenta of all particles, such a state is a single point. The space is denoted as the phase space, while the state is called the microstate. As in an arbitrary space, we can denote the phase space element

$$d\omega \equiv d^{3N}q d^{3N}p = dq_1 dq_2 \ldots dq_N dp_1 dp_2 \ldots dp_N\ . \tag{7.11}$$

If the system is isolated, the total energy of the system is conserved during the propagation of the microstate. The total energy is an experimental observable, that is, it is a macroscopically accessible parameter, while the microstate (all coordinates and momenta of particles) is not observed. In the classical thermodynamics described in the previous section, the system is characterized by macroscopic state parameters. From these arguments we can thus associate the observable quantities derived from the microscopic dynamics with the macroscopic thermodynamic parameters.

For a given macroscopic state defined by a finite set of parameters there are a large number of microstates. As an example we recall that the whole microscopic trajectory is characterized by the same macroscopic energy. We denote the number of microstates corresponding to a specified macroscopic state by Ω. Of course Ω depends on a set of macroscopic parameters that define the given macroscopic state. Below we form a relation between macroscopic and microscopic parameters.

Let us consider a macroscopic system, described by energy, volume, and the number of particles: E, V, and N, respectively. If we divide the system into two parts, we get

$$E = E_1 + E_2\ , \tag{7.12}$$

$$V = V_1 + V_2 , \quad (7.13)$$

and

$$N = N_1 + N_2 , \quad (7.14)$$

respectively. Now the total number of macrostates of the total system $\Omega(E, V, N)$ is the product of the number of microscopic states of the two subsystems:

$$\Omega(E, V, N) = \Omega_1(E, V, N)\Omega_2(E, V, N) . \quad (7.15)$$

The main principle of statistical physics affirms that all microscopic states of the isolated system have the same realization probability. Then the macroscopic state probabilities are completely defined by the number of microscopic states comprising the macrostate. According to the set theory of Section 3.1, the most probable macroscopic state should have the largest number of microstates when each microstate has equal probability.

Differentiation of (7.15) yields

$$\mathrm{d}\Omega = \mathrm{d}\Omega_1 \cdot \Omega_2 + \Omega_1 \cdot \mathrm{d}\Omega_2 , \quad (7.16)$$

or

$$\mathrm{d}\ln \Omega = \mathrm{d}\ln \Omega_1 + \mathrm{d}\ln \Omega_2 . \quad (7.17)$$

In the extremum we should have

$$\mathrm{d}\ln \Omega = 0 . \quad (7.18)$$

As Ω is a microscopic parameter, this result describes the most probable state of the system, which is equilibrium. From the thermodynamics, the equilibrium state is defined by the extremum of entropy, so

$$\mathrm{d}S = 0 . \quad (7.19)$$

Thus, we can postulate that

$$S(E, V, N) = k_\mathrm{B} \ln \Omega(E, V, N) . \quad (7.20)$$

The entropy defined by this equation is a proper extensive parameter. The proportionality constant is taken as the Boltzmann constant k_B, which properly normalizes the units.

To get this result we have assumed the concept of a closed system, where all microstates are equally likely. This assumption is the basic postulate of statistical mechanics. However, in the next section we consider open systems, where different microstates are not equally probable, and some weighting factors have to be taken into account.

7.3 Ensembles

In this section we describe three types of statistical ensembles. Therefore, we introduce the probability density, or the phase space density $\rho(q_j, p_j)$. $\rho(q_j, p_j)\mathrm{d}\omega$ denotes the probability of a microstate, defined by q_j, and p_j in volume element $\mathrm{d}^{3N} q \cdot \mathrm{d}^{3N} p \equiv \mathrm{d}\omega$. The probability density is normalized so that

$$\int \mathrm{d}\omega \rho(q_j p_j) = 1 . \tag{7.21}$$

If $f(q_j, p_j)$ is some observable function of all particle coordinates and momenta, the experimentally measurable value is the average quantity:

$$\langle f \rangle = \int \mathrm{d}\omega \rho(q_j, p_j) f(q_j, p_j) . \tag{7.22}$$

Since each point in a phase space can be identified with the copy of the system being in that state, the average here can be associated with the ensemble average.

7.3.1 Microcanonical Ensemble

Let us consider the closed system shown in Figure 7.1a. The phase space density of a closed system is characterized by constant energy. The copies of such a system may be in different microstates, but all of them have the same average energy. Such systems compose the *microcanonical ensemble.*

Let us introduce the energy uncertainty ΔE. In that case the system, described by the energy from E to $E + \Delta E$, is described by the phase space density of the form

$$\rho_{\mathrm{mc}} = \begin{cases} \text{const.,} & E \leq H(q_j, p_j) \leq E + \Delta E , \\ 0, & \text{otherwise} . \end{cases} \tag{7.23}$$

Above we denoted the number of microscopic states corresponding to the specified macroscopic state by Ω. This number can be used in the normalization of (7.23). The microcanonical phase space density is thus

$$\rho_{\mathrm{mc}} = \begin{cases} \frac{1}{\Omega}, & E \leq H(q_j, p_j) \leq E + \Delta E , \\ 0, & \text{otherwise} . \end{cases} \tag{7.24}$$

We can now use this density to write the entropy as an ensemble average. From (7.22) and (7.20) we assume that $S(E, V, N) = \langle S(q_i, p_i) \rangle$ and

$$\int \mathrm{d}\omega \rho_{\mathrm{mc}}(q_i, p_i) S(q_i, p_i) = k_{\mathrm{B}} \ln \Omega(E, V, N) , \tag{7.25}$$

where $S(q_i, p_i)$ is the "entropy" functional of a single microstate. Using the microcanonical phase space density, (7.24), we have

$$S(q_i, p_i) = -k_B \ln \rho_{mc}(q_i, p_i) . \tag{7.26}$$

This derivation can be considered as another form of (7.20); however, this expression in principle is not limited to the microcanonical phase space density. In general, we can postulate the entropy of a system is given by

$$S = -k_B \int d\omega \rho \ln(\rho) . \tag{7.27}$$

Extension to this expression is readily used in quantum physics to describe the *quantum entropy*.

Let us derive some useful property of a closed system density matrix. According to the classical description, all coordinates and momenta of the system under consideration evolve in time. The temporal evolution of the single system corresponds to a curve determined by the Hamilton equations of motion (which were defined in Chapter 2):

$$\dot{q}_i = \frac{\partial H}{\partial p_i} , \tag{7.28}$$

$$\dot{p}_i = -\frac{\partial H}{\partial q_i} . \tag{7.29}$$

In the closed system the Hamiltonian does not depend on time explicitly and thus the total energy does not depend on time. For an arbitrary quantity which is a function of the microstate, we can write $A = A(q_i(t), p_i(t), t)$. Its time dependence is described by

$$\frac{dA}{dt} = \frac{\partial A}{\partial t} + \sum_i \frac{\partial A}{\partial q_i}\dot{q}_i + \frac{\partial A}{\partial p_i}\dot{p}_i , \tag{7.30}$$

and taking into account the Hamilton equations, we have

$$\frac{dA}{dt} = \frac{\partial A}{\partial t} + \sum_i \frac{\partial A}{\partial q_i}\frac{\partial H}{\partial p_i} - \frac{\partial A}{\partial p_i}\frac{\partial H}{\partial q_i} , \tag{7.31}$$

or

$$\frac{dA}{dt} = \frac{\partial A}{\partial t} + \{A, H\} . \tag{7.32}$$

Here

$$\{f, g\} = \sum_i \frac{\partial f}{\partial q_i}\frac{\partial g}{\partial p_i} - \frac{\partial f}{\partial p_i}\frac{\partial g}{\partial q_i} \tag{7.33}$$

are the Poisson brackets. Equation (7.32) applied to the phase space density reads

$$\frac{d\rho}{dt} = \frac{\partial \rho}{\partial t} + \{\rho, H\} . \tag{7.34}$$

In the course of time the different points of the phase space propagate along different trajectories. As the dynamics is reversible, no points are gained and no points are lost; thus, the phase space density must be conserved. Mathematically this statement can be defined as

$$\frac{d\rho}{dt} \equiv 0 \tag{7.35}$$

or

$$\frac{\partial \rho}{\partial t} = -\{\rho, H\} . \tag{7.36}$$

This is the Liouville theorem for an arbitrarily closed system.

7.3.2
Canonical Ensemble

We next describe the system embedded in a thermal reservoir. This type of system is shown in Figure 7.1b. Given the system and the reservoir described by temperature T, the total energy of the whole system and the reservoir is

$$E = E_R + E_S . \tag{7.37}$$

The whole composite supersystem, that is, system plus reservoir, constitutes a closed system, whose energy E should not change and can be treated as a constant value. Different from the previous microcanonical ensemble, now the energy of the system part, E_S, is not fixed. We can thus calculate the probability p_i of the system being in microstate i of the system with energy E_i. When the system size is much smaller than the size of the reservoir, the number of degrees of freedom of the system is vanishingly small compared with that of the reservoir. The probability p_i is then proportional to the number of microstates *of the reservoir*. The reservoir has energy $E - E_i$; thus, its number of microstates is denoted by $\Omega_R(E - E_i)$ when the microstate of the system is i.

Using the relation $E_i \ll E_R$, we can expand the number of microstates of the reservoir Ω_R in powers of E_i. Let us consider the entropy of the reservoir, $S_R = k_B \ln \Omega_R$. Expansion up to the first power gives

$$S_R = k_B \ln \Omega_R(E - E_i) = k_B \ln \Omega_R(E) - \frac{d(k_B \ln \Omega_R(E))}{d E_R} E_i + \dots , \tag{7.38}$$

while higher powers can be disregarded. For the reservoir approaching infinite size, and keeping in mind that only heat exchange is happening between the system and the reservoir, so $\delta Q = d E_R$, we get from (7.2)

$$\frac{d S_R(E)}{d E_R} = \frac{1}{T} ; \tag{7.39}$$

thus,

$$k_B \ln \Omega_R(E - E_i) = k_B \ln \Omega_R(E) - \frac{1}{T} E_i \tag{7.40}$$

or

$$p_i \propto \Omega_R(E - E_i) = \Omega_R(E) \exp\left(-\frac{E_i}{k_B T}\right) . \tag{7.41}$$

Thus, the system can now have arbitrary energy, but the probability decreases exponentially with increasing energy. Normalization of the probability gives the prefactor:

$$Z(T) = \sum_i \exp\left(-\frac{E_i}{k_B T}\right) . \tag{7.42}$$

This quantity is known as the partition function of the canonical ensemble. It is one of the central quantities of statistical physics, and allows us to calculate significant thermodynamic properties. The quantity

$$F(T) = -k_B T \ln(Z(T)) \tag{7.43}$$

can be associated with the free energy of the system (see Appendix A.4).

When the system is described by classical particles, we should write the probability density

$$\rho(q_i, p_i) = \frac{1}{Z} \exp\left(-\frac{H(q_i, p_i)}{k_B T}\right) , \tag{7.44}$$

where $H(q_i, p_i)$ is the classical Hamiltonian. The partition function is now given by

$$Z = \int \mathrm{d}^N q \mathrm{d}^N p \, \exp\left(-\frac{H(q_i, p_i)}{k_B T}\right) . \tag{7.45}$$

These equations constitute the so-called Boltzmann statistics and the probability distribution is the Boltzmann probability distribution. They are general in the sense that they are easily extended to quantum physics, where the probability density is replaced by the density matrix and all properties of the thermodynamics can be evaluated. Some applications of Boltzmann statistics will be given in the following sections. Note that in the relaxation problems of weakly interacting systems, use of Boltzmann statistics is usually sufficient.

7.3.3
Grand Canonical Ensemble

In previous subsections we described systems whose number of particles is fixed. In this subsection we consider the open system, where the exchange of energy and matter is allowed as shown in Figure 7.1c.

Similar to the case of the canonical ensemble, we again consider the system embedded in a reservoir characterized by constant temperature and constant chemical potential for constituent particles. The system size is much smaller than that of the reservoir; thus,

$$E = E_\mathrm{S} + E_\mathrm{R} \,, \tag{7.46}$$

$$N = N_\mathrm{S} + N_\mathrm{R} \,, \tag{7.47}$$

$E_\mathrm{S}/E_\mathrm{R} \ll 1$, and $N_\mathrm{S}/N_\mathrm{R} \ll 1$. The system is now open, so it can have an arbitrary number of particles and arbitrary energy. We now consider the probability of the system being in a state with N_S particles and energy E_S. As the whole system plus reservoir supersystem is a microcanonical ensemble, the particular state of this ensemble is given by the number of microstates. Since the size of the reservoir is much larger than the size of the system, the number of microstates is determined by the microstates of the reservoir. The probability will thus be proportional to the number $\Omega_\mathrm{R}(E_\mathrm{R}, N_\mathrm{R})$.

Let us construct the entropy of the reservoir similarly to (7.38):

$$S_\mathrm{R}(E_\mathrm{R}, N_\mathrm{R}) = k_\mathrm{B} \ln \Omega\,(E_\mathrm{R}, N_\mathrm{R}) \,. \tag{7.48}$$

We can expand this quantity around $E_\mathrm{R} \approx E$:

$$S_\mathrm{R}(E_\mathrm{R}, N_\mathrm{R}) \approx S_\mathrm{R}(E, N) - \frac{\partial S_\mathrm{R}(E, N)}{\partial E} E_\mathrm{S} - \frac{\partial S_\mathrm{R}(E, N)}{\partial N} N_\mathrm{S} \,. \tag{7.49}$$

From (7.8) we have

$$\frac{\partial S_\mathrm{R}(E, N)}{\partial E} = \frac{1}{T} \tag{7.50}$$

and

$$\frac{\partial S_\mathrm{R}(E, N)}{\partial N} = -\frac{\mu}{T} \,, \tag{7.51}$$

which gives

$$k_\mathrm{B} \ln \Omega\,(E_\mathrm{R}, N_\mathrm{R}) = S_\mathrm{R}(E, N) - \frac{1}{T} E_\mathrm{S} + \frac{\mu}{T} N_\mathrm{S} \tag{7.52}$$

or

$$p_\mathrm{S} = \frac{1}{Z_\mathrm{g}} \exp\left(-\frac{1}{k_\mathrm{B} T}(E_\mathrm{S} - \mu N_\mathrm{S})\right) \,. \tag{7.53}$$

The normalization factor is denoted as the partition function of the grand canonical ensemble: it can be defined from

$$Z_\mathrm{g} = \sum_{N_\mathrm{S}=1}^{N} \int \mathrm{d}q_1 \ldots \mathrm{d}q_N \mathrm{d}p_1 \ldots \mathrm{d}p_N \exp\left(-\frac{1}{k_\mathrm{B} T}(E_\mathrm{S} - \mu N_\mathrm{S})\right) \,. \tag{7.54}$$

The probability density is thus given by

$$\rho(N, q_i p_i) = \frac{\exp\left(-\frac{H(q_i p_i)}{k_\mathrm{B} T} + \frac{\mu}{k_\mathrm{B} T} N\right)}{\sum_N \int \mathrm{d}q_1 \ldots \mathrm{d}q_N \mathrm{d}p_1 \ldots \mathrm{d}p_N \exp\left(-\frac{H(q_i p_i)}{k_\mathrm{B} T} + \frac{\mu}{k_\mathrm{B} T} N\right)} \,. \tag{7.55}$$

7.4 Canonical Ensemble of Classical Harmonic Oscillators

Let us consider an ensemble of N harmonic oscillators at temperature T having the same frequency ω and mass m. As shown in Section 2.1.3 the system Hamiltonian can then be given by

$$H(q_\alpha, p_\alpha) = \sum_{\alpha=1}^{N} \left(\frac{p_\alpha^2}{2m} + m\omega^2 \frac{q_\alpha^2}{2} \right) . \tag{7.56}$$

First we calculate the free energy of this system. In continuous space we use the dimensionless phase space element $(1/h)\mathrm{d}q\mathrm{d}p$, where h is the Planck constant (see Appendix A.5). The partition function in the case of continuous space is then a function of temperature and the number of oscillators, so

$$Z(T, N) = \frac{1}{h^N} \left(\prod_{\alpha}^{N} \int \mathrm{d}q_\alpha \mathrm{d}p_\alpha \right) \exp\left(-\frac{H(q_\alpha p_\alpha)}{k_\mathrm{B} T} \right) . \tag{7.57}$$

All integrals are of the Gaussian type:

$$\int \mathrm{d}x \exp\left(-\frac{(x-s)^2}{D} \right) = \sqrt{\pi D} , \tag{7.58}$$

which yields

$$Z(T, N) = \left(\frac{k_\mathrm{B} T}{\hbar\omega} \right)^N . \tag{7.59}$$

The phase space probability density, which is equivalent to the probability of a microstate, is given by

$$\rho(q_\alpha p_\alpha) = Z^{-1} \exp\left(-\frac{H(q_\alpha p_\alpha)}{k_\mathrm{B} T} \right) . \tag{7.60}$$

The entropy of the system is now

$$S = \frac{1}{h^N} \left(\prod_{\alpha}^{N} \int \mathrm{d}q_\alpha \mathrm{d}p_\alpha \right) \rho(q_\alpha p_\alpha) \left[-k_\mathrm{B} \ln \rho(q_\alpha p_\alpha) \right] , \tag{7.61}$$

which for our system is

$$S = k_\mathrm{B} N \left[1 + \ln\left(\frac{k_\mathrm{B} T}{\hbar\omega} \right) \right] . \tag{7.62}$$

This expression for the entropy is actually of the form

$$S = k_\mathrm{B} \ln Z + \frac{1}{T} \langle H \rangle . \tag{7.63}$$

Now the mean energy of the system $\langle H \rangle \equiv U$ is the internal energy of the system. For the free energy of the harmonic oscillators, from (7.43) we get

$$F(T, N) = -k_B T \ln Z = -N k_B T \ln\left(\frac{k_B T}{\hbar\omega}\right) \tag{7.64}$$

and the internal energy (7.9)

$$U = F + TS = N k_B T \,, \tag{7.65}$$

which is consistent with the ideal gas expression. Thus, the harmonic oscillators represent the bosonic ideal gas.

7.5 Quantum Statistics

Having in mind previous sections where we defined the statistical properties of thermodynamic quantities and introduced the elements of the phase space, we can straightforwardly extend the theory to the quantum density matrix formalism.

As shown in preceding sections, the phase space probabilities essentially depend on the system Hamiltonian. It is thus convenient to use the eigenenergy basis. In this case the energies of quantum states are properly defined. The system density matrix in the eigenenergy basis in the stationary state is diagonal:

$$\rho_{mn} = \rho_n \delta_{mn} \,. \tag{7.66}$$

Now the quantity

$$\rho_n = \langle \varphi_n | \hat{\rho} | \varphi_n \rangle \tag{7.67}$$

is the probability that a system assumes state n.

In the microcanonical ensemble we measure the probability of a system being in a state defined by the energy interval from E to $E + \Delta E$. All states with energy inside this interval have equal probability, which is a fundamental assumption of equal probability of microstates. In that case the probabilities must be defined as

$$\rho_n = \begin{cases} \frac{1}{\Omega}, & E \le E_n \le E + \Delta E \,, \\ 0, & \text{otherwise} \,. \end{cases} \tag{7.68}$$

Similarly to the classical description, the factor Ω should be obtained from the normalization condition of the total ensemble.

All states in the canonical ensemble are allowed, while their probabilities depend on their energy. In the eigenenergy basis these quantities are given by

$$\rho_n = \frac{\exp(-\beta E_n)}{\sum_i \exp(-\beta E_i)} \,, \tag{7.69}$$

where $\beta = (k_B T)^{-1}$. The partition function is similarly given by

$$Z = \sum_n \exp(-\beta E_n) . \tag{7.70}$$

Using the notation of the function of an operator, we can write the canonical density matrix in an arbitrary representation:

$$\hat{\rho} = \frac{\exp(-\beta \hat{H})}{\mathrm{tr}(\exp(-\beta \hat{H}))} , \tag{7.71}$$

where $\hat{H}$ is the quantum mechanical Hamiltonian of the system. Now a physical observable of the system represented by operator $\hat{A}$ is given by

$$A = \langle \hat{A} \rangle = \mathrm{tr}(\hat{A}\hat{\rho}) . \tag{7.72}$$

The equations are thus equivalent to the ensemble averaging in the classical case, just the probability densities are replaced by the density matrix and the integrals over the phase space are replaced by the trace operation.

7.6 Canonical Ensemble of Quantum Harmonic Oscillators

In Section 7.3.2 we calculated probabilities and thermodynamic properties of the classical harmonic oscillator. The quantum harmonic oscillator is described by the same Hamiltonian as in the classical system, but the coordinates and momenta of the phase space are replaced by the corresponding operators. The Hamiltonian then generates a ladder of states $E_n = \hbar\omega(n + 1/2)$ separated by the harmonic energy quantum $\hbar\omega$, as described in Section 4.6.1.

Once the state energies are known, the density matrix in the second-quantization (number of quanta) representation is diagonal. The diagonal values are simply given by

$$\rho_{nn} = \frac{\exp(-\beta\hbar\omega/2)}{Z} \exp(-\beta\hbar\omega n) \tag{7.73}$$

and the partition function

$$Z = \exp\left(-\frac{\beta\hbar\omega}{2}\right) \sum_n \exp(-\beta\hbar\omega n) = \frac{1}{2\sinh(\beta\hbar\omega/2)} , \tag{7.74}$$

so the complete expression for the density matrix is

$$\rho_{nn} = 2\sinh\left(\frac{\beta\hbar\omega}{2}\right) \exp\left(-\beta\hbar\omega\left(n + \frac{1}{2}\right)\right) . \tag{7.75}$$

We will now obtain the canonical density matrix in the coordinate representation. The wavefunctions in the coordinate representation are given by (4.105). Note that the density matrix is no longer diagonal in this coordinate representation since the coordinate operator does not commute with the Hamiltonian and the energy eigenstates are not eigenstates of the coordinate operator. The canonical density matrix in the coordinate representation is then given by

$$\begin{aligned}\langle x'|\hat{\rho}|x\rangle &\equiv \sum_{nn'} \phi_{n'}(x')\rho_{nn}\phi_n^*(x) \\ &= N \exp\left(-\frac{x^2+x'^2}{2} - \frac{\beta\omega}{2}\right) \sum_{n=0}^{\infty} \frac{\exp(-\beta\omega n)}{2^n n!} \mathcal{H}_n(x)\mathcal{H}_n(x') ,\end{aligned} \tag{7.76}$$

where N is the normalization constant, which we determined from the normalization of the diagonal of the density matrix – the probability density – and $\mathcal{H}(x)$ is the Hermite polynomial (see (4.104)). We also set $m = \hbar = 1$. Using the integral representation of the Hermite polynomials, we get

$$\mathcal{H}_n(x) = \frac{\exp(x^2)}{\sqrt{\pi}} \int \mathrm{d}y(-2\mathrm{i}y)^n \exp(-y^2 + 2\mathrm{i}xy) , \tag{7.77}$$

and calculating the Gaussian integrals, we get with proper normalization

$$\begin{aligned}\langle x'|\hat{\rho}|x\rangle &= 2\sinh\left(\frac{\beta\omega}{2}\right)\left(\frac{\omega}{2\pi\sinh(\beta\omega)}\right)^{1/2} \\ &\quad\times \exp\left(-\frac{\omega}{4}\left((x+x')^2\tanh\frac{\beta\omega}{2} + (x-x')^2\coth\frac{\beta\omega}{2}\right)\right) .\end{aligned} \tag{7.78}$$

This yields the Gaussian distribution for the coordinate along the diagonal:

$$\langle x|\hat{\rho}|x\rangle = 2\sinh\left(\frac{\beta\omega}{2}\right)\left(\frac{\omega}{2\pi\sinh(\beta\omega)}\right)^{1/2} \exp\left(-\omega x^2\tanh\frac{\beta\omega}{2}\right) . \tag{7.79}$$

In the high-temperature limit we obtain the classical distribution function

$$\langle x|\hat{\rho}|x\rangle \to \left(\frac{\beta\omega^2}{2\pi}\right)^{1/2} \exp\left(-\frac{\beta\omega^2}{2}x^2\right) . \tag{7.80}$$

An interesting quantity is also the quantum coherence given by the off-diagonal elements of the density matrix. Assuming $x \approx x'$ but $x \neq x'$, we denote $\Delta x = x - x'$, and in the high-temperature limit we have

$$\langle x'|\hat{\rho}|x\rangle \propto \exp\left(-\frac{1}{2\beta\hbar^2}\Delta x^2\right) . \tag{7.81}$$

The coherence thus decays as the coordinates separate, and this dependence becomes sharper as the temperature increases. Thus, in the high temperature of the classical regime, the density matrix becomes the classical probability distribution obeying Gaussian statistics.

In the case of N independent oscillators, we will have each oscillator in a specific state n_α, so the density matrix is

$$\rho_{\{nn\}_\alpha} = \frac{\exp(-N\beta\hbar\omega/2)}{Z} \exp\left(-\sum_{\alpha=1}^{N} \beta\hbar\omega n_\alpha\right) , \tag{7.82}$$

where

$$Z = \left[2\sinh\left(\frac{\beta\hbar\omega}{2}\right)\right]^{-N} . \tag{7.83}$$

Thus, the density matrix completely factorizes into products of individual density matrices of different oscillators:

$$\rho_{\{nn\}_\alpha} = \prod_{\alpha=1}^{N} 2\sinh\left(\frac{\beta\hbar\omega}{2}\right) \exp\left(-\beta\hbar\omega\left(n_\alpha + \frac{1}{2}\right)\right) . \tag{7.84}$$

7.7 Symmetry Properties of Many-Particle Wavefunctions

When we describe particles in quantum mechanics we formulate the problem using operators instead of observables. The convenient operators are usually the coordinate and momentum operators. The calculated wavefunction contains so many spatial coordinates as there are particles; however, these coordinates should not be understood as the coordinates of particles – these are only wavefunction parameters. The quantum particles are indistinguishable. The wavefunction is thus a multidimensional *field* of the quantum amplitude in real space. We often refer to the charge cloud or density.

The wavefunctions possess some specific symmetry with respect to particle permutation. Consider the many-particle wavefunction $\Psi(r_1 r_2 \ldots r_N)$. Let us assume there exists an operator which interchanges coordinates i and j:

$$\hat{P}_{ij}\Psi(r_1 \ldots r_i \ldots r_j \ldots r_N) = s\Psi(r_1 \ldots r_j \ldots r_i \ldots r_N) , \tag{7.85}$$

where s is some constant. As the Hamiltonian must be invariant to a change of the parameter order (interchange of identical particles), the wavefunction of the Hamiltonian is also the eigenfunction of the permutation operator. Note that application of the operator $\hat{P}_{ij}$ twice on the wavefunction brings about the same original wavefunction. From (7.85) then it follows that $s^2 = 1$, or $s = \pm 1$. We thus find that there may be two types of particles: particles characterized by a symmetric wavefunction, where $s = 1$, and particles with $s = -1$, which have an antisymmetric wavefunction.

Consider noninteracting quantum particles. The total Hamiltonian is given by

$$\hat{H}(r_1 r_2 \ldots p_1 p_2 \ldots) = \sum_i \hat{h}(r_i p_i) , \tag{7.86}$$

where $\hat{h}(r_i p_i)$ is the Hamiltonian of the ith system. Let us assume that we can solve the eigenvalue problem for a single system, that is,

$$\hat{h}(r_i p_i)\phi_k(r_i) = \varepsilon_k \phi_k(r_i) . \tag{7.87}$$

The total wavefunction can then be taken as the product of all wavefunctions of individual systems:

$$\Psi_{k_1 k_2 \ldots k_N}(r_1 r_2 \ldots r_N) = \Pi_i^N \phi_{k_i}(r_i) \tag{7.88}$$

with energy

$$E = \sum_i^N \varepsilon_{k_i} . \tag{7.89}$$

However, this product state is the eigenstate of the Hamiltonian, but it does not have a proper symmetry. Therefore, it has to be additionally symmetrized. We can create the symmetric and antisymmetric wavefunctions as a superposition of all possible permutations:

$$\Psi^{(S)}_{k_1 k_2 \ldots k_N}(r_1 r_2 \ldots r_N) = K_S \begin{vmatrix} \phi_{k_1}(r_1) & \phi_{k_2}(r_1) & \ldots & \phi_{k_N}(r_1) \\ \phi_{k_1}(r_2) & \phi_{k_2}(r_2) & \ldots & \phi_{k_N}(r_2) \\ \vdots & \vdots & \ddots & \vdots \\ \phi_{k_1}(r_N) & \phi_{k_2}(r_N) & \ldots & \phi_{k_N}(r_N) \end{vmatrix}_+ , \tag{7.90}$$

$$\Psi^{(A)}_{k_1 k_2 \ldots k_N}(r_1 r_2 \ldots r_N) = K_A \begin{vmatrix} \phi_{k_1}(r_1) & \phi_{k_2}(r_1) & \ldots & \phi_{k_N}(r_1) \\ \phi_{k_1}(r_2) & \phi_{k_2}(r_2) & \ldots & \phi_{k_N}(r_2) \\ \vdots & \vdots & \ddots & \vdots \\ \phi_{k_1}(r_N) & \phi_{k_2}(r_N) & \ldots & \phi_{k_N}(r_N) \end{vmatrix} . \tag{7.91}$$

Here K_S and K_A are the normalization constants, $|\ldots|_+$ denotes the *permanent*: the sum of all possible products as of the *determinant* where the signs of all terms are "+."

For the antisymmetric wavefunction $K_A = (N!)^{-1/2}$ and the sum of the products is a determinant, denoted the *Slater determinant*. This form of the determinant guarantees that the two particles cannot have the same quantum numbers: the determinant is equal to zero if any two coordinates, as quantum numbers, become identical. This demand is expressed by the *Pauli exclusion principle*, which is valid for half-spin quantum particles, called *fermions*. The symmetric wavefunction does not have this requirement – many particles can have the same state and the normalization becomes

$$K_S = \frac{1}{\sqrt{N! n_1! n_2! \ldots}} , \tag{7.92}$$

where n_i is the number of particles in state i. Particles obeying this type of symmetry have integer spin and are called *bosons*.

It is constructive to switch to the occupation number representation of one-particle states. Let the one-particle states be labeled by $|k\rangle$ so that

$$\hat{h}|k\rangle = \varepsilon_k |k\rangle . \tag{7.93}$$

The occupation numbers of these states are n_k. They obey the restriction

$$\sum_k n_k = N . \tag{7.94}$$

We can now denote the many-particle state as $|n_1, n_2, \ldots\rangle$, and it is an eigenfunction of the total Hamiltonian

$$\hat{H}|n_1, n_2, \ldots\rangle = \left[\sum_k \varepsilon_k n_k\right] |n_1, n_2, \ldots\rangle \tag{7.95}$$

as well as of the number operator

$$\hat{n}_k |n_1, n_2, \ldots\rangle = n_k |n_1, n_2, \ldots\rangle . \tag{7.96}$$

In this representation the fermions and bosons are identified by a set of possible occupation numbers. In the bosonic case the occupations $n_1, n_2, \ldots$ in a specific state $|n_1, n_2, \ldots\rangle$ can be arbitrary positive integers (of course (7.94) has to be satisfied). In the fermionic case, an additional restriction applies, so each n_k must be 0 or 1.

7.7.1
Bose–Einstein Statistics

Let us assume we have a set of bosons. The canonical density operator is diagonal in the occupation number representation. We can write the diagonal term as

$$(\rho_C)_{n_1 n_2 \ldots} = \frac{1}{Z_C} \exp\left(-\beta \sum_k n_k \varepsilon_k\right) , \tag{7.97}$$

and the partition function is

$$Z_C = \sum_{\{n_k\}} \exp\left(-\beta \sum_k n_k \varepsilon_k\right) . \tag{7.98}$$

Here the sum denotes the summation over all possible configurations of the occupation numbers with the condition of (7.94). In the same way the grand canonical ensemble is given by

$$(\rho_G)_{n_1 n_2 \ldots} = \frac{1}{Z_C} \exp\left(-\beta \sum_k n_k (\varepsilon_k - \mu)\right) \tag{7.99}$$

with the grand partition function

$$Z_G = \sum_{\{n_k\}} \exp\left(-\beta \sum_k n_k(\varepsilon_k - \mu)\right) . \tag{7.100}$$

This expression can be rewritten as

$$Z_G = \prod_{k=1}^{\infty} \sum_{n_k=0}^{\infty} \exp\left(-n_k \beta(\varepsilon_k - \mu)\right) , \tag{7.101}$$

while the resulting sum of exponents is

$$\sum_{n_k=0}^{\infty} \exp(-n_k \beta(\varepsilon_k - \mu)) = \frac{1}{1 - \exp(-\beta(\varepsilon_k - \mu))} . \tag{7.102}$$

The thermodynamic quantities can now be conveniently calculated from the partition function. For instance, the mean energy (for the derivation, see Appendix A.4)

$$U \equiv \langle \hat{H} \rangle = -\frac{\partial}{\partial \beta} \ln Z_G = \sum_k \frac{\varepsilon_k}{\exp(\beta(\varepsilon_k - \mu)) - 1} , \tag{7.103}$$

while the mean number of particles

$$N \equiv \langle \hat{N} \rangle = kT \frac{\partial}{\partial \mu} \ln Z_G = \sum_k \frac{1}{\exp(\beta(\varepsilon_k - \mu)) - 1} . \tag{7.104}$$

Considering (7.94), we have for the mean occupation of state k

$$n_k = \frac{1}{\exp(\beta(\varepsilon_k - \mu)) - 1} , \tag{7.105}$$

which is the famous Bose–Einstein distribution. This distribution is plotted in Figure 7.2. As the temperature becomes lower, the real distribution becomes sharper and all bosons occupy the lowest-energy state. This transition is denoted Bose condensation. At low occupation numbers the distribution approaches the exponential Boltzmann distribution.

Note that quanta of electromagnetic radiation, the photons, do not interact with each other and so their chemical potential $\mu = 0$. The photons have integer spin, and thus they obey Bose–Einstein statistics. Each of the photons has energy $\hbar\omega$ at frequency ω. The mean number of photons at this frequency is then given by

$$n(\omega) = \frac{1}{\exp(\beta\omega) - 1} . \tag{7.106}$$

As $k_B T$ at 300 K is approximately 25 meV, the number of photons in the microwave region ($\hbar\omega \ll k_B T$) is

$$n(\omega, T)|_{\text{long wavelength}} \propto \frac{T}{\omega} , \tag{7.107}$$

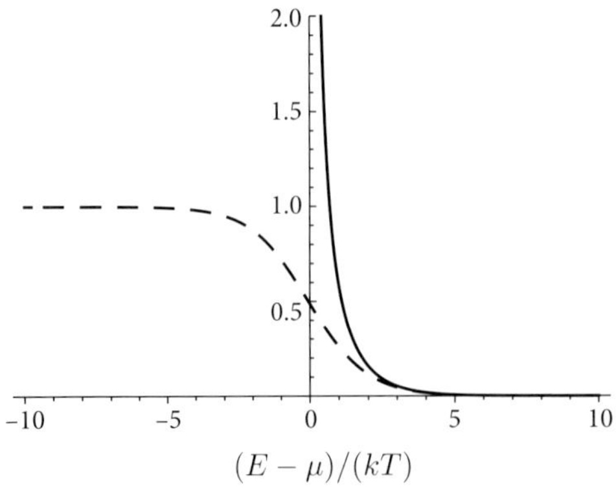

Figure 7.2 The distribution of the Bose-type occupation probability is represented by the solid line and that corresponding to the Fermi-type probability is represented by the dashed line.

while in the visible and UV regions $\hbar\omega \gg k_B T$, and these modes are mostly in the vacuum state, giving

$$n(\omega, T)|_{\text{short wavelength}} = \exp\left(-\frac{\omega}{T}\right) \to 0\,. \tag{7.108}$$

7.7.2
Pauli–Dirac Statistics

For fermions we can write the same set of equations as for bosons. Consider the grand canonical ensemble. Its density matrix is diagonal and the diagonal values are

$$(\rho_G)_{n_1 n_2 \ldots} = \frac{1}{Z_G} \exp\left(-\beta \sum_k n_k(\varepsilon_k - \mu)\right), \tag{7.109}$$

with the grand partition function

$$Z_G = \sum_{\{n_k\}} \exp\left(-\beta \sum_k n_k(\varepsilon_k - \mu)\right). \tag{7.110}$$

However, now different from bosons, the numbers n_k are either 0 or 1. The partition function thus takes a simple form:

$$Z_G = \prod_k (1 + \exp(-\beta(\varepsilon_k - \mu)))\,. \tag{7.111}$$

We follow the same procedure as for bosons and calculate the mean energy

$$U = \sum_k \frac{\varepsilon_k}{\exp(\beta(\varepsilon_k - \mu)) + 1}\,. \tag{7.112}$$

The mean number of particles is given by

$$N = \sum_k \frac{1}{\exp(\beta(\varepsilon_k - \mu)) + 1} . \tag{7.113}$$

This gives the mean occupation of state k:

$$n_k = \frac{1}{\exp(\beta(\varepsilon_k - \mu)) + 1} , \tag{7.114}$$

which is the famous Fermi–Dirac distribution.

This distribution (together with the Bose distribution) is shown in Figure 7.2. In the case of fermions the distribution becomes steplike as the temperature decreases. This is the state where all fermions fill all available states up to the Fermi level $E = \mu$. This is because no two fermions can occupy the same state.

7.8 Dynamic Properties of an Oscillator at Equilibrium Temperature

The canonical density matrix is a static property of the quantum system which can be used to describe the thermodynamic properties of classical and quantum systems. However, in Section 3.8 we encountered the quantity of the stochastic force, which was not identified explicitly. Statistical physics also allows an explicit identification of the dynamic stochastic parameters, which at equilibrium constant temperature can be treated as stochastic coordinates or stochastic forces acting on systems.

For this purpose we consider the harmonic oscillator once again. The quantum mechanical properties of a harmonic oscillator were considered in Section 4.6.1 and the Hamiltonian can be given by:

$$\hat{H} = \frac{\hat{p}^2}{2m} + \frac{mw^2}{2}\hat{x}^2 \tag{7.115}$$

Here $\hat{p}$ is the momentum operator, m is its mass, w is the frequency (we use w instead of ω as to be distinct from the frequency parameter of the Fourier transform), and $\hat{x}$ is its coordinate operator. It is convenient to reformulate the problem in terms of creation/annihilation operators by using (4.243) and (4.244) in the forms

$$\hat{x} = l\frac{1}{\sqrt{2}}\left(\hat{a}^\dagger + \hat{a}\right) \tag{7.116}$$

and

$$\hat{p} = \frac{i\hbar}{l}\frac{1}{\sqrt{2}}\left(\hat{a}^\dagger - \hat{a}\right) , \tag{7.117}$$

where we introduced the length scale

$$l = \sqrt{\frac{\hbar}{mw}} \,. \tag{7.118}$$

Having coordinate–momentum commutator identities, (4.4), we get the boson commutation relation (4.250) and the Hamiltonian in the form defined by (4.251). Now instead of coordinate and momentum dimensional operators, we have a dimensionless representation in terms of the creation and annihilation operators.

The equilibrium canonical density matrix of the oscillator at constant temperature has been calculated exactly, (7.75). This diagonal form of the density operator constitutes the complete statistical mixture of quantum states as all coherences are completely zero. We now turn to the dynamic properties of the oscillator. Consider the coordinate operator $\hat{x}$. The characteristics of the coordinate at equilibrium are the mean value and its variance. The mean value is

$$\bar{x} = \mathrm{tr}(\hat{x}\,\hat{W}) = \sum_{nn'} \hat{x}_{nn'} \left(\hat{W}_{\mathrm{eq}}\right)_{n'n} = 0 \tag{7.119}$$

as should be expected since in (7.116) $\hat{x}$ is given as a single power of creation and annihilation operators, while the density matrix is diagonal. Now the variance can also be calculated:

$$\sigma^2 = \bar{x^2} - \bar{x}^2 = \frac{l^2}{2} \coth\left(\frac{\beta\hbar w}{2}\right) \,. \tag{7.120}$$

We thus get a strong dependence on temperature, $\beta = (k_{\mathrm{B}} T)^{-1}$. This σ dependence on the temperature is shown in Figure 7.3. We find that at low temperature, when $\beta\hbar w \gg 1$, only the lowest-energy state of the oscillator is occupied and the coordinate is dispersed in a finite region ($\sigma \to l^2/2$). However, at high temperature, $\beta\hbar w \ll 1$, higher states of the oscillator are occupied and $\sigma^2 \to l^2 k_{\mathrm{B}} T/(\hbar w)$.

This result is intimately related to the probability density for finding the oscillator at coordinate x. Quantum mechanics dictates that this is given by the square of the

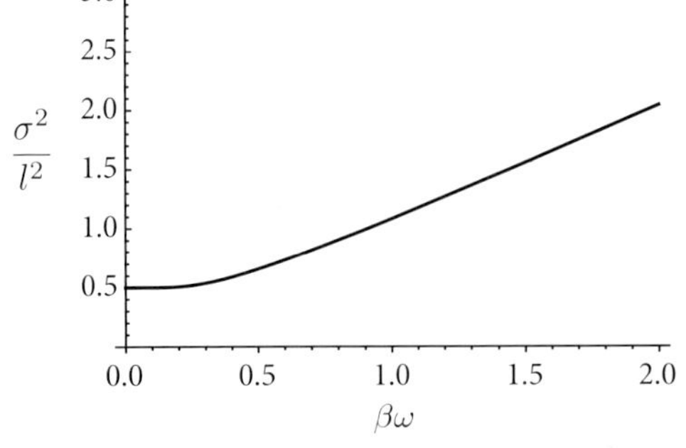

Figure 7.3 Variance of the density distribution as a function of temperature as defined in (7.120).

amplitude of the wavefunction, $|\varphi_n(x)|^2$, when the oscillator is in state $|n\rangle$. Having the equilibrium density matrix in (7.75), we can perform the sum over all states. This result is essentially the coordinate representation of the density matrix given by (7.78). For the probability density we have a Gaussian shape for the coordinate along the diagonal:

$$\langle x|\hat{\rho}|x\rangle = \frac{1}{l\sqrt{\pi}\coth(\beta w/2)} \exp\left(-\frac{(x/l)^2}{\coth(\beta w/2)}\right) . \tag{7.121}$$

The harmonic oscillator in thermal equilibrium is thus a Gaussian system. At low temperature the distribution does not contract to the Dirac delta distribution,

$$\langle x|\hat{\rho}|x\rangle \to \frac{1}{l\sqrt{\pi}} \exp\left(-\left(\frac{x}{l}\right)^2\right) , \tag{7.122}$$

because of quantum uncertainty.

An equivalent distribution can be calculated for a classical harmonic oscillator using the classical canonical statistical probability distribution. The potential energy of the classical harmonic oscillator is given by the parabolic function of the coordinate, $V(x) = m w^2 x^2/2$, and the probability of reaching a certain energy at finite temperature is governed by the Boltzmann law. The coordinate distribution function is therefore

$$\rho^{(\mathrm{cl})}(x) \propto \exp\left(-x^2\hbar\frac{w}{2k_\mathrm{B}T}\right) , \tag{7.123}$$

which is equivalent to (7.121) when $k_\mathrm{B}T \gg \hbar w$. Thus, the quantum harmonic oscillator behaves classically at high temperature, while its quantum properties emerge only at low temperature. This transition is defined by comparing the thermal energy, $k_\mathrm{B}T$, with the oscillator quantum level splitting, $\hbar w$.

As the coordinate of an oscillator (classical or quantum) is defined only through the distribution function, we interpret this as the result of stochastic fluctuations. The fluctuations have zero mean and nonzero variance. The other, dynamic property of the equilibrium oscillator, which is relevant to further chapters, is the coordinate correlation function. The equilibrium oscillator is thus a dynamic system executing Brownian-like motion. These fluctuations correspond to the thermal noise only at high temperature. When the temperature is low, $k_\mathrm{B}T \ll \hbar w$, we found that the coordinate distribution remained broad in the quantum case with temperature-independent variance. In this regime we should attribute fluctuations to the quantum vacuum or noise.

The correlation function will be defined in the Heisenberg representation, where

$$\hat{x}(t) = \exp\left(\frac{\mathrm{i}}{\hbar}\hat{H}t\right)\hat{x}\exp\left(-\frac{\mathrm{i}}{\hbar}\hat{H}t\right) , \tag{7.124}$$

and similarly

$$\hat{p}(t) = \exp\left(\frac{\mathrm{i}}{\hbar}\hat{H}t\right)\hat{p}\exp\left(-\frac{\mathrm{i}}{\hbar}\hat{H}t\right) . \tag{7.125}$$

The coordinate correlation function is given by

$$C_{xx}(t) \equiv \langle x(t)x(0)\rangle \equiv \mathrm{tr_B}\left(\hat{x}(t)\hat{x}(0)\, W_{\mathrm{eq}}\right) .$$

Using (7.116) and noting that the equilibrium density matrix is diagonal, we get

$$C_{xx}(t) \equiv \frac{\hbar}{2mw}\mathrm{tr_B}\left(\hat{a}^\dagger(t)\hat{a}(0)\, W_{\mathrm{eq}} + \hat{a}(t)\hat{a}^\dagger(0)\, W_{\mathrm{eq}}\right) , \tag{7.126}$$

where we defined the time-dependent annihilation operator

$$\hat{a}(t) = \exp\left(\frac{\mathrm{i}}{\hbar}\hat{H}t\right)\hat{a}\exp\left(-\frac{\mathrm{i}}{\hbar}\hat{H}t\right) \tag{7.127}$$

and a similar expression for the time-dependent creation operator.

According to the definition of the correlation function, let us define

$$\begin{aligned} C_{a^\dagger a}(t) &\equiv \mathrm{tr_B}\left(\hat{a}^\dagger(t)\hat{a}(0)\, W_{\mathrm{eq}}\right) \\ &= \exp(\mathrm{i}wt)[1-\exp(-\beta w)]\sum_n n\exp(-\beta\hbar w n) \end{aligned} \tag{7.128}$$

and

$$C_{aa^\dagger}(t) \equiv \mathrm{tr_B}\left(\hat{a}(t)\hat{a}^\dagger(0)\, W_{\mathrm{eq}}\right) = C^*_{a^\dagger a}(t) + \exp(-\mathrm{i}wt) . \tag{7.129}$$

After performing the infinite summation of exponential functions, we get

$$C_{a^\dagger a}(t) = \exp(\mathrm{i}wt)\frac{\exp(-\beta\hbar w)}{1-\exp(-\beta\hbar w)} , \tag{7.130}$$

thus giving the following expression for the coordinate–coordinate correlation function:

$$C_{xx}(t) = \frac{\hbar}{2mw}\left[\cos(wt)\coth\left(\frac{\beta\hbar w}{2}\right) - \mathrm{i}\sin(wt)\right] . \tag{7.131}$$

This function describes the very basic fluctuation properties of a harmonic oscillator at temperature $T \propto \beta^{-1}$.

Note that the correlation function is a complex function with properties

$$C_{xx}(t) = C^*_{xx}(-t) . \tag{7.132}$$

The dependence of the correlation function on temperature is not trivial: the imaginary part of the correlation function does not depend on temperature, while the real part is highly nonlinear. In the low-temperature limit $\beta\hbar w \gg 1$, only the lowest state of the oscillator is occupied and both real and imaginary parts have the same magnitudes, so

$$C^{(q)}_{xx}(t) = \frac{\hbar}{2mw}\mathrm{e}^{-\mathrm{i}wt} . \tag{7.133}$$

This is the regime of quantum noise. In the high-temperature limit $\beta\hbar w \ll 1$, the real part becomes proportional to temperature and the temperature-independent imaginary part can be disregarded. We then get the classical result:

$$C_{xx}^{(c)}(t) = \frac{k_B T}{m w^2} \cos(w t) . \tag{7.134}$$

The correlation function as it is for the single oscillator is thus a continuous periodic function as should be expected for a harmonic oscillator. At zero time in the quantum case $C_{xx}(0) = \hbar/(2mw)\coth(\beta\hbar w/2)$ we get the finite variance of the coordinate, while at later time we have the complete correlation of the harmonic motion.

Let us take the Fourier transform of the correlation function as described in Appendix A.5. We get

$$C_{xx}(\omega) \equiv \int \mathrm{d}t e^{\mathrm{i}\omega t} C_{xx}(t) = \frac{\hbar\pi}{2mw} \left\{ \coth\left(\frac{\beta\hbar w}{2}\right) \left[\delta(\omega + w) + \delta(\omega - w)\right] \right.$$
$$\left. + \left[\delta(\omega - w) - \delta(\omega + w)\right] \right\} . \tag{7.135}$$

The Fourier transform is thus a real function and it has two components: the even part,

$$C'_{xx}(\omega) = \frac{\hbar\pi}{2mw} \coth\left(\frac{\beta w}{2}\right) \left[\delta(\omega + w) + \delta(\omega - w)\right] , \tag{7.136}$$

and the odd part,

$$C''_{xx}(\omega) = \frac{\hbar\pi}{2mw} \left[\delta(\omega + w) - \delta(\omega - w)\right] . \tag{7.137}$$

As we find, the odd part does not depend on temperature and is thus the pure characterization of the oscillator. We can also form the general relation

$$C'_{xx}(\omega) = C''_{xx}(\omega) \coth\left(\frac{\beta\hbar\omega}{2}\right) , \tag{7.138}$$

and we can write the full correlation function as

$$C_{xx}(\omega) = C''_{xx}(\omega) \left[\coth\left(\frac{\beta\hbar\omega}{2}\right) + 1\right] \tag{7.139}$$

or

$$C_{xx}(t) = \int \frac{\mathrm{d}\omega}{2\pi} C''_{xx}(\omega) \left[\cos(\omega t) \coth\left(\frac{\beta\hbar\omega}{2}\right) - \mathrm{i}\sin(\omega t)\right] . \tag{7.140}$$

We thus see that in principle all properties of the harmonic oscillator are contained in a single time-independent entity $C''_{xx}(\omega)$. We can denote the function $C''_{xx}(\omega)$ in (7.137) as the *spectral density* of the harmonic system, which in our specific case is just a single delta peak. However, (7.138)–(7.140) imply general relations where the spectral density can be taken as an arbitrary real-value continuous function satisfying

$$C''_{xx}(\omega) = -C''_{xx}(-\omega) \tag{7.141}$$

and thus

$$C''(\omega = 0) \equiv 0 \,. \tag{7.142}$$

As we find later, the same type of relations are obtained for a collection of harmonic oscillators, but the spectral density is an infinite sum of delta peaks, and thus simulates the continuous function.

7.9
Simulation of Stochastic Noise from a Known Correlation Function

In Section 3.8 we assumed that a specific stochastic process such as Brownian motion generates the stochastic trajectory and that the trajectory can then be characterized by a correlation function. In the previous section we developed a very delicate procedure for how the statistical physics of a harmonic oscillator is related to the stochastic dynamics. If we want to propagate the stochastic Langevin-like equation, it is necessary to have the stochastic trajectory of a specific kind, characterized by a specific correlation function. In this section we describe the procedure to obtain the stochastic noise trajectory with the requested statistical properties.

Consider a stochastic process $z(t)$. We assume that its Fourier transform

$$Z(\omega) = \int \mathrm{d}t \exp(\mathrm{i}\omega t) z(t) \tag{7.143}$$

is integratable, so $Z(\omega)$ is also a stochastic function of frequency (in practice finite sets of data are used, and thus the integration can be always performed). The inverse transform is then given by

$$z(t) = \int \frac{\mathrm{d}\omega}{2\pi} \exp(-\mathrm{i}\omega t) Z(\omega) \,. \tag{7.144}$$

When the process is ergodic, the correlation function that appears in the stochastic Schrödinger equation apart from the normalization factor can be calculated as the following integral:

$$C(t) \equiv \langle z^*(t) z(0) \rangle = \int \mathrm{d}\tau z^*(t+\tau) z(\tau) \,. \tag{7.145}$$

Now we insert the Fourier transforms of the stochastic processes and get

$$C(t) = \int \frac{\mathrm{d}\omega}{2\pi} \exp(\mathrm{i}\omega t) Z^*(\omega) Z(\omega) \,. \tag{7.146}$$

Finally, the Fourier transform of the correlation function

$$C(\omega) = \int \mathrm{d}t \exp(\mathrm{i}\omega t) C(t) \tag{7.147}$$

leads to

$$C(\omega) = |Z(\omega)|^2 . \tag{7.148}$$

This relation is the form in the Wiener–Khinchin theorem [40]. It also states that the frequency image of the autocorrelation function is always a real function.

This theorem can be employed to generate the stochastic trajectory itself. Let us define the amplitude and the phase of the stochastic process in the Fourier space:

$$Z(\omega) = \zeta(\omega) \exp(\mathrm{i}\phi(\omega)) . \tag{7.149}$$

Both ζ and ϕ are real functions of frequency. From the Wiener–Khinchin theorem it follows that

$$\zeta(\omega) = \sqrt{C(\omega)} . \tag{7.150}$$

$\zeta(\omega)$ is thus fully defined by the correlation function of the process and therefore is not a stochastic function. The phase of the process is the stochastic function: it remains undefined by the correlation function and it can be chosen as some known stochastic process. From the Wiener–Khinchin theorem it does not affect the correlation and in practice it can be taken as a zero-correlated white noise, for example, it can be taken as a linearly distributed random number for each value of the frequency. However, in order to have the proper case that the imaginary-valued fluctuations become negligible at high temperature when the correlation function becomes symmetric, we should have $\phi(\omega) = -\phi(-\omega)$. The properly correlated stochastic process is then generated by

$$z(t) = \int \frac{\mathrm{d}\omega}{2\pi} \exp(-\mathrm{i}\omega t + \mathrm{i}\phi(\omega)) \sqrt{C(\omega)} . \tag{7.151}$$

This type of the random process is defined by the correlation function: if $C(\omega)$ is an even function of the frequency, the real $z(t)$ process is generated. However, that is not necessary. As is shown in the previous section, the realistic quantum thermal harmonic oscillator is described by a correlation function which is neither odd nor even. The stochastic process representing such behavior is essentially a complex one. However, the complex stochastic trajectory is still fully defined using (7.151). Substituting (7.139) into (7.151), we get the realistic stochastic trajectories. For the model spectral densities

$$C_1''(\omega) = \frac{\omega}{\omega^2 + \gamma^2} \tag{7.152}$$

and

$$C_2''(\omega) = \frac{1}{(\omega - 1)^2 + \gamma^2} - \frac{1}{(\omega + 1)^2 + \gamma^2} \tag{7.153}$$

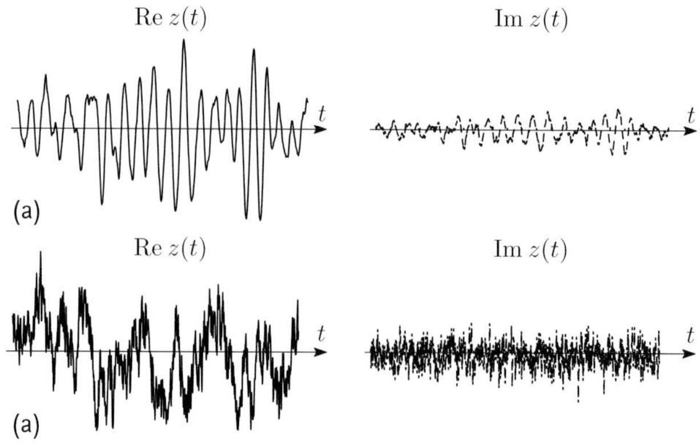

Figure 7.4 Modeled stochastic trajectories based on spectral densities (7.153) (a) and (7.152) (b) at temperature $kT = \sqrt{10}\gamma$. The real part (solid lines) and the imaginary part (dashed lines) are shown separately.

we get the fluctuating trajectories as shown in Figure 7.4. We thus find that the imaginary part in both cases is much smaller than the real part, which signifies finite temperature, but quantum effects may still be significant. Second, the spectral density of (7.152) describes the stochastic process without clearly visible oscillations, which is similar to the overdamped Brownian oscillator (Wiener process). Opposite to that, the spectral density, (7.153), which is a peaked function, shows perturbed oscillations, and thus signifies the presence of oscillatory dynamics. In both cases, however, the dynamics is chaotic.

8
An Oscillator Coupled to a Harmonic Bath

In the previous chapter we reviewed the basic properties of three types of systems and applied the results to the harmonic oscillator. The harmonic oscillator systems are the basic models used for a thermal bath – a reservoir at constant temperature. In this chapter we will describe the nonequilibrium properties of the oscillator systems.

8.1
Dissipative Oscillator

Let us now describe a general form of quantum oscillator, characterized by coordinate $\hat{Q}$ and momentum $\hat{P}$. This oscillator is our system of interest. The potential energy function is $U(\hat{Q})$. Let us assume that $U(Q)$ has a minimum at $Q = 0$. This oscillator weakly interacts with the set of harmonic oscillators, which determine the thermal bath. This part of the system is not of interest – it is not observed – and we thus study the observable system as it interacts with the bath.

The Hamiltonian of such a system is given by

$$\hat{H} = \frac{\hat{P}^2}{2M} + U(\hat{Q}) + \sum_k \left(\frac{\hat{p}_k^2}{2m_k} + \frac{m_k \omega_k^2}{2} \hat{x}_k^2 \right) + \hat{H}_{\mathrm{SB}}(\hat{Q}, \{\hat{x}\}) \,. \tag{8.1}$$

Here the first two terms denote the kinetic and potential energies of the observable oscillator described by the momentum $\hat{P}$ and the coordinate $\hat{Q}$ operators. The following sum represents the set of harmonic oscillators composing the bath. Their momenta are $\hat{p}_k$ and the coordinates are $\hat{x}_k$. m_k and ω_k are the mass and frequency, respectively. The coupling of the system with the bath is assumed to be linear:

$$\hat{H}_{\mathrm{SB}}(\hat{Q}, \{\hat{x}\}) = -\sum_k c_k \hat{Q} \hat{x}_k \,. \tag{8.2}$$

The sign in front of the sum is taken as negative for convenience.

In terms of these definitions we get the quadratic form of the parameter x_k in the Hamiltonian. We can thus combine the system–bath coupling term with the

Molecular Excitation Dynamics and Relaxation, First Edition. L. Valkunas, D. Abramavicius, and T. Mančal.

potential energy term of the bath oscillator. Let us write

$$\frac{m_k \omega_k^2}{2} \hat{x}_k^2 - c_k \hat{Q} \hat{x}_k = \frac{m_k \omega_k^2}{2} \left(\hat{x}_k - \frac{c_k \hat{Q}}{m_k \omega_k^2} \right)^2 - \frac{c_k^2 \hat{Q}^2}{2 m_k \omega_k^2} . \tag{8.3}$$

The last term is the system operator and thus can be merged into the bath potential. As a result, we get

$$U(\hat{Q}) - \frac{c_k^2 \hat{Q}^2}{2 m_k \omega_k^2} = V(\hat{Q}) . \tag{8.4}$$

The coupling with the bath thus shifts the potential surface; however, as we have formulated the problem for an arbitrary potential, we can take $V(\hat{Q})$ as the starting potential for the problem. So for convenience the Hamiltonian is redefined by using

$$\hat{H} = \frac{\hat{P}^2}{2M} + V(\hat{Q}) + \sum_k \left[\frac{\hat{p}_k^2}{2 m_k} + \frac{m_k \omega_k^2}{2} \left(\hat{x}_k - \frac{c_k}{m_k \omega_k^2} \hat{Q} \right)^2 \right] . \tag{8.5}$$

This model is usually referred to as the Caldeira–Leggett model [41, 42].

This system now has one advantage compared with the original problem. The global minimum of the bath oscillators can be determined from the condition

$$\frac{\partial}{\partial x_k} H = 0 , \tag{8.6}$$

which gives $m_k \omega_k^2 x_k^{(0)} = 0$. So we have the potential minimum at zero and this does not depend on the coordinate of the oscillator, which was present in the original Hamiltonian. Thus, under equilibrium conditions if the coordinate of the observable oscillator is characterized by zero mean, the mean of the bath oscillators is then not affected.

8.2 Motion of the Classical Oscillator

In the case when all oscillators are classical, all operators in (8.5) are regular coordinates and momenta. The equations of motion follow from the Hamilton principle and give essentially the set of Newton equations:

$$M \ddot{Q} + \frac{\partial}{\partial Q} V(Q) + \sum_k \frac{c_k^2}{m_k \omega_k^2} Q = \sum_k c_k x_k , \tag{8.7}$$

$$m_k \ddot{x}_k + m_k \omega_k^2 x_k = c_k Q . \tag{8.8}$$

We thus get the set of coupled equations where on the left-hand side we have the independent systems and on the right-hand side we have the driving force coming from the interaction. We can solve this system starting with (8.8) and using the frequency-domain Green's function method (see Appendix A.7). The solution of the homogeneous (h) part defined by the kinetic equation

$$m_k \ddot{x}_k^{(h)} + m_k \omega_k^2 x^{(h)} = 0 \tag{8.9}$$

for initial conditions $x_k(0) = x_k^{(0)}$ and $p_k(0) = p_k^{(0)}$ is given by

$$x_k^{(h)}(t) = x_k^{(0)} \cos(\omega_k t) + \frac{p_k^{(0)}}{m_k \omega_k} \sin(\omega_k t) . \tag{8.10}$$

For the inhomogeneous (i) part we take the solution via the Green's function:

$$x_k^{(i)}(t) = c_k \int_0^{\infty} G_k(t-\tau) Q(\tau) \mathrm{d}\tau . \tag{8.11}$$

We have used that $G_k(t < 0) = 0$ and $Q(t \leq 0) = 0$. The latter condition effectively turns off the system–bath coupling for negative times. This may seem artificial; however, when the bath performs stochastic fluctuations as described in the previous chapter the effects of the initial condition die out in a finite time. Thus, the expression for the Green's function holds for times longer than the bath correlation time. The Green's function satisfies the equation

$$m_k \ddot{G}_k(t) + m_k \omega_k^2 G_k(t) = \delta(t) , \tag{8.12}$$

which can be solved by using the Fourier transformation (see Appendix A.5). The Green's function is given by

$$G_k(\omega) = \frac{1}{m_k} \lim_{\eta \to 0} \frac{1}{\omega^2 - \omega_k^2 + \mathrm{i}\omega\eta} \tag{8.13}$$

and in the time domain we have

$$G_k(t) = \frac{1}{m_k \omega_k} \theta(t) \sin(\omega_k t) . \tag{8.14}$$

For (8.8) in the frequency space we get a simple solution:

$$x_k(\omega) = c_k G_k(\omega) Q(\omega) . \tag{8.15}$$

The full solution for the bath oscillator coordinate is then obtained using the Cauchy contour integration technique, which gives

$$x_k(t) = x_k^{(0)} \cos(\omega_k t) + \frac{p_k^{(0)}}{m_k \omega_k} \sin(\omega_k t) + \frac{c_k}{m_k \omega_k} \int_0^t \sin(\omega_k (t-\tau)) Q(\tau) \mathrm{d}\tau . \tag{8.16}$$

This expression shows that in fact not only the bath affects the system, but also that the system drives the bath through the last term. We now insert (8.16) into (8.7) and change the time integral as

$$\int_0^t \sin(\omega_k(t-\tau))\,Q(\tau)\mathrm{d}\tau$$

$$= \frac{1}{\omega_k}\left[Q(t) - Q(0)\cos(\omega_k t) - \int_0^t \dot{Q}(\tau)\cos(\omega_k(t-\tau)\mathrm{d}\tau)\right]. \tag{8.17}$$

We then obtain the Langevin equation for an oscillator of the form

$$M\ddot{Q} + M\int_0^t \mathrm{d}\tau\gamma(t-\tau)\dot{Q}(\tau) + \frac{\partial}{\partial Q}V(Q) = \zeta(t) - M\gamma(t)Q(0)\,, \tag{8.18}$$

where we have the time-dependent friction function

$$\gamma(t) = \frac{1}{M}\sum_k \frac{c_k^2}{m_k\omega_k^2}\cos(\omega_k t) \tag{8.19}$$

and the fluctuating force

$$\zeta(t) = \sum_k c_k\left[x_k^{(0)}\cos(\omega_k t) + \frac{p_k^{(0)}}{m_k\omega_k}\sin(\omega_k t)\right]. \tag{8.20}$$

These are properties of the bath fluctuations. We see that there is an initial condition on the right-hand side of (8.18). However, it should be "forgotten" if $\gamma(t) \to 0$ for long times.

Considering that $x_k = 0$ and $Q = 0$ correspond to the global potential minimum, the equilibrium dynamics should be characterized by $\langle x_k\rangle = 0$ and $\langle Q\rangle = 0$. We can then assume that the bath in equilibrium at temperature T has the canonical distribution as described in Section 7.5. On that basis we can define the initial condition $x_k^{(0)}$ and $p_k^{(0)}$ of bath oscillators. The classical canonical density matrix is given by

$$W_\mathrm{B} = \frac{1}{Z^{(c)}}\exp\left(-\frac{1}{2k_\mathrm{B}T}\sum_k\left(\frac{p_k^2}{m_k} + m_k\omega_k^2x_k^2\right)\right). \tag{8.21}$$

Here $Z^{(c)}$ is the classical partition function. It is given by

$$Z^{(c)} = \prod_k \int \mathrm{d}x_k \int \mathrm{d}p_k \exp\left(-\frac{1}{2k_\mathrm{B}T}\sum_k\left(\frac{p_k^2}{m_k} + m_k\omega_k^2x_k^2\right)\right). \tag{8.22}$$

All bath oscillators are independent; thus, the partition function is a product of partition functions of separate oscillators. The integrals in the partition function are all Gaussian integrals and can be easily calculated. We then get

$$Z^{(c)} = (2\pi k_B T)^N \prod_k \frac{1}{\omega_k} \; . \tag{8.23}$$

The fluctuating force can now be described by its moments. The first moment – the average value – is equal to zero,

$$\langle \zeta(t) \rangle = 0 \; , \tag{8.24}$$

and its correlation function is given by

$$C_{\zeta\zeta}(t) \equiv \langle \zeta(t)\zeta(0) \rangle = k_B T \sum_k \frac{c_k^2}{m_k \omega_k^2} \cos(\omega_k t) \; . \tag{8.25}$$

Checking (8.19), we can write the classical fluctuation–dissipation theorem:

$$C_{\zeta\zeta}(t) = M k_B T \gamma(t) \; . \tag{8.26}$$

Comparing (8.25) with (7.134), we find that

$$C_{\zeta\zeta}(t) = \sum_k c_k^2 C_{xx}^{(c)}(t) \; . \tag{8.27}$$

Thus, as should be expected, the fluctuating force, which originates from classical oscillators, shows the same correlation function as the harmonic oscillator itself. The effect of fluctuations on the system is twofold. First, fluctuations are created due to $\zeta(t)$. The fluctuations constitute input of thermal energy into the system. Second, the bath introduces dissipation due to $\gamma(t)$. This effect causes energy output. When both effects are in place, the system can reach thermal equilibrium with its environment. We thus find that all properties of the bath contract into the single correlation function $C_{\zeta\zeta}(t)$.

The realistic surroundings of a system contain infinitely many degrees of freedom. Thus, the number of oscillators representing the bath should approach infinity. Let us consider the definition of the damping parameter $\gamma(t)$ as given by (8.19). We can write it in the form

$$\gamma(t) = \frac{1}{M} \int_0^\infty \frac{d\omega}{2\pi} J(\omega) \cos(\omega t) \; , \tag{8.28}$$

where

$$J(\omega) = \sum_k \frac{c_k^2}{m_k \omega_k^2} \delta(\omega - \omega_k) \tag{8.29}$$

is the classical spectral density of the bath. It describes how many oscillators are in a specific frequency interval. As the total number of oscillators increases, the number of oscillators in the frequency interval should increase, and we have to switch to a continuous density function.

This spectral density is a positively defined function. In a specific problem it can now be defined as a model function. The simplest model of fluctuations is

$$J(\omega) = 2\Gamma \ . \tag{8.30}$$

Then

$$\gamma(t) = \frac{\Gamma}{M}\delta(t) \ . \tag{8.31}$$

From the fluctuation–dissipation relation we can write the correlation function

$$C_{\zeta\zeta}(t) = \Gamma k_{\mathrm{B}} T \delta(t) \ , \tag{8.32}$$

and finally we have the Langevin equation:

$$\ddot{Q} + \Gamma \dot{Q}(\tau) + \frac{1}{M}\frac{\partial}{\partial Q}V(Q) = \frac{1}{M}\zeta(t) \ . \tag{8.33}$$

This is an equation of a damped harmonic oscillator, driven by a fluctuating force. The stochastic process $\zeta(t)$ can be generated as described in Section 7.9.

8.3 Quantum Bath

In the previous section we showed that all properties of the bath which affect the system are contained in a single fluctuation correlation function. We can now introduce the quantum bath using these insights.

The damping process of the Langevin equation thus does not vanish at zero temperature. However, since the coordinate is essentially a real parameter, we find inconsistency of the classical Langevin equation due to the imaginary part of the damping parameter. This shows that the classical Langevin equation cannot account for the quantum bath. The quantum Langevin equation is obtained using the Heisenberg equation of motion for the coordinate operator in the Heisenberg representation. We use

$$[\hat{x}^n, \hat{p}] = n i\hbar \hat{x}^{n-1} \tag{8.34}$$

and

$$[\hat{p}^n, \hat{x}] = -n i\hbar \hat{p}^{n-1} \tag{8.35}$$

and we define

$$[V(\hat{x}), \hat{p}] = i\hbar V'(\hat{x}) \ . \tag{8.36}$$

Then from the Heisenberg equation of motion

$$\frac{\mathrm{d}\hat{A}}{\mathrm{d}t} = \frac{\mathrm{i}}{\hbar}[\hat{H}, \hat{A}] \tag{8.37}$$

for the Caldeira–Leggett Hamiltonian, (8.5), we obtain the equations for operators

$$M\ddot{\hat{Q}} + V'(\hat{Q}) + \sum_k \frac{c_k^2}{m_k\omega_k^2}\hat{Q} = \sum_k c_k\hat{x}_k \,, \tag{8.38}$$

$$m_k\ddot{\hat{x}}_k + m_k\omega_k^2\hat{x}_k = c_k\hat{Q} \,. \tag{8.39}$$

These equations are equivalent to the set of (8.7) and (8.8) and, thus the solution is equivalent to (8.16), which gives the quantum Langevin equation:

$$M\ddot{\hat{Q}} + M\int_0^t \mathrm{d}\tau\gamma(t-\tau)\dot{\hat{Q}}(\tau) + V'(\hat{Q}) = \zeta(t) - M\gamma(t)\hat{Q}(0) \,. \tag{8.40}$$

Now the time-dependent friction is a number as in (8.19), but the fluctuating force has the form given in (8.20) and is an operator of the bath variables.

To describe the force we have to use the quantum statistical physics description given in Section 7.8, where the full correlation function was calculated. The full quantum correlation function of a harmonic oscillator is given by (7.131). Inserting it into (8.27), we have the quantum form of the correlation function:

$$C_{\zeta\zeta}(t) = \sum_k \frac{c_k^2\hbar}{2m_k\omega_k}\left[\cos(\omega_k t)\coth\left(\frac{\beta\hbar\omega_k}{2}\right) - \mathrm{i}\sin(\omega_k t)\right] . \tag{8.41}$$

It has a symmetric (real) part with respect to time and an antisymmetric (imaginary) part. The symmetric part

$$\frac{1}{2}\left(C_{\zeta\zeta}(t) + C_{\zeta\zeta}(-t)\right) = \sum_k \frac{c_k^2\hbar}{2m_k\omega_k}\cos(\omega_k t)\coth\left(\frac{\beta\omega_k}{2}\right) \tag{8.42}$$

depends on temperature, while the antisymmetric part, which reflects quantum fluctuations, is temperature independent:

$$\frac{1}{2}(C_{\zeta\zeta}(t) - C_{\zeta\zeta}(-t)) = -\mathrm{i}\sum_k \frac{c_k^2\hbar}{2m_k\omega_k}\sin(\omega_k t) \,. \tag{8.43}$$

Comparing these expressions with the damping function defined in (8.19), we find that the symmetric part has the same symmetry as the damping function, that is, $\gamma(-t) = \gamma(t)$. At high temperature these results should turn into the classical result given by

$$\frac{1}{2}(C_{\zeta\zeta}(t) + C_{\zeta\zeta}(-t))|_{\mathrm{HT}} \rightarrow k_{\mathrm{B}}T\gamma(t) \,. \tag{8.44}$$

As we showed in Section 7.8, the Fourier transform of the full quantum correlation function allows us to define the spectral density. In light of the present problem we have the odd part of the Fourier transform of the correlation function as

$$C''_{\xi\xi}(\omega) = \pi \sum_k \frac{c_k^2 \hbar}{2 m_k \omega_k} \left[\delta(\omega - \omega_k) - \delta(\omega + \omega_k)\right] . \tag{8.45}$$

This function can be used as a spectral density extended to negative frequencies. As the classical form of the spectral density in (8.29) is one sided, we have the relation

$$C''_{\xi\xi}(\omega) = \hbar\omega\pi J(\omega) \tag{8.46}$$

when $\omega > 0$, while at negative frequencies

$$C''_{\xi\xi}(\omega) = \hbar\omega\pi J(-\omega), \quad \omega < 0 . \tag{8.47}$$

All relaxation properties are now described by the spectral density. The force fluctuation correlation function, coming from the quantum bath, is given by (7.140). The symmetric part is then

$$\frac{1}{2}(C_{\xi\xi}(t) + C_{\xi\xi}(-t)) = \int \frac{d\omega}{2\pi} C''_{\xi\xi}(\omega) \cos(\omega t) \coth\left(\frac{\beta\omega}{2}\right) \tag{8.48}$$

and the classical fluctuation–dissipation theorem then turns into

$$\gamma(t) = \int \frac{d\omega}{2\pi} \frac{C''_{\xi\xi}(\omega)}{\hbar\omega} \cos(\omega t) . \tag{8.49}$$

That relation is one of the forms of the *quantum fluctuation–dissipation theorem.* The spectral density or the correlation function is thus the main property of the fluctuating environment. In practice the spectral density for the realistic environment (it contains infinite number of harmonic oscillators) is a continuous function and various models for spectral densities are used.

8.4 Quantum Harmonic Oscillator and the Bath: Density Matrix Description

In the previous section we obtained the quantum Langevin equation for the coordinate operator. In this section we consider the density matrix of a quantum harmonic oscillator coupled to a bath of harmonic oscillators. This consideration can be connected with the Langevin equation derived in the previous section in the high-temperature limit.

To start, for convenience let us assume we have the harmonic oscillator model and disregard the Q^2 term in the Caldeira–Leggett Hamiltonian. The Hamiltonian can then be written in the following form:

$$\hat{H} = \hbar\varepsilon\left(\hat{a}^\dagger\hat{a} + \frac{1}{2}\right) + \sum_k \left[\frac{\hat{p}_k^2}{2m_k} + \frac{m_k\omega_k^2}{2}\hat{x}_k^2 + f_k\hat{x}_k(\hat{a}^\dagger + \hat{a})\right] . \quad (8.50)$$

Here we replaced the operators $\hat{Q}$ and $\hat{P}$ by the creation and annihilation operators $\hat{a}^\dagger$ and $\hat{a}$. The density matrix of the total system plus bath supersystem in the basis of observable eigenstates of the system can be written as $\hat{W}_{a,b}$, where a and b represent the states of the system. $\hat{W}$ is still an operator for the bath coordinates. The Liouville equation for this density matrix gives

$$\begin{aligned}\frac{\mathrm{d}\hat{W}_{a,b}}{\mathrm{d}t} &= -\mathrm{i}(a-b)\varepsilon\,\hat{W}_{a,b} - \mathrm{i}\mathcal{L}_\mathrm{B}\,\hat{W}_{a,b} \\ &\quad -\mathrm{i}\sum_k f_k\hat{x}_k\left(\sqrt{a}\,\hat{W}_{a-1,b} + \sqrt{a+1}\,\hat{W}_{a+1,b}\right) \\ &\quad +\mathrm{i}\sum_k f_k\left(\sqrt{b+1}\,\hat{W}_{a,b+1} + \sqrt{b}\,\hat{W}_{a,b-1}\right)\hat{x}_k . \end{aligned} \quad (8.51)$$

Here we introduced the superoperator notation

$$\mathcal{L}_\mathrm{B}\,\hat{W} = \left[\sum_k\left(\frac{\hat{p}_k^2}{2m_k} + \frac{m_k\omega_k^2}{2}\hat{x}_k^2\right), \hat{W}\right] . \quad (8.52)$$

We can now reduce the time dependence by switching to the interaction picture:

$$\hat{W}_{a,b}(t) = \exp(-\mathrm{i}(a-b)\varepsilon t)\exp(-\mathrm{i}\hat{H}_\mathrm{B}t)\hat{w}_{a,b}(t)\exp(\mathrm{i}\hat{H}_\mathrm{B}t) , \quad (8.53)$$

which for the density matrix gives

$$\begin{aligned}\frac{\mathrm{d}\hat{w}_{a,b}(t)}{\mathrm{d}t} &= -\mathrm{i}\sum_k f_k\hat{x}_k(t)\sqrt{a}\exp(\mathrm{i}\varepsilon t)\hat{w}_{a-1,b}(t) \\ &\quad -\mathrm{i}\sum_k f_k\hat{x}_k(t)\sqrt{a+1}\exp(-\mathrm{i}\varepsilon t)\hat{w}_{a+1,b}(t) \\ &\quad +\mathrm{i}\sum_k f_k\sqrt{b+1}\exp(\mathrm{i}\varepsilon t)\hat{w}_{a,b+1}(t)\hat{x}_k(t) \\ &\quad +\mathrm{i}\sum_k f_k\sqrt{b}\exp(-\mathrm{i}\varepsilon t)\hat{w}_{a,b-1}(t)\hat{x}_k(t) , \end{aligned} \quad (8.54)$$

where we denote

$$\exp(\mathrm{i}\hat{H}_\mathrm{B}t)\hat{x}_k\exp(-\mathrm{i}\hat{H}_\mathrm{B}t) = \hat{x}_k(t) \quad (8.55)$$

as the interaction representation of the bath coordinate. It evolves with respect to the bath Hamiltonian. We can now formally integrate the resulting Liouville equation

$$\hat{w}_{a,b}(t) = \hat{w}_{a,b}(0)$$
$$-\mathrm{i}\sum_k f_k \int_0^t \mathrm{d}\tau\, \hat{x}_k(\tau)\sqrt{a}\exp(\mathrm{i}\varepsilon\tau)\hat{w}_{a-1,b}(\tau)$$
$$-\mathrm{i}\sum_k f_k \int_0^t \mathrm{d}\tau\, \hat{x}_k(\tau)\sqrt{a+1}\exp(-\mathrm{i}\varepsilon\tau)\hat{w}_{a+1,b}(\tau)$$
$$+\mathrm{i}\sum_k f_k \int_0^t \mathrm{d}\tau\, \sqrt{b+1}\exp(\mathrm{i}\varepsilon\tau)\hat{w}_{a,b+1}(\tau)\hat{x}_k(\tau)$$
$$+\mathrm{i}\sum_k f_k \int_0^t \mathrm{d}\tau\, \sqrt{b}\exp(-\mathrm{i}\varepsilon\tau)\hat{w}_{a,b-1}(\tau)\hat{x}_k(\tau)\,, \tag{8.56}$$

and insert the result into the right-hand side of (8.54). This yields a somewhat complicated expression which can be considerably simplified. We first disregard the first powers of $\hat{x}_k(t)$ as they perform stochastic fluctuations with zero mean, and we also disregard nonresonant terms of the form

$$\int_0^t \mathrm{d}\tau \exp(\mathrm{i}\varepsilon(t+\tau))A(t,\tau) \to 0 \tag{8.57}$$

as $A(t,\tau)$ are slowly varying functions. This approach is denoted the rotating-wave approximation. Finally we obtain

$$\frac{\mathrm{d}\hat{w}_{a,b}(t)}{\mathrm{d}t} = \sum_k f_k^2 \int_0^t \mathrm{d}\tau$$
$$-\,\mathrm{e}^{\mathrm{i}\varepsilon(t-\tau)}\left(a\hat{x}_k(t)\hat{x}_k(\tau)\hat{w}_{a,b}(\tau) - \sqrt{(a+1)(b+1)}\hat{x}_k(\tau)\hat{w}_{a+1,b+1}(\tau)\hat{x}_k(t)\right)$$
$$-\,\mathrm{e}^{-\mathrm{i}\varepsilon(t-\tau)}\left((a+1)\hat{x}_k(t)\hat{x}_k(\tau)\hat{w}_{a,b}(\tau) - \sqrt{ab}\hat{x}_k(\tau)\hat{w}_{a-1,b-1}(\tau)\hat{x}_k(t)\right)$$
$$+\,\mathrm{e}^{-\mathrm{i}\varepsilon(t-\tau)}\left(\sqrt{(a+1)(b+1)}\hat{x}_k(t)\hat{w}_{a+1,b+1}(\tau)\hat{x}_k(\tau) - b\hat{w}_{a,b}(\tau)\hat{x}_k(\tau)\hat{x}_k(t)\right)$$
$$+\,\mathrm{e}^{\mathrm{i}\varepsilon(t-\tau)}\left(\sqrt{ab}\hat{x}_k(t)\hat{w}_{a-1,b-1}(\tau)\hat{x}_k(\tau) - (b+1)\hat{w}_{a,b}(\tau)\hat{x}_k(\tau)\hat{x}_k(t)\right)\,. \tag{8.58}$$

This is now a convenient form to perform averaging over the bath variables. First, the averaging over the bath of the density matrix yields the reduced density matrix of the oscillator:

$$\rho_{ab}(t) = \mathrm{tr}_\mathrm{B}(\hat{w}_{a,b}(t))\,. \tag{8.59}$$

Next, we assume that the bath is in the canonical equilibrium form for all times, so

$$\hat{w}_{a,b}(t) = \rho_{ab}(t)\rho_{\mathrm{B}} \,, \tag{8.60}$$

and this allows us to introduce the bath correlation function

$$C(t) = \sum_k f_k^2 \langle \hat{x}_k(t)\hat{x}_k(\tau)\rho_{\mathrm{B}}\rangle \,. \tag{8.61}$$

For the reduced density matrix we then obtain a non-Markovian master equation which describes the relaxation properties:

$$\begin{aligned}
\frac{\mathrm{d}}{\mathrm{d}t}\rho_{ab}(t) = \int_0^t \mathrm{d}\tau \\
&- \mathrm{e}^{\mathrm{i}\varepsilon(t-\tau)}\left(aC(t-\tau)\rho_{ab}(\tau) - \sqrt{(a+1)(b+1)}\,C(t-\tau)\rho_{a+1,b+1}(\tau)\right) \\
&- \mathrm{e}^{-\mathrm{i}\varepsilon(t-\tau)}\left((a+1)C(t-\tau)\rho_{ab}(\tau) - \sqrt{ab}\,C(t-\tau)\rho_{a-1,b-1}(\tau)\right) \\
&+ \mathrm{e}^{-\mathrm{i}\varepsilon(t-\tau)}\left(\sqrt{(a+1)(b+1)}\,C^*(t-\tau)\rho_{a+1,b+1}(\tau) - bC^*(t-\tau)\rho_{ab}(\tau)\right) \\
&+ \mathrm{e}^{\mathrm{i}\varepsilon(t-\tau)}\left(\sqrt{ab}\,C^*(t-\tau)\rho_{a-1,b-1}(\tau) - (b+1)C^*(t-\tau)\rho_{ab}(\tau)\right) \,.
\end{aligned} \tag{8.62}$$

This may seem to be a very complicated expression, but in fact it has a simple form. Various terms on the right-hand side denote specific physical phenomena, such as the transfer of energy between levels and coherence transfer. It also shows that the populations are in fact independent of coherences. This is the consequence of constant splitting between the nearest-neighbor energy levels.

We obtained this expression specifically for the harmonic oscillator. In coming chapters we will develop a general formalism for an arbitrary quantum system with countable energy levels.

When the interaction between the system and bath oscillators is weak, the time evolution of the density matrix in the interaction representation is slow (it is constant as the interaction is switched off). In this limit we can use the Markov approximation. Let us consider the relaxation of populations. From (8.62) we have a master equation

$$\frac{\mathrm{d}}{\mathrm{d}t}\rho_{aa} = \left[-ak_1 + (a+1)\tilde{k}_1\right]\rho_{aa} + (a+1)k_1\rho_{a+1,a+1} + a\tilde{k}_1\rho_{a-1,a-1} \,, \tag{8.63}$$

where the population rates

$$k_1 = k_- + k_-^* \tag{8.64}$$

and

$$\tilde{k}_1 = k_+ + k_+^* \,. \tag{8.65}$$

Here

$$k_- = \int_0^t d\tau e^{i\varepsilon(t-\tau)} C(t-\tau) \tag{8.66}$$

and

$$k_+ = \int_0^t d\tau e^{-i\varepsilon(t-\tau)} C(t-\tau) \,. \tag{8.67}$$

The equation for coherences is also very simple:

$$\begin{aligned} \frac{d}{dt}\rho_{ab} &= -\left[ak_- + bk_-^* + (a+1)k_+ + (b+1)k_+^*\right]\rho_{ab} \\ &+ \sqrt{(a+1)(b+1)}(k_-^* + k_-)\rho_{a+1,b+1} \\ &+ \sqrt{ab}(k_+ + k_+^*)\rho_{a-1,b-1} \,. \end{aligned} \tag{8.68}$$

We thus find that the whole relaxation process is characterized by two complex quantities k_+ and k_-. The first row now describes the decay of the coherence, the second and the third rows characterize the transfer of coherences. These nontrivial effects are only possible for the harmonic oscillator within the range of the approximations used.

However, the bath properties can be reduced to a single quantity. Consider the integral of k_- and let us change the integration variables $t - \tau = t_1$ so that we get the integral

$$k_- = \int_0^t dt_1 e^{i\varepsilon t_1} C(t_1) \,. \tag{8.69}$$

As the correlation function decays with time and if we disregard the initial short-time region, we can take the upper limit t to infinity. It is convenient then to introduce an auxiliary spectral function

$$M(\omega) = \int_0^\infty dt e^{i\omega t} C(t) \tag{8.70}$$

and write the correlation function in terms of the spectral density using (7.140):

$$C(t) = \int \frac{d\omega}{2\pi} C''(\omega) \left[\cos(\omega t)\coth\left(\frac{\beta\omega}{2}\right) - i\sin(\omega t)\right] . \tag{8.71}$$

The real part of the function M

$$\text{Re}\, M(\varepsilon) = \frac{1}{2} C''(\varepsilon) \left[\coth\left(\frac{\beta\varepsilon}{2}\right) + 1\right] \tag{8.72}$$

provides the population relaxation rates. We then get in (8.64) and (8.65)

$$k_1 = 2C''(\varepsilon)\frac{e^{\frac{\beta\varepsilon}{2}}}{e^{\frac{\beta\varepsilon}{2}} - e^{-\frac{\beta\varepsilon}{2}}} \tag{8.73}$$

and

$$\tilde{k}_1 = 2C''(\varepsilon)\frac{e^{-\frac{\beta\varepsilon}{2}}}{e^{\frac{\beta\varepsilon}{2}} - e^{-\frac{\beta\varepsilon}{2}}} . \tag{8.74}$$

The spectral density of the bath is now the only necessary function of the bath, and characterizes all relaxation properties. From the rate expressions given by (8.73) and (8.74), we can write the relation

$$k_1 = e^{\beta\varepsilon}\tilde{k}_1 . \tag{8.75}$$

This condition thus guarantees the proper Boltzmann equilibrium distribution of the occupation numbers.

The imaginary part of the function $M(\omega)$ denotes the Lamb shifts of energy levels, which become important for density matrix coherences. However, this part of the function has no analytical expression for an arbitrary spectral density and has to be calculated numerically, unless the spectral density has a simple form.

8.5 Diagonal Fluctuations

The quantum harmonic oscillator described by Hamiltonian (8.50) shows population and coherence relaxation. However, the interaction term in the Hamiltonian describes only the off-diagonal fluctuations due to the first powers of $\hat{a}$ and $\hat{a}^\dagger$. In this section we briefly repeat the derivation, but instead of off-diagonal fluctuations, we consider the diagonal or energy fluctuations. The Hamiltonian is then given by

$$\hat{H} = \hbar\varepsilon\left(\hat{a}^\dagger\hat{a} + \frac{1}{2}\right) + \sum_k\left(\frac{\hat{p}_k^2}{2m_k} + \frac{m_k\omega_k^2}{2}\hat{x}_k^2 + f_k\hat{x}_k\hat{a}^\dagger\hat{a}\right) . \tag{8.76}$$

Following the same procedure, we obtain the Liouville equation for the total density in the Schrödinger picture:

$$\begin{aligned}\frac{d\hat{W}_{a,b}}{dt} &= -i(a-b)\varepsilon\,\hat{W}_{a,b} - i\mathcal{L}_B\hat{W}_{a,b} \\ &\quad - i\sum_k f_k\hat{x}_k a\,\hat{W}_{a,b} + i\sum_k f_k b\,\hat{W}_{a,b}\hat{x}_k .\end{aligned} \tag{8.77}$$

Using the interaction representation, (8.53), we obtain

$$\frac{d\hat{w}_{a,b}(t)}{dt} = -i\sum_k f_k\hat{x}_k(t)a\hat{w}_{a,b}(t) + i\sum_k f_k b\hat{w}_{a,b}(t)\hat{x}_k(t) .$$

After the same procedure as in the previous section, we obtain the second-order expression

$$\frac{d\hat{w}_{a,b}(t)}{dt} = \sum_k f_k^2 \int_0^t d\tau \left[-aa\hat{x}_k(t)\hat{x}_k(\tau)\hat{w}_{a,b}(\tau) + ab\hat{x}_k(t)\hat{w}_{a,b}(\tau)\hat{x}_k(\tau)\right]$$
$$- \sum_k f_k^2 \int_0^t d\tau \left[-ab\hat{x}_k(\tau)\hat{w}_{a,b}(\tau)\hat{x}_k(t) + bb\hat{w}_{a,b}(\tau)\hat{x}_k(\tau)\hat{x}_k(t)\right] . \tag{8.78}$$

Now the averaging over the bath can be performed, which yields

$$\frac{d\rho_{ab}(t)}{dt} = -\int_0^t d\tau \left[a(a-b)C(t-\tau) - b(a-b)C^*(t-\tau)\right]\rho_{ab}(\tau) .$$

Let us now separate the correlation function into its real and imaginary parts:

$$C(t) = C'(t) + iC''(t) . \tag{8.79}$$

Using (7.140), we have

$$C'(t) = \int \frac{d\omega}{2\pi} C''(\omega)\cos(\omega t)\coth\left(\frac{\beta\omega}{2}\right) , \tag{8.80}$$

$$C''(t) = -\int \frac{d\omega}{2\pi} C''(\omega)\sin(\omega t) . \tag{8.81}$$

We then get

$$\frac{d\rho_{ab}(t)}{dt} = -\int_0^t d\tau \left[(a-b)^2 C'(t-\tau) + i(a^2-b^2)C''(t-\tau)\right]\rho_{ab}(\tau) .$$

We thus find that the diagonal fluctuations affect only coherences of the density matrix, that is, if $a = b$, the right-hand side of the Liouville equation is zero. In the Markovian limit we have the solution

$$\rho_{ab}(t) = e^{-(a-b)^2\gamma t + i(a^2-b^2)\eta t}\rho_{ab}(0) , \tag{8.82}$$

where

$$\gamma = \int_0^\infty d\tau\, C'(\tau) , \tag{8.83}$$

$$\eta = -\int_0^\infty d\tau\, C''(\tau) . \tag{8.84}$$

We thus find that the decay rate γ and the frequency shift η are both related to the function $M(\omega)$, (8.70); however, this spectral density may not be the same as that of the off-diagonal fluctuations. In a general case we may have one set of oscillators coupled to one type of phenomenon, and other oscillators coupled to the other type of phenomenon. We then have a set of spectral densities that control the relaxation dynamics.

8.6
Fluctuations of a Displaced Oscillator

In the previous sections we considered simple model systems. In this section we describe a more realistic case of a molecule coupled to the vibrational bath and describe some practical models of the spectral density. A molecule at equilibrium is in the electronic ground state. Absorption or emission of a photon transfers the molecule from one electronic state to another. As the molecule also has vibrational modes, during the optical process the (slow) nuclear degrees of freedom remain frozen and the molecular nuclear configuration becomes nonequilibrium with respect to the new electronic state. That is the Franck–Condon transition. The common assumption is that the nuclear potential energy is parabolic in the vicinity of the equilibrium. So for two electronic states – the ground state and the excited state – we obtain a displaced (harmonic) oscillator model.

The energy diagram of the potential energy of such a system is depicted in Figure 8.1. In one dimension the electronic potential of the ground state is $V_g(q) = \hbar\omega_0 q^2/2$, and the displaced electronic excited state is described by potential $V_e(q) = \omega_{eg} + \hbar\omega_0(q - d)^2/2$. Here ω_0 is the vibrational frequency, ω_{eg} is the energy gap, and d is a dimensionless displacement parameter. The Huang–Rhys factor $s = 1/2d^2$ characterizes the strength of the electron–phonon coupling.

Vibrational dynamics in the harmonic potentials is exactly described by the theory of the quantum harmonic oscillator. It gives an infinite set of wavefunctions ψ_m with quantum numbers $m = 1, \ldots, \infty$ and corresponding energies $E_m = \hbar\omega_0(m + 1/2)$ with respect to the bottom of the corresponding potential surface. Transitions between the ladder of the electronic ground state and the ladder of the

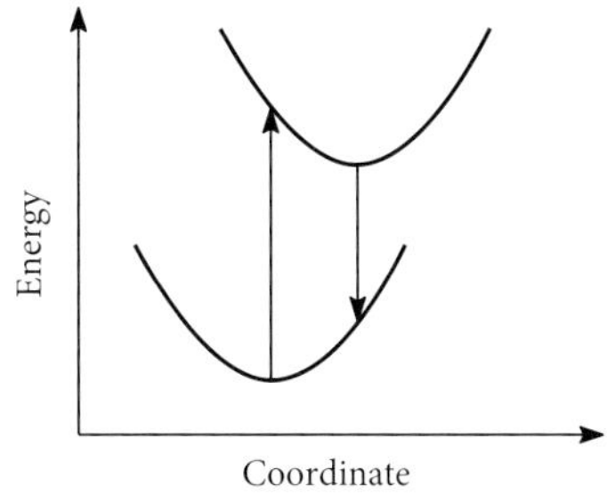

Figure 8.1 Energy levels of a two-state molecule – the displaced oscillator model.

electronic excited state determine the vibronic progression in the absorption spectrum. The intensity of each vibronic peak is scaled by the overlap of vibrational wavefunctions in the ground-state and excited-state potentials.

This description considers the electronic and vibrational system as a closed system. However, the nuclear degrees of freedom face damping due to the interaction with the rest of the vibrational degrees of freedom – the bath. A more general model of the electronic two-level system coupled to a phonon bath captures both the vibrational-type ladder of the energy spectrum and the homogeneous broadening at the same level of sophistication. Additionally, other vibrational degrees of freedom can be simply included. These can represent various types of vibrations.

The total Hamiltonian of such a system is given by

$$\hat{H} = 0|g\rangle\langle g| + (\omega_{eg} + \lambda)|e\rangle\langle e| - \sum_{\alpha} \hbar\omega_0^{(\alpha)} \sqrt{s_\alpha}\,(\hat{a}_\alpha^\dagger + \hat{a}_\alpha)\,|e\rangle\langle e| + \sum_{\alpha} \hbar\omega_0^{(\alpha)} \left(\hat{a}_\alpha^\dagger \hat{a}_\alpha + \frac{1}{2}\right) . \tag{8.85}$$

Here the ground state is the reference state with energy 0, the excited state is shifted by the reorganization energy $\lambda = \sum_\alpha \hbar\omega_0^{(\alpha)} s_\alpha$, and the bath is represented by a set of harmonic oscillators in terms of creation and annihilation operators. This form of system–bath coupling introduces fluctuations into the energy of the electronic excited state as described in the previous section. Therefore, we cannot now have population relaxation between the excited state and the bath. The only relaxation is inside the excited state or the ground state. This is essentially the vibrational relaxation.

The most convenient form to describe nuclear fluctuations is the correlation function of equilibrium system–bath fluctuations at constant temperature:

$$C(t) = \mathrm{Tr}_B \left\{ \hat{Q}_e(t)\hat{Q}_e(0)\rho_{eq} \right\} . \tag{8.86}$$

Here

$$\rho_{eq} = Z^{-1} \exp\left(-\sum_\alpha \beta\hbar\omega_0^{(\alpha)} \left(\hat{a}_\alpha^\dagger \hat{a}_\alpha + \frac{1}{2} \right) \right) \tag{8.87}$$

is the thermally equilibrated density operator ($\beta^{-1} = k_B T$), and

$$\hat{Q}_e(t) = \sum_\alpha \hbar\omega_0^{(\alpha)} \sqrt{s_\alpha}\,(\hat{a}_\alpha^\dagger(t) + \hat{a}_\alpha(t)) \tag{8.88}$$

is the fluctuating collective coordinate in the Heisenberg representation. Averaging in (8.86) gives

$$C(t) = \sum_\alpha \left(\hbar\omega_0^{(\alpha)}\right)^2 s_\alpha \left(\coth\frac{\beta\hbar\omega_\alpha}{2} \cos\omega_\alpha t - \mathrm{i}\sin\omega_\alpha t \right) , \tag{8.89}$$

which is a form of the two-point correlation function of the bath described in Section 7.8. From Section 7.8 we can calculate the spectral density using the correlation function. From (7.140) we have

$$C''(\omega) = -\int \mathrm{d}t \sin \operatorname{Im} C(t) \,. \tag{8.90}$$

Let us now discuss the types of correlation functions (or spectral densities) most relevant to such molecular systems. Assuming that the system is coupled to a continuous distribution of bath frequencies, we can obtain the integral form of (8.89) for a predefined coupling strength distribution $s(\omega)\mathrm{d}\omega \leftarrow s_\alpha$. Then the spectral density is given by

$$C''(\omega) = \pi(\hbar\omega)^2(s(\omega) - s(-\omega)) \,. \tag{8.91}$$

When we have a dominant mode frequency ω_0, while the mode is damped, the vibrational frequency distribution may be modeled using a Gaussian coupling:

$$s_\mathrm{G}(\omega) = \frac{1}{\sqrt{2\pi}\gamma} \mathrm{e}^{-\frac{(\omega-\omega_0)^2}{2\gamma^2}} \,. \tag{8.92}$$

This yields

$$C(t) \propto \omega_0^2 \mathrm{e}^{-\frac{1}{2}\gamma^2 t^2} \left(\coth \frac{\beta\hbar\omega_0}{2} \cos \omega_0 t - \mathrm{i} \sin \omega_0 t\right) \tag{8.93}$$

and the spectral density

$$C''_\mathrm{G}(\omega) = \lambda\mu \frac{\sqrt{\pi}}{\gamma\sqrt{2}} \left[\mathrm{e}^{-\frac{(\omega-\omega_0)^2}{2\gamma^2}} - \mathrm{e}^{-\frac{(\omega+\omega_0)^2}{2\gamma^2}}\right] , \tag{8.94}$$

where λ and μ are the added scaling prefactors of the energy dimension ensuring the right units for the spectral density. Also, the expression for μ can be chosen so that λ is equal to the reorganization energy $\pi^{-1}\int_0^\infty \mathrm{d}\omega/\omega \, C''(\omega)$.

Another model of Lorentzian-type coupling

$$s_\mathrm{L}(\omega) = \frac{1}{\pi} \frac{1}{(\omega-\omega_0)^2 + (2\ln 2)\,\gamma^2} \tag{8.95}$$

yields the correlation function

$$C(t) \propto \omega_0^2 \mathrm{e}^{-\gamma t} \left(\coth \frac{\beta\hbar\omega_0}{2} \cos \omega_0 t - \mathrm{i} \sin \omega_0 t\right) \tag{8.96}$$

and the spectral density

$$C''(\omega) = \lambda\mu \cdot \frac{4\omega\omega_0\gamma}{\left(\omega^2 - \omega_0^2 - \gamma^2\right)^2 + 4\omega^2\gamma^2} \,. \tag{8.97}$$

Equation (8.97) is the most general form of the spectral density. It includes both the damping parameter γ and the vibrational frequency ω_0. By taking various limits with respect to these parameters, we can obtain different damping regimes representing different baths. *Undamped, damped,* and *overdamped* regimes are discussed in the following.

For the motion of the *undamped* system ($\gamma = 0$) the spectral density function is obtained as a Fourier transform of a single nondecaying term of (8.89). It results in the spectral density which couples the system to a single mode of vibrations via the δ function:

$$C''_{\mathrm{u}}(\omega) = \pi s \omega_0^2 \left[\delta(\omega - \omega_0) - \delta(\omega + \omega_0)\right] . \tag{8.98}$$

The reorganization energy of such spectral density is $s\omega_0$.

However, the spectral density given by the δ functions is not realistic because in solutions any vibrational motion experiences dissipation. In that case the correlation function decays over time and the spectral density becomes a smooth function. Such a *damped* regime corresponds to the limit of the system being neither unaffected nor overdamped by the bath motion. This limit is achieved by taking $\gamma < \omega_0$ in (8.97) and redefining the damping strength $\gamma \to 2^{-1/2}\gamma$,

$$C''_{\mathrm{d}}(\omega) = \frac{4\lambda\omega\omega_0^2\gamma}{\left(\omega^2 - \omega_0^2\right)^2 + \gamma^2\omega^2} . \tag{8.99}$$

Here λ is the reorganization energy. This type of spectral density has a peak at ω_0; however, the peak width is defined by the damping strength γ, differently from (8.98), where the peak is the δ function (Figure 16.5a).

The *overdamped* bath regime is usually represented by the spectral density of a Brownian oscillator, which is derived *semiclassically*. In that case the classical correlation function of an overdamped oscillator is assumed (see Section 3.8):

$$C_{\mathrm{cl}}(t) = 2\lambda k_{\mathrm{B}} T \exp(-\gamma|t|) . \tag{8.100}$$

Its Fourier transform is given by

$$C_{\mathrm{cl}}(\omega) = 4k_{\mathrm{B}} T \frac{\lambda\gamma}{\omega^2 + \gamma^2} . \tag{8.101}$$

As expected, the function is real, so the spectral density cannot be defined from the relation given by (8.90). Instead we can use general relation (7.138). Note that in the high-temperature limit the classical correlation function coincides with the quantum correlation function. We must then have that

$$C'(t)|_{\mathrm{HT}} = C_{\mathrm{cl}}(t) \tag{8.102}$$

or

$$C'(\omega)|_{\mathrm{HT}} = C_{\mathrm{cl}}(\omega) . \tag{8.103}$$

From relation (7.138), we then have

$$C''(\omega) = C'(\omega)\frac{\beta\omega}{2} = C_{\mathrm{cl}}(\omega)\frac{\beta\omega}{2} \tag{8.104}$$

and we then get the well-known spectral density of an overdamped Brownian particle

$$C''_{\text{o-sc}}(\omega) = 2\lambda \frac{\gamma\,\omega}{\omega^2 + \gamma^2}, \tag{8.105}$$

with reorganization energy λ and relaxation rate γ. The full *quantum* correlation function in the frequency domain is later constructed by the direct application of the fluctuation–dissipation relation:

$$C_{\text{o-sc}}(\omega) = 2\lambda \frac{\gamma\,\omega}{\omega^2 + \gamma^2}\left[1 + \coth\left(\frac{\beta\omega}{2}\right)\right]. \tag{8.106}$$

The bath described by such spectral density is denoted as the *overdamped semiclassical bath.*

Alternatively, the overdamped bath can also be described by assuming the most general form of the spectral density. Taking (8.97) at the limit $\gamma \gg \omega_0$, we obtain

$$C''_{\text{o-q}}(\omega) = \frac{4\lambda\omega\gamma^3}{(\omega^2 + \gamma^2)^2}. \tag{8.107}$$

We denote this regime as *quantum overdamped* since it is introduced from the quantum definition of the correlation function rather than derived from classical assumptions. The reorganization energies are equal to λ for both types of overdamped spectral density. The semiclassical spectral density function is equal to λ at its maximum at $\omega = \gamma$, while the quantum function has its maximum at $\omega = \sqrt{3}/3\gamma \approx 0.58\gamma$ and the maximum value $C''_{\text{o-q}}(\sqrt{3}/3\gamma) = 3\sqrt{3}/4\lambda \approx 1.3\lambda$.

In Figure 8.2 we show how the spectral density of the damped quantum harmonic oscillator is interpolated from the undamped to the overdamped case. The undamped model $\gamma \to 0$ shows a well-resolved resonance at $\omega = \omega_0 = 1$, while the overdamped model $\gamma \gg \omega_0 = 1$ shows a steep rise at low frequencies and later decays as ω^{-3}.

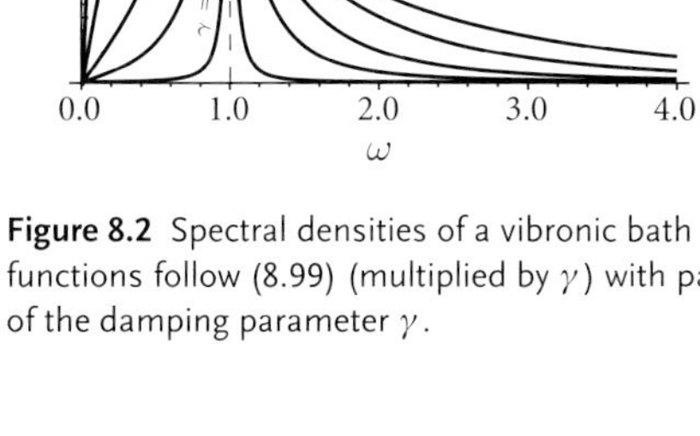

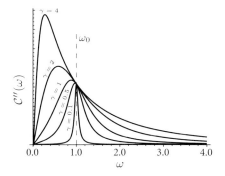

Figure 8.2 Spectral densities of a vibronic bath based on the damped Brownian oscillator. These functions follow (8.99) (multiplied by γ) with parameters $4\lambda = 1$ and $\omega_0 = 1$ for various values of the damping parameter γ.

In practice the molecules are usually coupled to several vibrational modes and a spectral density may be described by a multipeak function. For the model with several modes, the spectral density is then composed of several contributions:

$$C''(\omega) = C''_{\text{o}}(\omega) + C''_{\text{vib}}(\omega) . \tag{8.108}$$

Here the overdamped mode $C''_{\text{o}}(\omega)$ corresponds to the fast overdamped (*semiclassical* $C''_{\text{o-sc}}(\omega)$ or *quantum* $C''_{\text{o-q}}(\omega)$) bath and is responsible for the homogeneous broadening. The second term, $C''_{\text{vib}}(\omega)$, represents the spectral density of molecular vibrations (*undamped* $C''_{\text{u}}(\omega)$ or *damped* $C''_{\text{d}}(\omega)$) . This composition of the system–bath correlation is the commonly used approach in simulations of the spectroscopic results and is discussed in detail in the literature [43, 44].

9
Projection Operator Approach to Open Quantum Systems

In previous chapters we introduced the wavefunction and the density operator descriptions of quantum mechanical systems. They were first presented as equivalent alternatives in Chapter 4. Later in the same chapter, we recognized the flexibility of the density operator formalism, and we introduced the concept of mixed states and the concept of the reduced density operator. In Chapter 6 we examined the consequences of the quantum mechanical formalism for large systems. We introduced the concept of decoherence, which explains the emergence of the classical world from the quantum mechanical description. We are often dealing with systems with a small number of degrees of freedom, which interact with a large environment poorly characterized due to our limited knowledge of their state. Although we have been formally assigning wavefunctions to these environmental states, often the only information we were really using was that the overlap of the environmental wavefunctions corresponding to different states of the subsystem tends to zero. In other words, we have seen the coherence between different states of the subsystem dephase. We were thus interested in the dynamics of a small quantum mechanical subsystem embedded in some large environment. The total system (the subsystem and the environment) was itself described by a single wavefunction and it can therefore be said to be closed, because no interaction with the outside is taken into account. The small subsystem, on the other hand, exhibits the properties of an *open system*, because it exchanges energy and sometimes even matter with the external environment. The environment can thus be described by the concepts of statistical physics (Chapter 7). Here, the concept of the reduced density operator finds its most important application. We have already encountered the reduced density matrix in Chapter 8. Instead of following the precise time evolution of a wavefunction of a large closed quantum system, we are able to develop a reduced description of the subsystem evolution only, with some effective description of the rest of the environment and its influence on the subsystem dynamics.

The situation where we observe a microscopic subsystem on a macroscopic level is very common. Most spectroscopic measurements are done on molecules dissolved in a solution, or they are done on some specific (e.g., electronic) degrees of freedom that are embedded in a sea of other degrees of freedom (e.g., nuclear). The information that we extract from such experiments is specific to the microscopic subsystems, yet it depends on the unknown state of the large number of degrees

Molecular Excitation Dynamics and Relaxation, First Edition. L. Valkunas, D. Abramavicius, and T. Mančal.

of freedom that our experiment sees only indirectly. These degrees of freedom are often termed *irrelevant* as opposed to those of the subsystem that we are interested in, which are termed *relevant*. Because the irrelevant degrees of freedom are actually quite important for the evolution of the relevant part of the system, it is often better to speak of a (sub)system and its *environment, bath*, or *reservoir*. Because the bath is often something macroscopic, finding itself in equilibrium corresponding to some temperature, we often speak of the *thermodynamic bath* or *thermodynamic reservoir*. When talking about the subsystem, we will from now on use the shorter word *system* wherever it does not lead to confusion with the total closed system. In this chapter we will discuss methods for systematic derivation of evolution equations for the reduced density operator of the relevant system.

9.1
Liouville Formalism

Let us first introduce a suitable formalism for manipulations with the density operators. The equation of motion for a system described by the Hamiltonian H and the density operator ρ, the Liouville–von Neumann equation, was derived in Chapter 4 (4.82) and we give it again here for convenience:

$$\frac{\mathrm{d}}{\mathrm{d}t}\rho = -\frac{\mathrm{i}}{\hbar}(\hat{H}\rho - \rho\hat{H}) . \tag{9.1}$$

Represented in the selected basis $|a\rangle$, this equation would be an equation for a matrix, and the right-hand side would be obtained by matrix multiplication. We introduced the superoperator notation in (4.86). Equations (4.86) and (9.1) can be formally rewritten in a superoperator form:

$$\frac{\mathrm{d}}{\mathrm{d}t}\rho = -\mathrm{i}\mathcal{L}\rho . \tag{9.2}$$

The superoperator $\mathcal{L}$ is usually called the *Liouvillian* or *Liouville superoperator*. In the Dirac bra-ket formalism, we can represent it by the following scheme where $\mathcal{L}$ acts on an arbitrary operator $\hat{A}$:

$$\mathcal{L}\hat{A} = \sum_{ab}\sum_{cd}|a\rangle\langle b|\mathcal{L}_{abcd}\langle c|\hat{A}|d\rangle . \tag{9.3}$$

As in the case of operators, the superoperators can be specified by listing their matrix elements, which form tensors of rank 4.

In this chapter we introduce a shorthand summation notation – the Einstein summation convention – where a repeated index in matrix manipulation is understood as a summation, for example, $A_{ab}B_{bc}$ represents $\sum_b A_{ab}B_{bc}$. Thus, for quantum mechanical operators $\hat{A} = \sum_{ab} A_{ab}|a\rangle\langle b|$ and $\hat{B} = \sum_{ab} B_{ab}|a\rangle\langle b|$ and state vector $|\psi\rangle = \sum_a \psi_a|a\rangle$, we have

$$|\phi\rangle = \hat{A}|\psi\rangle \rightarrow A_{ab}\psi_b = \phi_a \tag{9.4}$$

and

$$\hat{C} = \hat{A}\hat{B} \rightarrow A_{ab}B_{bc} = C_{ac} , \tag{9.5}$$

where $|\phi\rangle$ is a new vector obtained by the action of $\hat{A}$ on $|\psi\rangle$ and $\hat{C}$ is obtained by the product of $\hat{A}$ and $\hat{B}$.

The action of a superoperator, on the other hand, requires summation of two indices. A superoperator acts on an operator as follows:

$$\hat{B} = \mathcal{L}\hat{A} \rightarrow \sum_{cd} \mathcal{L}_{abcd}A_{cd} \equiv \mathcal{L}_{abcd}A_{cd} = B_{ab} . \tag{9.6}$$

To see that there is close analogy between the operator–state vector formalism and the superoperator–operator formalism, we can group the pairs of indices into one superindex. This can be done in many equivalent ways. One can, for instance, start with

$$\begin{aligned} 1 &\equiv (11) \\ 2 &\equiv (12) \\ &\vdots \end{aligned} \tag{9.7}$$

to assign a unique number to all pairs of values by some rule. Relation (9.6) will then read

$$\mathcal{L}_{IJ}A_J = B_I , \tag{9.8}$$

where I and J are *superindices*. Equation (9.8) is completely analogous to relation (9.4). In other words, by reordering the density matrix elements into a vector, and by reordering the tensor elements of the Liouvillian into a matrix, we obtain the equation of motion for the density matrix, (9.2), in the same matrix form as the Schrödinger equation.

The space on which the superoperators act and which is formed by the operators is sometimes termed the Liouville space. In principle, both the Hilbert space formulation and the Liouville space formulation are equivalent for a closed system and they can be chosen depending on convenience. The advantage of the Liouville space becomes apparent when turning to open systems and the reduced density matrix.

In the Liouville space a formal solution of the Liouville–von Neumann equation can be given as

$$\rho(t) = \exp(-\mathrm{i}\mathcal{L}t)\rho(0) . \tag{9.9}$$

The forward propagator, or Green's function, $\mathcal{G}(t) = \theta(t)\exp(-\mathrm{i}\mathcal{L}t)$ is analogous to the corresponding evolution operator for the state vector in Hilbert space (see Chapter 4). For a density operator representing a pure state $|\psi\rangle$, that is, for $\rho = |\psi\rangle\langle\psi|$, the evolution superoperator $\mathcal{G}(t)$ can be expressed in terms of an action of two Hilbert space evolution operators $\hat{U}(t)$ (defined in Chapter 4) as

$$\rho(t) = \mathcal{G}(t)\rho(0) \Longleftrightarrow \hat{U}(t)\rho(0)\hat{U}^{\dagger}(t) . \tag{9.10}$$

As long as the density matrix ρ represents a closed quantum mechanical system, its time evolution can be expressed equivalently by the evolution superoperator $\mathcal{G}(t)$ and the evolution operator $\hat{U}(t)$. When the system is open, however, only the superoperator formalism has enough flexibility to fully describe the reduced system dynamics.

9.2 Reduced Density Matrix of Open Systems

In this section we introduce the reduced density matrix description of an open system interacting with some environment. We denote the relevant system by the letter S, the thermodynamic reservoir – the environment – by R, and any terms that correspond to their interaction by SR. Thus, the total Hamiltonian of the composite system consisting of the relevant system and its environment is given by

$$\hat{H} = \hat{H}_{\mathrm{S}} + \hat{H}_{\mathrm{R}} + \hat{H}_{\mathrm{SR}} , \tag{9.11}$$

and the corresponding Liouville superoperator reads

$$\mathcal{L} = \mathcal{L}_{\mathrm{S}} + \mathcal{L}_{\mathrm{R}} + \mathcal{L}_{\mathrm{SR}} . \tag{9.12}$$

The total density matrix of the composite system is denoted by $W(t)$, and in general it cannot be split into parts corresponding to subsystems S and R in a simple way. When the interaction term $\hat{H}_{\mathrm{SR}}$ is equal to zero, and subsystems S and R never interacted, we can write the total density operator as a product of two density operators: $\rho(t)$ for the system S and $w(t)$ for the environment R. This is our starting density operator:

$$W_0(t) = \rho(t) w(t) . \tag{9.13}$$

The density matrix reduced to subsystem S can be obtained by averaging W over the degrees of freedom of subsystem R. The reduction of the total density operator is achieved by applying the trace operation in the Hilbert space of subsystem R, which in the noninteracting case leads to

$$\mathrm{tr}_{\mathrm{R}}\{W_0(t)\} \equiv \mathrm{tr}_{\mathrm{R}}\{\rho(t) w(t)\} = \rho(t)\mathrm{tr}_{\mathrm{R}}\{w(t)\} = \rho(t) . \tag{9.14}$$

Here we used the fact that the density operator is normalized. If the systems interact, one can only define the reduced density operator formally by

$$\rho(t) = \mathrm{tr}_{\mathrm{R}}\{W(t)\} , \tag{9.15}$$

because there is no *a priori* density operator of the subsystem S. Subsystems S and R are in general entangled due to interaction (see Chapter 6). If we are interested just in observables related to subsystem S, the reduced density matrix still possesses all the necessary information.

It is useful to remind ourselves of the way the trace operation would be applied to the total density matrix operator expressed in some particular basis of the total Hilbert space. We chose some basis of vectors $|a\rangle$ of the Hilbert space of the system S (Latin indices) and some basis of vectors $|\alpha\rangle$ of the environment R (Greek indices). The total density matrix will be then represented by the rank-four matrix

$$W_{a\alpha,b\beta}(t) = \langle a|\langle\alpha|\,W(t)|b\rangle|\beta\rangle\ . \tag{9.16}$$

The trace operation over the environmental degrees of freedom amounts to summation of the elements of the density matrix diagonal in Greek indices:

$$\rho_{ab}(t) = \langle a|\mathrm{tr}_{\mathrm{R}}\{\,W(t)\}|b\rangle = \sum_{\alpha} W_{a\alpha b\alpha}(t)\ . \tag{9.17}$$

Here we reduced the density operator (represented by a matrix here) and we contracted it from the total Hilbert space to the smaller Hilbert space of the system.

9.3 Projection (Super)operators

The projection operator approach (see, e.g., [40]) allows us to stay in the Hilbert space of the total system, but it projects the total density matrix onto the degrees of freedom of the system. For the relevant system the total density matrix is replaced by the product of the reduced density matrix $\rho_{ab}(t)$ and some prescribed density matrix $w_{\alpha\beta}$ in the Hilbert space of R to represent the "*unknown*" bath state.

This operation (reduction and replacement $W_{a\alpha b\beta}(t) \to \rho_{ab}(t)w_{\alpha\beta}$) can be expressed by the action of a certain superoperator $\mathcal{P}$ on $W(t)$:

$$\mathcal{P}\,W(t) = \rho(t)w\ . \tag{9.18}$$

Expressed in the matrix elements, (9.18) is given by

$$\mathcal{P}_{a\alpha b\beta,a'\alpha'b'\beta'}\,W_{a'\alpha'b'\beta'}(t) \equiv \rho_{ab}(t)w_{\alpha\beta}\ , \tag{9.19}$$

which is fulfilled by

$$\mathcal{P}_{a\alpha b\beta,a'\alpha'b'\beta'} = \delta_{aa'}\delta_{bb'}w_{\alpha\beta}\delta_{\alpha'\beta'}\ . \tag{9.20}$$

where

$$\mathrm{tr}\{w\} = \sum_{\alpha} w_{\alpha\alpha} = 1\ . \tag{9.21}$$

Otherwise the operator w can be arbitrary. The superoperator $\mathcal{P}$ fulfills the relation

$$\mathcal{P}^2 = \mathcal{P}\ , \tag{9.22}$$

which qualifies it as a projection superoperator. The superoperator $\mathcal{P}$ thus performs a trace over the bath, producing the reduced density matrix. Then it represents the state of the bath by a predefined matrix w.

One can also always define a complementary superoperator $\mathcal{Q}$ as

$$\mathcal{Q} = 1 - \mathcal{P} . \tag{9.23}$$

It can be shown easily that $\mathcal{Q}$ is also a projector and that $\mathcal{PQ} = 0$. In the next few sections we will show how to use superoperators $\mathcal{P}$ and $\mathcal{Q}$ to our advantage together with the Liouville superoperators. This will significantly simplify the derivation of approximate equations of motion for the reduced density matrix $\rho(t)$.

9.4
Nakajima–Zwanzig Identity

Our final aim is to derive a suitable equation of motion for the reduced density operator $\rho(t)$. Such a derivation can be conveniently formalized if we first try to find the equation of motion of the density matrix $\mathcal{P}\,W(t)$, which implicitly contains $\rho(t)$ (see (9.18)). As we show below, one can derive a formal equation in a closed form. Solving this equation is as difficult as solving the full equation for $W(t)$, but it can be used as a starting point for a subsequent approximative treatment.

We start with the Liouville equation for the total density matrix:

$$\frac{\partial}{\partial t} W(t) = -\mathrm{i}\left(\mathcal{L}_\mathrm{S} + \mathcal{L}_\mathrm{R} + \mathcal{L}_\mathrm{SR}\right) W(t) . \tag{9.24}$$

It is useful to work in the interaction picture with respect to Liouvillians $\mathcal{L}_\mathrm{S}$ and $\mathcal{L}_\mathrm{R}$ because this will later allow us to perform a certain perturbation expansion with respect to $\mathcal{L}_\mathrm{SR}$ only. The interaction picture is used here in full analogy with the interaction picture explained in Section 4.4.3 for the state vector. The analogy is complete, just the operators are replaced by superoperators and the state vectors are replaced by density operators. Equation (9.24) in the interaction picture reads

$$\frac{\partial}{\partial t} W^{(\mathrm{I})}(t) = -\mathrm{i}\mathcal{L}_\mathrm{SR}(t)\, W^{(\mathrm{I})}(t) . \tag{9.25}$$

The projector operators $\mathcal{P}$ and $\mathcal{Q}$ introduced allow us to write the following two coupled equations:

$$\frac{\partial}{\partial t}\mathcal{P}\, W^{(\mathrm{I})}(t) = -\mathrm{i}\mathcal{P}\mathcal{L}_\mathrm{SR}(t)\mathcal{P}\, W^{(\mathrm{I})}(t) - \mathrm{i}\mathcal{P}\mathcal{L}_\mathrm{SR}(t)\mathcal{Q}\, W^{(\mathrm{I})}(t) , \tag{9.26}$$

$$\frac{\partial}{\partial t}\mathcal{Q}\, W^{(\mathrm{I})}(t) = -\mathrm{i}\mathcal{Q}\mathcal{L}_\mathrm{SR}(t)\mathcal{Q}\, W^{(\mathrm{I})}(t) - \mathrm{i}\mathcal{Q}\mathcal{L}_\mathrm{SR}(t)\mathcal{P}\, W^{(\mathrm{I})}(t) . \tag{9.27}$$

In (9.27) we introduced the new quantity $\mathcal{Q}\, W_0^{(\mathrm{I})}(t)$ (an interaction picture):

$$\mathcal{Q}\, W_0^{(\mathrm{I})}(t) = \exp_-\left(\mathrm{i}\int_{t_0}^{t} \mathrm{d}\tau\, \mathcal{Q}\mathcal{L}_\mathrm{SR}(\tau)\mathcal{Q}\right)\mathcal{Q}\, W^{(\mathrm{I})}(t) . \tag{9.28}$$

This transforms (9.27) into

$$\frac{\partial}{\partial t}\mathcal{Q}W_0^{(\mathrm{I})}(t) = -\mathrm{i}\exp_-\left(\mathrm{i}\int_{t_0}^{t} \mathrm{d}\tau\,\mathcal{Q}\mathcal{L}_{\mathrm{SR}}(\tau)\mathcal{Q}\right)\mathcal{Q}\mathcal{L}_{\mathrm{SR}}(t)\mathcal{P}W^{(\mathrm{I})}(t)\,. \tag{9.29}$$

The solution of (9.29) can be easily found by integration:

$$\mathcal{Q}W_0^{(\mathrm{I})}(t) = \mathcal{Q}W_0^{(\mathrm{I})}(t_0) - \mathrm{i}\int_{t_0}^{t}\mathrm{d}\tau\,\exp_-\left(\mathrm{i}\int_{t_0}^{\tau}\mathrm{d}\tau'\,\mathcal{Q}\mathcal{L}_{\mathrm{SR}}(\tau')\mathcal{Q}\right)\mathcal{Q}\mathcal{L}_{\mathrm{SR}}(\tau)\mathcal{P}W^{(\mathrm{I})}(\tau)\,. \tag{9.30}$$

After returning from the second interaction picture to $\mathcal{Q}W^{(\mathrm{I})}(t)$ by inverting (9.28), we obtain

$$\begin{aligned}\mathcal{Q}W^{(\mathrm{I})}(t) &= \exp_+\left(-\mathrm{i}\int_{t_0}^{t}\mathrm{d}\tau\,\mathcal{Q}\mathcal{L}_{\mathrm{SR}}(\tau)\mathcal{Q}\right)\mathcal{Q}W^{(\mathrm{I})}(t_0)\\ &\quad - \mathrm{i}\int_{t_0}^{t}\mathrm{d}\tau\,\exp_+\left(-\mathrm{i}\int_{t_0}^{t-\tau}\mathrm{d}\tau'\,\mathcal{Q}\mathcal{L}_{\mathrm{SR}}(\tau')\mathcal{Q}\right)\mathcal{Q}\mathcal{L}_{\mathrm{SR}}(\tau)\mathcal{P}W^{(\mathrm{I})}(\tau)\,.\end{aligned} \tag{9.31}$$

We next insert (9.31) into (9.26) and obtain a closed equation of motion for the relevant part of the density matrix $\mathcal{P}W^{(\mathrm{I})}(t)$:

$$\begin{aligned}\frac{\partial}{\partial t}\mathcal{P}W^{(\mathrm{I})}(t) &= -\mathrm{i}\mathcal{P}\mathcal{L}_{\mathrm{SR}}(t)\mathcal{P}W^{(\mathrm{I})}(t)\\ &\quad - \mathrm{i}\mathcal{P}\mathcal{L}_{\mathrm{SR}}(t)\exp_+\left(-\mathrm{i}\int_{t_0}^{t}\mathrm{d}\tau\,\mathcal{Q}\mathcal{L}_{\mathrm{SR}}(\tau)\mathcal{Q}\right)\mathcal{Q}W^{(\mathrm{I})}(t_0)\\ &\quad - \int_{t_0}^{t}\mathrm{d}\tau\,\mathcal{P}\mathcal{L}_{\mathrm{SR}}(t)\exp_+\left(-\mathrm{i}\int_{t_0}^{t-\tau}\mathrm{d}\tau'\,\mathcal{Q}\mathcal{L}_{\mathrm{SR}}(\tau')\mathcal{Q}\right)\\ &\quad \times \mathcal{Q}\mathcal{L}_{\mathrm{SR}}(\tau)\mathcal{P}W^{(\mathrm{I})}(\tau)\,.\end{aligned} \tag{9.32}$$

This equation, called the *Nakajima–Zwanzig identity*, is exact. No approximations were used in its derivation. We could have formulated it without the interaction picture, in which case its form would be simpler (ordinary instead of time-ordered exponentials). The Nakajima–Zwanzig identity is, however, useful mostly in expanding the exponentials to a low order. This would mean that we would expand the equations in the total Liouvillian, not just the interaction Liouvillian.

We can identify three terms in (9.32): a term which seems to contain an effective Liouvillian

$$\mathcal{L}_{\mathrm{NZ}}(t)\mathcal{P}\,W^{(\mathrm{I})}(t) = \mathcal{P}\mathcal{L}_{\mathrm{SR}}(t)\mathcal{P}\,W^{(\mathrm{I})}(t)\,, \tag{9.33}$$

the so-called initial correlation term

$$\mathcal{I}_{\mathrm{NZ}}(t) = \mathcal{P}\mathcal{L}_{\mathrm{SR}}(t)\exp_+\left(-\mathrm{i}\int\limits_{t_0}^{t}\mathrm{d}\tau\,\mathcal{Q}\mathcal{L}_{\mathrm{SR}}(\tau)\mathcal{Q}\right)\mathcal{Q}\,W^{(\mathrm{I})}(t_0)\,, \tag{9.34}$$

and a relaxation and dissipation term

$$\mathcal{K}_{\mathrm{NZ}}(t;\mathcal{P}\,W^{(\mathrm{I})}) = \int\limits_{t_0}^{t}\mathrm{d}\tau\,\mathcal{P}\mathcal{L}_{\mathrm{SR}}(t)\exp_+\left(-\mathrm{i}\int\limits_{t_0}^{t-\tau}\mathrm{d}\tau'\,\mathcal{Q}\mathcal{L}_{\mathrm{SR}}(\tau')\mathcal{Q}\right) \times \mathcal{Q}\mathcal{L}_{\mathrm{SR}}(\tau)\mathcal{P}\,W^{(\mathrm{I})}(\tau)\,. \tag{9.35}$$

The equation for the projected density matrix then reads

$$\frac{\partial}{\partial t}\mathcal{P}\,W^{(\mathrm{I})}(t) = -\mathrm{i}\mathcal{I}_{\mathrm{NZ}}(t) - \mathrm{i}\mathcal{L}_{\mathrm{NZ}}(t)\mathcal{P}\,W^{(\mathrm{I})}(t) - \mathcal{K}_{\mathrm{NZ}}(t;\mathcal{P}\,W^{(\mathrm{I})})\,. \tag{9.36}$$

We will interpret these terms after we will deal with an alternative use of the projector superoperators in the section that follows .

9.5
Convolutionless Identity

The exact equation, (9.32), is an integrodifferential equation. The time derivative of the state (the left-hand side of (9.32)) depends on the history of the system, that is, the equation is time nonlocal. However, this dependence is not fundamental. As we will demonstrate, it is possible to rewrite (9.32) in a time-local form.

We start with the full formal solution of the problem in the interaction representation, (9.25), that is, with

$$W^{(\mathrm{I})}(t) = \mathcal{G}(t,t_0)\,W^{(\mathrm{I})}(t_0)\,. \tag{9.37}$$

The evolution superoperator

$$\mathcal{G}(t,t_0) = \exp\left(-\mathrm{i}\int\limits_{t_0}^{t}\mathrm{d}\tau\,\mathcal{L}_{\mathrm{SR}}(\tau)\right)\,, \tag{9.38}$$

which solves (9.25), can be used to convert (9.32) into an ordinary differential equation. We will start with (9.31), that is, one step before we completed the derivation of the Nakajima–Zwanzig identity. We express all density operators in terms of $W^{(I)}(t)$, for example, $W^{(I)}(\tau) = \mathcal{G}(\tau, t)\, W^{(I)}(t)$. So for (9.31) we write

$$\begin{aligned}\mathcal{Q} W^{(I)}(t) = {} & \exp_+\left(-\mathrm{i}\int_{t_0}^{t} \mathrm{d}\tau \mathcal{Q}\mathcal{L}_{SR}(\tau')\mathcal{Q}\right) \mathcal{Q} W^{(I)}(t_0) \\ & - \mathrm{i}\int_{t_0}^{t} \mathrm{d}\tau \exp_+\left(-\mathrm{i}\int_{t_0}^{t-\tau} \mathrm{d}\tau' \mathcal{Q}\mathcal{L}_{SR}(\tau')\mathcal{Q}\right) \mathcal{Q}\mathcal{L}_{SR}(\tau) \\ & \times \mathcal{P} G(\tau, t)\, (\mathcal{Q} + \mathcal{P})\; W^{(I)}(t)\,. \end{aligned} \tag{9.39}$$

$\mathcal{Q} W^{(I)}(t)$ appears on both sides of (9.39). Collecting both terms containing it, and dividing by the factor in front of it, we obtain

$$\begin{aligned}\mathcal{Q} W^{(I)}(t) = {} & (1 - \Sigma(t))^{-1} \exp_+\left(-\mathrm{i}\int_{t_0}^{t} \mathrm{d}\tau \mathcal{Q}\mathcal{L}_{SR}(\tau')\mathcal{Q}\right) \mathcal{Q} W^{(I)}(t_0) \\ & + (1 - \Sigma(t))^{-1}\Sigma(t)\mathcal{P} W^{(I)}(t)\,, \end{aligned} \tag{9.40}$$

where the superoperator $\Sigma(t)$ reads

$$\Sigma(t) = -\mathrm{i}\int_{t_0}^{t} \mathrm{d}\tau \exp_+\left(-\mathrm{i}\int_{t_0}^{t-\tau} \mathrm{d}\tau' \mathcal{Q}\mathcal{L}_{SR}(\tau')\mathcal{Q}\right) \mathcal{Q}\mathcal{L}_{SR}(\tau)\mathcal{P} G(\tau, t)\,. \tag{9.41}$$

The new closed-form equations for $\mathcal{P} W^{(I)}(t)$ contain only $W^{(I)}(t)$ and $W^{(I)}(t_0)$ on the right-hand side. They are therefore ordinary differential equations and the history term was eliminated. The final convolutionless form of the equations of motion reads

$$\frac{\partial}{\partial t}\mathcal{P} W^{(I)}(t) = -\mathrm{i}\mathcal{I}_{CL}(t) - \mathcal{K}_{CL}(t)\mathcal{P} W^{(I)}(t)\,, \tag{9.42}$$

where

$$\mathcal{I}_{CL} = \mathrm{i}\mathcal{P}\mathcal{L}_{SR}(t)(1 - \Sigma(t))^{-1} \exp_+\left(-\mathrm{i}\int_{t_0}^{t} \mathrm{d}\tau \mathcal{Q}\mathcal{L}_{SR}(\tau')\mathcal{Q}\right) \mathcal{Q} W^{(I)}(t_0) \tag{9.43}$$

and

$$\mathcal{K}_{CL}(t) = \mathrm{i}\mathcal{P}\mathcal{L}_{SR}(t)(1 - \Sigma(t))^{-1}\mathcal{P}\,. \tag{9.44}$$

So we obtained again equations of motion of the projected density operator $\mathcal{P} W^{(I)}(t)$, this time in a convolutionless form. The term $\mathcal{K}_{CL}$ contains a contribution equivalent to $\mathcal{L}_{NZ}$ of (9.33), and a contribution replacing $\mathcal{L}_{NZ}$ of (9.35). Equations (9.32) and (9.42) represent two equivalent expressions for the dynamics of the projected density operator. These types of expressions can be efficiently used in relaxation theory, as we show in Section 11.3.

9.6 Relation between the Projector Equations in Low-Order Perturbation Theory

Identities (9.32) and (9.42) are not practical, unless we expand their terms in low order of $\mathcal{L}_{SR}$. In order to simplify the following approximate treatment, we will use a particular form of the projection superoperator, (9.18), and assume that the initial condition for the total density matrix reads

$$W(t_0) = \rho(t_0)w \, . \tag{9.45}$$

This choice leads to

$$\mathcal{Q}W(t_0) = 0 \, , \tag{9.46}$$

and thus the terms $\mathcal{I}_{NZ}$ and $\mathcal{I}_{CL}$ containing $W(t_0)$ are eliminated from all equations. Interestingly, this type of initial equation is quite common in spectroscopy, as we will discuss in Part Two (see Chapter 14).

Another point worth noting is the fact that we derived our identities in the interaction picture. We have to find the relation between the projected density operator $\mathcal{P}W^{(I)}(t)$ and the reduced density operator $\rho(t)$. Using the definition of the interaction picture, we have

$$\mathcal{P}W^{(I)}(t) = \hat{U}_S^\dagger(t)\mathrm{tr}_R\left\{\hat{U}_R^\dagger(t)\,W(t)\,\hat{U}_R(t)\right\} w\,\hat{U}_S(t) = \rho^{(I)}(t)w \, . \tag{9.47}$$

Here we used the fact that under the trace we can perform cyclical permutation of the operators, and correspondingly the bath evolution has no effect.

In the second order, the Nakajima–Zwanzig identity, (9.32), becomes (we omit the initial term because of (9.46))

$$\frac{\partial}{\partial t}\rho^{(I)}(t) = -\mathrm{i}\mathrm{tr}_R\{\mathcal{L}_{SR}(t)w\}\rho^{(I)}(t) - \int_0^{t-t_0} \mathrm{d}\tau \mathrm{tr}_R\{\mathcal{L}_{SR}(t)\mathcal{Q}\mathcal{L}_{SR}(t-\tau)w\}\rho^{(I)}(t-\tau) \, . \tag{9.48}$$

At the same time, the second-order expansion of the convolutionless identity leads to

$$\frac{\partial}{\partial t}\rho^{(I)}(t) = -\mathrm{i}\mathrm{tr}_R\{\mathcal{L}_{SR}(t)w\}\rho^{(I)}(t) - \int_0^{t-t_0} \mathrm{d}\tau \mathrm{tr}_R\{\mathcal{L}_{SR}(t)\mathcal{Q}\mathcal{L}_{SR}(t-\tau)w\}\rho^{(I)}(t) \, . \tag{9.49}$$

Equations (9.48) and (9.49) represent two different master equations for the reduced density operator in the interaction picture. The only difference between (9.48) and (9.49) is the retardation in (9.48). The convolutionless version of the equation can be obtained from (9.48) by an approximation which reminds us of the procedure by which we derived the convolutionless identity itself. Let us postulate an evolution superoperator $\mathcal{U}_{\mathcal{P}}(t)$ which solves (9.32) so that

$$\rho(t-\tau) = \mathcal{U}_{\mathcal{P}}(-\tau)\rho(t) . \tag{9.50}$$

Inserting this relation into (9.48), we obtain an ordinary differential equation for $\hat{\rho}(t)$. Because we want to keep the master equation in the second order of perturbation theory with respect to $\mathcal{L}_{\text{SR}}(t)$, we have to approximate $\mathcal{U}_P(-\tau)$, which now appears in the integration kernel, in zeroth order. This, however, means that we have to assume it is identity $\mathcal{U}_P(-\tau) \approx 1$. This leads to a second-order master equation which is identical to (9.49). Thus, after the replacement $\hat{\rho}(t-\tau) \rightarrow \hat{\rho}(t)$ in the second-order Nakajima–Zwanzig master equation, we obtain the second-order convolutionless master equation. This replacement is usually considered an approximation, but it certainly does not represent an approximation with respect to the interaction Liouvillian $\mathcal{L}_{\text{SR}}$.

The question of which of the two-second order equations, (9.48) or (9.49), is better has to be answered on an individual basis. Practical reasons limit us to these second-order representations, while at the same time, going to higher orders does not necessarily bring about an improvement [45]. In some cases, the second-order convolutionless generalized master equation can be shown to be exact [46] because it corresponds to the so-called second cumulant expansion [26].

9.7 Projection Operator Technique with State Vectors

The projection (super)operator technique is not exclusively reserved for density operator problems. We can apply it to state vector problems as well. We used the superoperator formulation in the previous sections, and we can now apply its results to the analogous problem with the state vector. The application to state vectors amounts to replacement of the superoperators by operators and the replacement of density operators by state vectors. In this section we use a slightly unusual look at open quantum systems. We will consider a multistate system, in which two levels will be considered explicitly, and the rest of the states will form a "bath" with respect to which the explicitly considered part of the system Hilbert space will be open. Let us consider an N-level system. The two lowest levels $|g\rangle$ and $|e\rangle$ will be the levels of interest, while the existence of a remaining band of a large (potentially infinite) number of states $|f_n\rangle$ is also assumed.

We will study the feasibility of exciting state $|e\rangle$ by light with frequency ω when the system is initially in state $|g\rangle$. In the case when the transition from $|g\rangle$ to $|e\rangle$ is dipole allowed, that is, the dipole moment operator has the form

$$\mu = d_{eg}(|e\rangle\langle g| + |g\rangle\langle e|) + \dots , \tag{9.51}$$

the situation is simple. Let the Hamiltonian have the simple form of

$$H = \epsilon_e |e\rangle\langle e| + \sum_n \epsilon_{f_n} |f_n\rangle\langle f_n| \,, \tag{9.52}$$

where we assume $\epsilon_g = 0$. The state vector $|\psi^{(I)}(t)\rangle$ in the interaction picture with respect to H and its expansion in terms of the basis states can be written as

$$|\psi^{(I)}(t)\rangle = c_g^{(I)}(t)|g\rangle + c_e^{(I)}|e\rangle + \sum_n c_{f_n}^{(I)}(t)|f_n\rangle \,. \tag{9.53}$$

We are interested in the excited state $|e\rangle$ and therefore we consider the Schrödinger equation for the coefficient $c_e^{(I)}(t)$. Let the interaction with the external monochromatic electric field $E(t) = \mathcal{E}(t)\mathrm{e}^{-\mathrm{i}\omega t}$ + c.c. be weak so that we can set $c_g^{(I)}(t) \approx c_g^{(I)}(0) = 1$. From the Schrödinger equation

$$\frac{\partial}{\partial t}|\psi^{(I)}(t)\rangle = \mathrm{i}\mu(t)E(t)|\psi^{(I)}(t)\rangle \,, \tag{9.54}$$

we get

$$\frac{\partial}{\partial t}c_e^{(I)}(t) = \mathrm{i}d_{eg}\mathrm{e}^{\mathrm{i}\omega_{eg}t}\mathcal{E}(t)(\mathrm{e}^{-\mathrm{i}\omega t} + \mathrm{e}^{\mathrm{i}\omega t})c_g^{(I)}(t) \,. \tag{9.55}$$

The term $\mathrm{e}^{\mathrm{i}\omega_{eg}t}$ was obtained from the interaction picture of the transition dipole operator. The right-hand side of (9.55) contains terms oscillating with frequencies $\omega_{eg} - \omega$ and $\omega_{eg} + \omega$. The former difference becomes zero at resonance and the corresponding term leads to a steady increase of $c_e^{(I)}(t)$. The second term oscillates with twice the frequency at resonance. Its contribution can therefore be disregarded. An effective equation for the expansion coefficient $c_e^{(I)}(t)$ is therefore

$$\frac{\partial}{\partial t}c_e^{(I)}(t) \approx \mathrm{i}d_{eg}E_0 c_g^{(I)}(t) \,. \tag{9.56}$$

This corresponds to a normal absorption of light.

The situation is more complicated when the direct transition from $|g\rangle$ to $|e\rangle$ is not allowed. Let us assume that at least some transitions from $|g\rangle$ to $|f_n\rangle$ and from $|f_n\rangle$ to $|e\rangle$ are allowed, that is,

$$\mu = d_{ef_n}|e\rangle\langle f_n| + \sum_n d_{gf_n}|g\rangle\langle f_n| + \text{h.c.} \tag{9.57}$$

We are still interested only in what happens in the subspace given by the projector $\mathcal{P} = |g\rangle\langle g| + |e\rangle\langle e|$, that is, the details of the system evolution concerning states $|f_n\rangle$ are to be considered by some effective theory. Equation (9.54) requires all states be known. However, applying the projector operator $\mathcal{P}$ to it would lead to an effective equation projected just onto the two states we are interested in.

By analogy with the previous section, we should expect the equation of motion for $\mathcal{P}|\psi^{(I)}(t)\rangle$ in the form of (9.42), with $\mathcal{K}_{\mathrm{CL}}^{(1)}(t)$ and $\mathcal{K}_{\mathrm{CL}}^{(2)}(t)$ in the corresponding operator forms:

$$\mathcal{K}_{\mathrm{CL}}^{(1)}(t) = \mathrm{i}\mathcal{P}\mu(t)E(t)\mathcal{P} \,, \tag{9.58}$$

$$\mathcal{K}_{\mathrm{CL}}^{(2)}(t) = \mathrm{i}\mathcal{P} \int_{t_0}^{t} \mathrm{d}\tau \mu(t) E(t) \mathcal{Q} \mu(\tau) E(\tau) \mathcal{P} \,. \tag{9.59}$$

Because the transition dipole moment μ contains $|f_n\rangle$ states in all terms (see (9.57)) the projection $\mathcal{P}\mu\mathcal{P}$ and correspondingly $\mathcal{K}_{\mathrm{CL}}^{(1)}$ are equal to zero. The only term that is able to bring the excitation to state $|e\rangle$ is therefore the term $\mathcal{K}_{\mathrm{CL}}^{(2)}$. The projected Schrödinger equation reads

$$\frac{\partial}{\partial t} \mathcal{P} |\psi^{(\mathrm{I})}(t)\rangle = -\mathcal{K}_{\mathrm{CL}}^{(2)}(t) \mathcal{P} |\psi^{(\mathrm{I})}(t)\rangle \,. \tag{9.60}$$

We will again assume that the population of states $|e\rangle$ and $|f_n\rangle$ is negligible with respect to the population of $|g\rangle$. Disregarding $c_e^{(\mathrm{I})}(t)$ on the right-hand side, we obtain

$$\frac{\partial}{\partial t} c_e^{(\mathrm{I})}(t) = -\int_{t_0}^{t} \mathrm{d}\tau \sum_n \mathrm{e}^{\mathrm{i}\epsilon_e/\hbar t} d_{e f_n} \mathrm{e}^{-\mathrm{i}\epsilon_{f_n}/\hbar (t-\tau)} d_{f_n g} \mathcal{E}(t) \mathcal{E}(\tau)$$
$$\times \left(\mathrm{e}^{-\mathrm{i}\omega t} + \mathrm{e}^{\mathrm{i}\omega t}\right)\left(\mathrm{e}^{-\mathrm{i}\omega\tau} + \mathrm{e}^{\mathrm{i}\omega\tau}\right) c_g^{(\mathrm{I})}(t) \,. \tag{9.61}$$

The integration can be performed, leading to

$$\frac{\partial}{\partial t} c_e^{(\mathrm{I})}(t) = -\mathrm{i} \sum_n \left(\mathrm{e}^{-\mathrm{i}(\omega - \omega_{eg} + \omega_{f_n g})t} + \mathrm{e}^{\mathrm{i}(\omega + \omega_{eg} - \omega_{f_n g})t}\right) d_{e f_n} d_{f_n g} \mathcal{E}(t) \mathcal{E}(t)$$
$$\times \left(\frac{\mathrm{e}^{-\mathrm{i}(\omega - \omega_{f_n g})t}}{\omega - \omega_{f_n g}} - \frac{\mathrm{e}^{\mathrm{i}(\omega + \omega_{f_n g})t}}{\omega + \omega_{f_n g}} \right) c_g^{(\mathrm{I})}(t) \,. \tag{9.62}$$

Here we assumed the slowly varying envelope approximation, and we assumed that $\mathcal{E}(t_0) = 0$.[1] There are now a number of conditions under which transition to state $|e\rangle$ can happen. We can identify four terms on the right-hand side of (9.62) which have to be integrated to obtain $c_e^{(\mathrm{I})}(t)$. The resonance conditions in the corresponding four cases are as follows:

$$\omega - \omega_{eg} + \omega_{f_n g} + \omega - \omega_{f_n g} = 2\omega - \omega_{eg} \approx 0 \,, \tag{9.63}$$

$$\omega - \omega_{eg} + \omega_{f_n g} - \omega - \omega_{f_n g} = \Delta\omega - \omega_{eg} \approx 0 \,, \tag{9.64}$$

$$\omega + \omega_{eg} - \omega_{f_n g} - \omega + \omega_{f_n g} = \Delta\omega + \omega_{eg} \approx 0 \,, \tag{9.65}$$

$$\omega + \omega_{eg} - \omega_{f_n g} + \omega + \omega_{f_n g} = 2\omega + \omega_{eg} \approx 0 \,. \tag{9.66}$$

Here, $\Delta\omega \approx 0$ represents a possible mismatch of the two appearences of ω due to finite width of the exciting pulses. First of all, condition (9.66) cannot be satisfied with positive frequencies, and it does not correspond to any physical process.

1) Integrating by parts, we obtain $\int_{t_0}^{t} \mathrm{d}\tau \mathcal{E}(\tau) \mathrm{e}^{\mathrm{i}\omega\tau} = [\mathcal{E}(\tau)\mathrm{e}^{\mathrm{i}\omega\tau}/\mathrm{i}\omega]_{t_0}^{t} - \int_{t_0}^{t} \mathrm{d}\tau \mathcal{E}'(\tau)(\mathrm{e}^{\mathrm{i}\omega\tau})/(\mathrm{i}\omega)$. The derivative $\mathcal{E}'(t)$ of the envelope $\mathcal{E}(t)$ is small and smooth so its integral with a fast exponential is negligible. Considering $\mathcal{E}(t_0) = 0$, we get $\int_{t_0}^{t} \mathrm{d}\tau \mathcal{E}(\tau) \mathrm{e}^{\mathrm{i}\omega\tau} \approx \mathcal{E}(t)(\mathrm{e}^{\mathrm{i}\omega t})/\mathrm{i}\omega$.

In conditions (9.63) and (9.66), the light frequency ω does not cancel out, while it completely cancels in conditions (9.64) and (9.65). In this latter case, we arrive at the condition $\omega_{eg} = 0$, which cannot be satisfied (we have no control over the transition frequency). However, if the excitation pulse carries a spectrum of frequencies with width $\Delta\omega$, the two occurrences of ω in (9.64) and (9.65) do not have to cancel exactly and the resonance condition can be fulfilled. We denote this by the resonance condition $\Delta\omega \pm \omega_{eg} \approx 0$. Alternatively, we could consider excitation by two pulses with different frequencies (different by the value of $\Delta\omega$), which would lead to the same resonance condition.

In (9.63) we arrived at the theory of *two-photon absorption*. Interestingly, although states $|f_n\rangle$ were instrumental in providing the channel for excitation, their transition frequencies do not appear in the resonance condition. Considering condition (9.63), we get from (9.62)

$$\frac{\partial}{\partial t} c_e^{(1)}(t) = -\mathrm{i}\left(\sum_n \frac{d_{\mathrm{e}f_n} d_{f_n\mathrm{g}}}{\omega - \omega_{f_n\mathrm{g}}}\right) \mathcal{E}(t)\mathcal{E}(t) c_g^{(1)}(t)\,, \tag{9.67}$$

which corresponds to the normal linear absorption process described by (9.56) only with the replacement $\mathcal{E}(t) \to \mathcal{E}(t)\mathcal{E}(t)$ and $d_{eg} \to \sum_n (d_{\mathrm{e}f_n} d_{f_n\mathrm{g}})/(\omega_{f_n\mathrm{g}} - \omega)$. The effective two-photon transition dipole moment,

$$D_{eg}^{(2)} = \sum_n \frac{d_{\mathrm{e}f_n} d_{f_n\mathrm{g}}}{\omega_{f_n\mathrm{g}} - \omega}\,, \tag{9.68}$$

contains the transition dipole moments and transition frequencies of states $|f_n\rangle$.

The conditions specified by (9.64) and (9.65) are even less specific when it comes to the value of the light frequency ω. They state that if the excitation pulse contains a spread of frequencies large enough to be equal to the transition frequency ω_{eg}, state $|e\rangle$ can be excited. Again, the transition frequencies $\omega_{f_n g}$ have no role in the resonance condition. As in two-photon absorption, they influence the effective transition dipole moment. The two conditions (9.64) and (9.65) correspond to the *Raman scattering* process.

10
Path Integral Technique in Dissipative Dynamics

In this chapter we make a short detour from the master-type equations describing the relaxation problems. The path integral approach (see, e.g., [47], and also [42], where its application to quantum dissipative systems is extensively described) is one of the techniques which applies quantum dynamics to classical trajectories. It can be applied in a wide variety of systems and it provides a very elegant extension to statistical physics. In the problem of quantum dissipative dynamics the path integral technique allows an easy description of mixed quantum–classical dissipative systems.

10.1
General Path Integral

Consider the coordinate and momentum operators. In the general spirit of quantum mechanics, the eigenstates of the position operator $\hat{q}$ can be defined using

$$\hat{q}|q\rangle = q|q\rangle \,. \tag{10.1}$$

They create the complete and orthogonal basis set in the sense $\langle q|q'\rangle = \delta(q-q')$, and

$$\int \mathrm{d}q|q\rangle\langle q| = 1 \,. \tag{10.2}$$

In the same way, we introduce the momentum operator eigenstates

$$\hat{p}|p\rangle = p|p\rangle \tag{10.3}$$

with the same properties $\langle p|p'\rangle = \delta(p-p')$, and

$$\int \mathrm{d}p|p\rangle\langle p| = 1 \,. \tag{10.4}$$

As momentum and coordinate states are equivalent, we can use either representation for quantum problems.

In the coordinate representation, the coordinate operator $\hat{q}_q = q$, and the canonical commutation relation, (Chapter 4), is satisfied when $\hat{p}_q = -\mathrm{i}\hbar\partial/\partial q$. In the momentum representation the result is similar, for example, $\hat{p}_p = p$, and the canonical relation is satisfied when $\hat{q}_p = \mathrm{i}\hbar\partial/\partial p$. The quantity of interest is now the scalar product of coordinate and momentum eigenstates denoted by $\langle q|p\rangle$. For this purpose we can write the eigenequation for the momentum operator in the coordinate representation:

$$\hat{p}\phi_p(q) = p\phi_p(q) , \tag{10.5}$$

which describes the wavefunction $\phi_p(q)$ of the momentum operator with momentum eigenvalue p. The normalized solution reads

$$\phi_p(q) = \frac{1}{\sqrt{2\pi\hbar}} \exp\left(\frac{\mathrm{i}}{\hbar} p q\right) , \tag{10.6}$$

with normalization

$$\int \mathrm{d}q \phi_p^*(q)\phi_{p'}(q) = \delta(p - p') . \tag{10.7}$$

In the coordinate representation, the wavefunction of the coordinate operator is given by

$$\hat{q}\psi_{q'}(q) = q'\psi_{q'}(q) , \tag{10.8}$$

which is satisfied by $\psi_{q'}(q) = \delta(q - q')$. Then,

$$\langle q|p\rangle \equiv \int \mathrm{d}q'\delta(q - q')\phi_p(q') = \frac{1}{\sqrt{2\pi\hbar}} \exp\left(\frac{\mathrm{i}}{\hbar} p q\right) . \tag{10.9}$$

This is an important relation for the path integral formulation in the coordinate representation.

Consider a particle with mass m moving in a one-dimensional space along coordinate q in a potential $V(q)$. The Hamiltonian describing this motion is

$$\hat{H} = \frac{\hat{p}^2}{2m} + V(\hat{q}) , \tag{10.10}$$

where $\hat{p}$ and $\hat{q}$ are the momentum and coordinate operators. The Schrödinger equation describes the time evolution of a state vector (see Chapter 4):

$$\frac{\partial}{\partial t}|\psi\rangle = -\frac{\mathrm{i}}{\hbar}\hat{H}|\psi\rangle . \tag{10.11}$$

We can now introduce the time-evolution operator

$$|\psi(t)\rangle = U(t)|\psi(0)\rangle , \tag{10.12}$$

where

$$U(t) = \exp\left(-\frac{\mathrm{i}}{\hbar}\hat{H}t\right) . \tag{10.13}$$

From (10.12) in the coordinate representation we write for the wavefunction

$$\begin{aligned}\psi(q,t) &= \int dq' \langle q|U(t)|q'\rangle \psi(q',0) \\ &\equiv \int dq' K(q,t;q',0)\psi(q',0)\,,\end{aligned} \qquad (10.14)$$

where we denote

$$K(q_f,t_f;q_0,t_0) \equiv \langle q_f|\exp\left(-\frac{i}{\hbar}\hat{H}(t_f-t_0)\right)|q_0\rangle\,. \qquad (10.15)$$

This value denotes the amplitude of the transition from system coordinate q_0 at time t_0 to coordinate q_f at later time t_f. It is a complex value, so it cannot be associated with the probability and is denoted by a probability amplitude. The probability is given by $|K(q_f,t_f;q_0,t_0)|^2$.

Due to the completeness of the coordinate representation (10.2), we can write the composition law of propagators:

$$K(q_2,t_2|q_0t_0) = \int dq_1 K(q_2,t_2|q_1t_1)K(q_1,t_1|q_0t_0)\,, \quad t_0 < t_1 < t_2\,. \qquad (10.16)$$

This bears strong similarity with the stochastic Markovian dynamics described by the equivalent Chapman–Kolmogorov equation for transition probabilities as in Chapter 3. Let us divide the time interval into many equidistant N intervals as shown in Figure 10.1:

$$t_f - t_0: \quad t_n = t_0 + n\epsilon\,, \quad \epsilon = (t_f - t_i)/N\,, \qquad (10.17)$$

and $n = 0\ldots N$. $n = 0$ denotes the initial point t_0, while $n = N$ denotes the final point t_f. The full propagator is then

$$\begin{aligned}K(q_f,t_f|q_0,t_0) = &\int dq_{N-1}\cdots\int dq_2\int dq_1 \\ &K(q_f,t_f|q_{N-1},t_{N-1})\cdots K(q_2,t_2|q_1,t_1)K(q_1,t_1|q_0,t_0)\end{aligned} \qquad (10.18)$$

(notice that the initial and final positions are not integrated over), where

$$K(q_n,t_n|q_{n-1},t_{n-1}) = \langle q_n|\exp\left(-\frac{i}{\hbar}\hat{H}\epsilon\right)|q_{n-1}\rangle \qquad (10.19)$$

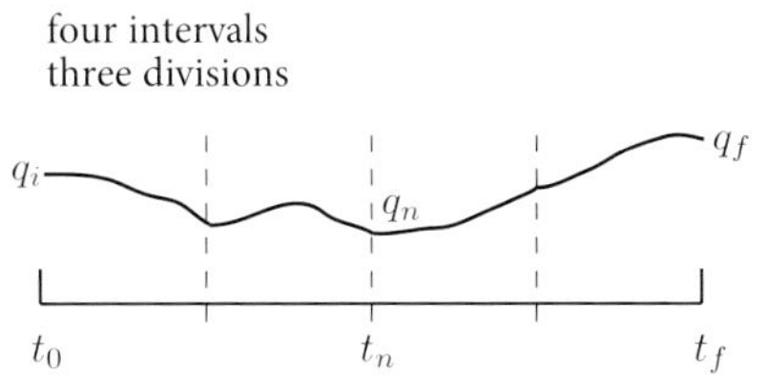

Figure 10.1 Construction of the path integral. The total interval is divided into $N = 4$ intervals by three divisions. The coordinate is followed at each interval.

is the elementary propagator over the infinitesimal interval ϵ.

The infinitesimal propagator can be easily calculated. If we take $N \to \infty$ and $\epsilon \to 0$, we can expand the exponent in (10.19) as

$$K(q_n, t_n | q_{n-1} t_{n-1}) = \langle q_n | 1 - \frac{\mathrm{i}}{\hbar} \hat{H} \epsilon + \ldots | q_{n-1} \rangle \,. \tag{10.20}$$

Using the properties of the Fourier integral, we obtain from the first term

$$\langle q_n | q_{n-1} \rangle = \delta(q_n - q_{n-1}) \equiv \int \frac{\mathrm{d}p}{2\pi\hbar} \exp\left(\mathrm{i}\frac{p}{\hbar}(q_n - q_{n-1})\right) \,. \tag{10.21}$$

The second term from the Hamiltonian (10.10) reads

$$\langle q_n | \frac{\mathrm{i}}{\hbar} \hat{H} \epsilon | q_{n-1} \rangle = \frac{\mathrm{i}}{\hbar} \epsilon \langle q_n | \frac{\hat{p}^2}{2m} | q_{n-1} \rangle + \frac{\mathrm{i}}{\hbar} \epsilon \langle q_n | V(\hat{q}) | q_{n-1} \rangle \,. \tag{10.22}$$

For the kinetic energy part we insert the full set of momentum operators and use (10.9):

$$\begin{aligned} \langle q_n | \frac{\hat{p}^2}{2m} | q_{n-1} \rangle &= \int \mathrm{d}p_{n-1} \langle q_n | \frac{\hat{p}^2}{2m} | p_{n-1} \rangle \langle p_{n-1} | q_{n-1} \rangle \\ &= \int \frac{\mathrm{d}p_{n-1}}{2\pi\hbar} \exp\left(\frac{\mathrm{i}}{\hbar} p_{n-1}(q_n - q_{n-1})\right) \frac{p_{n-1}^2}{2m} \,. \end{aligned} \tag{10.23}$$

The potential energy from (10.21) gives

$$\begin{aligned} \langle q_n | V(\hat{q}) | q_{n-1} \rangle &= V(q_{n-1}) \delta(q_n - q_{n-1}) \\ &= \int \frac{\mathrm{d}p_{n-1}}{2\pi\hbar} \exp\left(\frac{\mathrm{i}}{\hbar} p_{n-1}(q_n - q_{n-1})\right) V(q_{n-1}) \,. \end{aligned}$$

Inside the integral the sum of kinetic and potential energy terms gives

$$\frac{p^2}{2m} + V(q) \equiv H(p, q) \,, \tag{10.24}$$

which is the classical Hamiltonian function (not an operator). The elementary propagator is then reconstructed by taking the small terms into the exponent:

$$K(q_n, t_n | q_{n-1} t_{n-1}) = \int \frac{\mathrm{d}p_{n-1}}{2\pi\hbar} \mathrm{e}^{\mathrm{i}\frac{p_{n-1}(q_n - q_{n-1})}{\hbar} - \frac{\mathrm{i}}{\hbar}\epsilon H(p_{n-1}, q_{n-1})} \,. \tag{10.25}$$

We can now combine (10.18) to get the full propagator:

$$\begin{aligned} K(q_\mathrm{f}, t_\mathrm{f} | q_0 t_0) &= \left(\prod_{j=1}^{N-1} \int \mathrm{d}q_j\right) \left(\prod_{k=0}^{N-1} \int \frac{\mathrm{d}p_k}{2\pi\hbar}\right) \\ &\quad \times \exp\left(\frac{\mathrm{i}}{\hbar} \sum_{l=0}^{N-1} \epsilon (p_l \dot{q}_l - H(p_l, q_l))\right) \,, \end{aligned} \tag{10.26}$$

where $\dot{q}_l = (q_{l+1} - q_l)/\epsilon$. Notice that there is one p integral for each interval (N intervals), while there is one less q integral over points connecting these intervals. In the continuous limit we write *the general path integral*

$$K(q_f, t_f|q_0 t_0) = \int_{q_0}^{q_f} \mathcal{D}q \int \mathcal{D}p \exp\left(\frac{i}{\hbar}\int_{t_0}^{t_f} d\tau\, L(p(\tau)q(\tau))\right), \tag{10.27}$$

where

$$\int \mathcal{D}p = \prod_{k=0}^{N-1} \int \frac{dp_k}{2\pi\hbar}, \tag{10.28}$$

$$\int \mathcal{D}q = \prod_{k=1}^{N-1} \int dq_k . \tag{10.29}$$

Here and in the following the absence of integration limits implies the integration over the whole space, i.e. from $-\infty$ to $+\infty$. In the exponent we get the classical Lagrangian (not an operator) as a function of the paths in the phase space defined by $q(t)$ and $p(t)$:

$$L(p(\tau)q(\tau)) = p(\tau)\dot{q}(\tau) - H(p(\tau), q(\tau)) . \tag{10.30}$$

For the Hamiltonian form as in (10.10) we have an independent momentum-related term. So in the path integral we have independent Gaussian-type integrals over the momentum. These can be integrated by using

$$\int dp \exp\left(a\left(p\dot{q} - \frac{p^2}{2m}\right)\right) = \exp\left(a\frac{m\dot{q}^2}{2}\right)\sqrt{\frac{2\pi m}{a}} \tag{10.31}$$

by extending it to imaginary space ($a = i\epsilon/\hbar$). We then get the path integral in the sole coordinate space as a set of N trajectories and $N-1$ integrals:

$$K(q_f, t_f|q_0 t_0) = \left(\frac{m}{2\pi\hbar i\epsilon}\right)^{N/2}\left(\prod_{j=1}^{N-1}\int dq_j\right)\exp\left(\frac{i\epsilon}{\hbar}\sum_{l=0}^{N-1}\left(\frac{m\dot{q}_l^2}{2} - V(q_l)\right)\right) \tag{10.32}$$

or in the continuum version

$$K(q_f, t_f|q_0 t_0) = \int_{q_0}^{q_f} \mathcal{D}q \exp\left(\frac{i}{\hbar} S(q(\tau))\right), \tag{10.33}$$

where we assume the normalization constants are absorbed in $\mathcal{D}q$. The functional

$$S(q(\tau)) = \int_{t_0}^{t_f} d\tau \left(\frac{m\dot{q}^2(\tau)}{2} - V(q(\tau))\right) \tag{10.34}$$

denotes the classical action functional given for a path $q(\tau)$ starting at q_i at t_0 and ending at q_f at t_f. The path is continuous in time and is forward in time (i.e., it has no loops or parts with back-propagation).

Note that the path integral is the symbolical expression whose explicit simulation would require us to calculate the discrete expression (10.26). On the other hand, when the action is calculated, the proper normalization can be defined using some specific known-limit cases.

10.1.1 Free Particle

Let us now consider a free particle. In this case $V(q) = 0$ and the propagator is given by

$$K(q_\mathrm{f}, t|q_0 0) = \langle q_\mathrm{f}| \exp\left(-\frac{\mathrm{i}}{\hbar}\hat{H}t\right)|q_0\rangle \,. \tag{10.35}$$

The Hamiltonian is

$$\hat{H} = \frac{\hat{p}^2}{2m} \,. \tag{10.36}$$

We then have

$$K(q_\mathrm{f}, t|q_0 0) = \langle q_\mathrm{f}| \exp\left(-\frac{\mathrm{i}}{\hbar}\frac{\hat{p}^2 t}{2m}\right)|q_0\rangle \,. \tag{10.37}$$

Inserting the momentum eigenstate complete basis, we get the Gaussian integral

$$K(q_\mathrm{f}, t|q_0 0) = \int \frac{\mathrm{d}p}{2\pi\hbar} \exp\left(\frac{\mathrm{i}}{\hbar}p(q_\mathrm{f} - q_0) - \frac{\mathrm{i}t}{\hbar}\frac{p^2}{2m}\right) , \tag{10.38}$$

which finally gives

$$K(q_\mathrm{f}, t|q_0 0) = \sqrt{\frac{m}{2\pi\mathrm{i}\hbar t}} \exp\left(\frac{\mathrm{i}m}{2\hbar t}(q_\mathrm{f} - q_0)^2\right) . \tag{10.39}$$

The same result should be available from the path integral. The action over the paths is given by

$$S(q(\tau)) = \frac{m}{2}\int_0^t \dot{q}(\tau)\mathrm{d}(q(\tau)) = \frac{m}{2}(\dot{q}_f q_\mathrm{f} - \dot{q}_0 q_0) \,. \tag{10.40}$$

Now we partition the path into the classical $\bar{q}$ part and the deviation y:

$$q(\tau) = \bar{q}(\tau) + y(\tau) \,. \tag{10.41}$$

For the classical paths the straight paths contribute

$$\bar{q}(\tau) = q_0\frac{t-\tau}{t} + q_\mathrm{f}\frac{\tau}{t} \,. \tag{10.42}$$

These give

$$S(\bar{q}(\tau)) = \frac{m}{2t}(q_\mathrm{f} - q_0)^2 \,. \tag{10.43}$$

For the quantum paths we can take the Fourier series

$$y(\tau) = \sum_{\alpha=1}^{\infty} \xi_\alpha \sin(\nu_\alpha \tau) \,. \tag{10.44}$$

Here $\nu_\alpha = \pi\alpha/t$ guarantees boundary conditions $y(0) = y(\tau) = 0$. The action functional over the quantum paths is given by

$$S(y(\tau)) = \int_0^t \mathrm{d}\tau \frac{m\dot{y}^2(\tau)}{2} = \sum_{\alpha=1}^{\infty} \xi_\alpha^2 \frac{m\nu_\alpha^2}{2} \int_0^t \mathrm{d}\tau \cos^2(\nu_\alpha \tau) \,, \tag{10.45}$$

which gives

$$S(y(\tau)) = \frac{m\pi^2}{4t} \sum_{\alpha=1}^{\infty} \xi_\alpha^2 \alpha^2 \,. \tag{10.46}$$

The path integral over the quantum paths is

$$\int \mathcal{D}y \exp(\ldots) \propto \prod_\alpha \left(\int_{-\infty}^{\infty} \mathrm{d}\xi_\alpha \right) \exp\left(\frac{\mathrm{i}}{\hbar} S(y(\tau)) \right) \,, \tag{10.47}$$

which gives the Gaussian integrals over the ξ_α. However, note that this quantum contribution does not affect the dependence on the endpoints q_0 and q_f. This path integral thus affects only the normalization of the probability amplitude. The normalization factor can be easily obtained from

$$\int \mathrm{d}q_\mathrm{f} |K(q_\mathrm{f}, t|q_0 0)|^2 = 1 \,, \tag{10.48}$$

which finally gives the same result as (10.39).

10.1.2 Classical Brownian Motion

In this chapter we are describing quantum path integrals. However, we can use the path integral to describe a simple stochastic process – Brownian motion. In the simplest model, Brownian motion is described using the diffusion equation

$$\frac{\partial w(x,t)}{\partial t} = D \frac{\partial^2 w(x,t)}{\partial x^2} \,. \tag{10.49}$$

The solution for the initial condition is

$$w(x,t) = \frac{1}{\sqrt{4\pi D t}} \exp\left(-\frac{x^2}{4Dt} \right) \,. \tag{10.50}$$

We can introduce the propagator

$$w(x,t) = \int \mathrm{d}x_i \, K(x,t|x_i t_i) w(x_i t_i) \,. \tag{10.51}$$

The propagator is then given by

$$K(x,t|x't') = \frac{1}{\sqrt{4\pi D(t-t')}} \exp\left(-\frac{(x-x')^2}{4D(t-t')}\right) \,. \tag{10.52}$$

Let us now divide the time interval $t - t'$ into N intervals and consider the compound probability that the particle is at point x_n at time t_n in the interval $\mathrm{d}x$. The probability of such a path $x(\tau)$ is

$$\begin{aligned} &\left(\frac{\mathrm{d}x}{\sqrt{4\pi D\epsilon}}\right)^N \exp\left(-\frac{(x_1-x_0)^2}{4D\epsilon} - \frac{(x_2-x_1)^2}{4D\epsilon} \cdots - \frac{(x_N-x_{N-1})^2}{4D\epsilon}\right) \\ &= \left(\frac{\mathrm{d}x(\tau)}{\sqrt{4\pi D \mathrm{d}\tau}}\right)^N \exp\left(-\frac{1}{4D}\int_{t'}^{t} \mathrm{d}\tau \dot{x}^2(\tau)\right) \,. \end{aligned} \tag{10.53}$$

Now if we sum up over all possible trajectories, or integrate over all intermediate points x_n, we will get the full propagator in terms of the path integral:

$$K(x_2,t_2|x_1 t_1) = \int_{x_1}^{x_2} \mathcal{D}x \, \exp\left(-\frac{1}{4D}\int_{t'}^{t} \mathrm{d}\tau \dot{x}^2(\tau)\right) \,. \tag{10.54}$$

This result also introduces the weight of a single trajectory. All trajectories start at x_1 and end at x_2 and

$$\int_{x_1}^{x_2} \mathcal{D}x \equiv \lim_{N\to\infty} \left(\int \frac{\mathrm{d}x}{\sqrt{4\pi D\epsilon}}\right)^N \,. \tag{10.55}$$

Similarly, the path integral can be applied to the overdamped oscillator motion in a harmonic potential (see Section 3.6). In that case we will get a slightly different weight for a single trajectory. The stochastic process $\xi(\tau)$ is then described by the path integral of the form

$$K_{\text{osc}}(\xi, t|\xi_0 t_0) = \int_{\xi_0}^{\xi} \mathcal{D}\xi \, P_{\text{osc}}(\xi(\tau)) \,. \tag{10.56}$$

The probability weight of each path

$$P_{\text{osc}}(\xi(\tau)) = \exp\left(-\frac{1}{4D}\int_{t_i}^{t_f} \mathrm{d}\tau (\dot{\xi}(\tau) + D\xi(\tau))^2\right) \,. \tag{10.57}$$

Classical stochastic processes can thus be easily described by the simple path integrals and these can be used in the mixed semiclassical models as we show later in this chapter.

10.2
Imaginary-Time Path Integrals

In the previous section we showed how the quantum mechanical transition probability amplitude can be given in terms of the paths in space. We will use this result to derive equations of motion for the density matrix. In this section we present an elegant connection between the path integral expression of the probability amplitude and the partition function of statistical physics.

The partition function of the canonical ensemble was described in Chapter 7 and is given by

$$Z = \mathrm{tr}(\mathrm{e}^{-\beta \hat{H}}) , \tag{10.58}$$

where $\beta = (k_B T)^{-1}$. Consider the operator

$$\hat{Z}(\beta) = \mathrm{e}^{-\beta \hat{H}} . \tag{10.59}$$

It is equivalent to the evolution operator defined by (10.13) if we extend the time axis into the imaginary dimension by taking $\beta = \mathrm{i}t\hbar^{-1}$. Then $\hat{Z}(\beta) = \hat{U}(t = -\mathrm{i}\beta\hbar)$. In the coordinate representation we can then calculate the amplitude

$$\langle q_\mathrm{f}|\hat{Z}(\beta)|q_i\rangle = \langle q_\mathrm{f}|\mathrm{e}^{-\beta \hat{H}}|q_i\rangle , \tag{10.60}$$

and by taking the trace, we find the partition function

$$Z = \int \mathrm{d}q \langle q|\mathrm{e}^{-\beta \hat{H}}|q\rangle . \tag{10.61}$$

Knowledge of how to calculate the propagators can be easily translated to the problem of the partition function. Using the analogy with the path integral of the previous section, taking $t_0 = 0$ and $t_\mathrm{f} = -\mathrm{i}\beta\hbar$, as well as $q_0 = q_\mathrm{f} = q_N$, we find the imaginary-time path integral expression for the partition function:

$$Z = \int \mathrm{d}q \int_q^q \mathcal{D}q' \exp\left(-\frac{1}{\hbar} S^{(\mathrm{E})}(q'(\tau))\right) , \tag{10.62}$$

where now

$$S^{(\mathrm{E})}(q(\tau)) = \int_0^{\beta\hbar} \mathrm{d}\tau \left[\frac{m\dot{q}(\tau)^2}{2} + V(q(\tau))\right] . \tag{10.63}$$

The functional $S^{(\mathrm{E})}$ is called the *Euclidean action* for a path starting and ending at the same coordinate q. The starting time is 0 and the time propagates along the imaginary axis. The integrals

$$\int \mathrm{d}q \int_q^q \mathcal{D}q \equiv \int \mathcal{D}_1 q \tag{10.64}$$

can be denoted as a single path integral, whose endpoints are additionally integrated over the whole space.

In the same way we can now write the canonical density matrix, which in the coordinate representation reads

$$W(q_2 q_1) = Z^{-1} \langle q_2 | \hat{Z}(\beta) | q_1 \rangle = Z^{-1} \int_{q_1}^{q_2} \mathcal{D}q \exp\left(-\frac{1}{\hbar} S^{(E)}(q(\tau))\right) . \tag{10.65}$$

Now here the initial and final propagation coordinates of the trajectory $q(\tau)$ are different, leading to the canonical density matrix in the coordinate representation.

When the total system consists of an observable system and a bath we have the additive Hamiltonian

$$\hat{H} = \hat{H}_S + \hat{H}_B + \hat{H}_I . \tag{10.66}$$

We denote the system coordinates as q and the bath coordinates as x. In the same way, we get for the Euclidean action

$$S^{(E)} = S_S^{(E)}(q(\tau)) + S_B^{(E)}(x(\tau)) + S_I^{(E)}(q(\tau), x(\tau)) . \tag{10.67}$$

For the observable system we can easily define the reduced canonical density matrix by tracing over the bath coordinates:

$$\rho(q_2 q_1) \equiv \text{tr}_B W(q_2 x, q_1 x) = Z^{-1} \int_{q_1}^{q_2} \mathcal{D}q \int \mathcal{D}_1 x \times \exp\left(-\frac{1}{\hbar} S_S^{(E)}(q(\tau)) - \frac{1}{\hbar} S_B^{(E)}(x(\tau)) - \frac{1}{\hbar} S_I^{(E)}(q(\tau), x(\tau))\right) . \tag{10.68}$$

Note that the action functionals in the exponent are classical functions (not operators). The exponent thus factorizes. Additionally we define the reduced partition function of the system $Z_S Z_R = Z$ and we get

$$\rho(q_2 q_1) = Z_S^{-1} \int_{q_1}^{q_2} \mathcal{D}q \exp\left(-\frac{1}{\hbar} S_S^{(E)}(q(\tau))\right) \mathcal{F}^{(E)}(q(\tau)) , \tag{10.69}$$

where we introduced the *influence functional*

$$\mathcal{F}^{(E)}(q(\tau)) = Z_R^{-1} \int \mathcal{D}_1 x \exp\left(-\frac{1}{\hbar} S_B^{(E)}(x(\tau)) - \frac{1}{\hbar} S_I^{(E)}(q(\tau), x(\tau))\right) . \tag{10.70}$$

In the influence functional all paths run over loops with the same initial and final coordinates. That is not the case in the expression for the system density matrix in (10.69). In this path integral, the density matrix is defined as a regular path integral of system quantities. However, the bath affects the system through the influence functional.

10.3
Real-Time Path Integrals and the Feynman–Vernon Action

In the previous sections we described a general formulation of the path integrals. In this section we consider dissipative dynamics of open systems. We thus need to consider the time dependence of the reduced density matrix. The time dependence of the full density matrix of the global system is given by

$$W(t) = \mathrm{e}^{-\mathrm{i}\hat{H}t/\hbar}\,\hat{W}(0)\mathrm{e}^{\mathrm{i}\hat{H}t/\hbar}\,. \tag{10.71}$$

In the coordinate representation we have

$$\begin{aligned}\langle q_\mathrm{f}, x_\mathrm{f}|\,\hat{W}(t)|q'_\mathrm{f}x'_\mathrm{f}\rangle = \int \mathrm{d}q_\mathrm{i} \int \mathrm{d}q'_\mathrm{i} \int \mathrm{d}x_\mathrm{i} \int \mathrm{d}x'_\mathrm{i}\\ K(q_\mathrm{f}x_\mathrm{f}, t|q_\mathrm{i}x_\mathrm{i}, 0)\langle q_\mathrm{i}x_\mathrm{i}|\,\hat{W}(0)|q'_\mathrm{i}x'_\mathrm{i}\rangle\, K^*(q'_\mathrm{f}x'_\mathrm{f}, t|q'_\mathrm{i}x'_\mathrm{i}, 0)\,.\end{aligned} \tag{10.72}$$

$K(q_\mathrm{f}x_\mathrm{f}, t|q_\mathrm{i}x_\mathrm{i}, 0)$ is the coordinate representation of the time-evolution operator in the space of the system and bath coordinates,

$$K(q_\mathrm{f}x_\mathrm{f}, t|q_\mathrm{i}x_\mathrm{i}, 0) = \langle q_\mathrm{f}x_\mathrm{f}|\mathrm{e}^{-\mathrm{i}\hat{H}t/\hbar}|q_\mathrm{i}x_\mathrm{i}\rangle\,, \tag{10.73}$$

which we described in previous sections in this chapter. For the propagators we can now write the real-time path integral

$$K(q_\mathrm{f}x_\mathrm{f}, t|q_\mathrm{i}x_\mathrm{i}, 0) = \int_{q_\mathrm{i}}^{q_\mathrm{f}} \mathcal{D}q \int_{x_\mathrm{i}}^{x_\mathrm{f}} \mathcal{D}x \exp\left(\frac{\mathrm{i}}{\hbar} S(q(\tau), x(\tau))\right)\,. \tag{10.74}$$

The functional integration has endpoints $q(0) = q_\mathrm{i}$, $q(t) = q_\mathrm{f}$, $x(0) = x_\mathrm{i}$, and $x(t) = x_\mathrm{f}$. The total action is given by a sum of Lagrangians:

$$S = S_\mathrm{S} + S_\mathrm{R} + S_\mathrm{I} = \int_0^t \mathrm{d}\tau \left[L_\mathrm{S}(q(\tau)) + L_\mathrm{R}(x(\tau)) + L_\mathrm{I}(q(\tau), x(\tau))\right]\,. \tag{10.75}$$

The reduced density matrix of the system is then given by taking the trace over the bath:

$$\rho(q_\mathrm{f}q'_\mathrm{f}, t) = \int \mathrm{d}x_\mathrm{f}\langle q_\mathrm{f}, x_\mathrm{f}|\,\hat{W}(t)|q'_\mathrm{f}x_\mathrm{f}\rangle\,. \tag{10.76}$$

Since the action S is the classical functional, the conjugation amounts to complex conjugation; thus, the density matrix can be immediately described by the combined path integral of bra and ket parts.

Let us assume that the initial state is a product state of the system and the bath. The bath is in a canonical ensemble. We can then write

$$\hat{W}(0) = \hat{\rho}(0) \otimes \frac{1}{Z_\mathrm{R}} \exp\left(-\beta \hat{H}_\mathrm{R}\right)\,. \tag{10.77}$$

For the reduced density matrix we insert (10.77) into (10.72) and then into (10.76). For propagators we use the path integral expression (10.74) to finally get a compact expression for the reduced density matrix:

$$\rho\left(q_f q_f', t\right) = \int \mathrm{d}q_i \int \mathrm{d}q_i' \mathcal{J}\left(q_f, q_f', t | q_i, q_i', 0\right) \rho\left(q_i, q_i', 0\right) , \tag{10.78}$$

with the reduced density matrix propagator

$$\mathcal{J}(q_f, q_f', t | q_i, q_i', 0) = \int_{q_i}^{q_f} \mathcal{D}q \int_{q_i'}^{q_f'} \mathcal{D}q' \exp\left(\frac{i}{\hbar} S_S(q(\tau)) - \frac{i}{\hbar} S_S(q'(\tau))\right) \mathcal{F}(q(\tau), q'(\tau)) . \tag{10.79}$$

Here the functional

$$\mathcal{F}(q(\tau), q'(\tau)) = \frac{1}{Z_R} \int \mathrm{d}x_f \int \mathrm{d}x_i \int \mathrm{d}x_i' \langle x_i | e^{-\beta \hat{H}_R} | x_i' \rangle \int_{x_i}^{x_f} \mathcal{D}x \int_{x_i'}^{x_f} \mathcal{D}x' \times \exp\left(+\frac{i}{\hbar} S_R[x] + \frac{i}{\hbar} S_I[x, q] - \frac{i}{\hbar} S_R[x'] - \frac{i}{\hbar} S_I[x', q']\right) \tag{10.80}$$

is called the *Feynman–Vernon (FV) action functional* [48]. The FV functional accounts for the effect of the bath on the system during the propagation. This functional depends only on the bath and the system–bath interaction and it can be calculated separately. It consists of two conjugate propagators as functionals of paths $x(\tau)$ and $x'(\tau)$ and the equilibrium density matrix, which can be given in terms of the Euclidean path integral [49].

Next let us consider the bath of harmonic oscillators [49] as defined in the Caldeira–Leggett Hamiltonian (8.5). The bath-related Lagrangians are then given by

$$L_R = \sum_{\alpha=1}^{N} \frac{m_\alpha}{2} \left(\dot{x}_\alpha^2(t) - \omega_\alpha^2 x_\alpha^2(t)\right) , \tag{10.81}$$

$$L_I = \sum_{\alpha=1}^{N} \left(c_\alpha x_\alpha(t) q(t) - \frac{1}{2} \frac{c_\alpha^2}{m_\alpha \omega_\alpha^2} q^2(t)\right) . \tag{10.82}$$

In this form the exponents of the FV functional factorize into those for different bath oscillators. Now in the exponent of the influence functional we have Gaussian integrals, which can be evaluated exactly. From (10.80) we take the direct propagator for a single bath oscillator, given by

$$\int_{x_i}^{x_f} \mathcal{D}x \exp\left(\frac{i}{\hbar} S_R(x(\tau)) + \frac{i}{\hbar} S_I(q(\tau), x(\tau))\right) . \tag{10.83}$$

Let us take the path $x(\tau)$ for the bath oscillator as

$$x(\tau) = \tilde{x}(\tau) + y(\tau) . \tag{10.84}$$

Here $\tilde{x}$ is the classical path and y is the deviation from the classical path. After inserting the Lagrangian into the path integral, we get a single integral over the classical path and the path integral over the deviation:

$$\exp\left(\frac{i}{\hbar}\int_0^t d\tau \left(\frac{m_\alpha}{2}(\dot{\tilde{x}}^2(\tau) - \omega^2\tilde{x}^2(\tau)) + c\tilde{x}(\tau)q(\tau) - \frac{1}{2}\frac{c_\alpha^2}{m\omega^2}q^2(\tau)\right)\right)$$
$$\times \int \mathcal{D}y \exp\left(\frac{i}{\hbar}\int_0^t d\tau \frac{m_\alpha}{2}\left(\dot{y}^2(\tau) - \omega^2 y^2(\tau)\right)\right) . \tag{10.85}$$

The functional with respect to deviation $y(\tau)$ represents the quantum corrections. We have also dropped all terms linear in y. This is because $\tilde{x}(\tau)$ is the classical path that minimizes the action, while other terms of the classical paths do not contribute.

The action of the classical trajectory is computed by performing the integration by parts with respect to the $\dot{\tilde{x}}^2$ term and using the equation of motion, (8.8). We then have

$$\int_0^t d\tau \left[\frac{m}{2}(\dot{\tilde{x}}^2 - \omega^2\tilde{x}^2) + c\tilde{x}q - \frac{1}{2}\frac{c^2}{m\omega^2}q^2\right]$$
$$= \frac{1}{2}\left(\dot{x}_f x_f - \dot{x}_i x_i\right) + \frac{c}{2}\int_0^t d\tau \tilde{x}(\tau)q(\tau) - \frac{1}{2}\frac{c^2}{m\omega^2}\int_0^t d\tau q^2(\tau) . \tag{10.86}$$

We use $\tilde{x}(0) = x_i$ and $\tilde{x}(t) = x_f$.

The next part is the evaluation of the path integral over the fluctuations $y(\tau)$. These paths start and end at $y = 0$. The corresponding path integral can be evaluated by using the Fourier series

$$y(\tau) = \sum_{\alpha=1}^{\infty} \xi_\alpha \sin(\nu_\alpha \tau) . \tag{10.87}$$

Here $\nu_\alpha = \pi\alpha/t$ guarantees proper boundary conditions. The second row in (10.85) due to orthogonality of trigonometric functions is then (apart from the normalization)

$$\int \mathcal{D}y\,\exp(\ldots) \propto \prod_\alpha \left(\int_{-\infty}^{\infty} \mathrm{d}\xi_\alpha\right) \exp\left(\frac{imt}{4\hbar}\sum_\alpha \xi_\alpha^2\left(\nu_\alpha^2-\omega^2\right)\right) . \tag{10.88}$$

These are essentially the Gaussian independent integrals that give

$$\int_{-\infty}^{\infty} \mathrm{d}x\,\exp\left(\frac{imt}{4\hbar}\left(\nu_\alpha^2-\omega^2\right)x^2\right) = \sqrt{\frac{4\hbar\pi}{imt\left(\nu_\alpha^2-\omega^2\right)}} . \tag{10.89}$$

Noting that

$$\prod_{\alpha=1}^{\infty}\left[1-\left(\frac{x}{\pi\alpha}\right)^2\right] = \frac{\sin(x)}{x} , \tag{10.90}$$

we have

$$\int \mathcal{D}y\,\exp(\ldots) \propto \sqrt{\frac{\omega t}{\sin(\omega t)}} . \tag{10.91}$$

The normalization constant can be obtained by comparing the result with the free particle result ($\omega \to 0$) in Section 10.1.1. which finally gives

$$\int \mathcal{D}y\,\exp(\ldots) = \sqrt{\frac{m\omega}{2\pi i\hbar\sin(\omega t)}} . \tag{10.92}$$

The classical path for predefined initial conditions was given by (8.16). A solution with the boundary condition $\tilde{x}(0) = x_i$ and $\tilde{x}(t) = x_f$ can be obtained from the initial-value solution by choosing the appropriate $p^{(0)}$ value:

$$p^{(0)} = \frac{m\omega x_f}{\sin(\omega t)} - m\omega x_i\frac{\cos(\omega t)}{\sin(\omega t)} - c\int_0^t \frac{\sin(\omega(t-s))}{\sin(\omega t)}q(s)\mathrm{d}s , \tag{10.93}$$

which yields

$$\begin{aligned}\tilde{x}(\tau) = {} & x_i\frac{\sin(\omega(t-\tau))}{\sin(\omega t)} + x_f\frac{\sin(\omega\tau)}{\sin(\omega t)} + \frac{c}{m\omega}\int_0^\tau \sin(\omega(\tau-s))q(s)\mathrm{d}s \\ & - \frac{c}{m\omega}\frac{\sin(\omega\tau)}{\sin(\omega t)}\int_0^t \sin(\omega(t-s))q(s)\mathrm{d}s .\end{aligned} \tag{10.94}$$

Inserting this trajectory into the expression for the classical action, (10.86), we get the value of the classical action:

$$\begin{aligned} S^{(\mathrm{cl})} &= \frac{m\omega}{2\sin(\omega t)}\left[\left(x_\mathrm{f}^2 + x_\mathrm{i}^2\right)\cos(\omega t) - 2x_\mathrm{i}x_\mathrm{f}\right] \\ &+ \frac{c x_\mathrm{i}}{\sin(\omega t)}\int_0^t \mathrm{d}s \sin(\omega(t-s))q(s) + \frac{c x_\mathrm{f}}{\sin(\omega t)}\int_0^t \mathrm{d}s \sin(\omega s)q(s) \\ &- \frac{c^2}{m\omega\sin(\omega t)}\int_0^t \mathrm{d}s \int_0^s \mathrm{d}u \sin(\omega(t-s))\sin(\omega u)q(s)q(u) \\ &- \frac{c^2}{m\omega^2}\int_0^t \mathrm{d}s q^2(s)\,. \end{aligned} \tag{10.95}$$

The remaining task is to calculate the trace over the initial bath density matrix. In Chapter 7 we gave the equilibrium density matrix of a harmonic oscillator in the coordinate representation (7.78). We can rewrite it as

$$\begin{aligned} \rho_\mathrm{B}(x, x') &= \prod_\alpha 2\sinh\left(\frac{\beta\omega_\alpha}{2}\right)\left[\frac{m_\alpha\omega_\alpha}{2\pi\hbar\sinh(\beta\omega_\alpha)}\right]^{1/2} \\ &\times \exp\left(-\frac{m_\alpha\omega_\alpha}{2\hbar\sinh(\beta\omega_\alpha)}\left((x^2 + x'^2)\cosh\beta\omega_\alpha - 2xx'\right)\right)\,. \end{aligned} \tag{10.96}$$

The remaining integrals over the endpoints x_i and x_f are then again the Gaussian integrals, which finally give

$$\mathcal{F}_\mathrm{FV}(q(\tau), q'(\tau)) = \exp\left(-\frac{1}{\hbar}S_\mathrm{FV}(q(\tau), q'(\tau))\right)\,, \tag{10.97}$$

and the FV influence action is [48]

$$\begin{aligned} S_\mathrm{FV}(q(\tau), q'(\tau)) &= \int_0^t \mathrm{d}t' \int_0^{t'} \mathrm{d}t''(q(t') - q'(t')) \\ &\quad \times [C(t'-t'')q(t'') - C^*(t'-t'')q'(t'')] \\ &\quad + \mathrm{i}\lambda \int_0^t \mathrm{d}t'(q^2(t') - q'^2(t'))\,. \end{aligned} \tag{10.98}$$

Here

$$C(t) = \int \frac{\mathrm{d}\omega}{2\pi} C''(\omega)\left[\coth\left(\beta\hbar\frac{\omega}{2}\right)\cos(\omega t) - \mathrm{i}\sin(\omega t)\right] \tag{10.99}$$

is the equilibrium bath quantum correlation function given by the spectral density and

$$\lambda = \sum_\alpha \frac{c_\alpha^2}{2m_\alpha \omega_\alpha^2} = \int \frac{d\omega}{2\pi} \frac{C''(\omega)}{\omega}$$

is the potential renormalization term, usually denoted as the reorganization energy. The FV influence action has self-renormalization terms such as $q(t')C(t'-t'')q(t'')$ and $q(t')q(t')$ as well as cross-terms of the type $q'(t')C(t'-t'')q(t'')$. The bath thus convolutes the left and right elements of the density matrix, thus making the whole evolution entangled and convoluted.

10.4 Quantum Stochastic Process: The Stochastic Schrödinger Equation

Quantum mechanical systems are always in the presence of stochastic noise of the environment. This coupling introduces uncertainty and reversibility. This action of the environment in the Langevin meaning can be understood as the external time-dependent fluctuating force. In that case the classical system obtains fluctuating character. A quantum system, in principle, can be understood on the same basis. As described in Chapter 4, the wavefunction of an isolated system follows the Schrödinger equation. As we showed in previous sections in this chapter the propagator of the wavefunction can be given as a path integral. In this section we briefly review the derivation of the stochastic Schrödinger equation following [50, 51], which directly apply the FV result to describe a fluctuating wavefunction of the system coupled to the environment.

The common assumption is that there is a quantum system coupled to the harmonic bath. The reduced density matrix propagator of such a system is given by (10.79), while the FV influence functional has the form of (10.98).

Consider the real stochastic Gaussian process $\xi(t)$ and the statistical average of the integral [52]

$$G(f(\tau)) = \left\langle \exp\left(i\gamma \int_{t_i}^{t_f} d\tau \xi(\tau) f(\tau) \right) \right\rangle , \tag{10.100}$$

where $f(\tau)$ is a real function of time. The angle brackets here denote the statistical average over fluctuations. This average can be calculated by expanding the exponent in a Taylor series:

$$G(f(\tau)) = 1 + \sum_{n=1}^{\infty} \frac{(i\gamma)^n}{n!} \int_{t_i}^{t_f} d\tau_1 \ldots \int_{t_i}^{t_f} d\tau_n \langle \xi(\tau_1) \ldots \xi(\tau_n) \rangle f(\tau_1) \ldots f(\tau_n) . \tag{10.101}$$

As the stochastic process is Gaussian, the second-order cumulant completely describes the average (higher-order cumulants for the Gaussian process vanish). We thus get (see Appendix A.2)

$$\left\langle \exp\left(\mathrm{i}\gamma \int_{t_i}^{t_f} \mathrm{d}\tau \xi(\tau) f(\tau) \right) \right\rangle = \mathrm{e}^{-\frac{\gamma^2}{2} \int_{t_i}^{t_f} \mathrm{d}\tau_2 \int_{t_i}^{t_f} \mathrm{d}\tau_1 f(\tau_2) \langle \xi(\tau_2)\xi(\tau_1) \rangle f(\tau_1)} . \quad (10.102)$$

Consider now the FV influence functional (10.98), where the fluctuation correlation function is real and symmetric. The FV influence action is then

$$S_{\mathrm{FV}}(q(\tau), q'(\tau)) = \frac{1}{2} \int_0^t \mathrm{d}t' \int_0^t \mathrm{d}t'' (q(t') - q'(t')) C(t' - t'')(q(t'') - q'(t'')) . \quad (10.103)$$

Comparing this integral with (10.102), we find that the FV influence functional can be written as a statistical average of a single integral

$$S_{\mathrm{FV}}(q(\tau), q'(\tau)) = \left\langle \mathrm{i} \int_0^t \mathrm{d}\tau \xi(\tau)(q(\tau) - q'(\tau)) \right\rangle , \quad (10.104)$$

and the full propagator of the reduced density matrix becomes factorized into two conjugate parts. These can be associated with the propagators of the system wavefunction.

The propagator of the system wavefunction is then given as an ensemble average of

$$K(q_{\mathrm{f}}, t | q_{\mathrm{i}}, 0) = \int_{q_i}^{q_f} \mathcal{D}q \exp\left(\frac{\mathrm{i}}{\hbar} \int_0^t (L_{\mathrm{S}}(q(\tau)) + \mathrm{i}\xi(\tau) q(\tau)) \right) . \quad (10.105)$$

The corresponding *stochastic Schrödinger equation* in the case of the multimode bath can be given in the form

$$\frac{\mathrm{d}}{\mathrm{d}t} |\psi\rangle = -\frac{\mathrm{i}}{\hbar} \hat{H}_{\mathrm{S}} |\psi\rangle + \sum_k \hat{l}_k \xi_k(t) |\psi\rangle . \quad (10.106)$$

Here $\hat{H}_{\mathrm{S}}$ is the Hamiltonian of the isolated system, and $\hat{l}_k$ is the operator connecting the system to the kth mode of the bath, which produces the stochastic noise $\xi_k(t)$. The noise is described as the real-valued Gaussian noise. In that case the ensemble-averaged density matrix corresponds exactly to the reduced density matrix.

However, this approach produces the high-temperature result, which is due to the real-valued correlation function and the corresponding Gaussian stochastic noise (see Chapter 8). The realistic bath at equilibrium characterized by a constant temperature yields the fluctuation correlation function, which is a complex-valued function. It has been shown that this case of the FV influence functional can also be factorized into two conjugate parts using the complex-valued fluctuating Gaussian trajectories [51]. The stochastic propagator of the wavefunction that generates the proper FV influence functional is then given by

$$K(q_f, t|q_i, 0) = \int_{q_i(0)}^{q_f(t)} \mathcal{D}q \mathrm{e}^{\frac{i}{\hbar}\int_0^t \left[L_S(q(\tau)) + z(\tau)q(\tau) - \int_0^\tau \mathrm{d}s q(\tau) C(\tau - s) q(s)\right]} , \qquad (10.107)$$

where $z(t)$ is the complex-valued stochastic process with the correlation function $C(t)$, and the stochastic average of these propagators leads to the FV influence functional for a specific bath correlation function.

The stochastic wavefunction of the system under the action of such a bath satisfies the *non-Markovian stochastic Schrödinger equation* in the form [50]

$$\frac{\mathrm{d}}{\mathrm{d}t}|\psi\rangle = -\mathrm{i}\hat{H}_S|\psi\rangle + \sum_k \left[\hat{l}_k z_k(t)|\psi\rangle - \hat{l}_k \int_0^t \mathrm{d}s C_k(t-s)\frac{\delta}{\delta z_k(s)}|\psi\rangle\right] . \qquad (10.108)$$

The stochastic process $z_k(t)$ is characterized by the following properties:

$$\langle z_k(t)\rangle = 0 , \qquad (10.109)$$

$$\langle z_k(t) z_k(s)\rangle = 0 , \qquad (10.110)$$

and

$$\langle z_k^*(t) z_k(s)\rangle = C_k(t-s) ; \qquad (10.111)$$

thus, it simulates the stochastic action of the bath and the statistical average of the density matrix over different fluctuating trajectories, and then exactly generates the proper reduced density matrix.

10.5 Coherent-State Path Integral

In this section we derive the path integral representation of the system density operator in terms of coherent states. This representation is convenient for the class of problems where coordinate and momentum operators are not used explicitly. Instead we assume that the Hamiltonian is given in terms of bosonic creation and annihilation operators. Again we consider the time evolution of the density operator:

$$W(t) = \mathrm{e}^{-\mathrm{i}\hat{H}t/\hbar}\hat{W}(0)\mathrm{e}^{\mathrm{i}\hat{H}t/\hbar} . \qquad (10.112)$$

The density matrix in the coherent-state representation is:

$$\langle \alpha' | \hat{W}(t) | \alpha \rangle = \frac{1}{\pi^2} \int \mathrm{d}^2 \alpha_i' \int \mathrm{d}^2 \alpha_i K(\alpha', \alpha_i'; t) \langle \alpha_i' | \hat{W}(0) | \alpha_i \rangle K^*(\alpha_i, \alpha; t) \,, \tag{10.113}$$

where $K(\alpha, \alpha_i; t)$ is the coherent-state representation of the time-evolution operator

$$K(\alpha, \alpha_i; t) = \langle \alpha | \mathrm{e}^{-\mathrm{i}\hat{H}t/\hbar} | \alpha_i \rangle \,. \tag{10.114}$$

We will now calculate this quantity similarly to in the previous subsections.

First we insert the unity resolution by the coherent states:

$$K(\alpha, \alpha_i; t) = \int \frac{\mathrm{d}^2 \alpha_1}{\pi} K(\alpha, \alpha_1; t - t_1) K(\alpha_1, \alpha_i; t_1) \,, \quad 0 < t_1 < t \,, \tag{10.115}$$

and divide the time interval into N equidistant steps of size ϵ:

$$t : t_n = n\epsilon \,, \quad \epsilon = t/N \,, \tag{10.116}$$

and $n = 0, \ldots, N$. We then have

$$K(\alpha, \alpha_i; t) = \int \frac{\mathrm{d}^2 \alpha_1}{\pi} \int \frac{\mathrm{d}^2 \alpha_2}{\pi} \cdots \int \frac{\mathrm{d}^2 \alpha_{N-1}}{\pi} \\ K(\alpha, \alpha_{N-1}; \epsilon) \ldots K(\alpha_2, \alpha_1; \epsilon) K(\alpha_1, \alpha_i; \epsilon) \,, \tag{10.117}$$

where we have the infinitesimal propagator

$$K(\alpha_2, \alpha_1; \epsilon) = \langle \alpha_2 | \exp\left(-\frac{\mathrm{i}}{\hbar} \hat{H} \epsilon \right) | \alpha_1 \rangle \,. \tag{10.118}$$

In the limit $\epsilon \to 0$,

$$\exp\left(-\frac{\mathrm{i}}{\hbar} \hat{H} \epsilon \right) \approx 1 - \frac{\mathrm{i}}{\hbar} \hat{H} \epsilon \ldots \,, \tag{10.119}$$

and the integral is given by

$$\langle \alpha_2 | 1 | \alpha_1 \rangle = \exp\left(\alpha_2^* \alpha_1 - \frac{|\alpha_1|^2}{2} - \frac{|\alpha_2|^2}{2} \right) \,. \tag{10.120}$$

For the term $\hat{H}\epsilon$ we assume a Hamiltonian of the form that conserves the number of particles:

$$\hat{H} = \sum_m \Delta_m \hat{a}^{\dagger m} \hat{a}^m \,. \tag{10.121}$$

Expansion in the coherent states is trivial. Since

$$\hat{a}^{\dagger m} \hat{a}^m | n \rangle = n(n-1)(n-2) \ldots (n-m+1) | n \rangle \,, \tag{10.122}$$

we have

$$\langle\alpha_2|\hat{a}^{\dagger m}\hat{a}^m|\alpha_1\rangle = \alpha_2^{*m}\alpha_1^m \exp\left(\alpha_2^*\alpha_1 - \frac{|\alpha_1|^2}{2} - \frac{|\alpha_2|^2}{2}\right) . \tag{10.123}$$

The expansion of such a Hamiltonian in the coherent states then gives

$$\langle\alpha_2|1 - \frac{\mathrm{i}}{\hbar}\hat{H}\epsilon|\alpha_1\rangle = \exp\left(\alpha_2^*\alpha_1 - \frac{|\alpha_1|^2}{2} - \frac{|\alpha_2|^2}{2}\right) \times\left[1 - \frac{\mathrm{i}}{\hbar}\epsilon H(\alpha_2, \alpha_1)\right] , \tag{10.124}$$

where $H(\alpha_2, \alpha_1) = \sum_m \Delta_m \alpha_2^{*m}\alpha_1^m$: here all $\hat{a}^\dagger$ have been replaced by α_2^* and all $\hat{a}$ have been replaced by α_1. When ϵ is vanishing, by exponentiating the last term, we get

$$K(\alpha_2, \alpha_1; \epsilon) = \langle\alpha_2|\alpha_1\rangle \exp\left(-\frac{\mathrm{i}}{\hbar}\epsilon H(\alpha_2, \alpha_1)\right) . \tag{10.125}$$

We now can combine the elementary propagators to get

$$K(\alpha, \alpha_i t) = \lim_{N\to\infty} \langle\alpha|\alpha_{N-1}\rangle \left(\prod_{k=1}^{N-1} \int \frac{\mathrm{d}^2\alpha_k}{\pi} \langle\alpha_k|\alpha_{k-1}\rangle\right) \times \exp\left(\sum_{k=1}^{N}\left(-\frac{\mathrm{i}t}{\hbar N} H(\alpha_k, \alpha_{k-1})\right)\right) , \tag{10.126}$$

or in the continuous limit we can write the path integral

$$K(\alpha_2, \alpha_1 t) = \int_{\alpha_1}^{\alpha_2} \mathcal{D}\alpha \exp\left(\frac{\mathrm{i}}{\hbar} S(\alpha(\tau))\right) , \tag{10.127}$$

where the action in the coherent-state representation is

$$S(\alpha(\tau)) = \int_0^t \mathrm{d}\tau \left[\mathrm{i}\hbar\alpha^*(\tau)\dot{\alpha}(\tau) - H(\alpha(\tau))\right] . \tag{10.128}$$

Here $\alpha(\tau) = \alpha_1$ at $\tau = 0$ and $\alpha(\tau) = \alpha_2$ at $\tau = t$.

The final expression has some resemblance to the (qp) representation of the path integral of (10.30). According to the physical meaning, we can use the Lagrangian

$$L(\alpha(\tau)) = \mathrm{i}\hbar\alpha^*(\tau)\dot{\alpha}(\tau) - H(\alpha(\tau)) \tag{10.129}$$

and we can use all the formal theory of the coordinate representation. Now we can write the density matrix at time t:

$$\langle\alpha_2|\hat{W}(t)|\alpha_1\rangle \propto \int_{\alpha_{2i}}^{\alpha_2} \mathcal{D}\alpha' \int_{\alpha_{1i}}^{\alpha_1} \mathcal{D}\alpha \times \exp\left(\frac{\mathrm{i}}{\hbar} S(\alpha')\right) \langle\alpha_{2i}|\hat{W}(0)|\alpha_{1i}\rangle \exp\left(-\frac{\mathrm{i}}{\hbar} S^*(\alpha)\right) . \tag{10.130}$$

The equilibrium density matrix is given by the Euclidean action:

$$W(\alpha_2\alpha_1) \propto \int_{\alpha_1}^{\alpha_2} \mathcal{D}\alpha \exp\left(-\frac{1}{\hbar} S^{(E)}(\alpha(\tau))\right) , \tag{10.131}$$

where

$$S^{(E)}(\alpha(\tau)) \propto \int_0^{\beta\hbar} d\tau \, H(\alpha(\tau)) . \tag{10.132}$$

The proportionality signs here signify the overcompleteness of the coherent states. The normalization conditions can be determined at the level of the solution.

10.6 Stochastic Liouville Equation

In Section 10.4 we described the stochastic wavefunction approach consistently appearing from the open quantum system Hamiltonian. Here we obtain the equation for the reduced density matrix for the Gaussian overdamped bath [53]. Let us now consider the stochastic Hamiltonian

$$\hat{H}(\tau) = \hat{H}_S(\hat{a}^\dagger, \hat{a}) + \hat{l}(\hat{a}^\dagger, \hat{a})\xi(\tau) , \tag{10.133}$$

where the stochastic trajectory $\xi(\tau)$ is a Gaussian stationary process. The density matrix of the system with the specific stochastic trajectory will be given by $W(\xi, t)$

.

The stochastic noise is considered as the Gaussian process with the correlation function

$$\langle \xi(\tau)\xi(\tau')\rangle = e^{-\gamma|\tau-\tau'|} \tag{10.134}$$

and thus it satisfies the Fokker–Planck equation (3.84), which we rewrite here in the form

$$\frac{\partial}{\partial t} P(\xi, t|\xi_0 t_0) = \hat{\Gamma} P(\xi, t|\xi_0 t_0) . \tag{10.135}$$

The term $\hat{\Gamma} = \gamma\partial/\partial\xi(\xi + \partial/\partial\xi)$ can be understood as the operator. This process has also been described as a path integral by (10.56) and (10.57) (here γ replaces D). We now can write the total density matrix in terms of path integrals:

$$\begin{aligned} W\left(\alpha_f^*, \alpha_f', \xi, t\right) = & \int_{\alpha_i}^{\alpha_f} \mathcal{D}\alpha \int_{\alpha_i'}^{\alpha_f'} \mathcal{D}\alpha' \int d\xi_i \int_{\xi_i}^{\xi} \mathcal{D}\xi \\ & \times \exp\left(\frac{i}{\hbar} S(\alpha(\tau), \xi(\tau)) - \frac{i}{\hbar} S^*(\alpha'(\tau), \xi(\tau))\right) \\ & \times \rho_S\left(\alpha_i^*, \alpha_i', \xi_i, 0\right) \rho_B(\xi_i) P_{osc}(\xi(\tau)) . \end{aligned} \tag{10.136}$$

Here the factorized initial condition was assumed:

$$W_S\left(\alpha_i^*, \alpha_i', \xi_i, 0\right) = \rho_S\left(\alpha_i^*, \alpha_i, 0\right) \rho_B(\xi_i) . \quad (10.137)$$

Differentiation of the density matrix leads to the following equation:

$$\frac{\partial}{\partial t} \hat{W}(\xi, t) = -\frac{i}{\hbar}\left[\hat{H}_S, \hat{W}(\xi, t)\right] - \frac{i}{\hbar}\left[\hat{l}, \xi(t) \hat{W}(\xi, t)\right] + \hat{\Gamma} \hat{W}(\xi, t) . \quad (10.138)$$

This expression is sometimes denoted as Kubo's stochastic Liouville equation.

Note that the stochastic noise generated by this approach is due to the harmonic overdamped oscillator [53]. Therefore, it can be represented using the harmonic creation and annihilation operators $\hat{b}^\dagger$ and $\hat{b}$. In this approach

$$\hat{\Gamma} \to -\gamma \hat{b}^\dagger \hat{b} \quad (10.139)$$

and

$$\xi \to \hat{b}^\dagger + \hat{b} . \quad (10.140)$$

We then obtain the operator equation

$$\frac{\partial}{\partial t} \hat{W}(t) = -\frac{i}{\hbar}\left[\hat{H}_S, \hat{W}(t)\right] - \frac{i}{\hbar}\left[\hat{l}, (\hat{b}^\dagger + \hat{b})\hat{\rho}(t)\right] - \gamma \hat{b}^\dagger \hat{b} \hat{W}(t) . \quad (10.141)$$

Here we can introduce the number of quanta of the bath oscillator; thus, we expand the density matrix as

$$\hat{W} = \rho_n |n\rangle , \quad (10.142)$$

where $|n\rangle$ is the bath state. We thus get the *hierarchy of equations* for these density matrices:

$$\frac{\partial}{\partial t}\hat{\rho}_n(t) = -\frac{i}{\hbar}\left[\hat{H}_S, \hat{\rho}_n(t)\right] - n\gamma \hat{\rho}_n(t) - \frac{i\sqrt{n}}{\hbar}\left[\hat{l}, \hat{\rho}_{n-1}(t)\right] - \frac{i\sqrt{n+1}}{\hbar}\left[\hat{l}, \hat{\rho}_{n+1}(t)\right] . \quad (10.143)$$

Note that the density matrix $\hat{\rho}_0$ is described by the equation

$$\frac{\partial}{\partial t}\hat{\rho}_0(t) = -\frac{i}{\hbar}\left[\hat{H}_S, \hat{\rho}_0(t)\right] - \frac{i}{\hbar}\left[\hat{l}, \hat{\rho}_1(t)\right] , \quad (10.144)$$

and thus its time evolution is reminiscent of the system dynamics affected by the bath. The density matrix $\hat{\rho}_0$ is thus the reduced system density matrix and the other terms are then denoted as the auxiliary density matrices.

In this treatment the stochastic noise has been described by the classical stochastic process with the real correlation function, that is, the infinite-temperature limit. The finite-temperature effect can be introduced by taking the full FV influence functional in the density matrix propagator, (10.136), with respect to the harmonic bath. Then differentiating the propagator with respect to time yields equations of motion. We demonstrate the *hierarchical equations of motion* at finite temperature in Section 11.8.

11
Perturbative Approach to Exciton Relaxation in Molecular Aggregates

The most general model of a quantum dissipative system consists of a set of energy levels coupled to a set of fluctuating coordinates. This model is very general and covers a broad range of microscopic systems where the electronic degrees of freedom can be distinguished. Weak thermal fluctuations by nuclei comprising the bath can be easily identified. In the other context, we may address the problem of spin dephasing in the ferromagnetic or the current fluctuations in a superconductor. The electronic excitations in molecular aggregates are unique in this class of systems because their electronic excitations are localized on separate molecules as described in Section 5.4, while fluctuating nuclear vibrations originate from localized intramolecular nuclear normal modes. These vibrations are usually assumed to be isolated; thus, each molecule is then coupled to its own independent bath coordinates. Additionally, since all molecules in the aggregate are often of the same type, the different bath modes are of the same type. We will describe this model in this chapter.

The theoretical description of molecular excitations is usually given in terms of the perturbative exciton relaxation schemes. Within such an approach the interaction of electronic excitations with intramolecular and intermolecular vibrations causes a disruption of the phase relationship between excited states of the molecules. Such a type of interaction has a distinct influence on the coherence in the exciton dynamics and plays the dominant role by determining the exciton relaxation pathways, which is often described in terms of the Redfield theory. The simplest molecular aggregate – the molecular dimer – is an ideal model system disclosing effects caused by the excitonic quantum coherence [54, 55] and displaying energy relaxation and transport pathways. We present simulations on this type of system later in this chapter.

Recently developed nonlinear spectroscopies, such as two-dimensional photon echo spectroscopy [56–59], were a key tool in demonstrating a complex pathway of energy transfer in LH2 [60] and LH3 [61], the peripheral light-harvesting complexes in photosynthetic bacteria, and long-lasting coherence in Fenna–Matthews–Olson (FMO) complexes [57, 62] and in LHCII [63]. Recently, two-dimensional photon echo spectra were also recorded for the so-called reaction centers [64, 65] and for other molecular aggregates, for instance, for polymers [66] or for cylindrical (bitubular) *J* aggregates [67]. Photosynthetic molecular aggregates were described

Molecular Excitation Dynamics and Relaxation, First Edition. L. Valkunas, D. Abramavicius, and T. Mančal.

by using the exciton concept [24]. Apart from clear identification of exciton transfer between pigment molecules or their clusters, quantum coherence and population oscillations were also observed. All these new data boosted active research which aims to evaluate the importance of coherence, entanglement, and noise in the energy transport of biological molecular complexes [68–73]. This chapter reviews theoretical approaches based on the density matrix method used in molecular aggregates.

11.1 Quantum Master Equation

In this section we consider a quantum system coupled to a bath. Let us denote the system Hamiltonian by $\hat{H}_S$, the bath Hamiltonian by $\hat{H}_B$, and the coupling between the system and the bath as $\hat{H}_{SB}$. The total Hamiltonian is then given by

$$\hat{H} = \hat{H}_S + \hat{H}_B + \hat{H}_{SB} \,. \tag{11.1}$$

The time dependence for the density matrix of the system coupled with the bath, $\hat{W}$, follows the Liouville equation (we take $\hbar = 1$ throughout this chapter):

$$\frac{d\hat{W}}{dt} = -i\left[\hat{H}, \hat{W}\right] . \tag{11.2}$$

At this point we transform the problem into the interaction representation (we denote its operators by the time dependence):

$$\hat{W} = \exp\left(-i(\hat{H}_S + \hat{H}_B)t\right) \hat{W}(t) \exp\left(i\left(\hat{H}_S + \hat{H}_B\right)t\right) . \tag{11.3}$$

This leads to the following transformation of the Liouville equation

$$\frac{d}{dt}\hat{W}(t) = -i\left[\hat{H}_{SB}(t), \hat{W}(t)\right] , \tag{11.4}$$

where $\hat{H}_{SB}(t)$ is the interaction representation of the system–bath Hamiltonian:

$$\hat{H}_{SB}(t) = \exp\left(i\left(\hat{H}_S + \hat{H}_B\right)t\right) \hat{H}_{SB} \exp\left(-i\left(\hat{H}_S + \hat{H}_B\right)t\right) . \tag{11.5}$$

First we integrate (11.4) from some arbitrary initial time t_0:

$$\hat{W}(t) = \hat{W}(t_0) - i\int_{t_0}^{t} d\tau \left[\hat{H}_{SB}(\tau), \hat{W}(\tau)\right] . \tag{11.6}$$

Then we substitute this result into the right-hand side of (11.4):

$$\frac{d}{dt}\hat{W}(t) = -i\left[\hat{H}_{SB}(t), \hat{W}(t_0)\right] - \int_{t_0}^{t} d\tau \left[\hat{H}_{SB}(t), \left[\hat{H}_{SB}(\tau), \hat{W}(\tau)\right]\right] . \tag{11.7}$$

It is worthwhile mentioning that no approximation has been done so far, that is, (11.7) is exact.

A general simplification can be fulfilled at this point on the basis of two arguments. The first is that the term containing $\hat{W}(t_0)$ comes linearly into the equation of motion. If we assume that the system under the influence of the bath is dissipative, the initial condition should be "forgotten" after a sufficiently long time. The second argument is that the initial condition amounts to the point of reference of the problem. Assuming that the bath is composed of harmonic oscillators and the system–bath coupling is linear in the bath coordinates, we cause these terms to vanish when we perform averaging over the equilibrium bath when the density $\hat{W}(t_0)$ corresponds to the canonical ensemble for the bath coordinates. With these arguments in mind, we can safely disregard the initial condition, set t_0 to infinity, take the trace over the bath, and have the equation of motion for the reduced density matrix $\hat{\rho}(t) = \mathrm{Tr}_\mathrm{B}\,\hat{W}(\tau)$:

$$\frac{\mathrm{d}}{\mathrm{d}t}\hat{\rho}(t) = -\int_{-\infty}^{t} \mathrm{d}\tau \mathrm{tr}_\mathrm{B}\left[\hat{H}_\mathrm{SB}(t), \left[\hat{H}_\mathrm{SB}(\tau), \hat{W}(\tau)\right]\right] . \tag{11.8}$$

This equation of motion is nonperturbative and it is usually simplified by using various approximations [20]. The standard procedures involve assuming an equilibrium bath with its canonical form of the density matrix at constant temperature, denoted as the *Born approximation*: this gives $\hat{W}(\tau) \approx \hat{\rho}(\tau) \otimes \hat{\rho}_\mathrm{B}$. It allows us to define the relaxation operator, and the resulting equation is denoted as the *generalized master equation*. That form can be used to compute the time-dependent density matrix, while keeping the memory effects. When the bath dynamics is fast, the reduced density matrix of the system can be taken out of the integral, which implies the Markov approximation and the second-order perturbation expression for the relaxation operator. Using the harmonic bath model with a specific spectral density, we get the level of approximations which is denoted by the Redfield (or the modified Redfield) theory. Even more approximations can be involved by including only diagonal and zero-frequency terms in the relaxation operator. That gives the secular Redfield theory.

11.2
Second-Order Quantum Master Equation

In the Liouville space, (11.4) is given by

$$\frac{\mathrm{d}}{\mathrm{d}t} W_\mathrm{I}(t) = -\mathrm{i}\mathcal{V}(t)\, W_\mathrm{I}(t) , \tag{11.9}$$

where the system–bath interaction in the Liouville space is

$$\mathcal{V}(t)\hat{A} \Longrightarrow \left[\hat{H}_\mathrm{SB}(\tau), \hat{A}\right] . \tag{11.10}$$

The first approximation is the condition that the total density matrix can be factorized into the system $\rho(t)$ and the bath components (Born approximation). The second approximation is that the bath is in the equilibrium state, ρ_B, all the time:

$$W_I(t) = \rho_I(t) \otimes \rho_B \,. \tag{11.11}$$

The last step is to perform the trace operation over the equilibrium bath variables. This gives the generalized quantum master equation, (11.8), in the form

$$\frac{d}{dt}\rho_I(t) = -\int_{t_0}^{t} d\tau \, \mathrm{tr}_B \left(\mathcal{V}(t)\mathcal{V}(\tau)\rho_I(\tau) \otimes \rho_B\right) \,. \tag{11.12}$$

This Liouville space expression is very compact and captures the essential physical insight: the natural quantity which affects the density matrix dynamics is a two-times correlation function of the system–bath interactions. It is not very convenient for practical simulations, since one has to consider the complete microscopic bath dynamics inside the trace.

We next take a bilinear form of the system–bath interaction:

$$\hat{H}_{SB} = \sum_n \hat{S}_n \hat{q}_n \,, \tag{11.13}$$

where $\boldsymbol{n}$ is an index of expansion (i.e., not a state of the system), $\hat{S}_n$ is the system operator (usually a projector), and $\hat{q}_n$ is the associated bath coordinate operator. In the Hilbert space,

$$\mathcal{V}(t)\hat{A} \Longleftrightarrow \sum_n \hat{S}_n(t)\hat{q}_n(t)\hat{A} - \hat{A}\hat{S}_n(t)\hat{q}_n(t) \,. \tag{11.14}$$

The trace over the bath then yields coordinate–coordinate correlation functions and we get a compact form:

$$\frac{d}{dt}\rho_I(t) = -\int_{t_0}^{t} d\tau \, R_I(t,\tau)\rho_I(\tau) \,, \tag{11.15}$$

where the time-nonlocal rate superoperator is defined as

$$\begin{aligned} R_I(t,\tau)\rho_I(\tau) = \sum_{mn} \Big[& \hat{S}_m(t)\hat{S}_n(\tau)\rho_I(\tau)C_{mn}(t-\tau) \\ & - \hat{S}_m(t)\rho_I(\tau)\hat{S}_n(\tau)C_{nm}(\tau-t) \\ & - \hat{S}_n(\tau)\rho_I(\tau)\hat{S}_m(t)C_{mn}(t-\tau) \\ & + \rho_I(\tau)\hat{S}_n(\tau)\hat{S}_m(t)C_{nm}(\tau-t) \Big] \,. \end{aligned} \tag{11.16}$$

Here

$$C_{mn}(t) = \mathrm{tr}_B\left(\hat{q}_m(t)\hat{q}_n(0)\rho_B\right) \tag{11.17}$$

is the coordinate–coordinate correlation function. Note that in the bath equilibrium the relaxation operator is essentially a function of the time difference between t and τ, that is, $R_{\mathrm{I}}(t,\tau) \rightarrow R_{\mathrm{I}}(t-\tau)$.

The system operator $\hat{S}_n$ is understood here as the projection operator of the type $\hat{S}_n = |a\rangle\langle b|$, where $|a\rangle$ and $|b\rangle$ are electronic eigenstates. The bath coordinate is then understood as the specific coordinate or a whole set of coordinates, coupled to that specific Hamiltonian element. The model includes the effect of the system–bath interaction on the density matrix *nonperturbatively*, while the relaxation-inducing terms are calculated at the level of the Born approximation. This results in the memory-like effect of the bath through the bath correlation functions.

The rate operator is given by the second-order products of the system–bath interaction. Taking the initial condition $t_0 \rightarrow -\infty$, we have that the rate operator is a function of interaction delay times $t-\tau$. The evolution according to this rate is still infinite order in the system–bath interaction. It is convenient to introduce the delay time $s = t-\tau$ explicitly. In the Schrödinger representation we can then write

$$\frac{\mathrm{d}}{\mathrm{d}t}\rho(t) = -\mathrm{i}\left[\hat{H}_{\mathrm{S}}, \rho(t)\right] - \int_0^\infty \mathrm{d}s\, R(s)\rho(t-s) \,, \tag{11.18}$$

with the rate superoperator

$$\begin{aligned} R(s)\hat{A} = \sum_{mn} \Big(& \hat{S}_m \hat{G}(s)\hat{S}_n\hat{G}(-s)\hat{A}C_{mn}(s) \\ & - \hat{S}_m\hat{A}\hat{G}(s)\hat{S}_n\hat{G}(-s)C_{nm}(-s) \\ & - \hat{G}(s)\hat{S}_n\hat{G}(-s)\hat{A}\hat{S}_m C_{mn}(s) \\ & + \hat{A}\hat{G}(s)\hat{S}_n\hat{G}(-s)\hat{S}_m C_{nm}(-s)\Big) \,. \end{aligned} \tag{11.19}$$

Here we used the wavefunction propagator

$$\hat{G}(s) = \exp\left(-\mathrm{i}\hat{H}_{\mathrm{S}} s\right) \,. \tag{11.20}$$

The integrodifferential form of the quantum master equation obtained is very complicated. It can be simplified by using the *Redfield approximation*, that is, by assuming the *second-order approximation* for the relaxation kernel [74]. The second-order level is obtained by the Markov approximation. To that end we assume that the system–bath interaction is weak and the system density matrix *in the interaction picture* is a slowly evolving function, compared with the decay time of the relaxation tensor, that is, it can be taken out of the integral, since $R_{\mathrm{I}}(t-\tau)$ decays much faster than $\rho_{\mathrm{I}}(t)$ varies. Equation (11.15) can be simplified as

$$\int_{t_0}^{t} \mathrm{d}\tau'\, R_{\mathrm{I}}(t-\tau')\rho_{\mathrm{I}}(\tau') \approx \int_0^\infty \mathrm{d}\tau\, R_{\mathrm{I}}(\tau)\rho_{\mathrm{I}}(t) \,, \tag{11.21}$$

where we took $t_0 \rightarrow \infty$ and we introduced the integration over the delay time $\tau = t-\tau'$. The integration over the delay time τ can be performed and the time-local equation with the time-independent rate matrix is obtained. In the Schrödinger picture we get the *Redfield equation* is

$$\frac{\mathrm{d}}{\mathrm{d}t}\rho(t) = -\mathrm{i}[\hat{H}_\mathrm{S}, \rho(t)] - K\rho(t) \,, \tag{11.22}$$

where the *Redfield relaxation* superoperator is given by

$$K = \int_0^\infty \mathrm{d}\tau\, R(\tau) \,, \tag{11.23}$$

and the integral kernel is defined by (11.19).

The Redfield relaxation superoperator can be simplified considerably if the system operators are expanded into an orthogonal basis $|a\rangle$. The Hamiltonian is

$$\hat{H} = \sum_{ab} \left(h_{ab} + \tilde{h}_{ab}\hat{q}_{ab} \right) |a\rangle\langle b| + \hat{H}_\mathrm{B}(\hat{p}, \hat{q}) \,. \tag{11.24}$$

In general $h_{ab} = \delta_{ab}\epsilon_a + J_{ab}$. The remaining part can be partitioned out into the term $\hat{H}_\mathrm{B}(\hat{p}, \hat{q})$, representing the fluctuating environment, and the weak coupling amplitude $\tilde{h}_{ab}$ as a system–bath coupling amplitude and $\hat{q}_{ab}$ as the generalized coordinate of the bath, coupled to system Hamiltonian element ab. Inserting (11.24) into (11.23) and (11.19), we obtain the relaxation matrix defined by fluctuation correlation functions:

$$\begin{aligned} K_{ab,a'b'} = \sum_{cd} \int_0^\infty \mathrm{d}\tau \Bigg(& \delta_{bb'} \sum_e \tilde{h}_{ae}\tilde{h}_{dc} C_{ae,dc}(\tau) G_{ed}(\tau) G_{ca'}(-\tau) \\ & - \tilde{h}_{aa'}\tilde{h}_{cd} C_{cd,aa'}(-\tau) G_{b'c}(\tau) G_{db}(-\tau) \\ & - \tilde{h}_{dc}\tilde{h}_{b'b} C_{b'b,dc}(\tau) G_{ad}(\tau) G_{ca'}(-\tau) \\ & + \delta_{aa'} \sum_e \tilde{h}_{cd}\tilde{h}_{eb} C_{cd,eb}(-\tau) G_{b'c}(\tau) G_{de}(-\tau) \Bigg) \,. \end{aligned} \tag{11.25}$$

The correlation function $C_{ab,cd}(\tau)$ describes fluctuations of the Hamiltonian elements ab and cd. The functional form of the correlation function and the coefficients $\tilde{h}_{ab}$ can be defined for a specific system and the bath model.

A natural choice for the basis set for the Redfield relaxation superoperator is the eigenstate basis of the system Hamiltonian. This choice makes simulations much simpler, and it allows us to introduce the *secular approximation* and to define the requirement for the long-time limit. Let us assume that states $|a\rangle$ are eigenstates of the system Hamiltonian (the exciton states in the case of the exciton Hamiltonian). In that case

$$h_{ab} = \delta_{ab}\epsilon_a \tag{11.26}$$

is diagonal and the Redfield equation reduces to

$$\frac{\mathrm{d}}{\mathrm{d}t}\rho_{ab}(t) = -\mathrm{i}\omega_{ab}\rho_{ab}(t) - K_{ab,cd}\rho_{cd}(t) \,, \tag{11.27}$$

where $\omega_{ab} = \epsilon_a - \epsilon_b$.

If the system–bath coupling is weak compared with the splitting of energy levels, the relaxation effect is a small perturbation to the natural system evolution – quantum phase rotation. The free-system solution is

$$\rho_{ab}^{(0)}(t) = \exp(-\mathrm{i}\omega_{ab}t)\rho_{ab}^{(0)}(0) . \tag{11.28}$$

The relaxation effect can be included approximately by a slowly varying amplitude of the form

$$\rho_{ab}(t) = \rho_{ab}^{(1)}(t)\exp(-\mathrm{i}\omega_{ab}t)\rho_{ab}^{(0)}(0) . \tag{11.29}$$

Here we assume $|\dot{\rho}_{ab}^{(1)}| \ll \omega_{ab}$, where the dot denotes the time derivative. The Redfield equation in the interaction picture then gives

$$\frac{\mathrm{d}}{\mathrm{d}t}\rho_{ab}^{(1)}(t) = -K_{ab,cd}\rho_{cd}^{(1)}(t)\exp(\mathrm{i}(\omega_{ab} - \omega_{cd})t)\frac{\rho_{cd}^{(0)}(0)}{\rho_{ab}^{(0)}(0)} , \tag{11.30}$$

whose solution is

$$\rho_{ab}^{(1)}(t) = \rho_{ab}^{(1)}(0) - K_{ab,cd}\frac{\rho_{cd}^{(0)}(0)}{\rho_{ab}^{(0)}(0)}\int_0^t \mathrm{d}\tau\rho_{cd}^{(1)}(\tau)\exp(\mathrm{i}(\omega_{ab} - \omega_{cd})\tau) . \tag{11.31}$$

The integral

$$\int_0^t \mathrm{d}\tau\rho_{cd}^{(1)}(\tau)\exp(\mathrm{i}(\omega_{ab} - \omega_{cd})\tau) \tag{11.32}$$

is the essential quantity which affects the dynamics. When the system–bath coupling is smaller than the energy-level splitting, $\rho_{cd}^{(1)}(\tau)$ is a slowly varying function compared with the density matrix oscillation frequency. Inside the integral we have a difference of two frequencies. If that difference is of the order of the typical oscillation frequency, then the integral kernel becomes highly oscillatory; the integral thus vanishes. The terms which do not vanish are those where $\omega_{ab} = \omega_{cd}$. In general, all energy gaps are different and there are only two general cases where $|\omega_{ab} - \omega_{cd}| \ll |\dot{\rho}_{cd}^{(1)}|$: (1) when $a = c$ and $b = d$, and (2) when $a = b$ and $c = d$. We thus keep only those terms in the original Redfield equation, and we disregard all other terms. That is the essence of the secular approximation.

The secular approximation is thus different from the more general rotating-wave approximation (RWA). The RWA may include coherence transfer terms in the case when energy splittings of two different coherences are the same. Such a case is possible, for example, for the harmonic oscillator as was shown in Chapter 8. The secular approximation is thus more restrictive. As demonstrated, this approximation is well defined in the eigenstate basis. In some other basis one cannot define natural frequencies for different density matrix elements; thus, the secular approximation cannot be defined.

The *secular Redfield relaxation equation* can be written in the form

$$\frac{\mathrm{d}}{\mathrm{d}t}\rho_{ab}(t) = -\mathrm{i}(\omega_{ab} - \mathrm{i}\gamma_{ab})\rho_{ab}(t) - \delta_{ab}\sum_{b} k_{ab}\rho_{bb}(t)\,, \tag{11.33}$$

where $\gamma_{aa} \equiv 0$. In the eigenstate basis we have the population transport rate ($a \neq b$)

$$k_{ab} = -2|\tilde{h}_{ab}|^2 \mathrm{Re} \int_0^\infty \mathrm{d}\tau\, C_{ab,ba}(\tau)\mathrm{e}^{\mathrm{i}\omega_{ab}\tau}\,, \tag{11.34}$$

while for the diagonal we have

$$k_{aa} = -\sum_{b}^{b\neq a} k_{ba}\,. \tag{11.35}$$

The dephasing rate of the coherences ($a \neq b$) is

$$\gamma_{ab} = \int_0^\infty \mathrm{d}\tau \left(\sum_e |\tilde{h}_{ae}|^2 C_{ae,ea}(\tau)\mathrm{e}^{-\mathrm{i}\omega_{ea}\tau} + \sum_e |\tilde{h}_{be}|^2 C_{be,eb}(-\tau)\mathrm{e}^{-\mathrm{i}\omega_{be}\tau} \right.$$
$$\left. - 2\,\mathrm{Re}\,\tilde{h}_{aa}\tilde{h}_{bb}C_{aa,bb}(\tau) \right)\,. \tag{11.36}$$

All these rates are given by one-sided Fourier transforms of the coordinate–coordinate correlation function. As we showed in (8.71), these can be given in terms of the fluctuation spectral densities.

11.3
Relaxation Equations from the Projection Operator Technique

In an alternative approach, relaxation equations of a similar type can be obtained directly from the projection operator approach described in Chapter 9. Equation (9.49) already contains the second-order approximation in the system–bath interaction and it is time local (time convolutionless). For a more specific derivation we require the form of the interaction Hamiltonian $\hat{H}_{\mathrm{SB}}$ of (11.13). We assume an initial condition, $W(t) = \rho(t) \otimes \rho_{\mathrm{B}}$, where ρ_{B} is the equilibrium bath statistical operator. Such an initial condition leads to elimination of the initial term $\mathcal{I}_{\mathrm{CL}}$ if we choose a projector in the form

$$\mathcal{P}\hat{A} = \mathrm{tr}_{\mathrm{B}}\{\hat{A}\}\rho_{\mathrm{B}}\,. \tag{11.37}$$

Equation (9.49) leads to

$$\frac{\partial}{\partial t}\rho^{(\mathrm{I})}(t) = -\frac{\mathrm{i}}{\hbar}\left[\mathrm{tr}_\mathrm{B}\left\{\hat{H}_\mathrm{SB}(t)\rho_\mathrm{B}\right\}, \rho^{(\mathrm{I})}(t)\right]$$
$$-\frac{1}{\hbar^2}\int\limits_0^{t-t_0} \mathrm{d}\tau \left(\mathrm{tr}_\mathrm{B}\left\{\left[\hat{H}_\mathrm{SB}(t), \left[\hat{H}_\mathrm{SB}(t-\tau), \rho^{(\mathrm{I})}(t)\rho_\mathrm{B}\right]\right]\right\}\right.$$
$$\left. - \mathrm{tr}_\mathrm{R}\left\{\left[\hat{H}_\mathrm{SB}(t), \mathrm{tr}_\mathrm{B}\left\{\left[\hat{H}_\mathrm{SB}(t-\tau), \rho^{(\mathrm{I})}(t)\rho_\mathrm{B}\right]\right\}\rho_\mathrm{B}\right]\right\}\right) \ . \tag{11.38}$$

The first term on the right-hand side of (11.38) can be handled easily:

$$\mathrm{tr}_\mathrm{B}\left\{\hat{H}_\mathrm{SR}(t)\rho_\mathrm{B}\right\} = \sum_n \mathrm{tr}_\mathrm{B}\left\{U_\mathrm{R}^\dagger(t)\hat{q}_n U_\mathrm{R}(t)\rho_\mathrm{B}\right\} S_n^{(\mathrm{I})}(t) = \sum_n \langle q_n\rangle S_n^{(\mathrm{I})}(t) \ , \tag{11.39}$$

where we defined $\langle q_n\rangle = \mathrm{tr}_\mathrm{B}\{\hat{q}_n\rho_\mathrm{B}\}$ and we assumed that the equilibrium density operator ρ_B does not evolve in time due to the influence of $\hat{H}_\mathrm{B}$, that is, $U_\mathrm{R}(t)\rho_\mathrm{B} U_\mathrm{R}^\dagger(t) = \rho_\mathrm{B}$.

The four terms of the double commutator term of (11.38) have to be handled separately. For the sake of brevity, let us denote $H_t \equiv \hat{H}_\mathrm{SB}^{(\mathrm{I})}(t)$, $H_\tau \equiv \hat{H}_\mathrm{SB}^{(\mathrm{I})}(t-\tau)$, and $\rho \equiv \rho^{(\mathrm{I})}(t)$ for now. We have terms $H_t H_\tau \rho$ and $\rho H_\tau H_t$, which are Hermite conjugate to each other, and terms $H_t \rho H_\tau$ and $H_\tau \rho H_t$, which are also Hermite conjugate to each other. We then get

$$H_t H_\tau \rho \rightarrow \sum_{n,m}\langle q_n(t) q_m(t-\tau)\rangle \hat{S}_n^{(\mathrm{I})}(t)\hat{S}_m^{(\mathrm{I})}(t-\tau)\rho^{(\mathrm{I})}(t) \ , \tag{11.40}$$

$$H_t \rho H_\tau \rightarrow \sum_{n,m}\langle q_m(t-\tau) q_n(t)\rangle \hat{S}_n^{(\mathrm{I})}(t)\rho^{(\mathrm{I})}(t)\hat{S}_m^{(\mathrm{I})}(t-\tau) \ . \tag{11.41}$$

The third term of (11.38) contains terms similar to (11.40) and (11.41), where correlation functions $\langle q_a(t) q_b(t')\rangle$ are replaced by $\langle q_a\rangle\langle q_b\rangle$ (see (11.39)). We now define the bath correlation functions in the form

$$C_{nm}(t) = \langle q_n(\tau) q_m\rangle - \langle q_n\rangle\langle q_m\rangle \ , \tag{11.42}$$

and we can write (11.38) in the form

$$\frac{\partial}{\partial t}\rho^{(\mathrm{I})}(t) = -\frac{\mathrm{i}}{\hbar}\sum_n\left[\langle q_n\rangle \hat{S}_n^{(\mathrm{I})}(t), \rho^{(\mathrm{I})}(t)\right]$$
$$-\frac{1}{\hbar^2}\int\limits_0^{t-t_0}\mathrm{d}\tau\sum_{n,m}\left[C_{nm}(\tau)\hat{S}_n^{(\mathrm{I})}(t)\hat{S}_m^{(\mathrm{I})}(t-\tau)\rho^{(\mathrm{I})}(t)\right.$$
$$- C_{mn}(-\tau)\hat{S}_n^{(\mathrm{I})}(t)\rho^{(\mathrm{I})}(t)\hat{S}_m^{(\mathrm{I})}(t-\tau)$$
$$- C_{mn}(\tau)\hat{S}_n^{(\mathrm{I})}(t-\tau)\rho^{(\mathrm{I})}(t)\hat{S}_m^{(\mathrm{I})}(t)$$
$$\left. + C_{nm}(-\tau)\rho^{(\mathrm{I})}(t)\hat{S}_n^{(\mathrm{I})}(t-\tau)\hat{S}_m^{(\mathrm{I})}(t)\right] \ . \tag{11.43}$$

We can now transform the resulting equations to the Schrödinger picture, $\rho(t) = U_S(t)\rho^{(I)}(t)U_S^\dagger(t)$. This leads to

$$\frac{\partial}{\partial t}\rho(t) = -\frac{i}{\hbar}\left[\hat{H}_S + \sum_n \langle q_n\rangle \hat{S}_n, \rho(t)\right] - \frac{1}{\hbar^2}\int_0^{t-t_0} d\tau \sum_{n,m}\Big(C_{nm}(\tau)\hat{S}_n\hat{S}_m^{(I)}(-\tau)\rho(t) - C^*_{nm}(\tau)\hat{S}_n\rho(t)\hat{S}_m^{(I)}(-\tau) - C_{mn}(\tau)\hat{S}_n^{(I)}(-\tau)\rho(t)\hat{S}_m + C^*_{mn}(\tau)\rho(t)\hat{S}_n^{(I)}(-\tau)\hat{S}_m\Big)\,, \tag{11.44}$$

where we used $C^*_{nm}(t) = C_{mn}(-t)$, which can be derived from the definition given by (11.42). This form is, in principle, equivalent to (11.18), except for the first term $\langle S_n\rangle$, which was disregarded in (11.18).

An even more compact form of the master equation can be obtained by defining

$$\Lambda_n(t) = \frac{1}{\hbar^2}\int_0^{t-t_0} d\tau \sum_m C_{nm}(\tau)\hat{S}_m^{(I)}(-\tau)\,. \tag{11.45}$$

The final form of the master equation then reads

$$\frac{\partial}{\partial t}\rho(t) = -\frac{i}{\hbar}\left[\hat{H}_S + \sum_n \hat{S}_n(\langle q_n\rangle - i\hbar\Lambda_n(t)), \rho(t)\right] + \sum_n\left(\hat{S}_n\rho(t)\Lambda_n^\dagger(t) - \Lambda_n(t)\rho(t)\hat{S}_n\right)\,. \tag{11.46}$$

Here we can define an effective Hamiltonian $\hat{H}_{\text{eff}} = \hat{H}_S + \sum_n \hat{S}_n(\langle q_n\rangle - i\hbar\Lambda_n(t))$ which is not Hermitian. The commutator in (11.46) is defined as $[\hat{H}_{\text{eff}}, \rho] = \hat{H}_{\text{eff}}\rho - \rho\hat{H}^\dagger_{\text{eff}}$.

11.4
Relaxation of Excitons

The excitons of molecular aggregates are quantum systems with one key property: they have an isolated well-defined ground state $|g\rangle$, and the bath in equilibrium is in thermal equilibrium with respect to that state. The lifetime of state $|g\rangle$ is infinite. The next set of states can be reached by optical excitations. These are called the single-exciton states. The system in the single-exciton state lasts for several nanoseconds. However, the system may hop between different single-exciton states. Thus, the transport theory for excitons applies for the single-exciton manifold of states, which we label by $|e\rangle$.

The exciton model reflects the dipole–dipole-type intermolecular interaction. In addition, each molecule is coupled to the bath, which determines the transition energy fluctuations. These energy fluctuations are the key quantities in the description of the transport and dephasing rates.

Let us consider the secular Redfield equation. In this case states $|a\rangle$ and $|b\rangle$ are distinct single-exciton states $|e\rangle$ and $|e'\rangle$. They are related to molecular excitations $|m\rangle$ by the transformation

$$|e\rangle = \sum_m c_{em}|m\rangle \,, \tag{11.47}$$

where ψ_{em} is the exciton wavefunction. These wavefunctions transform the system Hamiltonian into a diagonal matrix

$$\varepsilon_e = \sum_{mn} (\hat{H}_{\mathrm{S}})_{mn} c_{em} c_{en} \,. \tag{11.48}$$

The system is affected by nuclear fluctuations of different molecules. The system–bath interaction is thus of the form

$$\hat{H}_{\mathrm{SB}} = \sum_n \tilde{h}_m q_m |m\rangle\langle m| \,. \tag{11.49}$$

The relaxation should be described in the eigenstate basis, while the fluctuation amplitudes of the molecular excitations are then transformed as follows:

$$\tilde{h}_{ee'} = \sum_m \tilde{h}_m c_{em} c_{e'm} \,. \tag{11.50}$$

The correlation functions for the Redfield superoperator in the exciton basis are given by

$$C_{e_1e_2,e_3e_4}(\tau) = \sum_{mn} \tilde{h}_m \tilde{h}_n \langle q_m(\tau) q_n(0)\rangle c_{e_1m} c_{e_2m} c_{e_3n} c_{e_4n} \,. \tag{11.51}$$

In the following we combine the fluctuation amplitudes into the energy–energy correlation function. Additionally we assume that all molecular fluctuations are of the same type and they are independent. We thus have

$$C_{e_1e_2,e_3e_4}(\tau) = \mathcal{C}(\tau) \sum_m c_{e_1m} c_{e_2m} c_{e_3m} c_{e_4m} \,, \tag{11.52}$$

where

$$\mathcal{C}(\tau) = |\tilde{h}_m|^2 \langle q_m(\tau) q_m(0)\rangle \,. \tag{11.53}$$

Other types of correlation functions vanish, $\langle q_m(\tau) q_n(0)\rangle = 0$ for $m \neq n$. With these definitions we have the Redfield rate of exciton transfer:

$$k_{e_1e_2} = -2\mathrm{Re}\, M_{e_1e_2,e_2e_1}(\omega_{e_1e_2}) \,,$$

where the function $M(\omega)$ is the one-sided Fourier transform of the correlation function in the exciton basis

$$M_{e_1e_2e_3e_4}(\omega) = \left(\sum_m c_{e_1m}c_{e_2m}c_{e_3m}c_{e_4m}\right)\int_0^\infty d\tau C(\tau)e^{i\omega\tau} .$$

The dephasing rate of the single-exciton coherences is also given in terms of the function $M(\omega)$:

$$\gamma_{e_1e_2} = \sum_e M_{e_1e,ee_1}(\omega_{e_1e}) + \sum_e M^*_{e_2e,ee_2}(\omega_{e_2e})$$
$$- 2\mathrm{Re}\, M_{e_1e_1,e_2e_2}(0) . \quad (11.54)$$

This theory thus allows us to evaluate all model parameters using the single spectral density (or their family). As we additionally find, all fluctuations in this model become correlated to some degree. So a theory which includes some additional correlations may be desirable.

11.5 Modified Redfield Theory

In this section we consider only the population transfer, which is the most important component in the exciton relaxation and energy transfer problems. The Redfield approach assumes that the bath is Markovian and is in thermal equilibrium. However, it is important that the bath is often affected by the system; thus, the bath equilibrium for different system states can be slightly shifted. The *modified Redfield theory* includes these effects: it is nonperturbative with respect to diagonal fluctuations and it includes correlations between diagonal and off-diagonal fluctuations [75] (the term "modified" was not introduced in the original paper of Zhang *et al.*).

The exact result for the relaxation part of the generalized quantum master equation in the Born approximation, as shown above, in the Liouville space is given by the relaxation kernel:

$$R_I(t,\tau)\rho_I(\tau) \equiv \mathrm{tr_B}\left(\mathcal{V}(t)\mathcal{V}(\tau)\rho_I(\tau)\otimes\rho_B\right) . \quad (11.55)$$

In the exciton relaxation problem, the initial state is created by absorption of a photon. Thus, ρ_B should not be considered in equilibrium, since the equilibrium is for the electronic ground state.

Let us consider the Hamiltonian defined by (11.24), where the system-part matrix h_{ab} is diagonal. Thus, states $|a\rangle$ and $|b\rangle$ are the system eigenstates in the absence of the bath. In the Schrödinger representation for the system the population transfer integral kernel from state a to state b is

$$[R(t-\tau)\rho(\tau)]_{bb} = \mathrm{tr_B}\left\{\left[\mathrm{e}^{-\mathrm{i}\hat{H}_0(t-\tau)}\right]_{bb} \tilde{h}_{ba} q_{ba} W_{aa}(\tau)\left[\mathrm{e}^{\mathrm{i}\hat{H}_0(t-\tau)}\right]_{aa} \tilde{h}_{ab} q_{ab}\right\}, \tag{11.56}$$

where $\hat{H}_0 = \hat{H}_\mathrm{S} + \hat{H}_\mathrm{B}$, and $W_{aa}(\tau)$ is the matrix element of the total density matrix in the eigenstate basis, still an operator in the bath subspace. Let us assume that the bath performs stochastic fluctuations, so in the Heisenberg representation for the bath Hamiltonian $q_{ab}(t)$ is the time-dependent fluctuation. Now keeping the second order in the off-diagonal fluctuations q_{ab} in the eigenstate basis, we write the adiabatic propagator with respect to electronic state $|a\rangle$ as

$$\left[\mathrm{e}^{\mathrm{i}\hat{H}_0(t-\tau)}\right]_{aa} = \exp\left(\mathrm{i}\varepsilon_a(t-\tau) + \mathrm{i}\int_\tau^t \mathrm{d}\tau' q_{aa}(\tau')\right). \tag{11.57}$$

For the initial state a and the final state b we then get

$$\begin{aligned}[R(t-\tau)\rho(\tau)]_{bb} &= |\tilde{h}_{ba}|^2 \exp(\mathrm{i}\omega_{ab}(t-\tau)) \\ &\quad\times \mathrm{tr_B}\left[\exp\left(-\mathrm{i}\int_\tau^t \mathrm{d}\tau' q_{bb}(\tau')\right) q_{ba}(\tau)\right. \\ &\quad\left.\times \exp\left(-\mathrm{i}\int_t^\tau \mathrm{d}\tau' q_{aa}(\tau')\right) W_{aa}(t) q_{ab}(t)\right]. \end{aligned} \tag{11.58}$$

The last step in formulating the problem is to determine W_{aa}. It is taken as the total density matrix when the bath is in thermal equilibrium with respect to exciton state $|a\rangle$. To obtain that limit we take the time limit for which the bath density matrix remained in state $|a\rangle$ as infinitely long time. We thus get

$$\begin{aligned}[R(t-\tau)\rho(\tau)]_{bb} &= |\tilde{h}_{ba}|^2 \rho_{aa}(t) \exp(\mathrm{i}\omega_{ab}(t-\tau)) \\ &\times \lim_{t'\to-\infty} \mathrm{tr_B}\left[\exp\left(-\mathrm{i}\int_\tau^t \mathrm{d}\tau' q_{bb}(\tau')\right) q_{ba}(\tau)\right. \\ &\left.\times \exp\left(-\mathrm{i}\int_{t'}^\tau \mathrm{d}\tau' q_{aa}(\tau')\right) \rho_\mathrm{B} \exp\left(-\mathrm{i}\int_t^{t'} \mathrm{d}\tau' q_{aa}(\tau')\right) q_{ab}(t)\right]. \end{aligned} \tag{11.59}$$

This trace can be calculated exactly using the cumulant expansion when the bath is harmonic (see Appendix A.8). We then get the following rate expression in terms of the so-called spectral lineshape $g(t)$ functions (A98):

$$k_{ba} = 2\mathrm{Re}\,|\tilde{h}_{ba}|^2 \int_0^\infty \mathrm{d}\tau \mathrm{e}^{\mathrm{i}\omega_{ab}\tau} \{\ddot{g}_{ab,ba}(\tau) - \left[\dot{g}_{bb,ba}(\tau) - \dot{g}_{aa,ba}(\tau) + 2\mathrm{i}\lambda_{ba,aa}\right]\left[\dot{g}_{ab,aa}(\tau) - \dot{g}_{ab,bb}(\tau) + 2\mathrm{i}\lambda_{ab,aa}\right]\} \times \exp(-g_{aa,aa}(\tau) - g_{bb,bb}(\tau) + g_{aa,bb}(\tau) + g_{bb,aa}(\tau) + 2\mathrm{i}(\lambda_{aa,bb} - \lambda_{aa,aa})\tau) \,. \tag{11.60}$$

Here

$$g_{ab,cd}(t) = \int_0^t \mathrm{d}\tau \int_0^\tau \mathrm{d}s\, C_{ab,cd}(\tau - s) \tag{11.61}$$

is the so-called lineshape function, dots and double dots denote the time derivatives. and $\lambda_{ab,cd}$ are the reorganization energies (Stokes shifts) given in the limit

$$\lambda_{ab,cd} = -\lim_{t\to\infty} \dot{g}_{ab,cd}(t) \,. \tag{11.62}$$

Also note that $\ddot{g}(\tau) = C(\tau)$.

The modified Redfield rate expression is very suitable for excitons. It includes full equilibration in the excited initial state. The way this expression interpolates between Redfield and Förster theories has been demonstrated [31, 76]. It also includes correlations of diagonal and off-diagonal fluctuations. The modified rate formula thus explicitly includes the bath relaxation effects.

11.6 Förster Energy Transfer Rates

We next consider the Förster model for energy transfer between donor molecules (initial state – d) and acceptor molecules (final state – a) with weak electrostatic interactions. It is thus yet another population transfer model. In that case their transition energy and coupling fluctuations can be considered as independent. The intermolecular coupling is now considered as a weak perturbation due to the dipole–dipole interaction between transition charge densities, which is described in Section 5.4.

The Förster energy transfer rate expression follows directly from the modified Redfield rate expression. In the modified Redfield rate expression we then only need to consider autocorrelation functions, that is, the quantities $g_{aa,aa}$, $g_{dd,dd}$, and $g_{ad,da} = g_{da,ad}$; all other combinations of indices have zero amplitudes. We then have

$$F_{ad}(\tau) = F_{ad}^{(0)}(\tau)|U_{ad}|^2 \tag{11.63}$$

and

$$F^{(0)}_{ad}(\tau) = \exp(-\mathrm{i}\omega_{ad}\tau - g_{aa,aa}(\tau) - g_{dd,dd}(\tau) - 2\mathrm{i}\lambda_{dd,dd}\tau) \,. \tag{11.64}$$

Integrating these relations, we obtain the Förster energy transfer rate formula [76, 77]:

$$K^{(\mathrm{F})}_{ad} = |U_{ad}|^2 \int \frac{\mathrm{d}\omega}{2\pi} \mathcal{A}_a(\omega)\mathcal{F}_d(\omega) \,. \tag{11.65}$$

Here

$$\mathcal{A}_a(\omega) = \int \mathrm{d}\tau \exp(\mathrm{i}(\omega - \varepsilon_a)\tau - g_{aa,aa}(\tau)) \tag{11.66}$$

and

$$\mathcal{F}_d(\omega) = \int \mathrm{d}\tau \exp\left(\mathrm{i}(\omega - (\varepsilon_d - 2\lambda_{dd,dd}))\tau - g^*_{dd,dd}(\tau)\right) \tag{11.67}$$

are spectral lineshapes of the acceptor absorption and the donor fluorescence, normalized to unit area (these are derived in Part Two). The symmetry $g(\tau) = g^*(-\tau)$ ensures that $\mathcal{A}_a(\omega)$ and $\mathcal{F}_d(\omega)$ are real. Equation (11.65) is commonly used with experimental normalized absorption and emission, and the intermolecular coupling is calculated using the dipole–dipole model between transition densities.

11.7 Lindblad Equation Approach to Coherent Exciton Transport

The relaxation process of a quantum system is in general not time local, that is, it has some memory. The memory is present since in practice the energy transfer through the bath has a finite timescale. This is formally described by the bath correlation function. However, when the bath correlation time is short compared with the system dynamics time, the time-local equation well describes the exciton dynamics. The memory may still effectively remain due to the nonsecular nature of the equation, that is, the effect of excitation is transferred from population to coherence and back, leading to some effective phase delay.

We first assume that we can prepare the system at the initial time $t = 0$ in the system and bath product state, so the total system density matrix is of the form $W(0) = \rho(0) \otimes \rho_\mathrm{B}$. Such a condition is commonly realized in optical excitation of molecular aggregates. Following the description of the Markov processes, we can define the transformation of the system density matrix to some later time [9]:

$$\rho(t) = V(t)\rho(0) \equiv \mathrm{tr}_\mathrm{B}\left(U(t)\rho(0) \otimes \rho_\mathrm{B} U^\dagger(t)\right) \,, \tag{11.68}$$

where $U(t)$ is the Green's function of the total system. This transformation is sometimes denoted as a "dynamic map." If we write the bath operator in its eigenstate basis, $\rho_\mathrm{B} = \sum_n \lambda_n |\varphi_n\rangle\langle\varphi_n|$, we find

$$\rho(t) = V(t)\rho(0) \equiv \sum_{mn} \lambda_n \langle\varphi_m| U(t)|\varphi_n\rangle \rho(0) \langle\varphi_n| U^\dagger(t)|\varphi_m\rangle \,. \tag{11.69}$$

We can now denote $\mathcal{W}_{mn}(t) = \sqrt{\lambda_n}\langle\varphi_m|U(t)|\varphi_n\rangle$ and get

$$V(t)\rho(0) \equiv \sum_{mn} \mathcal{W}_{mn}(t)\rho(0)\mathcal{W}^{\dagger}_{mn}(t) . \tag{11.70}$$

Taking into account that there is no assumption about the dynamics involved, the dynamic map is a completely positive and trace-preserving operation; however, it is time irreversible.

As in Markov processes we may use Markov-type relations, that is,

$$V(t_1)V(t_2) = V(t_1 + t_2) . \tag{11.71}$$

This condition is satisfied by an exponential form,

$$V(t) = \exp(\mathcal{L}t) , \tag{11.72}$$

and the density matrix of the system then satisfies

$$\frac{\mathrm{d}}{\mathrm{d}t}\rho = \mathcal{L}\rho . \tag{11.73}$$

That is the generalization of the Liouville equation and it now includes the effects of the coupling with the bath, that is, the relaxation. This dissipative process formally composes the quantum dynamical semigroup and $\mathcal{L}$ is its generator.

It is possible to construct the most general form of the superoperator $\mathcal{L}$ by expanding it into an arbitrary set of operators. However, the whole mathematical derivation is beyond the scope of this book (see, e.g., [9]). The most important part is the final result:

$$\mathcal{L}\rho \Longrightarrow -\mathrm{i}\left[\hat{H}_{\mathrm{S}}, \rho\right] + \sum_{k=1}^{N^2-1} \gamma_k \left(\hat{A}_k\rho\hat{A}^{\dagger}_k - \frac{1}{2}\hat{A}^{\dagger}_k\hat{A}_k\rho - \frac{1}{2}\rho\hat{A}^{\dagger}_k\hat{A}_k\right) . \tag{11.74}$$

This is called the Lindblad form [78]. Here the first term on the right-hand side represents the unitary part of the dynamics, and $\hat{A}_k$ are Lindblad operators. While they are obtained as mathematical constructions, they have the meaning of various modes which couple the system with the bath (collective coordinates). Having taken the Lindblad operators as dimensionless, we have γ_k as the relaxation rates of different decay modes. The sum runs over the number of independent bath modes; N is the number of degrees of freedom (number of states) of the system.

Assuming that the number of bath modes is uncountable, we next write the Lindblad equation as follows:

$$\frac{\mathrm{d}}{\mathrm{d}t}\rho = -\mathrm{i}[H_{\mathrm{S}}, \rho] + \sum_k \left(\hat{L}_k\rho\hat{L}^{\dagger}_k - \frac{1}{2}\hat{L}^{\dagger}_k\hat{L}_k\rho - \frac{1}{2}\rho\hat{L}^{\dagger}_k\hat{L}_k\right) . \tag{11.75}$$

We expand the Lindblad operator $\hat{L}_k$ in a orthonormal basis set of system states:

$$\hat{L}_k = \sum_{ab} u^{(k)}_{ab}|a\rangle\langle b| . \tag{11.76}$$

Substituting this into the Lindblad equation, we have the following terms:

$$\left(\hat{L}_k \rho \hat{L}_k^\dagger\right)_{ab} = \sum_{a'b'} u_{bb'}^{(k)*} u_{aa'}^{(k)} \rho_{a'b'} , \tag{11.77}$$

$$\left(\hat{L}_k^\dagger \hat{L}_k \rho\right)_{ab} = \sum_{a'b'} u_{a'a}^{(k)*} u_{a'b'}^{(k)} \rho_{b'b} , \tag{11.78}$$

and

$$\left(\rho \hat{L}_k^\dagger \hat{L}_k\right)_{ab} = \sum_{a'b'} u_{b'a'}^{(k)*} u_{b'b}^{(k)} \rho_{aa'} . \tag{11.79}$$

By denoting

$$Z_{ab,cd} = \sum_k u_{ab}^{(k)*} u_{cd}^{(k)} , \tag{11.80}$$

we obtain the Redfield-like equation

$$\frac{\mathrm{d}}{\mathrm{d}t} \rho_{ab} = -\mathrm{i}\left[\hat{H}_\mathrm{S}, \rho\right]_{ab} + \sum_{a'b'} K_{ab,a'b'} \rho_{a'b'} , \tag{11.81}$$

with relaxation rates

$$K_{ab,a'b'} = Z_{bb',aa'} - \frac{\delta_{b'b}}{2} \sum_c Z_{ca,ca'} - \frac{\delta_{a'a}}{2} \sum_c Z_{cb',cb} . \tag{11.82}$$

That is a very important relation which has a direct connection with the Redfield equation and it defines the physical meaning of the correlation coefficients $Z_{ab,cd}$. Let us consider the population transport from state b to state a and $a \neq b$. The rate governing this process is

$$K_{aa,bb} = Z_{ab,ab} . \tag{11.83}$$

The dephasing rate of the density matrix coherence ρ_{ab} is

$$K_{ab,ab} = Z_{bb,aa} - \frac{1}{2} \sum_c [Z_{ca,ca} + Z_{cb,cb}] . \tag{11.84}$$

The commonly used expression can be written

$$K_{ab,ab} = -\bar{\gamma}_{ab} - \frac{1}{2}\left(\tau_a^{-1} + \tau_b^{-1}\right) , \tag{11.85}$$

where the lifetime of state a is given by the total rate of population escape $\tau_a = \sum_c^{c \neq a} Z_{ca,ca}$ and

$$\bar{\gamma}_{ab} = \frac{1}{2}(Z_{aa,aa} + Z_{bb,bb}) - Z_{bb,aa} \tag{11.86}$$

is known as the pure dephasing rate. Terms such as $Z_{aa,bb}$ can be taken as complex numbers, and then $\bar{\gamma}_{ab}$ might include a Lamb shift of the oscillation frequency. Note, however, that we must have $Z_{aa,bb} = Z^*_{bb,aa}$, so $Z_{aa,aa}$ is real and $\bar{\gamma}_{ab} = \bar{\gamma}^*_{ba}$.

However, this description leaves a lot of undefined "off-diagonal" parameters. They must be chosen in a specific way to lead to a physically reasonable result. We thus assume that the Lindblad equation yields the canonical equilibrium distribution of the isolated system $\rho^{(\infty)} = \exp(-\beta \hat{H}_S)$ at long times $t \to \infty$ [79]. The equilibrium exciton populations are then given by $\rho^{(\infty)}_{aa} \propto \exp(-\beta\varepsilon_a)$, where ε_a is the energy of state a, and all off-diagonal density matrix elements vanish: $\rho^{(\infty)}_{ab} = 0$ for $a \neq b$. For the equilibrated state at $t \to \infty$ the Lindblad equation for all a and b gives

$$0 = \sum_c \left[Z_{bc,ac} - \frac{1}{2} \sum_d (\delta_{ac} Z_{dc,db} + \delta_{bc} Z_{da,dc}) \right] \exp(-\beta\varepsilon_c) \,. \tag{11.87}$$

This equation is satisfied when the Lindblad operator matrix elements are completely uncorrelated, that is, $Z_{ab,cd} = \delta_{ac}\delta_{bd} Z_{ab,ab}$, thus leading to the secular relaxation equation which satisfies the detailed balance and the requirement for the equilibrium to be satisfied. However, that is not the only solution. Let us rearrange (11.87) in the form

$$0 = \sum_c \left[Z_{bc,ac} \exp(-\beta\varepsilon_c) - \frac{1}{2} Z_{ca,cb} \exp(-\beta\varepsilon_a) - \frac{1}{2} Z_{ca,cb} \exp(-\beta\varepsilon_b) \right] . \tag{11.88}$$

A sufficient condition for this equality is obtained when each term in the sum of (11.88) is required to be 0, which can be satisfied by

$$\frac{Z_{ab,cd}}{\exp(-\beta\varepsilon_a) + \exp(-\beta\varepsilon_c)} = \frac{Z_{dc,ba}}{\exp(-\beta\varepsilon_d) + \exp(-\beta\varepsilon_b)} \,. \tag{11.89}$$

The Lindblad equation determines all other off-diagonal rates responsible for coherence–coherence transfer and for population–coherence mixing. Terms such as $K_{aa,bb}$ and $K_{ab,ab}$ can be calculated microscopically from the Redfield rate expressions. This leads to fixing correlation coefficients $\langle |u_{ab}|^2 \rangle$ $(a \neq b)$.

Additional determination of coefficients is possible for excitons with respect to the special – ground – state. This state $|g\rangle$ has an infinitely long lifetime and it is a reference state for all fluctuating coordinates, that is, it is not fluctuating itself. Therefore, $\langle u^*_{ab} u_{cd} \rangle = 0$ if one of the indices a, b, c, or d coincides with g. The Lindblad equation then guarantees that there is no population transfer to the ground state. However, the coherence decay rates involving the ground state, which determine the homogeneous linewidth of the absorption spectrum, are as follows:

$$K_{ge,ge} = -\bar{\gamma}_{ge} - \frac{1}{2}\tau_e^{-1} \,, \tag{11.90}$$

$$\bar{\gamma}_{ge} = \frac{1}{2} Z_{ee,ee} \,. \tag{11.91}$$

With this model we are thus able to relate all coefficients of type $C_{ee',ee'}$ to the Redfield theory:

$$Z_{ee,ee} = 2\bar{\gamma}_{ge} , \tag{11.92}$$

$$Z_{ee',ee'} = K_{ee,e'e'} . \tag{11.93}$$

And finally

$$Z_{e'e',ee} = \frac{1}{2}(Z_{ee,ee} + Z_{e'e',e'e'}) - \bar{\gamma}_{ee'} . \tag{11.94}$$

The exciton concept allows us to suggest the relation between the Lindblad correlation coefficients and the exciton wavefunctions. First note that for correlation coefficients we can assume that

$$Z^2_{ab,cd} < Z_{ab,ab} Z_{cd,cd} \tag{11.95}$$

or

$$Z_{ab,cd} = \sqrt{Z_{ab,ab} Z_{cd,cd}} \cos(\alpha_{ab,cd}) , \tag{11.96}$$

where we define the mixing angle $\alpha_{ab,cd}$.

11.8 Hierarchical Equations of Motion for Excitons

Hierarchical equations of motion (HEOM) correspond to a nonperturbative theory describing the exciton dynamics in the open quantum systems introduced in Section 10.6. Because it is a full theory, it is computationally expensive. An additional complication for excitons is that we assume an independent bath for each molecular excitation. We therefore have to deal with N bath modes for N molecules denoted as a coordinate $\hat{Q}_m$, and we have N hierarchy dimensions. Here we present a modified HEOM theory usually termed the hierarchical quantum master equation (HQME) [80]. It is still nonperturbative, but is restricted to the approximate form of the bath correlation function.

Let us start with the semiclassical overdamped spectral density for the nth molecule (see Section 8.6):

$$C''_n(\omega) = 2\lambda_n \frac{\gamma_n \omega}{\omega^2 + \gamma_n^2} . \tag{11.97}$$

The fluctuation correlation function of the bath (7.140) for the nth molecule can be given in the form

$$C_n(t) = \frac{1}{\pi} \int \frac{1}{1 - \mathrm{e}^{-\beta\omega}} \mathrm{e}^{-\mathrm{i}\omega t} C''_n(\omega) \mathrm{d}\omega . \tag{11.98}$$

By expanding the Bose–Einstein function up to the $(\beta\omega)$ term

$$\frac{1}{1-e^{-\beta\omega}} = \frac{1}{\beta\omega} + \frac{1}{2} + \frac{\beta\omega}{12} + \mathcal{O}[(\beta\omega)^3] \tag{11.99}$$

(this expansion corresponds to [0/0] Padé decomposition of the spectrum [81]), we have

$$C_n(t) = \left(\frac{2\lambda_n}{\beta} - \frac{\beta\lambda_n\gamma_n^2}{6}\right) e^{-\gamma_n t} - i\lambda_n\gamma_n e^{-\gamma_n t} + \frac{\lambda_n\gamma_n\beta}{3}\delta(t)\,. \tag{11.100}$$

The high-temperature approximation schemes use only the first two terms of this expansion. Here, all three terms are used and the rest of the correlation function is accounted for by the Markovian-white-noise residue ansatz [80]. The criterion of applicability certainly depends on temperature: it was shown in [80] to represent adequate dynamics when

$$\min\{\Gamma_n(\gamma_n)/\Omega_S, \kappa_n\} \gtrsim 2\,, \tag{11.101}$$

where $\Gamma_n(\gamma_n) = [\sqrt{12 + (\beta\gamma_n)^2} + 6]/\beta$, Ω_S is the characteristic frequency of the system, and $\kappa_n = \sqrt{6\Gamma_n(\gamma_n)/(\beta\lambda_n\gamma_n)}$.

The HQME for the correlation function given in (11.100) is written in the Liouville space as a hierarchy of coupled differential equations for auxiliary density operators denoted by $|\rho_{\boldsymbol{n}}(t)\rangle\rangle$ [80]:

$$\begin{aligned}\frac{d}{dt}|\rho_{\boldsymbol{n}}(t)\rangle\rangle &= -i\hat{\mathcal{L}}_e|\rho_{\boldsymbol{n}}(t)\rangle\rangle - \sum_{m=1}^{N}\left(\gamma_m n_m + \delta\hat{\mathcal{R}}_m\right)|\rho_{\boldsymbol{n}}(t)\rangle\rangle \\ &\quad + \sum_{m=1}^{N} n_m\hat{\mathcal{A}}_m|\rho_{\boldsymbol{n}_m^-}(t)\rangle\rangle + i\sum_{m=1}^{N}\hat{\mathcal{Q}}_m^{\times}|\rho_{\boldsymbol{n}_m^+}(t)\rangle\rangle\,,\end{aligned} \tag{11.102}$$

where the auxiliary superoperators

$$\delta\hat{\mathcal{R}}_m = \frac{\lambda_m\gamma_m\beta}{3}\hat{\mathcal{Q}}_m^{\times}\hat{\mathcal{Q}}_m^{\times} \tag{11.103}$$

and

$$\hat{\mathcal{A}}_m = i\left[\left(\frac{2\lambda_m}{\beta} - \frac{\beta\lambda_m\gamma_m^2}{6}\right)\hat{\mathcal{Q}}_m^{\times} - i\lambda_m\gamma_m\hat{\mathcal{Q}}_m^{\circ}\right] \tag{11.104}$$

are introduced. Here $\hat{\mathcal{Q}}_m^{\times}\bullet \Leftrightarrow [\hat{Q}_m, \bullet]$ ($\bullet$ is an arbitrary operator) is the commutator and $\hat{\mathcal{Q}}_m^{\circ}\bullet \Leftrightarrow \{\hat{Q}_m, \bullet\}$ is the anticommutator. In (11.102) $|\rho_0(t)\rangle\rangle$ corresponds to the physical reduced density operator, while $\boldsymbol{n}$ is a vector of indices $\boldsymbol{n} \equiv (n_1, n_2, \ldots, n_N)$, and we use the notation $\boldsymbol{n}_m^{\pm} \equiv (n_1, n_2, \ldots, n_m \pm 1, \ldots, n_N)$. All indices of auxiliary density matrices represent the number of vibrational quanta and are positive numbers [53].

Formally the hierarchy in (11.102) is infinite; thus, the equations are nonperturbative and non-Markovian. Various truncation schemes can be used. The simplest one is based on the assumption that all auxiliary density matrices with tier level $L = \sum_{m=1}^{N} n_m$ greater than the truncation level L_{trunc} are simply discarded. The truncation level is usually chosen to guarantee convergence of the simulation results. Other truncation schemes are also possible [53, 82].

Since the HEOM theory is derived as an operator equation and makes no approximation for the bath, it is thus independent of the basis chosen for the solution of the problem. It can thus capture effects such as exciton delocalization, polaron formation, and collapse of a delocalized exciton onto a single chromophore.

11.9 Weak Interchromophore Coupling Limit

Energy transfer in molecular aggregates in the case of weak resonance coupling is usually considered within the framework of Förster resonance energy transfer (FRET) theory. This is a widely employed method that works remarkably well even in situations where the condition of weak chromophore–chromophore coupling might be questionable. However, for some applications FRET theory has several important closely related deficiencies. Namely, since FRET theory is an approach of the Fermi golden rule type, it gives the population transfer rates but no prescription for propagating the other elements of the density matrix, for example, coherences. By the same token, FRET theory intrinsically assumes that the excitations are localized on individual chromophores despite their mutual interaction. The latter deficiency can be overcome by using the modified Redfield theory, but in this case the former still persists, which renders the description of coherent phenomena impossible.

It is possible, however, to formulate a dynamical description of the whole reduced density matrix in the weak resonance coupling limit [83]. The derivation is similar to that of the generalized quantum master equation, except that this time the resonance coupling instead of the system–bath interaction is treated as a perturbation. Therefore, we split the system Hamiltonian in the following way:

$$\hat{H}_S = \hat{H}_\epsilon + \hat{H}_J \,. \tag{11.105}$$

Here $\hat{H}_\epsilon$ and $\hat{H}_J$ denote accordingly the diagonal and off-diagonal parts of the Frenkel exciton Hamiltonian. In the presence of the bath the reference part of the total Hamiltonian for the perturbation expansion is taken as

$$\hat{H}_0 = \hat{H}_\epsilon + \hat{H}_{SB} + \hat{H}_B \,, \tag{11.106}$$

which leaves out the resonance coupling part $\hat{H}_J$ as a perturbation. Switching to the Liouville space notation, we split the total Liouvillian in the same manner:

$$\mathcal{L} = \mathcal{L}_0 + \mathcal{L}_J \,. \tag{11.107}$$

We first write the solution of the Liouville equation in the usual way using the Green's function

$$W(t) = \mathcal{G}(t)\, W(0) \,. \tag{11.108}$$

We next define the interaction representation with respect to $\mathcal{L}_0$:

$$W_\mathrm{I}(t) = \mathcal{G}_0(-t)\, W(t) \tag{11.109}$$

and

$$\mathcal{V}(t) = \mathcal{G}_0(-t)\mathcal{L}_\mathrm{J}\mathcal{G}_0(t)\ . \tag{11.110}$$

Then, using the projection operator technique as introduced in Chapter 9, we get an equation for $\mathcal{P}\, W_\mathrm{I}(t)$ and truncate it at second order in $\hat{H}_\mathrm{J}$:

$$\frac{\mathrm{d}}{\mathrm{d}t}\mathcal{P}\, W_\mathrm{I}(t) = -\mathrm{i}\mathcal{P}\mathcal{V}(t)\mathcal{P}\, W_\mathrm{I}(t) - \int\limits_{t_0}^{t} \mathrm{d}\tau \mathcal{P}\mathcal{V}(t)\mathcal{P}\mathcal{Q}\mathcal{V}(\tau)\mathcal{P}\, W_\mathrm{I}(\tau)\ ; \tag{11.111}$$

Here the so-called initial term proportional to $\mathcal{Q}\, W_\mathrm{I}(t_0)$ is omitted. To justify this omission as well as to ensure the maximum quality of the second-order approximation, the choice of the operator $\mathcal{P}$ is essential, and it is best dictated by the physical situation in question. For instance, in the case of optical excitation, we can assume that the system is initially in the ground state, $\rho = |\mathrm{g}\rangle\langle\mathrm{g}|$, and the bath is in the canonical equilibrium, ρ_B, so the total density matrix is factorized as $W(t_0) = \rho \otimes \rho_\mathrm{B}$. Then according to the Franck–Condon principle, upon the excitation the bath part remains unchanged and the system–bath state remains factorized. In this way the appropriate projection operator reads

$$\mathcal{P}\hat{Z} = \mathrm{tr}_\mathrm{B}(\hat{Z})\rho_\mathrm{B}\ , \tag{11.112}$$

and therefore $\mathcal{Q}\, W_\mathrm{I}(t_0) = 0$, which eliminates the initial term.

Denoting $\bar{\rho}(t) = \mathrm{Tr}_\mathrm{B}(\, W_\mathrm{I}(t))$ and resolving the projectors and commutators (thus returning to the Hilbert space), we can rewrite (11.111) in terms of $\bar{\rho}(t)$:

$$\frac{\mathrm{d}}{\mathrm{d}t}\bar{\rho}(t)\rho_\mathrm{B} = -\mathrm{i}\mathrm{tr}_\mathrm{B}\left(\hat{H}_\mathrm{J}(t)\rho_\mathrm{B}\right)\bar{\rho}(t)\rho_\mathrm{B} + \mathrm{i}\bar{\rho}(t)\mathrm{tr}_\mathrm{B}\left(\rho_\mathrm{B}\hat{H}_\mathrm{J}(t)\right)\rho_\mathrm{B} - R(J^2)\ , \tag{11.113}$$

where

$$\begin{aligned} R(J^2) = \int\limits_{t_0}^{t} \mathrm{d}\tau \Big[& \mathrm{tr}_\mathrm{B}\left(\hat{H}_\mathrm{J}(t)\hat{H}_\mathrm{J}(\tau)\rho_\mathrm{B}\right)\bar{\rho}(\tau) - \mathrm{tr}_\mathrm{B}\left(\hat{H}_\mathrm{J}(t)\rho_\mathrm{B}\right)\mathrm{tr}_\mathrm{B}\left(\hat{H}_\mathrm{J}(\tau)\rho_\mathrm{B}\right)\bar{\rho}(\tau) \\ & - \mathrm{tr}_\mathrm{B}\left(\hat{H}_\mathrm{J}(t)\bar{\rho}(\tau)\rho_\mathrm{B}\hat{H}_\mathrm{J}(\tau)\right) + \mathrm{tr}_\mathrm{B}\left(\hat{H}_\mathrm{J}(t)\rho_\mathrm{B}\right)\bar{\rho}(\tau)\mathrm{tr}_\mathrm{B}\left(\rho_\mathrm{B}\hat{H}_\mathrm{J}(\tau)\right) \\ & - \mathrm{tr}_\mathrm{B}\left(\hat{H}_\mathrm{J}(\tau)\bar{\rho}(\tau)\rho_\mathrm{B}\hat{H}_\mathrm{J}(t)\right) + \mathrm{tr}_\mathrm{B}\left(\hat{H}_\mathrm{J}(\tau)\rho_\mathrm{B}\right)\bar{\rho}(\tau)\mathrm{tr}_\mathrm{B}\left(\rho_\mathrm{B}\hat{H}_\mathrm{J}(t)\right) \\ & + \bar{\rho}(\tau)\mathrm{tr}_\mathrm{B}\left(\rho_\mathrm{B}\hat{H}_\mathrm{J}(\tau)\hat{H}_\mathrm{J}(t)\right) - \bar{\rho}(\tau)\mathrm{tr}_\mathrm{B}\left(\rho_\mathrm{B}\hat{H}_\mathrm{J}(\tau)\right)\mathrm{tr}_\mathrm{B}\left(\rho_\mathrm{B}\hat{H}_\mathrm{J}(t)\right)\Big]\ . \end{aligned} \tag{11.114}$$

The traces can be evaluated as follows. Taking the matrix element of $\mathrm{tr_B}(\hat{H}_\mathrm{J}(t)\rho_\mathrm{B})$, we obtain the expression

$$\langle a|\mathrm{Tr_B}\left(\hat{H}_\mathrm{J}(t)\rho_\mathrm{B}\right)|b\rangle = J_{ab}\mathrm{e}^{\mathrm{i}\omega_{ab}t}\mathrm{tr_B}(G_a(-t)G_b(t)\rho_\mathrm{B}) \,, \tag{11.115}$$

where $G(t)$ is the Green's function corresponding to the Hamiltonians involving the nuclear coordinate $\hat{q}$: $\hat{H}_\mathrm{SB}(\hat{q}) + \hat{H}_\mathrm{B}(\hat{q}, \hat{p})$. The trace can now be evaluated by employing the second-order cumulant approximation:

$$\mathrm{Tr_B}(G_a(-t)G_b(t)\rho_\mathrm{B}) = \exp\left[-(1-\delta_{ab})\left(g_a^*(t) + g_b(t)\right)\right] \,, \tag{11.116}$$

where $g_a(t)$ is the lineshape function associated with the transition from state $|g_a\rangle$ to state $|e_a\rangle$. Thus,

$$\langle a|\mathrm{Tr_B}\left(\hat{H}_\mathrm{J}(t)\rho_\mathrm{B}\right)|b\rangle = J_{ab}\mathrm{e}^{\mathrm{i}\omega_{ab}t-(1-\delta_{ab})(g_a^*(t)+g_b(t))} \equiv J_{ab}(t) \,. \tag{11.117}$$

Similarly, the larger traces in (11.114) can be evaluated. Then changing the integration variable in (11.114) to $\tau' = t - \tau$, we can employ the Markov approximation $\bar{\rho}(t-\tau) \approx \bar{\rho}(t)$. Setting $t_0 = 0$ (e.g., the optical excitation giving the time reference), we obtain the equations of motion in the final form:

$$\begin{aligned} \frac{\mathrm{d}}{\mathrm{d}t}\bar{\rho}_{ab}(t) = &-\mathrm{i}\sum_c J_{ac}(t)\bar{\rho}_{cb}(t) + \mathrm{i}\sum \bar{\rho}_{ac}(t)J_{cb}(t) \\ &-\sum_{cd}\left[R_{accd}(t)\bar{\rho}_{db}(t) - R^*_{cabd}(t)\bar{\rho}_{cd}(t)\right. \\ &\left.- R_{dbac}(t)\bar{\rho}_{cd}(t) + R^*_{bddc}(t)\bar{\rho}_{ac}(t)\right] \,. \end{aligned} \tag{11.118}$$

The relaxation tensor reads

$$R_{abcd}(t) = \int_0^t \mathrm{d}\tau[J_{ab}J_{cd}M_{abcd}(t, t-\tau) - J_{ab}(t)J_{cd}(t-\tau)] \,, \tag{11.119}$$

where the auxiliary function is given as

$$M_{abcd}(t,\tau) = \mathrm{e}^{F_{abcd}(t,\tau)+\mathrm{i}\omega_{ab}t+\mathrm{i}\omega_{cd}\tau} \,, \tag{11.120}$$

and

$$\begin{aligned} F_{abcd}(t,\tau) = &-g_a^*(t) - g_b(t) - g_c^*(\tau) - g_d(\tau) \\ &-\delta_{ac}\left[g_a(t) - g_a(t-\tau) + g_a^*(\tau)\right] + \delta_{ad}\left[g_a(t) - g_a(t-\tau) + g_a^*(\tau)\right] \\ &+\delta_{bc}\left[g_b(t) - g_b(t-\tau) + g_b^*(\tau)\right] - \delta_{bd}\left[g_b(t) - g_b(t-\tau) + g_b^*(\tau)\right] \,. \end{aligned} \tag{11.121}$$

Here we disregarded the cross-correlations, that is, assumed $g_{ab}(t) = \delta_{ab}g_a(t)$.

Finally, the connection between the reduced density operator in the interaction picture $\bar{\rho}(t) = \mathrm{Tr}_{\mathrm{B}}(W_{\mathrm{I}}(t))$ and in the Schrödinger picture $\rho(t) = \mathrm{Tr}_{\mathrm{B}}(W(t))$ is given by the equation

$$\bar{\rho}_{ab}(t) = \mathrm{e}^{\mathrm{i}\omega_{ab}t+(1-\delta_{ab})[g_a(t)+g_b^*(t)]}\rho_{ab}(t) . \tag{11.122}$$

We thus get a dynamic relaxation picture through the $g(t)$ functions (these are introduced in Appendix A.2). The bath thus affects not only the excitation in a rigid frame of excitonic wavefunctions, but the frame itself becomes "flexible." The bath renormalized the intermolecular couplings dynamically, which is never captured by the Redfield equations.

11.10 Modeling of Exciton Dynamics in an Excitonic Dimer

As described in Section 5.3, an elementary molecular aggregate is a molecular dimer made up of two strongly coupled molecules. It was described in Section 5.3. It is characterized by two site energies ε_1 and ε_2 and the intermolecular coupling J. We next denote $\Delta = \varepsilon_2 - \varepsilon_1$.

The Hamiltonian of such a system is solvable analytically by introducing the mixing angle θ (5.57) so that the exciton eigenvectors are

$$\begin{pmatrix} \psi_{e_1 n_1} & \psi_{e_2 n_1} \\ \psi_{e_1 n_2} & \psi_{e_2 n_2} \end{pmatrix} = \begin{pmatrix} \cos(\theta) & -\sin(\theta) \\ \sin(\theta) & \cos(\theta) \end{pmatrix} . \tag{11.123}$$

The Schrödinger equation for the system gives the mixing angle

$$\tan(2\theta) = \frac{2J}{\Delta} \tag{11.124}$$

and then the eigenstate energies are given by

$$\varepsilon_{\pm} = \bar{\varepsilon} \pm \frac{\Delta}{2}\sqrt{1+\tan^2(2\theta)} , \tag{11.125}$$

where $2\bar{\varepsilon} = \varepsilon_1 + \varepsilon_2$.

To include the relaxation effects we couple each molecular excitation in the dimer (the excitation energy) to its own phonon bath characterized by the spectral density:

$$C''(\omega) = 2\lambda \frac{\omega\Lambda}{\omega^2 + \Lambda^2} . \tag{11.126}$$

In that case their energy fluctuations are uncorrelated.

We next present simulations of exciton dynamics in this system using several different models described in the various sections in this chapter. This section is thus devoted to the application to the model dimer. The set of simulation parameters for the dimer is as follows. The trial system Hamiltonian in reciprocal centimeters is taken as

$$\begin{pmatrix} 10\,200 & 100 \\ 100 & 10\,100 \end{pmatrix} . \tag{11.127}$$

This leads to a mixing angle of $\theta \approx 2.6$ rad. The eigenenergies are $10\,000 + 38.2$ and $10\,000 + 261.8$. The bath is characterized by $\Lambda = 0.53J$, which is taken as approximately 100 fs. These values of the parameters are typical of photosynthetic pigment–protein complexes [24].

For the initial condition we assume that the molecular dimer is excited by an ultrashort laser pulse. In that case the pulse bandwidth covers both exciton eigenstates, so both are excited simultaneously. Such laser excitation prepares the system in a highly nonequilibrium excited configuration described by the density matrix

$$\rho(t = 0) = \begin{pmatrix} 0.5 & 0.5 \\ 0.5 & 0.5 \end{pmatrix} . \tag{11.128}$$

We will next follow the exciton dynamics in this system using the propagation methods described below when this system is either weakly ($\lambda = 0.1J$) or strongly ($\lambda = J$) coupled to the bath.

Secular Redfield Dynamics The secular Redfield scheme serves as the simplest approach. The dynamics for the weak system–bath coupling strength are shown in Figure 11.1. As population and coherence dynamics are uncoupled, we can observe only exponential monotonous decay of nonequilibrium populations to the equilibrium. The coherences oscillate with their native frequencies dictated by the energy gaps between the eigenlevels. The decay of these coherences is limited only by the exciton lifetimes and by their pure dephasings. The strong system–bath coupling in Figure 11.2 leads to faster relaxation and almost overdamped relaxation of the coherence.

Modified Redfield Theory The modified Redfield theory is intrinsically similar to the secular Redfield theory, only the transport rates are calculated differently. In this case in Figure 11.1 the population transfer rates are different; thus, populations redistribute faster. Note that this weak coupling regime does not induce any noticeable relaxation in the excited state; thus the rates of the Redfield and modified Redfield theories almost coincide. In Figure 11.2 the strong coupling case is shown. We find that the rates in this case are higher and the population dynamics are faster. This change in rates is induced by exciton relaxation in the excited state. The coherences propagate in the same way as in the secular Redfield theory as the coherence is described by the same rates.

Full Redfield Theory The full Redfield theory as described by (11.27) is the most straightforward approach to tackle the nonequilibrium system dynamics as it is a direct outcome of second-order perturbation theory. All fluctuations are included

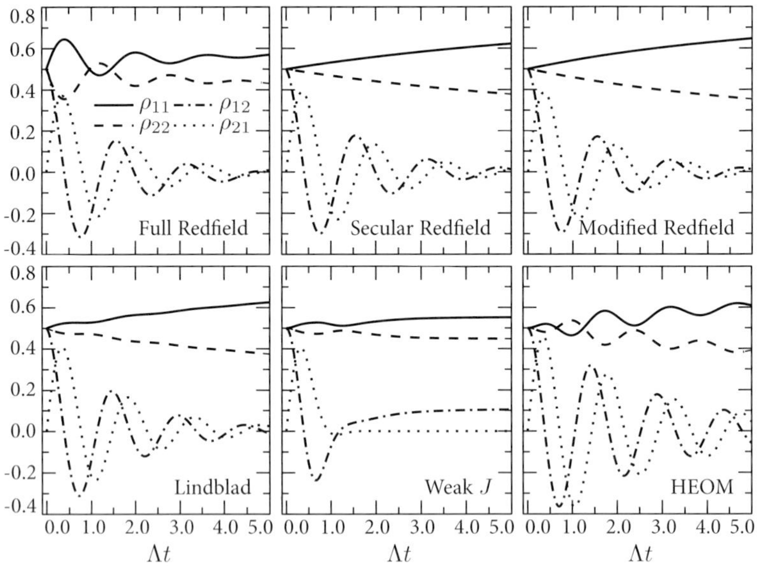

Figure 11.1 The density matrix dynamics of the dimer with initial conditions as in (11.128) in the case of weak system–bath coupling strength. "HEOM" means "hierarchical equation of motion."

up to second order. In the weak system–bath coupling case the system remains highly coherent as implied by oscillatory populations and density matrix coherences. After the initial coherent phase, the populations cease to oscillate and approach equilibrium values monotonically. In this state the coherences are almost zero. A completely different picture is obtained in the strong system–bath coupling case, where the Redfield equation parameters make the solution diverge. This demonstrates that the Redfield equation has a limited parameter space of validity. It must be noted that since the equation is up to second order in fluctuations, the error of the solution accumulates with time. At second order we have the fluctuation intensity, which perturbs the system dynamics, proportional to

$$\delta_c = \xi \left| \int_0^\infty C(t)\mathrm{d}t \right| , \tag{11.129}$$

where ξ characterizes the strength of off-diagonal fluctuations in the eigenstate basis. For the dimer $\xi = 1/2\sin^2(2\theta)$, while to estimate the integral we take the high-temperature limit and get

$$\delta_c = \frac{\lambda}{\beta\Lambda}\sin^2(2\theta) . \tag{11.130}$$

δ_c^{-1} provides the timescale where the Redfield solution should be trustworthy. Note that δ_c roughly coincides with the homogeneous linewidth of the absorption line.

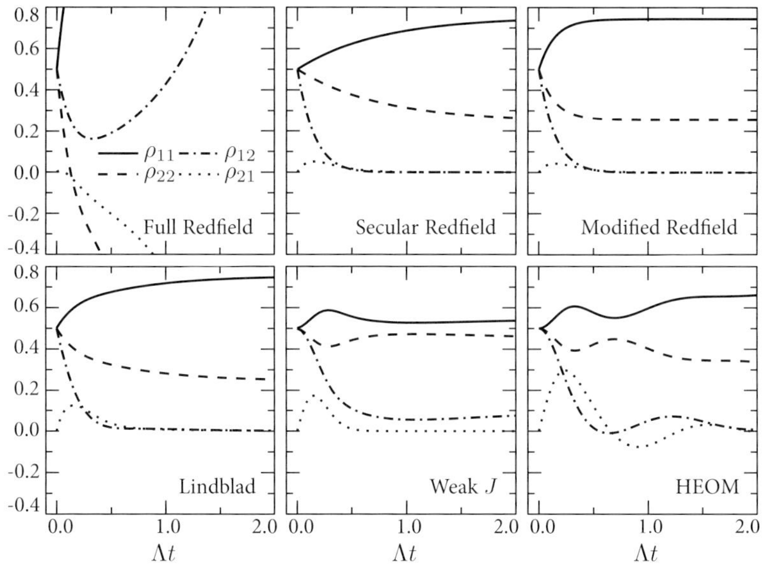

Figure 11.2 The density matrix dynamics of the dimer with initial conditions as in (11.128) in the case of strong system–bath coupling strength.

For the dimer parameters we get

$$\Lambda \cdot \left(t < \frac{\Lambda \beta}{\lambda} \right) . \tag{11.131}$$

Using the above parameters we roughly have $\Lambda t < 1$ for the weak coupling and $\Lambda t < 0.1$ for the strong coupling. In the weak system–bath coupling case the result is physically feasible also for $\Lambda t \sim 1$; however, it should be concluded that it is not trustworthy without relating it to a particular experiment. That is obvious in the case of the strong system–bath coupling, where the result diverges for $\Lambda t \sim 0.1$. The secular theories cure this problem; however their result is questionable as well.

Lindblad Dynamics The Lindblad equation allows for the quantum transport regime where the populations are coupled with density matrix coherences. The equation maintains the physical density matrix for all times. The Lindblad dynamics as presented in this chapter are parametrized with respect to the Redfield theory: we choose the Lindblad correlation cosine matrix to be [79]

$$\cos(\alpha_{e_4,e_3,e_2,e_1}) = \text{Max}(w_{e_1,e_2} \cdot w_{e_3,e_4}, w_{e_1,e_3} \cdot w_{e_2,e_4}, w_{e_1,e_4} \cdot w_{e_2,e_3}) , \tag{11.132}$$

where

$$w_{e_1,e_2} = \sum_n |c_{ne_1} c_{ne_2}| \tag{11.133}$$

is the exciton overlap integral with condition $0 < w_{e_1,e_2} < 1$.

The Lindblad equation now includes the same transport rates as in the Redfield theory. Additionally they are tuned so that the thermal equilibrium coincides with the secular Redfield theory for excitons. This results in the Lindblad equation being intermediate between the full Redfield theory and the secular Redfield theory. It thus captures the coherent dynamics as observed in the weak system–bath coupling case (oscillatory population evolution) as well as satisfactory relaxation to the equilibrium similar to the secular equation. The result remains physically reasonable in the strong system–bath coupling limit as well. In this case the coherences are overdamped and the system is driven to equilibrium monotonically. However, this construction of the parameters relies on heuristic arguments, so the result must be understood with respect to a specific experiment.

Weak Resonance Coupling Regime The weak-J theory captures the other limit of the strong system–bath coupling correctly. In this case when the coupling with the bath is strong and comparable to the interchromophore coupling J, the theory correctly reflects the density matrix dynamics since the diagonal bath fluctuations in the eigenstate basis are included through the cumulant expansion. This part can be understood as exact provided the diagonal and off-diagonal fluctuations are weakly correlated. This is in stark contrast to the Redfield theory, which fails and diverges. The long-time limit is however shifted from the correct thermal equilibrium, but the initial dynamics is properly recovered. The finite coherence elements at long times show that the excitonic basis is no longer the eigenbasis of the solution and the site basis is more appropriate. Thus, in the long-time limit the system relaxes to the site basis equilibrium.

HEOM (HQME) Dynamics The HQME method is a way to compute the aggregate dynamics without approximations at any temperature. However, the hierarchy is infinite and additional constrains are included with respect to the fluctuation correlation functions. It thus correctly includes the quantum transport effects as well as realistic coherence dephasing rates. In Figures 11.1 and 11.2 we show the cases of weak and strong couplings using the HEOM method. In the initial time it reflects the coherent dynamics, which transform into the dissipative ones. In Figures 11.1 and 11.2 only the short-time dynamics are shown. However, in the long-time limit, it can be shown that coherences in the strong system–bath coupling case do not relax to the equilibrium. This shows that the weak-J theory becomes more accurate in that limit and the site basis is more appropriate especially in the strong system–bath coupling limit. The exciton states thus decay and the dynamics should be described in the site basis.

11.11
Coherent versus Dissipative Dynamics: Relevance for Primary Processes in Photosynthesis

An essential step in developing a description of exciton relaxation in molecular aggregates is to make the right approximations. The most sophisticated HEOM theory is by no means the best tool to describe few-chromophore systems. However, it is an essentially exact description and its ultimate application amounts to the full solution of quantum dynamics. It thus scales badly with the system size. For instance, if the system consists of N chromophores and we assume that each site is coupled to an independent bath coordinate, we have to use an N-dimensional space for hierarchy of equations. If additionally we take M terms into expansion of the Bose–Einstein distribution with respect to the temperature, we add M more dimensions in the hierarchy. The problem becomes hardly tractable on a computer even for moderate system sizes (approximately 10). However, with use of a multicore supercomputer, aggregates as large as $N = 36$ have been processed [84] using the HEOM. Current applications are mostly limited to the overdamped model of the bath, while the colored bath is theoretically possible in HEOM [85]. Alternative exact methods including an arbitrary bath spectral density are available as well [71, 84, 86–89]. These are not considered in this book because of space limitations.

As the HEOM is numerically exact, it is now taken as a reference by comparing the approximate methods. As we showed, the approximate methods for density matrix can be used in specific parameter regions. For instance, in the case of weak system–bath coupling, the full Redfield theory is very suitable at short times. As the system reaches the incoherent relaxation phase, the secular theory can be used to propagate the populations to the thermal equilibrium. Surprisingly, the simplest secular Redfield theory gives proper relaxation timescales. As it is simple to implement and is very efficient in simulations, it is the best choice for estimating qualitative behavior. In the case of strong system–bath couplings, the weak-J theory gives the correct coherent dynamics phase and encapsulates the formation of polarons.

Special attention has to be paid to the Lindblad and modified Redfield theories. The Lindblad equation of motion guarantees a physically proper density matrix for all time delays; however, it does not provide a recipe to obtain the relaxation rates or to define the correlation coefficients using spectral densities of the harmonic bath. The method to relate the Lindblad parameters with the Redfield rates presented in this book gives a qualitatively important result as it allows us to capture the short-time coherent dynamics and the long-time dissipative dynamics correctly. The modified Redfield scheme is "an upgrade" of the secular Redfield theory, and promises better exciton transfer rates. However, as the theory relies on the exciton basis, its second-order approximation in the off-diagonal fluctuations limits the rates again to the weak coupling regime since in the case of strong system–bath couplings (strongly coupled systems) all fluctuations are large, become correlated, and cannot be treated at second order. The modified Redfield scheme is thus trustworthy when all fluctuations are weak or when the chromophores are weakly coupled. However, the weak-J theory may be a better choice in that case as it captures dynamic wavefunction relaxation.

The theories presented were demonstrated for a simple pair of two-level sites. The results and conclusions obtained can be extended to larger systems since the excitonic dimer is the simplest system having all ingredients as in larger systems.

Recent multidimensional spectroscopy experiments, which we describe in Part Two, done primarily on the FMO aggregate [24, 90] and later on other molecular aggregates, revealed a broad range of coherent processes that take place in the femtosecond time domain [62, 72]. As these experiments initiated the discussion of the importance of quantum coherent effects in the biological functioning of the light harvesting complex in photosynthesis, this initiated a revision of the quantum relaxation theories with a strong emphasis on the quantum coherence [68, 69, 71, 73]. According to estimations [91] the strongest J coupling in FMO aggregates is on the order of approximately $100\,\mathrm{cm}^{-1}$, while the homogeneous absorption linewidth is approximately $60\,\mathrm{cm}^{-1}$. In this case the exciton concept should hold and the theories which are based on the weak J coupling regime should be avoided. The theories which include coherent excitons are thus preferable. For instance, the full Redfield theory should be appropriate to account for the damped evolution of the excitonic coherences. The modified Redfield theory includes some additional concepts compared with the full Redfield theory. For instance, the correlations of the diagonal and off-diagonal fluctuations are included approximately in the modified Redfield population transport rates; however, this addition does not guarantee better agreement with experiments since a lot of model parameters are not well defined and are usually fitted by indirect experiments.

We note that the simulations of FMO aggregates based on the Redfield theory have never been able to reproduce the long-lived quantum coherence beats observed in the seminal two-dimensional spectroscopy experiment. Recently it has been shown that the observed spectral beats in the two-dimensional spectrum reflect the molecular vibrations, which have been ignored before [92–94]. The latter is not surprising knowing that the vibrational progression is very weakly expressed in the FMO absorption spectrum, while it may influence the energy transfer rates [95].

Part Two
Spectroscopy

12
Introduction

Electronic spectra of molecular aggregates are usually attributed to Frenkel-type excitons, which have been described in Chapter 5. Such attribution is based on a significant shift and narrowing of the absorption band in comparison with the absorption bands corresponding to separate molecules. It is noteworthy that the resonance inter-molecular interaction is the main parameter resulting in delocalization of the exciton states defined by a linear superposition of the excited states of the individual molecules.

The exciton phenomenon was well-resolved as early as 1936 when spectral changes of the pseudosocianine dyes in water solution were observed, while changing the dye concentration [96, 97], and later on in molecular crystals and polymers [23, 25, 27]. Excitonic features are also disclosed in stationary and time-resolved spectra of various photosynthetic pigment-protein complexes [24] that start the "cycle of life."

To harvest the solar light efficiently, photosynthetic organisms are equipped with pigment-protein antenna complexes. These complexes are involved in the initial stage of photosynthesis, starting with the absorption of the solar light by the pigment molecules and followed by the transfer of the accumulated energy to the reaction center, where this energy is stabilized as a chemical potential [98, 99]. In the latter process of the energy accumulation the charge transfer states are involved.

Variations of protein environment at different pigment molecules introduce distribution of transition energies and determine the timescale of their changes. Such interaction of the pigment molecules with their environment is usually qualified in two limiting cases corresponding to the static and dynamic disorder of their transition energies. The static disorder corresponds to the slow rearrangement of the environment in the vicinity of a particular pigment molecule while the dynamic disorder reflects the opposite limiting case corresponding to fast vibrations, thus determining the exciton dephasing and restraining the coherence in exciton transport. The entire set of such vibrations might be considered as the bath, while interaction of these vibrations with molecular excitations can be treated perturbatively. According to such a theoretical scheme the exciton dynamics can contain both coherent and incoherent behavior.

As elementary excitations carry no charge or spin, only energy, they cannot be tracked by electric measurements extensively used in semiconductor electronics.

Molecular Excitation Dynamics and Relaxation, First Edition. L. Valkunas, D. Abramavicius, and T. Mančal.

Spectroscopic methods are then efficiently applied. The traditional measurements include absorption, fluorescence (steady state and time resolved) and pump-probe, and various other types. In addition to the conventional spectroscopic methods used for studies of various molecular aggregates, the four-wave mixing measurement provides the set of simple nonlinear techniques available for isotropic systems. In the time domain these are performed by applying either two pulses (pump-probe) or three pulses (homodyne three-pulse photon echo), or four pulses (coherent heterodyned signals) to generate and detect the desired signal. The four-wave mixing signal is generated by the induced third order polarization, which is a parametric function of the delays between the adjacent laser pulses. The polarization dynamics with respect to these parameters reflect a wide variety of ultrafast molecular processes. Recent development of coherent two-dimensional photon echo spectroscopy disclosed the possibility to estimate the time scale of the exciton decoherence. Thus, two-dimensional spectroscopy is already a key tool demonstrating a complex pathway network of the energy transfer in various photosynthetic pigment-protein complexes, molecular aggregates, and polymers. It is evident that the interaction of electronic excitations with intra- and inter-molecular vibrations causes a disruption of the phase relationship between excited states of the molecules constituting the exciton wave functions. This type of interaction has a distinct influence on the coherence in the exciton dynamics and plays the dominant role by determining the exciton transport pathways. Thus, nonlinear optical techniques performed using ultrashort laser pulses are capable of probing various dynamical phenomena on microscopic/nanoscopic scale.

In this part of the book we discuss basic theoretical approaches for the description of various spectroscopic observables. We illustrate the basic steps which take from the formulation of the quantum mechanical problem in terms of equations of motion as described in Part One of this book to the formulation of the spectroscopic signal in terms of the response functions, which is the basis of Part Two. Indeed, most of the spectroscopic experimental observations can be well described in terms of the response theory by treating the excitation light perturbatively. The basics of the semiclassical response theory is presented in Chapter 13. Theoretical principles of the linear spectroscopy are presented in Chapter 14. The theoretical basis applicable for the description of the nonlinear spectroscopy is presented in Chapter 15. The 2D coherent spectroscopy approach is discussed by analyzing first the simple model systems (Chapter 16) and afterwards by considering the spectral changes of the photosynthetic pigment-protein complexes (Chapter 17). The basics of the single molecular spectroscopy is described in Chapter 18.

13
Semiclassical Response Theory

Spectroscopic experiments are usually described using the semi-classical approximation [26]. According to this approximation the incoming excitation fields, the outgoing signal field are assumed to be classical (electric) fields, and the molecular system is considered to be a quantum object. This scheme applies to most experimental situations, unless the quantum properties of light are explicitly used or investigated, for instance, by considering entangled photons or quantum photon statistics measurements. Because the lasers and coherent fields are used for the light generation in most experiments considered in this book, classical description is an appropriate approximation.

In the semiclassical model, the experiment is formally divided into two stages. In the first stage a quantum system interacts with the incoming field. As discussed in Section 2.4.1, this interaction is described by the polarization (or transition dipole moment) operator. The system gets displaced from the equilibrium due to the influence of the external field, and thus, creates the nonequilibrium time-dependent polarization. In the second stage the expectation value of the induced polarization becomes a source of the signal field. This stage can be treated as a problem of the classical electrodynamics, and it can be demonstrated by the Maxwell equations as demonstrated in Section 2.2.

Assuming that the polarization is ultimately originating in the microscopic quantum system, we need to consider the dynamics of this quantum system influenced by the external field. Thus, we have to deal with a coupled problem of time evolution of the fields $\boldsymbol{E}(\boldsymbol{r}, t)$ and the system state $\hat{W}(t)$. The whole process can be described by the following three Maxwell–Liouville equations:

$$\nabla \times \nabla \times \boldsymbol{E}(\boldsymbol{r}, t) + \frac{1}{c^2}\frac{\partial^2}{\partial t^2}\boldsymbol{E}(\boldsymbol{r}, t) = -\mu_0 \frac{\partial^2}{\partial t^2}\boldsymbol{P}(\boldsymbol{r}, t) , \tag{13.1}$$

$$\boldsymbol{P}(\boldsymbol{r}, t) = \mathrm{Tr}\left(\hat{\boldsymbol{P}}(\boldsymbol{r})\hat{W}(t)\right) , \tag{13.2}$$

$$\frac{\partial \hat{W}(t)}{\partial t} = -\frac{\mathrm{i}}{\hbar}\left[\hat{H}_{\mathrm{sc}}(\boldsymbol{E}_i(\boldsymbol{r}, t)), \hat{W}(t)\right] . \tag{13.3}$$

They have the following meaning: the first equation is the relationship between photo-induced polarization $\boldsymbol{P}(\boldsymbol{r}, t)$ of the system and the outgoing electric field.

Molecular Excitation Dynamics and Relaxation, First Edition. L. Valkunas, D. Abramavicius, and T. Mančal.

This expression will be greatly simplified by assuming the so-called phase-matching experimental geometry, which will be discussed in Section 13.3.1. The second equation is the definition of the physically observable polarization as the trace of the product of the polarization operator and the density matrix. The third equation determines the time evolution of the system density matrix under the influence of the excitation field $\boldsymbol{E}_i(\boldsymbol{r}, t)$. Here $\hat{H}_{sc}$ is the semi-classical Hamiltonian, which describes the quantum system under the influence of the classical electric field as an external force.

In the next few sections, we will describe the physics behind these three equations. One straightforward step which we can perform here is that (13.1) can be rewritten in a form:

$$-\nabla^2 \boldsymbol{E}(\boldsymbol{r}, t) + \frac{1}{c^2}\frac{\partial^2}{\partial t^2}\boldsymbol{E}(\boldsymbol{r}, t) = -\mu_0 \frac{\partial^2}{\partial t^2}\boldsymbol{P}(\boldsymbol{r}, t) , \tag{13.4}$$

which is valid for transverse radiation fields, $\boldsymbol{E} \equiv \boldsymbol{E}^{\perp}$ (in other words, $\nabla \cdot \boldsymbol{E} = 0$). This assumption usually applies to the dielectric medium as we study it here.

13.1
Perturbation Expansion of Polarization: Response Functions

We will now assume that the sample is made of some molecules, which are much smaller than the wavelength of light. In this case, the molecule-field interaction is assumed to be well described as a dipole-field interaction. The Hamiltonian describing the whole problem is then divided into the following terms:

$$\hat{H} = \hat{H}_S + \hat{H}_B + \hat{H}_{SB} + \hat{H}_{int} = \hat{H}_{mat} - \hat{\boldsymbol{\mu}} \cdot \boldsymbol{E}(t) . \tag{13.5}$$

Here, $\hat{H}_S$ is the molecular (or system) part of the Hamiltonian, containing all degrees of freedom (or states) of the molecular system, which have to be included explicitly. They are usually the electronic states that can be directly manipulated by light, but sometimes they might also include some selected vibrational or other levels. The second term, $\hat{H}_B$, represents the bath, causing dephasing and energy relaxation in the system. $\hat{H}_{SB}$ corresponds to the interaction between the system and the bath. These three terms of the Hamiltonian constitute the material part of the system, $\hat{H}_{mat}$. The last term, $\hat{H}_{int} = -\hat{\boldsymbol{\mu}} \cdot \boldsymbol{E}(t)$, is the dipolar system-field interaction (see Section 2.4).

The dynamics of the system is more conveniently described using superoperators, which were already introduced in Section 4.5.3 where they proved to be useful in deriving equations of motion of the reduced density matrix. Here the problem is of a similar complexity, and the abbreviation of the commutators (those from (13.3)) into superoperators will be of great convenience. We use the material Liouville superoperator and interaction superoperator

$$\mathcal{L}_{mat}\hat{A} = \frac{1}{\hbar}[\hat{H}_{mat}, \hat{A}] ,$$

$$\mathcal{L}_{\text{int}}\hat{A} = -\frac{1}{\hbar}[\hat{\mu}, \hat{A}]E(t) = -\mathcal{V}\hat{A}E(t) , \tag{13.6}$$

where $\mathcal{V}$ is the polarization (or transition dipole moment) superoperator. For the sake of simplicity, we introduce $E(t) = \boldsymbol{n} \cdot \boldsymbol{E}(t)$ and $\hat{\mu} = \boldsymbol{n} \cdot \hat{\boldsymbol{\mu}}$, where $\boldsymbol{n}$ is a vector of unit length and determines the direction of the transition dipole operator. The equation of motion for the total density matrix $\hat{W}(t)$ is then given by

$$\frac{\partial \hat{W}(t)}{\partial t} = -\mathrm{i}\mathcal{L}_{\text{mat}}\hat{W}(t) - \mathrm{i}\mathcal{L}_{\text{int}}\hat{W}(t) . \tag{13.7}$$

This equation cannot be solved exactly. The external field is a parameter under control by experimentalists, and usually it is weak in the sense of the perturbation theory. In other words, the spectroscopic signals can be sorted out by the power dependencies on the external perturbation. We will thus treat the field-system interaction perturbatively, while the material part will be taken as a reference, for which the dynamical problem can be solved, at least formally. This scheme of the time-dependent perturbation theory is typical for most of the spectroscopic calculations. At the next step we use the interaction picture, which represents the problem in "rotating frame" corresponding to the field-free evolution of the system. The reference material evolution operator is defined as

$$\mathcal{U}_{\text{mat}}(t) \equiv \exp(-\mathrm{i}\mathcal{L}_{\text{mat}}t) . \tag{13.8}$$

We transform the system density operator into the interaction picture:

$$\hat{W}^{(\text{I})}(t) = \mathcal{U}^{\dagger}_{\text{mat}}(t)\hat{W}(t) . \tag{13.9}$$

and we get accordingly a new equation of motion:

$$\frac{\partial \hat{W}^{(\text{I})}(t)}{\partial t} = \mathrm{i}\mathcal{V}(t)\hat{W}^{(\text{I})}(t)E(t) , \tag{13.10}$$

where $\mathcal{V}(t) = \mathcal{U}^{\dagger}_{\text{mat}}(t)\mathcal{V}\mathcal{U}_{\text{mat}}(t)$ is the polarization superoperator in the interaction representation. This expression can be formally integrated:

$$\hat{W}^{(\text{I})}(t) = \hat{W}^{(\text{I})}(t_0) + \mathrm{i}\int_{t_0}^{t} \mathrm{d}\tau \mathcal{V}(\tau)\hat{W}^{(\text{I})}(\tau)E(\tau) . \tag{13.11}$$

By repeating the iteration we obtain an infinite series:

$$\begin{aligned}\hat{W}^{(\text{I})}(t) &= \hat{W}^{(\text{I})}(t_0) + \mathrm{i}\int_{t_0}^{t} \mathrm{d}\tau \mathcal{V}(\tau)\hat{W}^{(\text{I})}(\tau)E(\tau)\\ &+ \mathrm{i}^2\int_{t_0}^{t}\mathrm{d}\tau\int_{t_0}^{\tau}\mathrm{d}\tau'\mathcal{V}(\tau)\mathcal{V}(\tau')\hat{W}^{(\text{I})}(\tau')E(\tau)E(\tau')\\ &+ \mathrm{i}^3\int_{t_0}^{t}\mathrm{d}\tau\int_{t_0}^{\tau}\mathrm{d}\tau'\int_{t_0}^{\tau'}\mathrm{d}\tau''\mathcal{V}(\tau)\mathcal{V}(\tau')\mathcal{V}(\tau'')\hat{W}^{(\text{I})}(\tau'')E(\tau)E(\tau')E(\tau'') + \ldots\end{aligned} \tag{13.12}$$

In the same form we can write the induced polarization. Taking the expansion:

$$P(t) = P^{(0)}(t) + P^{(1)}(t) + P^{(2)}(t) + \dots \tag{13.13}$$

and using the general expression for the polarization expectation value in the Schrödinger picture:

$$P(t) = \mathrm{Tr}\left[\hat{\mu}\,\hat{W}(t)\right] \tag{13.14}$$

we can now write an expression for an arbitrary order of the induced polarization:

$$P^{(n)}(t) = \int_{t_0}^{t} \mathrm{d}\tau_n \int_{t_0}^{\tau_n} \mathrm{d}\tau_{n-1} \dots \int_{t_0}^{\tau_2} \mathrm{d}\tau_1\, E(\tau_n) E(\tau_{n-1}) \dots E(\tau_1)$$
$$\times\, \mathrm{i}^n \mathrm{Tr}\left[\hat{\mu}\mathcal{U}_{\mathrm{mat}}(t)\mathcal{V}(\tau_n)\mathcal{V}(\tau_{n-1}) \dots \mathcal{V}(\tau_1)\mathcal{U}^{\dagger}_{\mathrm{mat}}(t_0)\,\hat{W}(t_0)\right]. \tag{13.15}$$

Here, from a pure mathematical point of view, the operator $\hat{\mu}$ inside trace is not appropriate as all other quantities are superoperators in the Liouville space. However, we interpret this notation as (read from right to left): take the density matrix $\hat{W}(t_0)$, map it to the Liouville space, apply all Liouville space propagators $\mathcal{U}_{\mathrm{mat}}$ and interactions $\mathcal{V}$ until we reach $\hat{\mu}$; then map the result back to the Hilbert space and act on the left by the operator $\hat{\mu}$.

We now make an assertion that the optical field is turned on at time $t = 0$, so at $t_0 < 0$ the system is in the equilibrium state. In the equilibrium the system state is preserved under the time evolution. We denote $\hat{W}(0) \equiv \hat{\rho}_{\mathrm{eq}}$ is the time-independent equilibrium density matrix. We also introduce time intervals between interactions instead of absolute interaction times and make a corresponding substitution in the integral, (13.15). This will lead to an expression where we integrate from zero to time $t - t_0$. The absolute starting time t_0 is arbitrary as long as it is before the light arrives, and it can, therefore, be formally set to $t_0 = -\infty$. The integration thus goes formally to infinity which is sometimes convenient for calculations. The polarization then reads as:

$$P^{(n)}(t) = \int_{0}^{\infty} \mathrm{d}\tau_n \int_{0}^{\infty} \mathrm{d}\tau_{n-1} \dots \int_{0}^{\infty} \mathrm{d}\tau_1\, S^{(n)}(t_n, t_{n-1}, \dots, t_1)$$
$$\times\, E(t-\tau_n)E(t-\tau_n-\tau_{n-1}) \dots E(t-\tau_n-\tau_{n-1}-\dots-\tau_1)\,, \tag{13.16}$$

where

$$S^{(n)}(t_n, t_{n-1}, \dots, t_1) = \mathrm{i}^n \mathrm{Tr}\left[\hat{\mu}\mathcal{U}_{\mathrm{mat}}(\tau_n)\mathcal{V}\mathcal{U}_{\mathrm{mat}}(\tau_{n-1})\mathcal{V} \dots \mathcal{U}_{\mathrm{mat}}(\tau_1)\mathcal{V}\hat{\rho}_{\mathrm{eq}}\right] \tag{13.17}$$

is the nth order response function of the system.

The optical field and the polarization are vector quantities. The response functions, hence, are tensorial quantities. For instance, the first order response function $S^{(1)}$ is the second rank tensor $S^{(1)}_{\nu_2\nu_1}$ connecting three incoming components of the optical field E_{ν_1} with the outgoing component of the induced polarization P_{ν_2}. In most of the cases these properties are not relevant since simple optical field configurations are realized; for example, all incoming fields are collinear and linearly polarized, and their electric vectors are all parallel, while the sample is homogeneous. In this case the tensorial properties can be disregarded. We will emphasize the tensorial properties when required.

The first order response function is given by:

$$S^{(1)}(t_1) = \mathrm{iTr}\left\{\hat{\mu}\mathcal{U}_{\mathrm{mat}}(t_1)\mathcal{V}\hat{\rho}_{\mathrm{eq}}\right\} = \frac{\mathrm{i}}{\hbar}\mathrm{Tr}\left\{\hat{\mu}\,U(t_1)\left[\hat{\mu},\hat{\rho}_{\mathrm{eq}}\right]U^{\dagger}(t_1)\right\}\,, \tag{13.18}$$

Switching back to the Hilbert space it can also be given through a time correlation function of the polarization operators in the interaction representation:

$$S^{(1)}(t_1) = \frac{\mathrm{i}}{\hbar}\theta(t)\left[J_1(t) - J_1^{*}(t)\right]\,, \tag{13.19}$$

$$J_1(t) \equiv \mathrm{Tr}\left\{\hat{\mu}(t)\hat{\mu}(0)\hat{\rho}_{\mathrm{eq}}\right\} \equiv \langle\hat{\mu}(t)\hat{\mu}(0)\rangle\,. \tag{13.20}$$

The linear response function is responsible for such effects as linear absorption, circular dichroism, and so on, as we will show in detail later.

If the optical experiments are performed on liquid state solutions, the observable molecular ensembles include all possible orientations of the relevant quantum systems with respect to the laboratory frame. As will be shown later in Section 15.2.6 the second order response is zero for such ensembles of molecules, so that the next contribution to the optical response is the third order response function, which is related to the third order polarization $P^{(3)}$.

From (13.16) and (13.17) we can write down the expression for the third order polarization:

$$\begin{aligned} P^{(3)}(t) = &\int_0^{\infty}\mathrm{d}t_3\int_0^{\infty}\mathrm{d}t_2\int_0^{\infty}\mathrm{d}t_1\,S^{(3)}(t_3,t_2,t_1)\\ &\times E(t-t_3)E(t-t_3-t_2)E(t-t_3-t_2-t_1)\,, \end{aligned} \tag{13.21}$$

where the response function is given by:

$$S^{(3)}(t_3,t_2,t_1) = \mathrm{i}^3\mathrm{Tr}\left\{\hat{\mu}_4\mathcal{U}_{\mathrm{mat}}(t_3)\mathcal{V}_3\mathcal{U}_{\mathrm{mat}}(t_2)\mathcal{V}_2\mathcal{U}_{\mathrm{mat}}(t_1)\mathcal{V}_1\hat{\rho}_{\mathrm{eq}}\right\}\,. \tag{13.22}$$

Expanding the commutators we get:

$$\begin{aligned} S^{(3)}(t_3,t_2,t_1) = &\left(\frac{\mathrm{i}}{\hbar}\right)^3\theta(t_1)\theta(t_2)\theta(t_3)\\ &\times\sum_{\alpha=1}^{4}\left(R_\alpha(t_3,t_2,t_1) - R_\alpha^{\star}(t_3,t_2,t_1)\right) \end{aligned} \tag{13.23}$$

where the four terms in the Hilbert space are given in the form of the four-point correlation functions:

$$R_1(t_3, t_2, t_1) = \mathrm{Tr}\left\{\hat{\mu}(t_1)\hat{\mu}(t_1 + t_2)\hat{\mu}(t_1 + t_2 + t_3)\hat{\mu}(0)\hat{\rho}_{eq}\right\}, \tag{13.24}$$

$$R_2(t_3, t_2, t_1) = \mathrm{Tr}\left\{\hat{\mu}(0)\hat{\mu}(t_1 + t_2)\hat{\mu}(t_1 + t_2 + t_3)\hat{\mu}(t_1)\hat{\rho}_{eq}\right\}, \tag{13.25}$$

$$R_3(t_3, t_2, t_1) = \mathrm{Tr}\left\{\hat{\mu}(0)\hat{\mu}(t_1)\hat{\mu}(t_1 + t_2 + t_3)\hat{\mu}(t_1 + t_2)\hat{\rho}_{eq}\right\}, \tag{13.26}$$

$$R_4(t_3, t_2, t_1) = \mathrm{Tr}\left\{\hat{\mu}(t_1 + t_2 + t_3)\hat{\mu}(t_1 + t_2)\hat{\mu}(t_1)\hat{\mu}(0)\hat{\rho}_{eq}\right\}. \tag{13.27}$$

Heaviside functions in (13.23) emphasize the principle of causality: as it is seen in (13.21), the third order polarization at time t depends on the electric field of earlier times. In other words, an electric field in the past (a cause) determines the polarization in the present time. If any of the arguments of the system response function is negative, the function must be zero. One can also notice that the system response function is always real. It is clear from the physical definition, since the polarization is a real observable quantity and has a corresponding hermitian quantum mechanical operator whose expectation value is always real. We have also used the fact that the operators inside the trace can be cyclically permuted.

The full set of response functions contains all properties of the observable system relevant to all possible optical measurements. The response function contains many contributions, for instance the nth order response function has 2^n terms due to the commutators. As we show later, due to high frequency of the laser field, resonant conditions can be established and only terms with oscillation frequencies equal to that of the laser field can be retained for a specific measurement. The number of relevant contributions for specific experiment reduces significantly. The polarization is found to oscillate with frequency, or the set of frequencies, similar to the excitation laser field. The oscillatory polarization then generates the new optical field as the oscillating dipole in classical electrodynamics [3]. The detector then measures that new field. In the coming chapters we apply this theory to specific optical experiments.

13.2
First Order Polarization

The first order polarization is tightly related to the simplest spectroscopic experiments. The best representative of the corresponding experiment is the linear absorption measurement, which we study in detail in the next chapter. We introduce the main ideas in this section.

13.2.1
Response Function and Susceptibility

The crucial quantity in the theory of linear absorption is the linear polarization $\boldsymbol{P}^{(1)}(\boldsymbol{r}, t)$. It is (by definition) linearly proportional to the electric field $\boldsymbol{E}$. As indicated in Section 13.1 it can depend on the field at other locations and at previous times. Thus, the most general relation between the two quantities can be described in the form of a combined space-time convolution [26]:

$$\boldsymbol{P}^{(1)}(\boldsymbol{r},t) = \int_V \mathrm{d}\boldsymbol{r}' \int_{-\infty}^{\infty} \mathrm{d}t' \, \boldsymbol{S}^{(1)}(\boldsymbol{r}-\boldsymbol{r}', t-t') \boldsymbol{E}(\boldsymbol{r}', t') \,, \tag{13.28}$$

where we require the response function to be identically equal to zero at time $t-t' < 0$ in order to allow only the past values of the field to influence the polarization at the present time. This means that $\boldsymbol{S}^{(1)}(\boldsymbol{r},t)$ is proportional to the *Heaviside step function* $\theta(t)$. The boldface character here denotes the tensor character of the response function. We will restrict ourselves to isotropic materials and the linear response function will be represented by a scalar quantity $S^{(1)}(\boldsymbol{r},t)$

$$\boldsymbol{S}^{(1)}(\boldsymbol{r},t) = S^{(1)}(\boldsymbol{r},t) \overleftrightarrow{\mathbb{I}} \,, \tag{13.29}$$

where $\overleftrightarrow{\mathbb{I}}$ is a unity tensor of rank 2 with respect to the electric vectors of the incoming field and of the induced polarization.

In the Hamiltonian (2.110), the material can be viewed as an ensemble of identical independent systems interacting with the incoming electric field $\boldsymbol{E}$. The space dependence of $S^{(1)}$, therefore, has the form of a spatial $\delta(\boldsymbol{r})$ function, and will be omitted in further discussion. For a given point in space, the relation between the polarization $\boldsymbol{P}(t)$ and the external field $\boldsymbol{E}(t)$ is given by:

$$\boldsymbol{P}(t) = \int_{-\infty}^{t} \mathrm{d}t' \, S^{(1)}(t-t') \boldsymbol{E}(t') = \int_{0}^{\infty} \mathrm{d}t_1 \, S^{(1)}(t_1) \boldsymbol{E}(t-t_1) \,, \tag{13.30}$$

where we used the substitution $t_1 = t - t'$.

It is often advantageous to work with Fourier transformed quantities (see Appendix A.5) such as, for example,

$$\boldsymbol{E}(\omega) = \int_{-\infty}^{\infty} \mathrm{d}t \, \boldsymbol{E}(t) \mathrm{e}^{\mathrm{i}\omega t} \,, \quad \boldsymbol{E}(t) = \frac{1}{2\pi} \int_{-\infty}^{\infty} \mathrm{d}\omega \, \boldsymbol{E}(\omega) \mathrm{e}^{-\mathrm{i}\omega t} \,. \tag{13.31}$$

In the frequency domain the nonlocal relation, (13.28), turns into a local relation:

$$\boldsymbol{P}^{(1)}(\omega) = S^{(1)}(\omega) \boldsymbol{E}(\omega) \,, \quad S^{(1)}(\omega) = \int_{-\infty}^{\infty} \mathrm{d}t \; S^{(1)}(t) \mathrm{e}^{\mathrm{i}\omega t} \,. \tag{13.32}$$

In the classical electrodynamics we introduce the *linear susceptibility* by the relation:

$$\boldsymbol{P}^{(1)}(\omega) = \epsilon_0 \chi(\omega) \boldsymbol{E}(\omega) \,, \tag{13.33}$$

from which it follows that:

$$\chi(\omega) = \frac{1}{\epsilon_0} S^{(1)}(\omega) \,. \tag{13.34}$$

The Fourier transformed linear response function is (up to a constant) equal to the linear susceptibility.

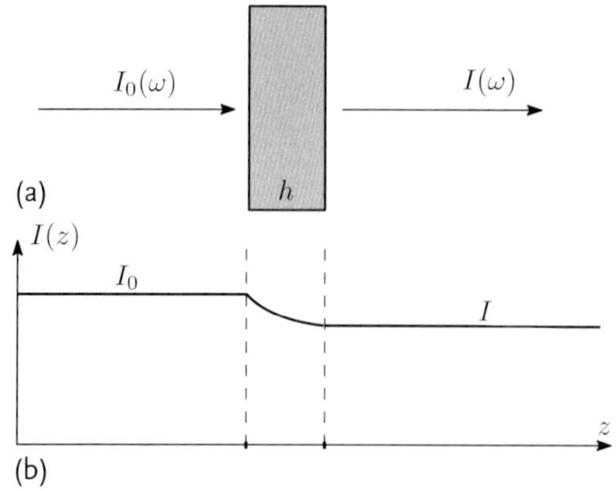

Figure 13.1 Scheme of the linear absorption experiment. (a) Light of certain frequency and intensity $I_0(\omega)$ is sent through a sample of the width h and its diminished intensity $I(\omega)$ is measured. (b) The light intensity as a function of the position z.

13.2.2
Macroscopic Refraction Index and Absorption Coefficient

In spectroscopy the spectral content of radiation is of highest interest. Often we actively send radiation with a known spectrum through a given macroscopic sample of matter containing a molecular system of interest. Then we measure the effect of the matter on the radiation (for example, changes in intensity), trying to deduce the properties of the investigated molecular system. The transmittance or the absorbance can thus be measured.

One of the simplest and most fundamental spectroscopic experiments is the measurement of the linear *absorption coefficient* $\kappa_a(\omega)$. It is based on the *Lambert–Beers law* which states that the intensity of light passing through the material, $\boldsymbol{I}(\boldsymbol{\omega})$, decreases exponentially with its thickness h. The absorption strenght can vary with the light frequency ω, and thus the law can be given by:

$$I(\omega) = I_0(\omega)\mathrm{e}^{-\kappa_a(\omega)h} \,. \tag{13.35}$$

A simple scheme of the absorption experiment is depicted in Figure 13.1.

Now we will relate the linear susceptibility with the linear absorption coefficient. We are interested in propagation of the electric field $\boldsymbol{E}(\boldsymbol{r},t)$ through a medium with no free charges. This situation is described by (13.4). In vacuum (where $\boldsymbol{P}(\boldsymbol{r},t) = 0$), any function of the argument $\boldsymbol{s}\cdot\boldsymbol{r} - ct$, where $\boldsymbol{s}$ is some unit vector, is a possible solution of (13.4). In particular, plain waves $\boldsymbol{E}(\boldsymbol{r},t) = \boldsymbol{e}\,E_0\mathrm{e}^{-\mathrm{i}\omega t + \mathrm{i}\omega/c\boldsymbol{s}\cdot\boldsymbol{r}}$ + c.c. are its solutions. The vector $(\omega/c)\boldsymbol{s}$ pointing in the direction of the plain wave propagation is called the *wave vector*, and it is usually denoted by $\boldsymbol{k} = k\boldsymbol{s}$, $k = \omega/c$. If we assume a slowly varying envelope $\mathcal{E}(t)$ instead of constant E_0, (13.4) will be still approximately satisfied as long as $|\partial/\partial t\mathcal{E}(t)| \ll \omega\mathcal{E}(t)$.

Inside a dielectric material, we might look for a new solution, again in the form of a plain wave with a slowly varying envelope, with an additional degree of freedom to account for changes in the spatial phase factor. The wave vector magnitude k will thus be allowed to depend on ω in a more general way than just linear. However, we still assume that $k(\omega)$ is real. We will start with the electric field in the form of a single beam with a given carrier frequency. Hence, we use the following *ansatz* for the positive frequency component (the negative part is complex conjugate):

$$\boldsymbol{E}^{(+)}(\boldsymbol{r}, t) = \boldsymbol{e}\mathcal{E}(\boldsymbol{r}, t)\mathrm{e}^{\mathrm{i}k(\omega)\boldsymbol{s}\cdot\boldsymbol{r}-\mathrm{i}\omega t} , \tag{13.36}$$

$$\boldsymbol{P}^{(+)}(\boldsymbol{r}, t) = \boldsymbol{e}\mathcal{P}(\boldsymbol{r}, t)\mathrm{e}^{\mathrm{i}k(\omega)\boldsymbol{s}\cdot\boldsymbol{r}-\mathrm{i}\omega t} . \tag{13.37}$$

Let the vector $\boldsymbol{s}$ point along the z axis. We might, therefore, replace $\boldsymbol{r} \rightarrow z = \boldsymbol{s}\cdot\boldsymbol{r}$ and reduce the spatial derivative to a derivative in z only, $\nabla^2 \rightarrow \partial^2/\partial z^2$. We neglect all time derivatives of both envelopes (of polarization and field), and we do the same with the second derivative of the field envelope according to z. We expect that some weak absorption will take place and consequently the changes of $\mathcal{E}(\boldsymbol{r}, t)$ along the z axis are allowed. The wave equation, (13.4), then yields:

$$\frac{\partial}{\partial z}\mathcal{E}(z, t) + \frac{\mathrm{i}}{2}\left(k(\omega) - \frac{\omega^2}{k(\omega)c^2}\right)\mathcal{E}(z, t) = \frac{\mathrm{i}\mu_0\omega^2}{2k(\omega)}\mathcal{P}(z, t) . \tag{13.38}$$

From the imaginary part of (13.38) we have:

$$k^2(\omega) - \frac{\omega^2}{c^2} = \mu_0\omega^2\mathrm{Re}\left[\frac{\mathcal{P}(z, t)}{\mathcal{E}(z, t)}\right] . \tag{13.39}$$

According to (13.33) the ratio $(P(z, t))/(\mathcal{E}(z, t))/\epsilon_0$ at the given frequency can be identified[1] with the susceptibility $\chi(\omega)$. We introduce the real and imaginary parts of the susceptibility such that $\chi(\omega) = \chi'(\omega) + \mathrm{i}\chi''(\omega)$. By defining the refraction index via $\omega n(\omega)/c = k(\omega)$ and using the relation $c^2 = 1/\epsilon_0\mu_0$ we arrive at:

$$n(\omega) = \sqrt{1 + \chi'(\omega)} . \tag{13.40}$$

The real part of the linear susceptibility, therefore, determines the refraction index $n(\omega)$.

Taking the real part of (13.38) we obtain:

$$\frac{\partial}{\partial z}\mathcal{E}(z, t) = -\frac{\omega}{2\epsilon_0 n(\omega)c}\mathrm{Im}\,\mathcal{P}(z, t) . \tag{13.41}$$

or multiplying with $\mathcal{E}(z, t)$ for the field intensity we get:

$$\frac{\partial}{\partial z}I(z, t) = -\frac{\omega}{n(\omega)c}\chi''(\omega)I(z, t) . \tag{13.42}$$

1) Strictly speaking, the ratio of the polarization and field is proportional to the susceptibility only in the frequency domain. For slowly varying envelopes, however, it can be assumed to be approximately valid also in the time domain because the field is assumed to be close to monochromatic, having just one frequency component.

The absorption coefficient $\kappa_a(\omega)$ describing the exponential decay of $I(z,t)$ is, therefore:

$$\kappa_a(\omega) = \frac{\omega}{n(\omega)c}\chi''(\omega) . \tag{13.43}$$

A typical *linear absorption* experiment is performed with a weak monochromatic continuous wave (CW) field and the polarization $P(z,t)$ can be identified with its linear component $P^{(1)}(z,t)$. The absorption coefficient κ_a of (13.43) then corresponds to the *linear* absorption coefficient. In principle, however, the validity of (13.38) is not restricted to the linear polarization and one can describe the intensity-dependant non-linear absorption.

13.3
Nonlinear Polarization and Spectroscopic Signals

We shall now generalize the discussion already made for the linear spectroscopy; that is, the relation between fields, polarizations and response functions, for the case of nonlinear process, of so-called N-wave mixing process.

13.3.1
N-wave Mixing

In (13.36) and (13.37) we have assumed that the electric field is composed of a single beam traveling in a single direction. Now let us consider a more general excitation by $N-1$ fields

$$\boldsymbol{E}(\boldsymbol{r},t) = \sum_{n=1}^{N-1} \boldsymbol{e}_n \mathcal{E}_n(\boldsymbol{r},t) \mathrm{e}^{\mathrm{i}\boldsymbol{k}_n\cdot\boldsymbol{r}-\mathrm{i}\omega_n t} + \text{c.c.} \tag{13.44}$$

From (13.15) we know that the mth order polarization will depend on the product of m fields of the form of (13.44), and can thus contain all possible combinations of m available wave vectors $\boldsymbol{k}_s = \pm\boldsymbol{k}_1 \pm \boldsymbol{k}_2 \pm \ldots$ Some of these combinations yield wave vectors different in direction from wave vectors of the incident light. And a corresponding new field (wave) is generated in the sample. The most prominent methods used in the nonlinear spectroscopy are the four wave mixing (FWM) methods where three fields with wave vectors $\boldsymbol{k}_1$, $\boldsymbol{k}_2$ and $\boldsymbol{k}_3$ are used to generate field into a fourth new direction. An arbitrary configuration from $\pm\boldsymbol{k}_1 \pm \boldsymbol{k}_2 \pm \boldsymbol{k}_3$ can be measured in an experiment. To account for this new field, our procedure of calculating electric fields has to be slightly modified. Instead of (13.36) we have (13.44), and (13.37) will be replaced by the third order polarization component with the desired wave vector $\boldsymbol{k}_s$ and carrier frequency ω_s that is with

$$\boldsymbol{P}^{(3)(+)}(\boldsymbol{r},t) = \boldsymbol{e}\mathcal{P}^{(3)}(\boldsymbol{r},t)\mathrm{e}^{\mathrm{i}\boldsymbol{k}_s\cdot\boldsymbol{r}-\mathrm{i}\omega_s t} . \tag{13.45}$$

We have chosen to look at the (+) component of the polarization, but the third order polarization apparently has both (+) and (−) components. In the next step we insert our new expression into the wave equation, (13.4), and obtain an equation analogical to (13.38). We split the total field into a part solving (13.38) with the linear polarization on the right hand side and the expected third order field $\boldsymbol{E}^{(3)}(\boldsymbol{r}, t)$, which will be generated by the third order polarization. Thus, we get equation analogical to (13.38), but containing purely the third order quantities.

We shall not forget that the third order field $\boldsymbol{E}^{(3)}$ generates its own polarization that causes its absorption and refraction. Since the third order field itself is weak and the sample is usually optically thin, we can stay to the first order with the description of its absorption, and denote the slow varying envelope of the first order polarization induced by the third order field by $\mathcal{P}^{(1[3])}(\boldsymbol{r}, t)$.

Another important point is that due to the dispersion, the third order electric field might be generated into a direction $\boldsymbol{k}$ different from $\boldsymbol{k}_s$. Hence, (13.38) leads to

$$\frac{\partial}{\partial z}\mathcal{E}^{(3)}(z,t) + \frac{\mathrm{i}}{2}\left(k(\omega) - \frac{\omega^2}{k(\omega)c^2}\right)\mathcal{E}^{(3)}(z,t) = \frac{\mathrm{i}\omega^2}{2k(\omega)\epsilon_0 c^2}\left(\mathcal{P}^{(1[3])}(z,t) + \mathcal{P}^{(3)}(z,t)\mathrm{e}^{\mathrm{i}\Delta \boldsymbol{k}\cdot\boldsymbol{s}z}\right). \tag{13.46}$$

The absolute value of the wave vector $k(\omega)$ can be assumed to satisfy (13.39) with the first order polarization $\mathcal{P}^{(1[3])}$ only, so that we could cancel its real part. Ignoring the reabsorption for now, we get from (13.46)

$$\frac{\partial}{\partial z}\mathcal{E}^{(3)}(z,t) = \mathrm{i}\frac{\omega}{2n(\omega)\epsilon_0 c}\mathcal{P}^{(3)}(z,t)\mathrm{e}^{\mathrm{i}\Delta \boldsymbol{k}\cdot\boldsymbol{s}z}. \tag{13.47}$$

Integration of (13.47) in a simple box geometry, Figure 13.2, with initial condition $\mathcal{E}^{(3)}(z = 0, t) = 0$ can be easily done. Let us assume that the thickness of the material sample in the direction $\boldsymbol{s}$ of the third order field propagation is h and let us set the origin of z at the start of the sample. The integration of (13.47) yields

$$\mathcal{E}^{(3)}(h,t) = \mathrm{i}\frac{\omega}{n(\omega)\epsilon_0 c}\mathcal{P}^{(3)}(t)h\frac{\sin(\Delta k h/2)}{(\Delta k h/2)}\mathrm{e}^{\mathrm{i}\Delta k h/2}. \tag{13.48}$$

When $h \to \infty$ the function $\sin(\Delta k h/2)/(\Delta k h/2) \to \delta(\Delta k h)$ and thus the function has a sharp maximum at $\Delta k = 0$. In other directions the signal becomes significantly weaker. If we measure the third order signal in a phase matching direction, such as $\boldsymbol{k}_s$, the third order field at its exit from the box of the thickness h reads:

$$\mathcal{E}^{(3)}(h,t) = \mathrm{i}\frac{\omega}{n(\omega)\epsilon_0 c}\mathcal{P}^{(3)}(t)h. \tag{13.49}$$

In most of the spectroscopic investigations we are not interested in absolute values of the nonlinear signal. Rather, we are interested in its frequency domain profile or relative time dependence. Often we can neglect the change of the refraction index in the spectral region of the experiment, and so the spectroscopically significant dependence of the generated field on the polarization is reduced to a change of

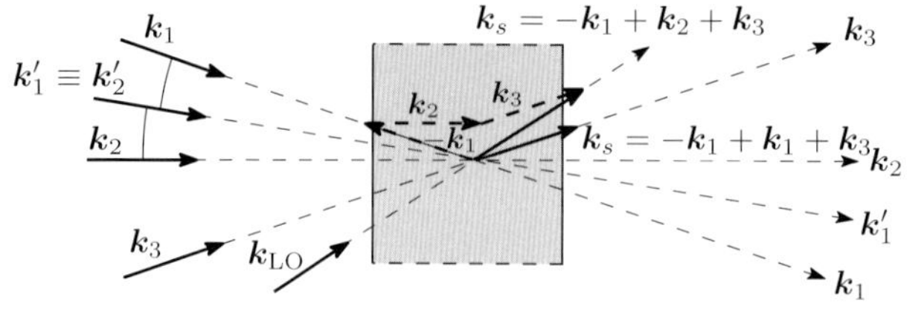

Figure 13.2 A two-dimensional representation of the geometry corresponding to the four-wave mixing experiment. The incoming electric fields with wavevectors k_1, k_3 and k_3 generate nonlinear signals in all possible wavevector combinations (only $-k_1+k_2+k_3$ is depicted). The local oscillator field with wavevector $k_{\text{LO}} = k_s$ is added corresponding to the heterodyne detection scheme.

phase by i and a distortion of the line shape by factor ω. Thus, the main result of this section is that the slowly varying envelope of the nonlinear signal reads:

$$\mathcal{E}^{(3)}(t) \approx \mathrm{i}\omega \mathcal{P}^{(3)}(t) . \tag{13.50}$$

From here it is obvious that $\mathcal{E}^{(n)}(t)$ is directly related to the response functions:

$$\mathcal{E}^{(n)}(t) \approx \mathrm{i} S^{(n)}(t, t_{n-1}, \ldots, t_1) , \tag{13.51}$$

an approximation which works well in the limit of ultrashort excitation pulses, as we will show in detail in Section 15.1.4.

13.3.2
Pump Probe

As an example we briefly introduce here the pump-probe experiment. Later we make quantitative relation with the response function in Section 15.3.2.

The pump-probe experiment is one of the standard or "classical" nonlinear experiments performed with short light pulses. The scheme is shown in Figure 13.3. Quantitatively the experiment can be understood as differential measurement of absorption. In this scheme two measurements of absorption of a probe laser pulse are performed. The first measurement reads out the absorption of the *probe* in the sample, that has been excited in advance by another short *pump* pulse. The pump pulse is responsible for creation of a nonequilibrium system state. This measurement yields absorption $A_{\text{p+}}$. The second measurement performs the regular measurement of the probe pulse absorption A_{p} as the reference. The pump probe measurement corresponds to the difference

$$A_{\text{pp}} = A_{\text{p+}} - A_{\text{p}} . \tag{13.52}$$

As both absorption measurements involve linear polarization, the non-linear part is highlighted in pump-probe.

The pump probe can be designed in different representations. The pump pulse can be tuned to a specific system resonance, while the probe can be tuned to the

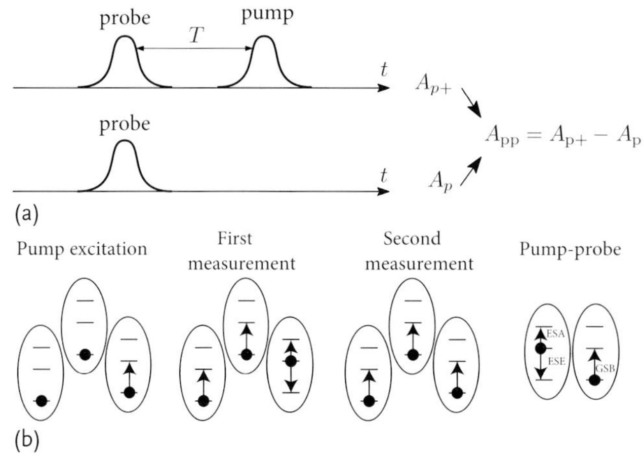

Figure 13.3 The scheme of a traditional pump probe measurement (a). Transitions in the ensemble of three-level molecules are shown in (b). ESA – excited state absorption, ESE – excited state emission, GSB – ground state bleach.

same or different resonance: the pump-probe intensity as a function of the delay time can then target excited state decay or transfer. In another setup the pump can again excite the specific resonance, but the absorption *spectrum* of the resulting nonequilibrium system can be measured in a broad frequency interval using white-light probe. This measurement yields correlations between various system resonances. In a third setup we can assume a broad band excitation, broad band detection, and we can imagine a correlation between the pump frequency and the probe frequency at a predefined delay time. Thus, the pump-probe is very versatile technique to target nonequilibrium system resonances and their dynamics.

The pump-probe experiment can be sufficiently easily visualized in terms of transitions between energy levels of the molecules under consideration as shown in Figure 13.3b. Initially the sample resides in the ground state. So, the regular absorption measurement detects all possible absorption events from the ground state into the upper energy levels, and thus the regular absorption spectrum is drawn. When the sample is affected by the pump pulse, some constituent molecules are resonantly transferred into their excited states. The consequent measurement of the probe absorption records fewer transitions from the ground state – this effect is termed as the *ground state bleach*. However, the measurement also records induced emission from the excited states (often termed as the *stimulated emission* or the *excited state emission*) and the absorption into even higher-lying excited states of the nonequilibrium molecules. This is termed as the *induced* (or the *excited state*) *absorption*, respectively. These processes introduce new bands into the pump probe spectrum as compared to the linear absorption.

In light of (13.50) the field measured in pump-probe is $\propto \mathrm{i}\omega \mathcal{P}^{(3)}(t)$. Below we describe the electrodynamic picture of the pump-probe measurement. From the point of view of the detection of the time domain polarization, the main difference between absorption described in Section 13.2.2 and the FWM described in Section 13.3.1 is that in the absorption, the new field component generated by polarization interferes with the original field, while in our case of the FWM it travels unaffected in its own direction. In the pump-probe the signal is emitted in the same direction as the probe pulse. We can imagine a third order polarization generated in a modified FWM scheme (see Figure 13.2) where the first two interactions happen along the same direction $\boldsymbol{k}_1 \equiv \boldsymbol{k}_2$ inside one pump pulse. The system is effectively excited by two pulses with different wave vectors $\boldsymbol{k}_1$ and $\boldsymbol{k}_3$, respectively. If the $\boldsymbol{k}_3$-pulse – the probe pulse – with envelope $\mathcal{E}_3(t)$ and frequency ω_3 follows the $\boldsymbol{k}_1$-pulse into the sample, they generate the third order polarization also into the direction $\boldsymbol{k}_s = \boldsymbol{k}_3 = -\boldsymbol{k}_1 + \boldsymbol{k}_1 + \boldsymbol{k}_3$. A detector can be used to measure the integrated intensity of the transmitted beam along $\boldsymbol{k}_3$ depending on the delay T between the two pulses. According to (13.41) the relative change of the probing light intensity with respect to the input intensity is:

$$S(\omega_3; T) = \frac{\omega_3}{n(\omega_3)\epsilon_0 c} \frac{\int_{-\infty}^{\infty} dt\, \mathrm{Im}\left[(\mathcal{P}^{(1)}(t) + \mathcal{P}^{(3)}(t; T))\mathcal{E}_3^*(t)\right]}{\int_{-\infty}^{\infty} dt |\mathcal{E}_3(t)|^2} , \tag{13.53}$$

where $\mathcal{P}^{(3)}(t; T) = 0$ for $T < 0$. Subtracting the absorption without the first pulse we can get to the nonlinear (T dependent) part of the absorption

$$\Delta S(\omega_3; T) \equiv S(\omega_3; T) - S(\omega_3; T < 0) . \tag{13.54}$$

Above, we have written the argument of the absorption coefficient as ω_3 – the carrier frequency of the second pulse. This is because we have assumed that the change of the total intensity of the second pulse is measured. This is analogous to the linear absorption measurement which is done by scanning excitation frequency.

Alternatively, one can measure a frequency dispersed signal $\Delta S_{\omega_3}(\omega; T)$ for which we have

$$\Delta S(\omega_3; T) = \int_{-\infty}^{\infty} d\omega \Delta S_{\omega_3}(\omega; T) . \tag{13.55}$$

The dispersed signal can be written as

$$\Delta S_{\omega_3}(\omega; T) = \frac{\omega_3}{n(\omega_3)\epsilon_0 c} \frac{\mathrm{Im}\,[\mathcal{P}^{(3)}(\omega; T)\mathcal{E}_3^*(\omega)]}{\int_{-\infty}^{\infty} d\omega |\mathcal{E}_3(\omega)|^2} , \tag{13.56}$$

where all quantities with argument ω are Fourier transforms of their time-dependent counterparts.

An important conclusion here is that the relative absorption is proportional to $\mathrm{Im}\,[\mathcal{P}^{(3)}(t)\mathcal{E}^*(t)]$. If we assume $\mathcal{E}$ to be real, the imaginary part of the product corresponds to the component of the polarization which is out of phase with respect to the field. This phase difference thus defines the absorption intensity.

13.3.3 Heterodyne Detection

From the above section we can see that the pump-probe signal is linearly proportional to the third order polarization. In terms of the signal strength, this is an advantage over direct detection of the third order field, which corresponds to the FWM experiment discussed in Section 13.3.1 where the intensity depends on the square of the weak third order field. To get the same advantage of detecting the signal linearly dependent on the third order field, we introduce the so-called *heterodyne detection* scheme. We add an extra electric field $E_{\text{LO}}(t)$, traditionally termed the *local oscillator*, along the selected nonlinear signal direction in the N-wave mixing experiment, and measure them together (see Figure 13.2). The total intensity is then

$$I_{\text{tot}} \approx |E_s(t) + E_{\text{LO}}(t)|^2 = |E_s(t)|^2 + |E_{\text{LO}}(t)|^2 + 2\text{Re}\left[E_s(t) E^*_{\text{LO}}(t)\right] . \quad (13.57)$$

The first term is of the second order in the weak field and can be neglected, while the second one is known, or can be measured separately and afterwards subtracted. The third term is linear in $E_s(t)$ and due to (13.50) it is proportional to $\text{Im}\left[\mathcal{P}_s^{(3)}(t) E^*_{\text{LO}}(t)\right]$. The signal that we thus measure is in some sense analogous to the pump probe. Or from a slightly different point of view, we can say that the pump probe is a *self-heterodyne detected* signal.

In the pump-probe method, the relative phase of the excitation field and the generated third order signal is very significant. It determines the differential absorption. The local oscillator, however, does not generate the signal E_s and the relative phase between E_s and E_{LO} can in principle be arbitrary in the experiment, unless it is phase-locked with the excitation field (this could be achieved if the LO pulse is obtained by branching the excitation pulse pathway and using a part of the excitation pulse as the LO pulse). The aim of the heterodyne detection is to determine the field E_s, but if the phase of the LO field is not locked with the excitation field, the signal field can be only determined up to arbitrary phases difference. One way of setting the phase of E_s to something meaningful is to compare the heterodyne detected intensity, (13.57), with the signal detected in pump probe, where the relation between the exciting field and E_{LO} is properly defined as it is self-heterodyned. More details on this issue are given in Appendix A.9.

14
Microscopic Theory of Linear Absorption and Fluorescence

To introduce basic theoretical ideas of spectroscopy we need to discuss a system of at least two quantum levels interacting resonantly with radiation. In any realistic situation, the two levels interact not only with the radiation, which is controlled in an experiment, but also with the electromagnetic vacuum and possibly other degrees of freedom (molecular electronic levels are inevitably coupled to nuclei of the molecule, for example) which form its environment (see Chapter 9). Linear absorption spectroscopy, and indeed any other type of spectroscopy, reveals some basic properties of both the two-level system and its environment.

14.1
A Model of a Two-State System

The interaction of the system with the environment can cause both the energy relaxation (its irreversible transfer to the environment) and pure dephasing. We will see later that in the linear absorption spectra (to which we will limit ourselves for a start) both effects appear as different forms of dephasing. The model system we study has a ground state $|g\rangle$ and an excited state $|e\rangle$ so that its Hamiltonian reads as:

$$\hat{H}_0 = \hat{H}_g|g\rangle\langle g| + \hat{H}_e|e\rangle\langle e| . \tag{14.1}$$

Here, the operators $\hat{H}_g$ and $\hat{H}_e$ represent the Hamiltonian operators of the environment when the system is in the states $|g\rangle$ and $|e\rangle$, respectively. The terms proportional to $|g\rangle\langle e|$ would inevitably cause the energy transfer to the ground state. Hence, this process is neglected.

We have studied the fluctuating properties of such a system in Section 8.6, where the energy fluctuations have been characterized by a set of spectral densities. In this section we continue to study this system and describe its optical responses.

In electronic spectroscopies one can often assume that the system in thermal equilibrium (before excitation by the external fields) occupies only the ground electronic state $|g\rangle$ and its total density matrix has a form:

$$\hat{\rho}_{\text{eq}} = w_{\text{B}}|g\rangle\langle g| , \tag{14.2}$$

Molecular Excitation Dynamics and Relaxation, First Edition. L. Valkunas, D. Abramavicius, and T. Mančal.

where the equilibrium density matrix of the bath is $w_B = \exp(-\beta \hat{H}_g)/Z_g$, $\beta^{-1} = k_B T$ (see Chapter 7). The system interacts with the classical external field $E(t)$ via the Hamiltonian (see (13.5)),

$$\hat{H}_{\text{int}} = -\hat{\mu} \cdot E(t) \,, \tag{14.3}$$

where

$$\hat{\mu} = d_{ge}|g\rangle\langle e| + d_{eg}|e\rangle\langle g| \tag{14.4}$$

is the polarization operator, and we disregard the vector notation. The optical field thus induces vertical transitions between the electronic ground and excited states. Otherwise, when the optical field is off, the system is adiabatic.

In coming sections we relate microscopic properties with the susceptibility $\chi(\omega)$, (13.34), and through it with the macroscopic absorption coefficient $\kappa_a(\omega)$, (13.43).

14.2 Energy Gap Operator

Based on the Liouville equation, (13.7) in Section 13.1, we derived an expression for the first order response function $S^{(1)}(t)$. The scalar first order response function of (13.18) reads as:

$$S^{(1)}(t) = \frac{\mathrm{i}}{\hbar}\theta(t)\mathrm{Tr}\left(\hat{U}_0^\dagger(t)\hat{\mu}\,\hat{U}_0(t)\hat{\mu}\,\hat{W}_{\text{eq}}\right) + \text{c.c.} \tag{14.5}$$

The microscopic expression for $S^{(1)}(t)$ together with (13.43), (13.33) and (13.32) form a closed quantum mechanical theory of the absorption spectrum. Now we apply it to the model of the two-level system introduced above.

First of all, the trace operation can be performed separately on the Hilbert spaces of the environment and of the relevant degrees of freedom so that

$$\mathrm{Tr}\hat{A} \equiv \mathrm{Tr}_B\left(\sum_{a=\mathrm{g,e}} \langle a|\hat{A}|a\rangle\right) , \tag{14.6}$$

for an arbitrary operator $\hat{A}$. The response function thus reads as

$$S^{(1)}(t) = \frac{\mathrm{i}}{\hbar}\theta(t)\mathrm{Tr}_B\left(\langle g|\,\hat{U}_0^\dagger(t)\hat{\mu}\,\hat{U}_0(t)\hat{\mu}|g\rangle w_B\right) + \text{c.c.} \tag{14.7}$$

The Hamiltonian $\hat{H}_0$ is diagonal in the basis of vectors $|g\rangle$ and $|e\rangle$, and correspondingly the evolution operator $\hat{U}_0(t)$ is diagonal. For the dipole moment operator only the off-diagonal elements $d_{ab} = \langle a|\hat{\mu}|b\rangle$ (where $a \neq b$) are nonzero and thus we have

$$S^{(1)}(t) = \frac{\mathrm{i}}{\hbar}\theta(t)|d_{eg}|^2\mathrm{Tr}_B\left(\hat{U}_g^\dagger(t)\,\hat{U}_e(t)w_B\right) + \text{c.c.} \tag{14.8}$$

Here, the elements of the evolution operator

$$\hat{U}_g(t) = \exp(-i\hat{H}_g t) , \quad \hat{U}_e(t) = \exp(-i\hat{H}_e t) , \tag{14.9}$$

contain the Hamiltonian operators of the environment. The operator $\hat{U}_{ge}(t) \equiv \hat{U}_g^\dagger(t)\hat{U}_e(t)$ composed of them can be found to obey a simple equation of motion

$$\frac{\partial}{\partial t}\left(\hat{U}_g^\dagger(t)\hat{U}_e(t)\right) = -i U_g^\dagger(t)(\hat{H}_e - \hat{H}_g)U_g(t)\left(\hat{U}_g^\dagger(t)\hat{U}_e(t)\right) , \tag{14.10}$$

and it can, therefore, be written in terms of the difference $\hat{H}_e - \hat{H}_g$. In (14.8) we take the expectation value of the operator $\hat{U}_{ge}(t)$ at the equilibrium state of the environment. It is advantageous to take the difference $\hat{H}_e - \hat{H}_g$ at this equilibrium. We define the so-called *energy gap operator*,

$$\Delta\hat{V}(t) = U_g^\dagger(t)\left(\hat{H}_e - \hat{H}_g\right)U_g(t) - \hbar_{eg} , \tag{14.11}$$

where $\omega_{eg} = \mathrm{Tr_B}[(\hat{H}_e - \hat{H}_g)w_\mathrm{B}]$ so that the solution of (14.8) is

$$\hat{U}_g^\dagger(t)\hat{U}_e(t) = e^{-i\omega_{eg}t}\exp_+\left(-i\int_0^t dt'\Delta\hat{V}(t')\right) . \tag{14.12}$$

The energy gap operator, (14.11), is an appropriate small parameter upon which one can base perturbative evaluation of the response function, (14.8). As ω_{eg} is subtracted in (14.11), the $\Delta\hat{V}(t)$ represents environment-induced fluctuations of the transition energy.

14.3
Cumulant Expansion of the First Order Response

Equation (14.8) can be evaluated by the perturbation theory with respect to $\Delta\hat{V}$. One might, for instance, try to approximate the time ordered exponential in (14.12) by its expansion up to a certain order. Ideally, one would like to evaluate such a series in all orders. It turns out that a partial summation to all orders is possible by the so-called cumulant expansion explained in Appendix A.8. We start with the second order Taylor expansion in (14.12)

$$\begin{aligned} e^{i\omega_{eg}t}\mathrm{Tr_B}\left(\hat{U}_g^\dagger(t)\hat{U}_e(t)w_\mathrm{B}\right) &= 1 - i\int_0^t dt'\mathrm{Tr_B}\left(\Delta\hat{V}(t')w_\mathrm{B}\right) \\ &\quad - \int_0^t dt'\int_0^{t'} dt''\mathrm{Tr_B}\left(\Delta\hat{V}(t')\Delta\hat{V}(t'')w_\mathrm{B}\right) + \ldots \end{aligned} \tag{14.13}$$

Here we notice that the first order term is zero, because of the invariance of w_B to the time evolution by $\hat{U}_g^\dagger(t)$, and thanks to (14.11) where we defined $\Delta\hat{V}(t)$ such that its equilibrium average is zero:

$$\mathrm{Tr_B}\left(\Delta\hat{V}(t')w_B\right) = \mathrm{Tr_B}\left(\Delta\hat{V}(0)\,\hat{U}_g(t')w_B\,\hat{U}_g^\dagger(t')\right)$$
$$= \mathrm{Tr_B}\left(\Delta\hat{V}(0)w_B\right) = 0\,. \tag{14.14}$$

The first nonzero order of the Taylor series is, therefore, quadratic in $\Delta\hat{V}(t)$. The second order term of (14.13)

$$g(t) = \frac{\mathrm{i}}{\hbar^2}\int_0^t \mathrm{d}\tau_2 \int_0^{\tau_2} \mathrm{d}\tau_1 \mathrm{Tr}\left(\Delta\hat{V}(\tau_2)\Delta\hat{V}(\tau_1)w_B\right) \tag{14.15}$$

is usually termed the *line shape function* (see Appendix A.8) [26, 91]. The function inside the double integration, is the energy gap correlation function

$$C(t) = \mathrm{Tr_B}\left(\Delta\hat{V}(t)\Delta\hat{V}(0)w_B\right)\,. \tag{14.16}$$

It plays an important role not only in determining the absorption spectrum, but forms a basis for the energy relaxation theory as described in Part One. As we show later, the nonlinear spectroscopic experiments involve also a set of correlation functions of the fluctuations of various energy levels. Using the result of Appendix A.8 we can write the ordinary exponential for the propagators

$$\mathrm{Tr_B}\left(\hat{U}_g^\dagger(t)\,\hat{U}_e(t)w_B\right) \approx \mathrm{e}^{-g(t)-\mathrm{i}\omega_{eg}t}\,. \tag{14.17}$$

So the linear response function is

$$S^{(1)}(t) = \frac{\mathrm{i}}{\hbar}\theta(t)|d_{eg}|^2\mathrm{e}^{-\mathrm{i}\omega_{eg}t-g(t)} + \mathrm{c.c.} \tag{14.18}$$

Combining (13.34) for the susceptibility $\chi(t)$ and (13.43) for the absorption coefficient $\kappa_a(\omega)$ we arrive at the final expression for the absorption spectrum in terms of the line shape function $g(t)$ (or the corresponding correlation function $C(t)$)

$$\kappa_a(\omega) = \frac{\omega}{n(\omega)c}|d_{eg}|^2\mathrm{Re}\int_0^\infty \mathrm{d}t\mathrm{e}^{\mathrm{i}(\omega-\omega_{eg})t-g(t)}\,. \tag{14.19}$$

As already pointed out at the beginning of Section 14.1, the absorption spectrum reveals both the properties of the transition (in terms of d_{eg} and ω_{eg}) and the properties of its coupling to the environment (in terms of the lineshape $g(t)$ function). While the transition dipole moment d_{eg} and the transition frequency ω_{eg} are quantities accessible in many cases from the first principles of quantum chemical calculations, it is much more difficult to obtain the energy gap correlation function $C(t)$ from such calculation. To gain insight into the physics of molecular systems it is sometimes more useful to investigate spectra using model correlation functions discussed in Section 8.6.

14.4 Equation of Motion for Optical Coherence

In the previous section, we have calculated the absorption spectrum of a two-level system by directly evaluating the first order response function, (14.5). What mattered for this calculation was the dynamics of the degrees of freedom of the environment which were reduced to the energy gap correlation function $C(t)$. It seems as if nowhere in our description did we refer to the density matrix of the system reduced to the degrees of freedom of the two-level system itself. It is, however, the first order response function that hides the reduced density matrix as we will show now. Using our assumption that $\hat{\mu}$ does not depend on the environmental degrees of freedom we can rewrite (14.5) in the following way:

$$S^{(1)}(t) = \frac{\mathrm{i}}{\hbar}\theta(t)\mathrm{Tr}\left(\hat{\mu}\hat{\sigma}(t)\right) = \mathrm{i}\theta(t) d_{ge}\sigma_{eg}(t) , \tag{14.20}$$

where $\hat{\sigma}$ is a reduced operator

$$\hat{\sigma}(t) = \mathrm{Tr}_{\mathrm{B}}\left(\hat{U}_0(t)\hat{\mu}\hat{W}_0\hat{U}_0^{\dagger}(t)\right) = |e\rangle d_{eg}\rho_{eg}(t)\langle g| . \tag{14.21}$$

Here, we dropped the part of the response function oscillating with $\mathrm{e}^{\mathrm{i}\omega_{eg}t}$ and in the last line we have introduced the matrix element $\rho_{eg}(t) = \langle e|\hat{\rho}(t)|g\rangle$ of the reduced density matrix

$$\hat{\rho}(t) = \mathrm{Tr}_{\mathrm{B}}(\hat{W}(t)) . \tag{14.22}$$

The initial condition for the reduced density matrix is (see (14.2))

$$\hat{\rho}(t_0) = |e\rangle\langle g| . \tag{14.23}$$

The response function $S^{(1)}(t)$ can thus be written in terms of a single element of the reduced density matrix

$$S^{(1)}(t) = \frac{\mathrm{i}}{\hbar}\theta(t)|d_{ge}|^2\rho_{eg}(t) . \tag{14.24}$$

The response function and the absorption spectrum could thus also be obtained from the dynamics of the reduced density matrix for which one could in principle derive equations of motion.

It is, however, very important to note at this point that the initial condition in (14.23) and, therefore, the reduced density matrix introduced by (14.22) do not fulfill the properties of the density matrix (see Section 4.5). Obviously, $|\rho_{eg}|^2 \not\leq \rho_{ee}\rho_{gg} = 0$. This is a consequence of the fact that we calculated the spectroscopic signal by the perturbation theory and the density matrix $\hat{\sigma}(t)$ in (14.21) is only the first order in the field correction. The total density matrix, represented by the complete series, should follow the equation of motion, (13.7), and the reduced density matrix properly derived from it by the trace operation fulfills the requirements of Section 4.5. For the Hamiltonian, (14.1), the evolutions of all reduced density matrix elements $\rho_{ab}(t) = \langle a|\hat{\rho}(t)|b\rangle$ remain independent of each other, and it is thus possible to augment the reduced density matrix initial conditions, (14.23) to

$$\hat{\rho}(t_0) = \frac{|e\rangle\langle g| + |g\rangle\langle e| + |g\rangle\langle g| + |e\rangle\langle e|}{2} , \tag{14.25}$$

After obtaining the solution of the Liouville equation for the reduced density matrix in the form of the time evolution $\hat{\rho}(t)$ one can calculate the absorption spectrum from a single element $\rho_{eg}(t)$. This issue underlines the uneasy relation between the response functions and the equations of motion of the reduced density matrix.

At present we can derive the equation of motion for the element $\rho_{eg}(t)$ directly from the known expression for the first order response function. For $t > 0$ we have by differentiating (14.24)

$$\frac{\partial}{\partial t}\rho_{eg}(t) = -i\omega_{eg}\rho_{eg}(t) - \int_0^t dt' C(t')\rho_{eg}(t) . \tag{14.26}$$

This master equation has been derived indirectly via derivation of the response function, and it has the same validity as the cumulant expansion. Notice that for the bath of harmonic oscillators with linear system–bath coupling the fluctuations of the canonical ensemble are Gaussian (see Section 7.8) and the cumulant expansion is then exact, so (14.26) is exact. It is, therefore, interesting to note that it contains $\Delta\hat{V}$ only to the second order. Its exactness is possible (but not guaranteed) by the fact that its solution actually contains all orders of $\Delta\hat{V}$.

14.5
Lifetime Broadening

As shown in the previous section the interaction between the optical transition and the environmental degrees of freedom enters the absorption spectrum via decay of the matrix element $\rho_{eg}(t)$. Such decay is also termed the *dephasing,* and it is not a result of population relaxation (which is not involved in linear response). If the system interacts with some degrees of freedom which can cause transition between the levels, and consequently the energy relaxation, such relaxation would nevertheless result in a decay of the *coherence element*$\rho_{eg}(t)$ *of the density matrix* as shown in Chapter 8. This can happen because of two sources: the fluctuations of the nuclear bath which causes radiationless transition between electronic states, and the interaction with the quantum electromagnetic field.

In Chapters 8 and 11 we have described the density matrix dynamics under the influence of the fluctuating bath coordinates. Equation (11.33) is the simplest model for the density matrix dynamics. Let us now assume that we have two electronic states $|g\rangle$ and $|e\rangle$ and all fluctuations are off diagonal (Section 8.4). In this case (11.33) turns into

$$\frac{d}{dt}\rho_{gg}(t) = -|k_{eg}|\rho_{gg}(t) + |k_{ge}|\rho_{ee}(t) \tag{14.27}$$

$$\frac{d}{dt}\rho_{ee}(t) = -|k_{ge}|\rho_{ee}(t) + |k_{eg}|\rho_{gg}(t) \tag{14.28}$$

and

$$\frac{d}{dt}\rho_{eg}(t) = -i(\omega_{eg} - i\gamma_{eg})\rho_{eg}(t) . \tag{14.29}$$

Here the parameters are

$$|k_{eg}| = 2|\tilde{h}_{eg}|^2 \mathrm{Re} \int_0^\infty d\tau\, C(\tau) e^{i\omega_{eg}\tau} , \tag{14.30}$$

$$|k_{ge}| = 2|\tilde{h}_{eg}|^2 \mathrm{Re} \int_0^\infty d\tau\, C(\tau) e^{-i\omega_{eg}\tau} , \tag{14.31}$$

and

$$\gamma_{eg} = |\tilde{h}_{eg}|^2 \int_0^\infty d\tau \left(C(\tau) e^{-i\omega_{ge}\tau} + C(-\tau) e^{-i\omega_{ge}\tau} \right) , \tag{14.32}$$

where $\tilde{h}_{eg}$ is the parameter determining the system-bath coupling strength (see Section 8.4). So the real part of the dephasing rate, which is responsible for the decay of the coherence is

$$\mathrm{Re}\, \gamma_{eg} = \frac{1}{2}(|k_{eg}| + |k_{ge}|) . \tag{14.33}$$

Notice that this may be included as an additional decay term to the $g(t)$ function, since the $g(t)$ function is a result of the diagonal fluctuations, which do not induce the decay of the excited state (that is correct when the diagonal and off diagonal fluctuations are independent). So the linear response function can then be given as

$$S^{(1)}(t) = i\theta(t)|d_{eg}|^2 e^{-i\omega_{eg}t - g(t) - \Gamma_{\mathrm{deph}}t} + \text{c.c.} \tag{14.34}$$

and the absorption is the Fourier transform of the $\rho_{eg}(t)$ density matrix element; we can write the absorption as

$$\kappa_a(\omega) = \frac{\omega}{n(\omega)c}|d_{eg}|^2 \mathrm{Re} \int_0^\infty dt e^{i(\omega - \omega_{eg})t - g(t) - \Gamma_{\mathrm{deph}}t} . \tag{14.35}$$

Here Γ_{deph} represents the lifetime induced dephasing part which can in general be given as

$$\Gamma_{\mathrm{deph}}|_{ab} = \frac{1}{2}\left(\tau_a^{-1} + \tau_b^{-1}\right) , \tag{14.36}$$

where the lifetime of state $|a\rangle$, τ_a, describes all causes of the decay of the excited state.

Even when the nuclear degrees of freedom are absent (so the environment fluctuations can be neglected) interaction with the electromagnetic field is yet another cause of the decay of an excited state. The Hamiltonian, (14.1), can then be extended by the free radiation Hamiltonian $\hat{H}_T$[1] and the corresponding interaction term. Thus, the new Hamiltonian reads as

$$\hat{H}_0 = \hat{H}_g|g\rangle\langle g| + \hat{H}_e|e\rangle\langle e| + \hat{H}_T - \hat{E}(d_{eg}|e\rangle\langle g| + d_{ge}|g\rangle\langle e|) \tag{14.37}$$

Now we want to evaluate the density matrix element $\rho_{eg}(t)$ when the vacuum contains the bath of the quantum electromagnetic field in addition to the nuclear bath. We are interested solely in the evolution (dephasing) of the density matrix element $\rho_{eg}(t)$. To this end, let us use the results of Chapter 9.

We use the projection operator technique to project the information onto the single matrix element of our interest, that is on $\rho_{eg}(t)$. The corresponding projector $\mathcal{P}$ can be defined as follows

$$\mathcal{P}\hat{W}(t) = \mathrm{Tr}_{B,T}(\langle e|\hat{W}(t)|g\rangle)\hat{\mathcal{W}}_0|e\rangle\langle g| = \rho_{eg}(t)\hat{\mathcal{W}}_0|e\rangle\langle g| , \tag{14.38}$$

where $\hat{\mathcal{W}}_0$ is some operator on the Hilbert space of the environment and the radiation. The trace is taken over the bath and the radiation field. It is useful to choose it in a form that eliminates the initial condition term so that $\mathcal{P}\hat{W}(t_0) = \hat{W}(t_0)$. In our present case we assume that the emission occurs from a thermally equilibrated state of the environment, and the radiation is in its vacuum state denoted by $|\varnothing\rangle$. It is important to note that the equilibrium state of the environment from which we start is taken with respect to the excited state $|e\rangle$ of the system. Therefore, we assume that before the emission the system had enough time to equilibrate in the excited state, which is usually the case for the electronic excitations [24]. Thus, we choose the initial density matrix

$$\hat{\mathcal{W}}_0 = w_B^{(e)}|\varnothing\rangle\langle\varnothing| , \tag{14.39}$$

where w_B is the canonical distribution of the vibrational environment with respect to the excited electronic state $|e\rangle$. The initial state of the system will then be assumed in a form of a product

$$\hat{W}(t_0) = |e\rangle\langle e|w_B^{(e)}|\varnothing\rangle\langle\varnothing| . \tag{14.40}$$

Let us start with (9.49) using the projector defined by (14.38). To this end we divide the Hamiltonian $\hat{H}_0$ into the three (system, reservoir, and their interaction) parts in the following way

$$\hat{H}_S = \hbar\omega_{eg}|e\rangle\langle e| , \quad \hat{H}_R = \hat{H}_T + \hat{H}_B , \tag{14.41}$$

1) The index T stands for *transversal* and relates to the fact that the radiation is formed only by the transverse field, that is with property $\mathrm{rot}\,\boldsymbol{E} = 0$.

$$\hat{H}_{\mathrm{SR}} = \hat{\Phi}\left(|e\rangle\langle g| + |g\rangle\langle e|\right) + \Delta\hat{V}|e\rangle\langle e| \,, \tag{14.42}$$

where for brevity we use

$$\hat{\Phi} = -d \cdot \hat{E} \,. \tag{14.43}$$

Two terms of $\mathcal{P}\mathcal{L}_{\mathrm{SR}}(t)\mathcal{P}\hat{\rho}(t) = \mathcal{P}\hat{H}_{\mathrm{SR}}(t)\mathcal{P}\hat{\rho}(t) - \mathcal{P}(\mathcal{P}\hat{\rho}(t))\hat{H}_{\mathrm{SR}}(t)$ vanish as they are linear in operators $\Delta\hat{V}$ and $\hat{\Phi}$. This result removes the first term in (9.49) and also simplifies the integration kernel of the same equation. The integration kernel has a form $\mathcal{P}\mathcal{L}_{\mathrm{SR}}(t)\mathcal{Q}\mathcal{L}_{\mathrm{SR}}(t-\tau)\mathcal{P}$ where $\mathcal{Q} = 1 - \mathcal{P}$. Because $\mathcal{P}\mathcal{L}_{\mathrm{SR}}(t)\mathcal{P} = 0$, we can write $\mathcal{P}\mathcal{L}_{\mathrm{SR}}(t)\mathcal{Q}\mathcal{L}_{\mathrm{SR}}(t-\tau)\mathcal{P} = \mathcal{P}\mathcal{L}_{\mathrm{SR}}(t)\mathcal{L}_{\mathrm{SR}}(t-\tau)\mathcal{P}$. The double commutator $\mathcal{L}_{\mathrm{SR}}(t)\mathcal{L}_{\mathrm{SR}}(t-\tau)$ results in four terms and so the integration kernel reads as

$$\begin{aligned}
\mathcal{P}\mathcal{L}_{\mathrm{SR}}(t)\mathcal{L}_{\mathrm{SR}}(t-\tau)\mathcal{P}\hat{\rho}(t) &= \rho_{eg}(t)|e\rangle\langle g|w_{\mathrm{B}}^{(e)}|\varnothing\rangle\langle\varnothing| \\
&\quad\times\frac{1}{\hbar^2}\Big[\mathrm{Tr}_{\mathrm{B}}\left(\langle\varnothing|\langle e|\hat{H}_{\mathrm{SR}}(t)\hat{H}_{\mathrm{SR}}|e\rangle|\varnothing\rangle\right) \\
&\quad-\mathrm{Tr}_{\mathrm{B}}\left(\langle\varnothing|\langle e|\hat{H}_{\mathrm{SR}}|e\rangle w_{\mathrm{B}}^{(e)}\langle g|\hat{H}_{\mathrm{SR}}(t)|g\rangle|\varnothing\rangle\right) \\
&\quad-\mathrm{Tr}_{\mathrm{B}}\left(\langle\varnothing|\langle e|\hat{H}_{\mathrm{SR}}(t)|e\rangle w_{\mathrm{B}}^{(e)}\langle g|\hat{H}_{\mathrm{SR}}|g\rangle|\varnothing\rangle\right) \\
&\quad+\mathrm{Tr}_{\mathrm{B}}\left(w_{\mathrm{B}}^{(e)}\langle\varnothing|\langle g|\hat{H}_{\mathrm{SR}}\hat{H}_{\mathrm{SR}}(t)|g\rangle|\varnothing\rangle\right)\Big] \,.
\end{aligned} \tag{14.44}$$

The third and fourth lines contain the element $\langle g|\hat{H}_{\mathrm{SR}}|g\rangle$, and they are, therefore, zero. The remaining matrix elements containing two occurrences of $\hat{H}_{\mathrm{SR}}$ can be easily separated into contributions of the nuclear bath and the electromagnetic field by using the definition, (14.42)

$$\langle\varnothing|\langle e|\hat{H}_{\mathrm{SR}}(t)\hat{H}_{\mathrm{SR}}|e\rangle|\varnothing\rangle = \langle\varnothing|\hat{\Phi}(t)\hat{\Phi}|\varnothing\rangle e^{i\omega_{eg}t} + \Delta\hat{V}(t)\Delta\hat{V} \,, \tag{14.45}$$

$$\langle\varnothing|\langle g|\hat{H}_{\mathrm{SR}}\hat{H}_{\mathrm{SR}}(t)|g\rangle|\varnothing\rangle = \langle\varnothing|\hat{\Phi}\hat{\Phi}(t)|\varnothing\rangle e^{-i\omega_{eg}t} \,. \tag{14.46}$$

The equation for the density matrix element $\rho_{eg}(t)$ now reads

$$\begin{aligned}
\frac{\partial}{\partial t}\rho_{eg}(t) &= -i\omega_{eg}\rho_{eg}(t) - \frac{1}{\hbar^2}\int_0^{t-t_0} d\tau\,\mathrm{Tr}_{\mathrm{B}}\left(\Delta\hat{V}(\tau)\Delta\hat{V}w_{\mathrm{B}}^{(e)}\right)\rho_{eg}(t) \\
&\quad-\left(\frac{2}{\hbar^2}\mathrm{Re}\int_0^{t-t_0} d\tau\langle\varnothing|\hat{\Phi}(\tau)\hat{\Phi}|\varnothing\rangle e^{i\omega_{eg}\tau}\right)\rho_{eg}(t) \,,
\end{aligned} \tag{14.47}$$

where we used $\langle\varnothing|\hat{\Phi}\hat{\Phi}(t)|\varnothing\rangle = \langle\varnothing|\hat{\Phi}(t)\hat{\Phi}|\varnothing\rangle^*$. The first term on the right hand side recovers the lineshape function discussed in Section 14.4 and the second term corresponds to the contribution of the light-matter interaction to the dephasing. The corresponding rate of dephasing due to radiation is

$$\Gamma(\omega_{eg}) = \int_{-\infty}^{\infty} dt\langle\varnothing|\hat{\Phi}(t)\hat{\Phi}|\varnothing\rangle e^{i\omega_{eg}t} \,. \tag{14.48}$$

Here we assumed, $t_0 \to -\infty$. The expression, (14.48), can be evaluated. First we evaluate the matrix element with the interaction Hamiltonians

$$\begin{aligned}\langle \varnothing | \hat{\Phi}(t) \hat{\Phi} | \varnothing \rangle &= \langle \varnothing | \boldsymbol{d}_{eg} \cdot \hat{\boldsymbol{E}}(t) \boldsymbol{d}_{eg} \cdot \hat{\boldsymbol{E}} | \varnothing \rangle \\ &= \sum_{k\lambda} \sum_{q\mu} \omega_k \omega_q (\boldsymbol{d}_{eg} \cdot \boldsymbol{e}_{q\mu})(\boldsymbol{d}_{eg} \cdot \boldsymbol{e}_{k\lambda}) \frac{1}{\Omega} \\ &\quad \times \sqrt{\frac{\hbar}{2\varepsilon_0 \omega_k}} \sqrt{\frac{\hbar}{2\varepsilon_0 \omega_q}} \langle \varnothing | a_{\lambda k}(t) a^{\dagger}_{\mu q} | \varnothing \rangle \,, \end{aligned} \tag{14.49}$$

where we applied the definition of the electric field operator, (4.273). In the summation over the field modes we perform a summation over scalar product of a single vector, $\boldsymbol{d}_{eg}$, with wavevectors corresponding to all possible directions of the field polarization. We can replace this summation with an integral over the density of modes on one hand, and an average of the expression $(\boldsymbol{d}_{eg} \cdot \boldsymbol{e}_{q\mu})(\boldsymbol{d}_{eg} \cdot \boldsymbol{e}_{k\lambda})$ over an isotropic distribution of orientations, on the other hand. The averaging leads to

$$\langle (\boldsymbol{d}_{eg} \cdot \boldsymbol{e}_{q\mu})(\boldsymbol{d}_{eg} \cdot \boldsymbol{e}_{k\lambda}) \rangle_{\text{orient.}} = \frac{1}{3} \delta_{\mu\lambda} |\boldsymbol{d}_{eg}|^2 \,. \tag{14.50}$$

In this intermediate step we obtain

$$\langle \varnothing | \hat{\Phi}(t) \hat{\Phi} | \varnothing \rangle = |\boldsymbol{d}_{eg}|^2 \sum_{k\lambda} \frac{\hbar \omega_k}{6\varepsilon_0 \Omega} \mathrm{e}^{-\mathrm{i}\omega_k t} \,. \tag{14.51}$$

Instead of summing over modes we will now integrate over the number of modes at frequency ω. The number of modes can be obtained by multiplying the mode density $n(\omega)\mathrm{d}\omega = \omega^3/(\pi^2 c^3)\mathrm{d}\omega$, (2.79), by the volume Ω and we get

$$\langle \varnothing | \hat{\Phi}(t) \hat{\Phi} | \varnothing \rangle = |\boldsymbol{d}_{eg}|^2 \int \mathrm{d}\omega \; n(\omega) \frac{\hbar \omega}{6\varepsilon_0} \mathrm{e}^{-\mathrm{i}\omega t} \,. \tag{14.52}$$

Now we perform the time integration in (14.48) which leads to a Dirac delta function, that expresses the energy conservation:

$$\int_{-\infty}^{\infty} \mathrm{d}t \mathrm{e}^{-\mathrm{i}(\omega - \omega_{eg})t} = \pi \delta(\omega - \omega_{eg}) \,. \tag{14.53}$$

Now we can easily perform the integration over ω leading to

$$\Gamma(\omega_{eg}) = \frac{|\boldsymbol{d}_{eg}|^2 \omega^3}{6\varepsilon_0 \hbar \pi c^3} \,. \tag{14.54}$$

This is consistent with excited state life time τ_e and the Einstein coefficient A_{if} for spontaneous emission

$$A_{if} = \frac{1}{\tau_e} = 2\Gamma(\omega_{eg}) = \frac{|\boldsymbol{d}_{eg}|^2 \omega^3}{3\varepsilon_0 \hbar \pi c^3} \,. \tag{14.55}$$

We thus find that irrespective of the source of off-diagonal fluctuations, be it the nuclear vibrations or the electromagnetic vacuum, they all contribute to the decay of the density matrix coherence in the same way. From the density matrix dynamics including the populations (see (11.33)) the off-diagonal fluctuations induce the decay of populations with the rates tightly connected with the dephasing rates. Our calculation, therefore, also provides a result for the lifetime due to electromagnetic vacuum $\tau_e = \Gamma(\omega_{eg})/2$.

Now we can update the first order response function by adding the dephasing rate in (14.24). The absorption spectrum will therefore also include an additional lifetime-relaxation terms.

14.6 Inhomogeneous Broadening in Linear Response

In the above sections we have always assumed that we work with a single representative molecular system, and that the total response is the response of this system multiplied by the number of the equivalent molecular systems in the sample. The situation where all molecules are exactly the same is, however, rather rare. Their parameters, such as the optical transitions energy, might depend, for example, on some very slowly changing parameters of the environment. This is referred to as the inhomogeneous *disorder*, and it leads to the so-called *inhomogeneous broadening* of the spectra. The word inhomogeneous refers to the fact that individual molecules exhibit different parameters from each other during the spectroscopic experiment, and it is in contrast to the *homogeneous broadening* which stems from the interaction of the molecule with some fast component of the environment, which averages out quickly so that it is essentially the same for all molecules. Our averaging that led to the line broadening function was of this latter type. One should often keep in the consideration two types of disorder. First, in this section we will look at the consequences of energetic disorder, that is the situation where transition frequencies of the molecules are distributed according to some simple law. Second, we will consider the fact that molecules or complexes are oriented randomly in the sample in Section 15.2.6.

For simplicity we assume that the disorder concerns only the energy levels of a system in question, and that the fast fluctuations of the environment responsible for the energy gap correlation function are independent of the slow fluctuations that cause the disorder. This enables us to perform two averaging procedures, over the fast and slow fluctuations, independently. We might assume that the total density matrix reads as $\hat{W}_{\mathrm{eq}} = \hat{W}_{\mathrm{eq}}^{(\mathrm{fast})} \otimes \hat{W}_{\mathrm{eq}}^{(\mathrm{slow})}$ and the evolution operator corresponding to the slow bath is $\hat{U}_{\mathrm{B}}^{(\mathrm{slow})}(t) \approx 1$. As the origins of the two types of fluctuations are essentially the same, we may add the slow component to the energy gap operator, (14.11) and write for its time dependence

$$\Delta V_{eg}(t) = \Delta V_{eg}^{(\mathrm{slow})} + \Delta V_{eg}^{(\mathrm{fast})}(t) . \tag{14.56}$$

The response functions can be now constructed in the same way as in Section 14.3, that is by using the cumulant expansion. Also, here we will assume that $\text{Tr}_\text{B}\{\Delta V_{eg}^{(\text{slow})} W_{\text{eq}}^{(\text{slow})}\} = 0$, and only term $\text{Tr}_\text{B}\{\Delta V_{eg}^{(\text{slow})} \Delta V_{eg}^{(\text{slow})} W_{\text{eq}}^{(\text{slow})}\} = \hbar^2 \Delta_{\text{inh}}^2$ is considered to be nonzero. The slow energy gap correlation function and the line shape function read as

$$C_{ee}^{(\text{slow})}(t) = \hbar^2 \Delta_{\text{inh}}^2 , \quad g_{ee}^{(\text{slow})}(t) = \frac{1}{2} \Delta_{\text{inh}}^2 t^2 . \tag{14.57}$$

Adding the slow component to our expression for the absorption coefficient, (14.35), we obtain

$$\kappa_\text{a}(\omega) = \frac{\omega}{n(\omega)c} |d_{eg}|^2 \text{Re} \int_0^\infty \text{d}t e^{\text{i}(\omega - \omega_{eg})t - g(t) - \frac{1}{2}\Delta_{\text{inh}} t^2 - \Gamma_{\text{deph}} t} . \tag{14.58}$$

This expression contains both inhomogeneous and lifetime broadening and combines thus the Gaussian and Lorenzian line shapes.

As we study third order nonlinear response in the next chapter we find the third order response function more complicated; however, we can still use the lineshape function for the slow degrees of freedom as we did in this section.

14.7 Spontaneous Emission

So far we have been only interested in the system properties during the experiment corresponding to the absorption or emission of radiation. Interaction with the electromagnetic vacuum as we described in Section 14.5 causes the decay of the excited state, but also radiation of new field, which we study in this section and results in a spontaneous emission. Here we will be interested in the spectrum of such an emitted radiation.

The Hamiltonian, (14.1), can be extended by the free radiation Hamiltonian $\hat{H}_\text{T}$ and an interaction term. Thus, the new Hamiltonian reads as

$$\hat{H}_0 = \hat{H}_g |g\rangle\langle g| + \hat{H}_e |e\rangle\langle e| + \hat{H}_\text{T} - \Phi |e\rangle\langle g| + \Phi^\dagger |g\rangle\langle e|) . \tag{14.59}$$

Here

$$\Phi^\dagger = \text{i}\mu_{ge} \sum_{\alpha k} \sqrt{\frac{\hbar \omega_k}{2\Omega \epsilon_0}} \hat{a}_{\alpha k}^\dagger \tag{14.60}$$

is the form of the electromagnetic field multiplied by the transition dipole (4.269) (we neglect vector notation, so μ_{ge} should be understood as the projection of the transition dipole into the electric vector of the quantum radiation field, greek alpha signifies the field polarization). This setup describes the system not affected by the external classical field, $\boldsymbol{E} = 0$, but the vacuum contains the bath of the quantum electromagnetic field. As a consequence, the excited matter radiates photons. When we measure the emission of photons into a given mode of the electromagnetic field (characterized by a polarization vector $\boldsymbol{e}_\alpha$ and a wave vector $\boldsymbol{k} = \boldsymbol{n}\omega/c$) we are interested in the rate of change of the population $n_{\alpha\boldsymbol{k}} = \mathrm{Tr}(\hat{a}^\dagger_{\alpha\boldsymbol{k}}\hat{a}_{\alpha\boldsymbol{k}}\hat{W}(t))$ of this mode with the initial condition that the material system is excited and there are no photons in the field, while the vibrational environment is equilibrated. This is all represented by the following initial density operator:

$$\hat{W}(t_0) = |e\rangle|\varnothing\rangle\langle\varnothing|\langle e|w_{\mathrm{B}}^{(e)} . \tag{14.61}$$

Here $|e\rangle$ denotes the excited state of the electronic system, $|\varnothing\rangle\langle\varnothing|$ denotes the vacuum of the electromagnetic field and $w_{\mathrm{B}}^{(e)}$ is the canonical ensemble of bath oscillators with respect to the excited state of the matter. $w_{\mathrm{B}}^{(e)}$, therefore, has to be invariant to the evolution operator $\hat{U}_e(t)$ and not to $\hat{U}_g(t)$. The Hamiltonian, (14.37), may be for our purposes rearranged into the standard form $\hat{H}_0 = \hat{H}_{\mathrm{S}} + \hat{H}_{\mathrm{R}} + \hat{H}_{\mathrm{SR}}$ in a way reflecting our preference for the excited state as a starting point of the dynamics:

$$\hat{H}_{\mathrm{S}} = -\hbar\omega_{eg}|g\rangle\langle g| , \quad \hat{H}_{\mathrm{R}} = \hat{H}_e + \hat{H}_{\mathrm{T}} , \tag{14.62}$$

$$\hat{H}_{\mathrm{SR}} = \hat{\Phi}(|e\rangle\langle g| + |g\rangle\langle e|) + \Delta\hat{V}|g\rangle\langle g| . \tag{14.63}$$

This time the energy gap operator reads

$$\Delta\hat{V}_{ge} = \hat{H}_g - \hat{H}_e + \hbar\omega_{eg} , \quad \hbar\omega_{eg} = \mathrm{Tr}_{\mathrm{B}}\left((\hat{H}_g - \hat{H}_e)w_{\mathrm{B}}^{(e)}\right) . \tag{14.64}$$

The difference between the vertical absorption and vertical fluorescence frequency

$$\omega_{eg}^{(\mathrm{abs})} - \omega_{eg}^{(\mathrm{fl})} = \mathrm{Tr}_{\mathrm{B}}\left((\hat{H}_e - \hat{H}_g)w_{\mathrm{B}}^{(e)}\right) - \mathrm{Tr}_{\mathrm{B}}\left((\hat{H}_e - \hat{H}_g)w_{\mathrm{B}}^{(g)}\right) = 2\lambda .$$

The rate of fluorescence into the given mode described by polarization α and wavevector $\boldsymbol{k}$ can be given as the rate of change of the number of photons

$$\begin{aligned}\sigma_{\alpha\boldsymbol{k}} &= \frac{\mathrm{d}}{\mathrm{d}t}n_{\alpha\boldsymbol{k}} = \mathrm{Tr}\left(\hat{a}^\dagger_{\alpha\boldsymbol{k}}\hat{a}_{\alpha\boldsymbol{k}}\frac{\mathrm{d}}{\mathrm{d}t}\hat{W}(t)\right) \\ &= -\mathrm{i}\mathrm{Tr}\left(\hat{a}^\dagger_{\alpha\boldsymbol{k}}\hat{a}_{\alpha\boldsymbol{k}}\mathcal{L}_{\mathrm{SR}}\hat{W}(t)\right) ,\end{aligned} \tag{14.65}$$

where in the second line we have written out explicitly the trace over the electronic as well as electromagnetic degrees of freedom, and we used the Liouville equation for the total density $\hat{W}(t)$ (the Liouvillians $\mathcal{L}_{\mathrm{S}}$ and $\mathcal{L}_{\mathrm{R}}$ do not contribute.

Now we will expand the total density matrix in (14.65) to the first order in the interaction Liouvillian (see (13.12)) which yields:

$$\sigma_{\alpha\boldsymbol{k}} = (-\mathrm{i})^2\mathrm{Tr}\left(\hat{a}^\dagger_{\alpha\boldsymbol{k}}\hat{a}_{\alpha\boldsymbol{k}}\int\limits_0^\infty \mathrm{d}t'\mathcal{L}_{\mathrm{SR}}\mathcal{U}_0(t')\mathcal{L}_{\mathrm{SR}}\mathcal{U}_0(t-t')\hat{W}(t_0)\right) . \tag{14.66}$$

Going back to the Hilbert space and using the condition that the electromagnetic field is initially in vacuum, we get

$$\sigma_{\alpha k} = -2\mathrm{i}\,\mathrm{Im}\,\mathrm{Tr}\left(\int_0^\infty \mathrm{d}t'\, U_0^\dagger(t')\hat{H}_{\mathrm{SR}}\, U_0(t')\hat{H}_{\mathrm{SR}}\,\hat{W}(t_0)\right) . \tag{14.67}$$

Using the explicit form of the density operator we have

$$\sigma_{\alpha k} = -2\mathrm{i}\,\mathrm{Im}\int_0^\infty \mathrm{d}t' \mathrm{e}^{\mathrm{i}(\omega_{eg}-2\lambda)t'} \times \mathrm{Tr}_{\mathrm{B}}\left(\Delta\hat{V}_{ge}(t')\Delta\hat{V}_{ge}(0)w_{\mathrm{B}}^{(e)}\right)\mathrm{Tr}_{\Phi}\left(\hat{\Phi}(t')\hat{\Phi}^\dagger(0)\hat{\rho}_{\Phi}\right) . \tag{14.68}$$

The trace over the field can be obtained using (7.131), which for the present case gives

$$\mathrm{Tr}_{\Phi}(\Phi(t)\Phi^\dagger(0)) = |\mu_{ge}|^2 \sum_{\alpha k} \frac{\hbar\omega_k}{2\Omega\epsilon_0}\mathrm{e}^{-\mathrm{i}\omega_k t} , \tag{14.69}$$

while for the nuclear bath correlation function we use the cumulant expansion that gives

$$\mathrm{Tr}_{\mathrm{B}}\left(\Delta\hat{V}_{ge}(t')\Delta\hat{V}_{ge}(0)w_{\mathrm{B}}^{(e)}\right) = \exp(-g^*(t)) . \tag{14.70}$$

Here the complex conjugation is because the bath is equilibrium with respect to the excited electronic state $|e\rangle$, while $g(t)$ is determined for the equilibrium with respect to the electronic $|g\rangle$ state. We thus get the electromagnetic field mode occupation number

$$\sigma_{\alpha k} \propto |\mu_{ge}|^2 \sum_k \omega_k \mathrm{Re}\int_0^\infty \mathrm{d}t \mathrm{e}^{\mathrm{i}\omega_{eg}t-2\mathrm{i}\lambda t-g^*(t)-\mathrm{i}\omega_k t} . \tag{14.71}$$

As can be seen inside the integral we find the resonance selection, where the system will emit the field in the range of frequencies ω_k close to $\omega_{eg} - 2\lambda$. In the case of uniform density of electromagnetic modes (within the range of the resonant frequency) we can write the spontaneous emission or the fluorescence spectrum as

$$\sigma(\omega) \propto \omega|\mu_{ge}|^2 \mathrm{Re}\int_0^\infty \mathrm{d}t \mathrm{e}^{-\mathrm{i}\omega t+\mathrm{i}\omega_{eg}t-2\mathrm{i}\lambda t-g^*(t)} . \tag{14.72}$$

The presence of the emission spectrum implies that the emergence of emitted photons is due to the electronic transition between the excited state and the ground state. The excited state lifetime broadening due to radiational transition should be included when the system absorption linewidths are extremely narrow, for instance, in cold gasses. Their atoms will show spectrum, whose linewidth will be influenced by the excited state lifetime as described in Section 14.5. However, the radiative lifetime is usually quite long in complex biological molecules and, therefore, the radiative lifetime-induced spectral broadening is much smaller than that induced by nuclear vibrations of the surrounding. In that case the radiative effects may be disregarded.

14.8
Fluorescence Line-Narrowing

In the previous section we have seen that both absorption and emission spectra crucially depend on the bath correlation function $C(t)$. As explained in Chapter 8, the time-dependent function $C(t)$ is in turn related to the spectral density $C''(\omega)$ or to the often used quantity $J(\omega)$ for which we have $C''(\omega) = \omega^2 J(\omega)$ (see e.g. [20]):

$$C(t) = \int_0^\infty \mathrm{d}\tau \omega^2 J(\omega) \left[(1 + n(\omega))\mathrm{e}^{-\mathrm{i}\omega t} + n(\omega)\mathrm{e}^{\mathrm{i}\omega t}\right] , \tag{14.73}$$

where $n(\omega)$ is the Bose-Einstein distribution. We will now discuss an experimental measurement, which allows us to directly estimate $J(\omega)$.

We will need the relation of $J(\omega)$ to the lineshape function $g(t)$, which is easy to obtain by integrating (14.73) according to (14.15). This can be given in a form

$$g(t) = G(0) - G(t) - \mathrm{i}\lambda t , \tag{14.74}$$

where λ is the reorganization energy $\lambda = \int_0^\infty \mathrm{d}\omega\omega J(\omega)$ and

$$G(t) = \int_0^\infty \mathrm{d}\omega J(\omega) \left[(1 + n(\omega))\mathrm{e}^{-\mathrm{i}\omega t} + n(\omega)\mathrm{e}^{\mathrm{i}\omega t}\right] . \tag{14.75}$$

We notice that at low temperatures, $T \to 0$, when we have $n(\omega) \to 0$, the line shape function simplifies significantly:

$$G_{T\to 0}(t) = \int_0^\infty \mathrm{d}\omega J(\omega)\mathrm{e}^{-\mathrm{i}\omega t} . \tag{14.76}$$

Consequently, the expressions for absorption and fluorescence simplifies too.

Let us introduce the so-called *fluorescence line narrowing* measurement. In this measurement the system is excited at a frequency ω_{exc} by the spectrally narrow light, and the spectrally resolved fluorescence is recorded. The recorded intensity $I(\omega, \omega_{\mathrm{exc}})$ will depend not only on the emission coefficient $\sigma(\omega)$, but also on the ability of the molecular system to absorb the excitation frequency, that is the absorption coefficient at the given frequency $\alpha(\omega_{\rm exc})$:

$$I(\omega, \omega_{\rm exc}) = \alpha(\omega_{\rm exc})\sigma(\omega)\,. \tag{14.77}$$

This holds for the case when all the molecules in the ensemble have the same absorption and fluorescence spectra.

The ensemble of molecules at low temperature will contain inhomogeneities, causing disorder of transition energies. To take into account the disorder, we have to integrate over some inhomogeneous distribution function $P_{\rm inh}(\omega_{eg} - \bar{\omega}_{eg})$ of transition energies around the mean value $\bar{\omega}_{eg}$. The absorption spectrum of a single molecule with the transition energy ω_{eg} will be denoted by $\alpha(\omega; \omega_{eg})$ and similarly, the fluorescence spectrum of the same molecule will be denoted $\sigma(\omega; \omega_{eg})$. Thus, in the case of disorder, the recorded fluorescence will read:

$$\bar{I}(\omega, \omega_{\rm exc}) = \int_{-\infty}^{\infty} d\omega_{eg}\, P_{\rm inh}(\omega_{eg} - \bar{\omega}_{eg})\alpha(\omega; \omega_{eg})\sigma(\omega; \omega_{eg})\,. \tag{14.78}$$

The results of the previous sections for the absorption and fluorescence can be written in a very symmetric fashion:

$$\alpha(\omega; \omega_{eg}) \approx e^{-G(0)} \int_{-\infty}^{\infty} dt e^{G(t)+i(\omega-\omega_{eg}^{0})t}\,, \tag{14.79}$$

$$\sigma(\omega; \omega_{eg}) \approx e^{-G(0)} \int_{-\infty}^{\infty} dt e^{G^*(t)-i(\omega-\omega_{eg}^{0})t}\,, \tag{14.80}$$

where we have introduced the purely electronic transition frequency $\omega_{eg}^{0} = \omega_{eg} - \lambda$ (ω_{eg} is the vertical Frank-Condon transition). For the absorption we can write

$$\begin{aligned}\alpha\left(\omega; \omega_{eg}^{0}\right) &\approx \int_{-\infty}^{\infty} dt e^{i(\omega-\omega_{eg}^{0})t} + \int_{-\infty}^{\infty} dt (e^{G(t)} - 1) e^{i(\omega-\omega_{eg}^{0})t} \\ &= 2\pi\delta\left(\omega - \omega_{eg}^{0}\right) + \phi\left(\omega - \omega_{eg}^{0}\right)\,,\end{aligned} \tag{14.81}$$

where we defined

$$\phi(\omega) = \int_{-\infty}^{\infty} dt (e^{G(t)} - 1) e^{i\omega t}\,. \tag{14.82}$$

Similarly for the fluorescence we have

$$\sigma\left(\omega; \omega_{eg}^{0}\right) \approx 2\pi\delta\left(\omega - \omega_{eg}^{0}\right) + \phi\left(\omega_{eg}^{0} - \omega\right)\,. \tag{14.83}$$

Now we can construct the averaged fluorescence intensity, (14.78), using the results of (14.81) and (14.83). We obtain:

$$\bar{I}(\omega, \omega_{\mathrm{exc}}) \approx \int_{-\infty}^{\infty} \mathrm{d}\omega_{eg}^{0} P_{\mathrm{inh}}\left(\omega_{eg}^{0} - \bar{\omega}_{eg}\right)\left[4\pi^2 \delta\left(\omega_{\mathrm{exc}} - \omega_{eg}^{0}\right)\delta\left(\omega - \omega_{eg}^{0}\right)\right.$$
$$+ 2\pi\delta\left(\omega_{\mathrm{exc}} - \omega_{eg}^{0}\right)\phi\left(\omega_{eg}^{0} - \omega\right) + 2\pi\delta\left(\omega - \omega_{eg}^{0}\right)$$
$$\left.\times\, \phi\left(\omega_{\mathrm{exc}} - \omega_{eg}^{0}\right) + \phi\left(\omega_{\mathrm{exc}} - \omega_{eg}^{0}\right)\phi\left(\omega_{eg}^{0} - \omega\right)\right] . \tag{14.84}$$

All four terms have a rather straightforward interpretation. The first term corresponds to the excitation of the zero phonon transition, $0 \to 0$ and the emission through the same transition $0 \leftarrow 0$. The second term corresponds to the excitation $0 \to 0$ and the emission from the sideband, and the third term is correspondingly the excitation to the sideband and emission through the zero-phonon line. The very last term corresponds to the situation when both excitation and emission occur throughout the sideband. The strongest signal of all is the first term, and it corresponds to a very narrow zero-phonon line. At low temperature, the side band only occurs on the higher energy side of the zero-phonon line so that the second term is weak, although the zero-phonon line is strongly allowed. The fourth term is even weaker because it is excited through a weak sideband.

The third term which leads to

$$\bar{I}_{\mathrm{side}}(\omega, \omega_{\mathrm{exc}}) \approx P_{\mathrm{inh}}(\omega - \bar{\omega}_{eg})\phi\left(\omega_{\mathrm{exc}} - \omega_{eg}^{0}\right) \tag{14.85}$$

is, therefore, the main focus in the experiment. The importance of this expression becomes clear when we realize that due to (14.76) and after an approximation $\mathrm{e}^{G(t)} \approx 1 + G(t)$ we can write

$$\phi(\omega) \approx \int_{-\infty}^{\infty} \mathrm{d}t \int_{0}^{\infty} \mathrm{d}\omega' J(\omega')\mathrm{e}^{\mathrm{i}(\omega-\omega')t} = J(\omega) . \tag{14.86}$$

The sideband of the fluorescence line-narrowing spectrum is, therefore, directly related to the spectral density $J(\omega)$. If the disorder is small in the fluorescence line-narrowing experiment, that is the distribution $P_{\mathrm{inh}}(\omega - \bar{\omega}_{eg})$ is sharply peaked around $\bar{\omega}_{eg}$, we scan the sideband and, therefore, we scan the spectral density by the excitation frequency ω_{exc}. This enables us to estimate the spectral density of real molecular systems directly.

14.9 Fluorescence Excitation Spectrum

Another relatively simple experimental method using the fluorescence signal is designed to give information about the excitation energy transfer phenomena in molecular aggregates. The so-called *fluorescence excitation spectrum* is designed to reveal the efficiency of the excitation energy transfer between various spectral regions of an aggregate when these regions are energetically distinct.

Let us assume a molecular donor and acceptor. Since the energy transfer occurs preferentially downwards, the donor molecule will absorb at higher-energy than the acceptor molecule. The two spectra can be overlapping; however, the high-energy region and low-energy region can be separately attributed. We denote the high energy region by D, because it represents the part of the aggregate which donates energy to the acceptor region which we denoted by A. If we now excite the system at frequency ω_{exc} in the D region, we can expect the system to fluoresce at some frequency ω_{A}, which will be in the acceptor region. The total recorded fluorescence from the acceptor, when excited at ω_{exc}, will be denoted as:

$$F(\omega_{\text{exc}}) \approx \int d\omega_{\text{fl}} \sigma(\omega_{\text{fl}}; \omega_{\text{exc}}) \,. \tag{14.87}$$

The dependence of the fluorescence on the excitation wavelength was not discussed in the above sections, because we started the theory at the point when the excitation has been predefined, and the excited state has relaxed to the thermal equilibrium from which the fluorescence occurs. In the case of excited energy transfer from the donor to the acceptor, it is evident, however, that when the excitation gets lost during the transfer between the original excited donor states and the fluorescing acceptor states, the total recorded fluorescence from the acceptor molecule will be diminished. This dependence of the fluorescence on the excitation frequency can be quite strong.

The spectrum defined by (14.87) already gives us an information from which spectral regions the energy flows to the fluorescing region. To turn it into a quantitative estimation of the efficiency we have to record the absorbance A, which is for low absorption similar to the absorption coefficient, (13.43),

$$A(\omega) = \frac{I(\omega) - I_0(\omega)}{I_0(\omega)} \approx \kappa_{\text{a}}(\omega) \,. \tag{14.88}$$

Now, when the system is excited directly in the acceptor region (in the region of the maximum fluorescence $F(\omega_{\text{A}})$), that is, no transfer is needed to achieve fluorescence, the situation is equivalent to the 100% transfer efficiency. If we had another region from which the efficiency were 100%, the fluorescence would change only by the difference in absorption. Consequently, by normalizing both the absorption and fluorescence excitation spectrum to one at the frequency ω_{A} we can estimate the fluorescence loss $\zeta(\omega)$ at a given frequency:

$$\zeta(\omega) = \frac{A(\omega)}{A(\omega_{\text{A}})} - \frac{F(\omega)}{F(\omega_{\text{A}})} \,. \tag{14.89}$$

We can consequently also define the efficiency of the energy transfer as:

$$Q(\omega) = (1 - \zeta(\omega)) \cdot 100\% \,. \tag{14.90}$$

This quantity tells us what percentage of the absorbed quanta of energy were lost during the transfer of excitation process and, therefore, this rather simple spectroscopic approach can provide us with an important information about connectivity between different regions of the aggregate absorption spectra. No expensive time-resolved experiments have to be performed if we want to know how much of the absorbed energy gets delivered to the lowest point of the energy landscape, provided the system is able to fluoresce from this point.

The above described spectroscopic techniques illustrate simple examples of usage of the theory described in the beginning of this chapter. We find, that the crucial quantity is again the correlation function of the environment fluctuations or its spectral density.

15
Four-Wave Mixing Spectroscopy

The previous chapter has dealt with the spectroscopic techniques whose theoretical description could be achieved within linear response theory. Absorption spectra could be understood as a result of a single perturbative interaction of light with the system originally in equilibrium. Spontaneous emission or fluorescence could also be understood in terms of a single perturbative interaction of the quantized light with the system, this time in an electronically excited state. The preparation stage of the excited state was not included in our description of fluorescence assuming that the system has enough time to equilibrate in the electronic excited state before it is likely to fluoresce. The more precise description of fluorescence should take into account details of preparation of the excited state, which is included in this chapter. Such a description is presented using the nonlinear spectroscopy formalism, what leads to the time-resolved fluorescence method. In the following sections we will discuss nonlinear techniques, which are mostly related to the four-wave mixing type of signals generated by the third order nonlinear response function.

15.1
Nonlinear Response of Multilevel Systems

We already know how the response functions relate to the spectroscopic signal which is measured (see (13.51)). The response functions have a certain structure that we will investigate and make use of in this section. We will first disregard the details of the system–bath interaction. This interaction will be formally included by evolution superoperators, but our discussion will be centered on the structure given to the response functions by a set of states that are interesting from the point of view of the spectroscopic experiment. These states are connected by optically allowed transitions. The rest of the system is then responsible for effects of energy relaxation and dephasing.

Molecular Excitation Dynamics and Relaxation, First Edition. L. Valkunas, D. Abramavicius, and T. Mančal.

15.1.1
Two- and Three-Band Molecules

The full third order response function expression was derived in Section 13.1 (see (13.22)) and is given by

$$S^{(3)}(t_3, t_2, t_1) = \mathrm{i}^3 \mathrm{Tr}\left\{\hat{\mu}_4 \mathcal{U}_{\mathrm{mat}}(t_3)\mathcal{V}_3\mathcal{U}_{\mathrm{mat}}(t_2)\mathcal{V}_2\mathcal{U}_{\mathrm{mat}}(t_1)\mathcal{V}_1\hat{\rho}_{\mathrm{eq}}\right\} . \quad (15.1)$$

Here we denoted the order of interactions by $\hat{\mu}_j$ or $\mathcal{V}_j$. $j = 4$ denotes the last emission step, $\hat{\rho}_{\mathrm{eq}}$ is the full equilibrium density matrix. The response function tells us that the third order nonlinear signal is constructed by acting three times by $\mathcal{V}$ at the initial density matrix corresponding to the electronic ground state, followed by the emission step through the last polarization operator $\hat{\mu}$ acting on the left in the Hilbert space, and finally taking the trace. In the Hilbert space using the Heisenberg representation we have the response in terms of the four-point correlation functions (13.24)–(13.27), (for convenience these are repeated here by additionally labeling the interaction order inside the trace)

$$R_1(t_3, t_2, t_1) = \mathrm{Tr}\left\{\hat{\mu}_2(t_1)\hat{\mu}_3(t_1 + t_2)\hat{\mu}_4(t_1 + t_2 + t_3)\hat{\mu}_1(0)\hat{\rho}_{\mathrm{eq}}\right\} , \quad (15.2)$$

$$R_2(t_3, t_2, t_1) = \mathrm{Tr}\left\{\hat{\mu}_1(0)\hat{\mu}_3(t_1 + t_2)\hat{\mu}_4(t_1 + t_2 + t_3)\hat{\mu}_2(t_1)\hat{\rho}_{\mathrm{eq}}\right\} , \quad (15.3)$$

$$R_3(t_3, t_2, t_1) = \mathrm{Tr}\left\{\hat{\mu}_1(0)\hat{\mu}_2(t_1)\hat{\mu}_4(t_1 + t_2 + t_3)\hat{\mu}_3(t_1 + t_2)\hat{\rho}_{\mathrm{eq}}\right\} , \quad (15.4)$$

$$R_4(t_3, t_2, t_1) = \mathrm{Tr}\left\{\hat{\mu}_4(t_1 + t_2 + t_3)\hat{\mu}_3(t_1 + t_2)\hat{\mu}_2(t_1)\hat{\mu}_1(0)\hat{\rho}_{\mathrm{eq}}\right\} . \quad (15.5)$$

and

$$S^{(3)}(t_3, t_2, t_1) = \mathrm{i}^3\theta(t_1)\theta(t_2)\theta(t_3)\sum_{\alpha=1}^{4}\left(R_\alpha(t_3, t_2, t_1) - R_\alpha^\star(t_3, t_2, t_1)\right) . \quad (15.6)$$

Now $\hat{\mu}_j(t)$ is the transition dipole in Heisenberg representation representing the jth interaction with the field with respect to real time. For the trace to be nonzero, the final matrix under the trace needs to have nonzero diagonal elements. The transition dipole moment operator $\hat{\mu}$ facilitates transitions between electronic states. Using this ordering of operators inside the trace in (15.2) to (15.5), the transitions (going from the right side) have to start from the ground state and have to finish in the electronic ground state where they started, because the density operator is at the right-most position under the trace. While this ordered form is convenient to reflect the form of the four-point correlation function, the proper reordering with respect to time implies that the action on the density matrix is either on the left or on the right at different time moments. The last $j = 4$ emission step occurs always in the left as indicated by (13.22).

In the following we study the systems whose levels construct several bands. The lowest band is the electronic ground state g, or a set of closely-lying states. There is another band of states separated by a wide energy gap above the g band. We call the this band the singly-excited levels e. And the third band is denoted as the

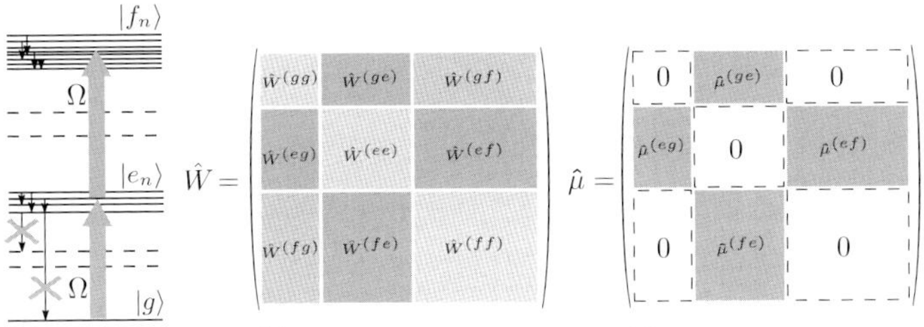

(a) (b) (c)

Figure 15.1 Energy level structure of a three-band electron system (a), the block structure of its statistical operator $\hat{W}$ (b), and the transition dipole operator $\hat{\boldsymbol{\mu}}$ (c). The three-band model consists of the ground state $|g\rangle$, a band of excited states $|e_n\rangle$ separated from the ground state by the wide gap possible to overcome only by the optical transition energy quantum $\hbar\Omega$, and another band of states $|f_n\rangle$ separated by another energy quantum $\hbar\Omega$ (all states depicted by full lines). Energy relaxation is allowed within the bands, but not between the bands or to other states with nonresonant transitions (dashed lines). The statistical operator $\hat{W}$, therefore, has a block structure (center of the figure) with nine independent blocks connected by optical coupling only.

double-excited band f (see Figure 15.1). This distinction becomes obvious if we consider an arbitrary experiment performed with the beam of light having a carrier frequency Ω that is close to the energy gap between the nearest bands. The light can then induce transitions between g and e and between e and f. In this type of experiment, states lying even higher or between the bands would not be able to contribute to the third order as they would not contribute to the final trace or would be off-resonant. We can, therefore, limit our consideration just to three-band systems.

An important example of a generic molecular system which fits the above description is represented by excitonic aggregates studied in Chapter 5. In these aggregates some relatively simple molecules (chromophores) are bound together noncovalently, and held at well defined positions by some specific geometrical constraints, be it self-organized structures or determined by protein scaffold. Because of the mutual electrostatic interaction, the complex has different spectroscopic properties compared to the individual chromophores (for example, in solution). The states of the aggregate can be well classified using the states of individual chromophores. The aggregate with N chromophores is said to be in its ground state $|g\rangle$ if all its constituting chromophores are in their ground states $|g_i\rangle$. The first excited band is formed of the states $|e_k\rangle$ which include one excited chromophore and the second excited state band is formed by the state $|f_{kl}\rangle$ with two excited chromophores. These levels represent three well defined bands as shown in Figure 15.1.

The evolution superoperator elements in (15.1) can be also organized into blocks according to the bands. As the light transfers the system between nearest bands, the density matrix breaks up into nine blocks (see Figure 15.1) to which we can assign corresponding blocks in the evolution superoperator. We further assume no direct relaxation between the bands, so no evolution superoperator connects two different blocks of the density matrix. Thus, for example, $\mathcal{U}^{(ee)}(\tau)$ will be the block of the evolution superoperator governing the evolution of the populations and intra-band coherences inside the e-band, which is in turn described by the density operator block $\hat{W}^{(ee)}$. Whenever we need to evaluate the evolution of the density operator block in detail, we can go back to the indices of the individual states, so we can write, for example,

$$W_{mn}^{(ee)}(\tau + t_0) = \sum_{kl} \mathcal{U}_{mn,kl}^{(ee)}(\tau)\, W_{kl}^{(ee)}(t_0) \,, \tag{15.7}$$

m, n, k, l run over the singly-excited band, or, using shorthand notation, we will use

$$\hat{W}^{(ee)}(\tau + t_0) = \mathcal{U}^{(ee)}(\tau)\, \hat{W}^{(ee)}(t_0) \tag{15.8}$$

with the same meaning. The superscript (ee), represents the time evolution which starts in the "diagonal" block, (ee), of the statistical operator, and finishes in the same block.

One important property of the statistical operator is that its off-diagonal blocks, for example $\hat{W}^{(ge)}(t)$ or $\hat{W}^{(ef)}(t)$, oscillate with a frequency similar to Ω and the block $\hat{W}^{(fg)}(t)$ oscillates with $\sim 2\Omega$. The evolution superoperator has to reflect this fact and corresponding frequencies are indeed imprinted in its time dependence. For further manipulations with these blocks we can explicitly subtract this fast frequency component and introduce slowly varying envelopes (the Rotating Frame) as in the following example

$$\mathcal{U}^{(eg)}(t) = \tilde{\mathcal{U}}^{(eg)}(t) e^{-i\Omega t} \,, \quad \mathcal{U}^{(fg)}(t) = \tilde{\mathcal{U}}^{(fg)}(t) e^{-i2\Omega t} \,. \tag{15.9}$$

From now on in this chapter, all superoperator blocks appearing with tilde as above will represent these slow envelopes.

The second quantity which appears in the response function in both the operator and superoperator form is the polarization superoperator $\mathcal{V}$ and the corresponding polarization operator $\hat{\mu}$. The polarization operator is essentially a dipole operator, because we assume that the molecular length-scale is much shorter than the optical wavelength. As this operator mediates transitions between bands, it has a block form as in Figure 15.1. Matrices can be multiplied in blocks, and thus it is easy to verify that the action of the transition dipole moment on the density operator transforms one block into another. With the transition dipole moment operator

$$\hat{\mu} = \mu^{(eg)}|e\rangle\langle g| + \mu^{(fe)}|f\rangle\langle e| + \text{h.c.} \,, \tag{15.10}$$

where h.c. stands for the Hermite conjugate, we have, for example, $\hat{\mu}\, \hat{W}^{(gg)} \equiv \hat{\mu}^{(eg)}\, \hat{W}^{(gg)} = \hat{W}^{(eg)}$. Similarly, we can show that off-diagonal blocks are transformed into diagonal ones by the action of the dipole moment operator from left or right. In the superoperator notation we have, for example,

$$\mathcal{V}\, \hat{W}^{(gg)} = \mathcal{V}^{(eggg)}\, \hat{W}^{(gg)} - \mathcal{V}^{(gegg)}\, \hat{W}^{(gg)} = \hat{W}^{(eg)} - \hat{W}^{(ge)} \,. \tag{15.11}$$

Here we also introduced a block notation for the dipole moment superoperator $\mathcal{V}$. It must have four indices as it transforms one block (gg) into another block (eg) or (ge). Action of the transition dipole moment on a block of the statistical operator yields contributions to its two different blocks, and the notation denotes the transformation performed by the superoperator on the band indices, for example, $(eggg) = (eg) \leftarrow (gg)$.

15.1.2 Liouville Space Pathways

Let us study the third order response function given in (15.1). As we have already discussed, the response functions consist of sets of transitions and propagations between them by evolution superoperators. In electronic spectroscopy we always start with only the electronic ground state populated because the optical transition energies are much larger than the thermal energy, $\hbar\Omega \gg k_B T$. The tracing requires that after all actions of the dipole operator, we end up at the same initial ground state in the time-unordered representation of (15.2)–(15.5) or the density matrix ends in a population state in the time-ordered representation of (15.1). This means that there are only certain *Liouville space pathways* through the states of the system that can contribute. These pathways connect specific density matrix elements or density matrix blocks.

Let us consider (15.2)–(15.5) and let a and b denote bands or individual energy levels and let us draw the following diagram (Figures 15.2 and 15.3). A density matrix element in Figure 15.2 is depicted by a circle with two letters denoting the element of the matrix. The action of a dipole moment operator from the left-hand side changes the left index of the density matrix and, similarly, its action from the right-hand side changes the right index. In the diagram we denote the action of

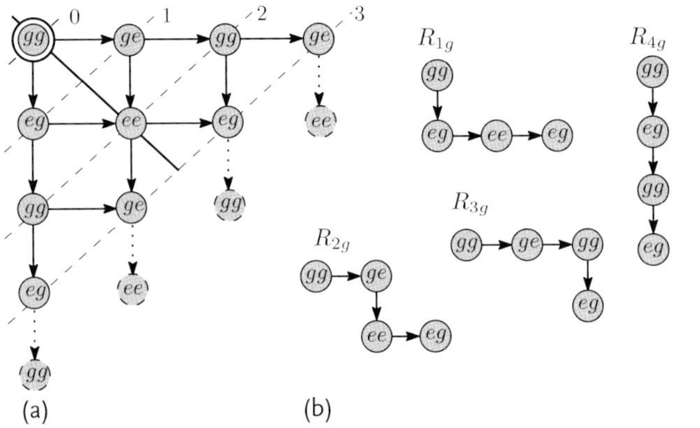

Figure 15.2 Liouville space pathway diagram: possible pathways starting with the electronic ground state $|g\rangle$ in a two-band system are shown in (a). The dashed lines connect the states after zero, one, two and three perturbative interactions with the field. Independent pathways with "+" overall sign are shown in (b).

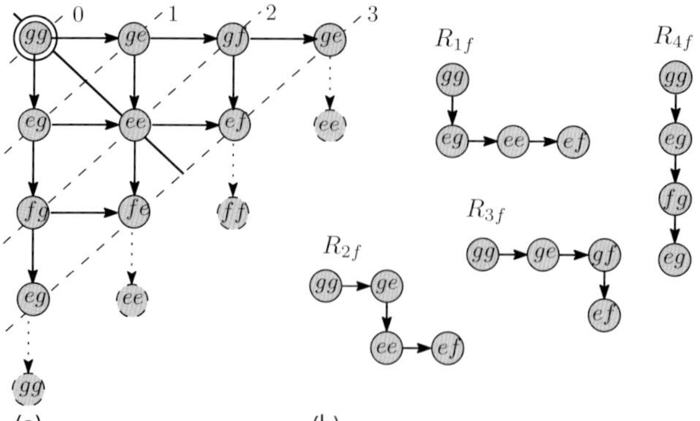

Figure 15.3 Liouville space pathways similar to Figure 15.2. Here we show additional pathways that become possible in a three-band system (a). Independent pathways with "+" overall sign are shown in (b).

the dipole operator from the left side by a downward pointing arrow and the corresponding change of the left index. The action on the right side is denoted by a right arrow that changes the right-hand side index. Each commutator $\mathcal{V}$ in (15.1) corresponds to a *fork* of the two arrows, to the right-hand side, or downwards. Dashed lines indicate the propagation intervals.

In a system with two bands, g and e, all possible pathways are shown in Figure 15.2a. There are only three possible configurations of system states possible: gg, ge, $eg \equiv (ge)^*$ and ee. In time intervals t_1 and t_3 the system is always at interband coherence while during t_2 we have populations gg or ee. In a three-band system, some additional pathways become possible since we can have transitions $e \rightarrow f$. These are shown in Figure 15.3a. We thus get possible fe and fg coherences that can be probed at time intervals t_3 and t_2, respectively. The last interaction in (15.1) is always on the left, so the last arrow must go down to some population state (gg or ee or ff).

The Figures (15.2) and (15.3) show that there are 16 possible Liouville pathways between the three bands that can be traveled by three actions of the dipole moment. The elements (ab) and (ba) correspond to mutual Hermite conjugated blocks of the density matrix, and, therefore, the pathways that are mirrors of each other with respect to the diagonal line of the diagram have mutually complex conjugated contributions. Moreover, those pathways that involve an odd number of dipole operator actions from the left-hand side carry a minus sign, which originates from the second term of the commutator. We have, therefore, eight independent contributions, four of which contain only bands g and e, and four of which also include the f band. There are four pairs of pathways with a plus sign that differ by exchanging g and f and which correspond to four different orderings of left- and right-hand side actions of the dipole moment. We denote the contributions to the total response function by a letter R with a lower index $1, \ldots, 4$ denoting the four independent orderings of interaction from the left and right, that is, the four independent shapes of the Liouville pathways in Figure 15.2. Compared to (15.6) we introduce a second index g or f to distinguish pathways which do not reach the f-band from those that do, respectively. Thus, instead of (15.6) we write the total third order response function in a form:

$$S^{(3)}(t_3,t_2,t_1) = \mathrm{i}^3\theta(t_3)\theta(t_2)\theta(t_1)\sum_{n=1}^{4}\sum_{a=g,f}\left[R_{na}(t_3,t_2,t_1) - R^*_{na}(t_3,t_2,t_1)\right]. \tag{15.12}$$

Now it is convenient to explicitly write down individual Liouville pathways and evaluate them using the evolution superoperators. In the third order to the field, following Figure 15.2 we then have

$$R_{1g}(t_3,t_2,t_1) = \mathrm{Tr}\left\{\hat{\mu}^{(ge)}\tilde{\mathcal{U}}^{(eg)}(t_3)\mathcal{V}^{(egee)}\tilde{\mathcal{U}}^{(ee)}(t_2)\right.$$
$$\left.\times\,\mathcal{V}^{(eeeg)}\tilde{\mathcal{U}}^{(eg)}(t_1)\mathcal{V}^{(eggg)}\,\hat{W}^{(gg)}\right\}\mathrm{e}^{-\mathrm{i}\Omega(t_1+t_3)}\,. \tag{15.13}$$

$$R_{2g}(t_3,t_2,t_1) = \mathrm{Tr}\left\{\hat{\mu}^{(ge)}\tilde{\mathcal{U}}^{(eg)}(t_3)\mathcal{V}^{(egee)}\tilde{\mathcal{U}}^{(ee)}(t_2)\right.$$
$$\left.\times\,\mathcal{V}^{(eege)}\tilde{\mathcal{U}}^{(ge)}(t_1)\mathcal{V}^{(gegg)}\,\hat{W}^{(gg)}\right\}\mathrm{e}^{-\mathrm{i}\Omega(t_3-t_1)} \tag{15.14}$$

$$R_{3g}(t_3,t_2,t_1) = \mathrm{Tr}\left\{\hat{\mu}^{(ge)}\tilde{\mathcal{U}}^{(eg)}(t_3)\mathcal{V}^{(eggg)}\tilde{\mathcal{U}}^{(gg)}(t_2)\right.$$
$$\left.\times\,\mathcal{V}^{(ggge)}\tilde{\mathcal{U}}^{(ge)}(t_1)\mathcal{V}^{(gegg)}\,\hat{W}^{(gg)}\right\}\mathrm{e}^{-\mathrm{i}\Omega(t_3-t_1)} \tag{15.15}$$

$$R_{4g}(t_3,t_2,t_1) = \mathrm{Tr}\left\{\hat{\mu}^{(ge)}\tilde{\mathcal{U}}^{(eg)}(t_3)\mathcal{V}^{(eggg)}\tilde{\mathcal{U}}^{(gg)}(t_2)\right.$$
$$\left.\times\,\mathcal{V}^{(ggeg)}\tilde{\mathcal{U}}^{(eg)}(t_1)\mathcal{V}^{(eggg)}\,\hat{W}^{(gg)}\right\}\mathrm{e}^{-\mathrm{i}\Omega(t_1+t_3)}\,. \tag{15.16}$$

This is the set of pathways for a two-band system where f is excluded. Each of these pathways is composed of a possibly large number of different sub-pathways between individual levels within the bands. Notice that during the interval t_2 of the response, the system evolves according to an evolution superoperator of either the excited state band or the ground state band. For bare excitonic systems (Chapter 5), the ground state band is composed of just one level; however, the dynamics of an arbitrary realistic system also involve evolution of the reservoir degrees of freedom. The evolution inside the bands can have a form of energy relaxation and transport, as well as coherence dephasing. This evolution can be accessed in some types of FWM experiments.

The existence of f band leads to an excitation of the system from the band e to a higher lying band f. This adds an extra set of pathways whose contributions are

$$R_{1f}(t_3,t_2,t_1) = \mathrm{Tr}\left\{\hat{\mu}^{(fe)}\tilde{\mathcal{U}}^{(ef)}(t_3)\mathcal{V}^{(efee)}\tilde{\mathcal{U}}^{(ee)}(t_2)\right.$$
$$\left.\times\,\mathcal{V}^{(eeeg)}\tilde{\mathcal{U}}^{(eg)}(t_1)\mathcal{V}^{(eggg)}\,\hat{W}^{(gg)}\right\}\mathrm{e}^{-\mathrm{i}\Omega(t_1-t_3)}\,. \tag{15.17}$$

$$R_{2f}(t_3, t_2, t_1) = \mathrm{Tr}\left\{\hat{\mu}^{(fe)}\tilde{\mathcal{U}}^{(ef)}(t_3)\mathcal{V}^{(efee)}\tilde{\mathcal{U}}^{(ee)}(t_2)\right.$$
$$\left. \times\, \mathcal{V}^{(eege)}\tilde{\mathcal{U}}^{(ge)}(t_1)\mathcal{V}^{(gegg)}\,\hat{W}^{(gg)}\right\} \mathrm{e}^{\mathrm{i}\Omega(t_1+t_3)} \tag{15.18}$$

$$R_{3f}(t_3, t_2, t_1) = \mathrm{Tr}\left\{\hat{\mu}^{(fe)}\tilde{\mathcal{U}}^{(ef)}(t_3)\mathcal{V}^{(efgf)}\tilde{\mathcal{U}}^{(gf)}(t_2)\right.$$
$$\left. \times\, \mathcal{V}^{(gfge)}\tilde{\mathcal{U}}^{(ge)}(t_1)\mathcal{V}^{(gegg)}\,\hat{W}^{(gg)}\right\} \mathrm{e}^{\mathrm{i}\Omega(t_1+2t_2+t_3)} \tag{15.19}$$

$$R_{4f}(t_3, t_2, t_1) = \mathrm{Tr}\left\{\hat{\mu}^{(ge)}\tilde{\mathcal{U}}^{(eg)}(t_3)\mathcal{V}^{(egfg)}\tilde{\mathcal{U}}^{(fg)}(t_2)\right.$$
$$\left. \times\, \mathcal{V}^{(fgeg)}\tilde{\mathcal{U}}^{(eg)}(t_1)\mathcal{V}^{(eggg)}\,\hat{W}^{(gg)}\right\} \mathrm{e}^{-\mathrm{i}\Omega(t_1+2t_2+t_3)}\,. \tag{15.20}$$

The eight functions $R_{1g}, \ldots, R_{4g}$ and $R_{1f}, \ldots, R_{4f}$ and their complex conjugate parts completely determine the third order response of a three-band system and thus describes all possible FWM experiments.

Individual contributions to the response function, the Liouville space pathways, can be conveniently visualized by the *double sided Feynman diagrams.* A double sided Feynman diagram carries at least the same information as the diagrams in Figure 15.2b. Additional information which completely determines the Liouville space pathway characteristics, for example, the frequencies of the transitions and the interaction field wavevectors, can be included after performing the rotating wave approximation (selection of resonances) with respect to specific optical fields. Figure 15.4 shows Feynman diagrams compared to the Liouville pathway diagram from Figure 15.2. The two vertical arrows denote the time evolution of the ket and bra of the system's density operator (time is running upwards). Each horizontal bar denotes the time when an interaction with the external field (in other words, multiplication of the statistical operator from left or right by the transition dipole operator) occurs. The ket and bra after each interaction are denoted. The last arrow (from the bottom to top) represents the action of the transition dipole moment operator $\hat{\mu}$, which we wrote to the left-most position under the trace in the response function. The position of the last arrow is convenient at this point, and we will keep it on the left for the pathways that yield relevant response.

15.1.3
Third Order Polarization in the Rotating Wave Approximation

In the previous section we have treated the response function separately from (13.16) which connects the response function with the third order polarization. One can easily verify that the product of three electric fields in (13.16) carries phase factors similar to those of the Liouville pathways. The polarization is obtained by integrating the product of the response functions and the fields. Because we have assumed that the optical frequency Ω is similar to the inter-band energy gap, the terms where optical frequencies match the response function frequencies and they overall cancel out, should yield a much larger contribution to the convolution integral in (13.16) than those where some highly oscillatory factors remain. All the

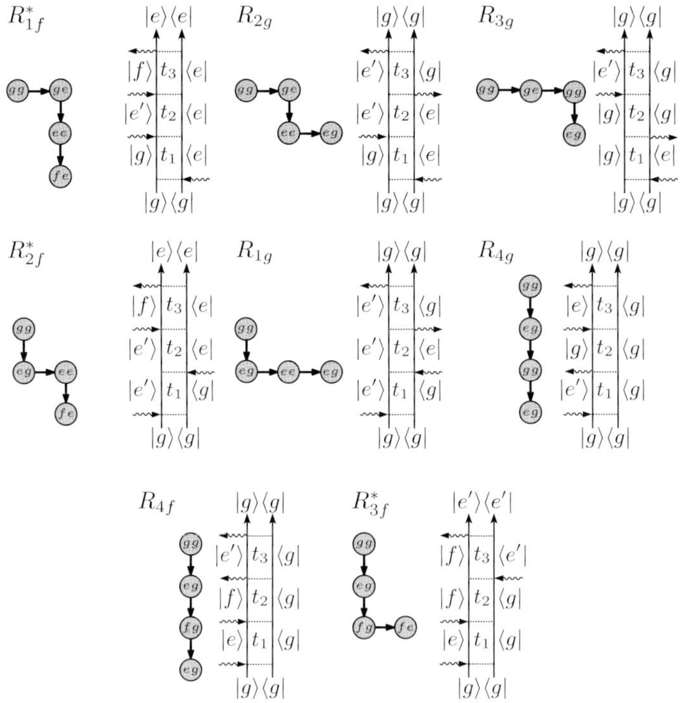

Figure 15.4 Feynman diagrams for resonant interaction configurations when all fields have approximately the same carrier frequency. Liouville pathways are indicated as well. Incoming/outgoing arrows denote absorption/emission events.

oscillating terms under the integral will, therefore, be neglected. This is usually denoted as the *rotating wave approximation*.

In the remainder of this book we will consider time-resolved techniques performed with several short laser pulses. In that case it is convenient to introduce a Gedankenexperiment (a thought experiment) as follows. Let us assume the *incoming* electric field to be formed of three pulses with envelopes depicted in Figure 15.5, that is

$$E(\boldsymbol{r}, t) = \sum_{j=1}^{3} \mathcal{E}_j(t - \tau_j)\mathrm{e}^{\mathrm{i}\boldsymbol{k}_j \cdot \boldsymbol{r} - \mathrm{i}\Omega(t-\tau_j)} + \mathrm{c.c.}$$

where c.c. denotes the complex conjugated term. Then each interaction between the system and the field, represented by $\mathcal{V}$ in the response function, happens with a distinct optical pulsed field. Later we can associate these "virtual" pulses with the physical pulses of the laser when we describe a realistic spectroscopic measurement.

The cubic form of electric fields

$$E(\boldsymbol{r}, t - t_3 - t_2 - t_1)E(\boldsymbol{r}, t - t_3 - t_2)E(\boldsymbol{r}, t - t_3) \tag{15.21}$$

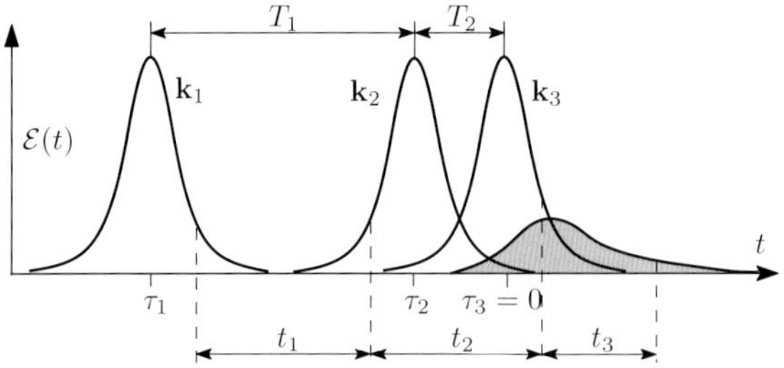

Figure 15.5 Pulse scheme of the time-ordered four-wave mixing experiment when pulse 1 is first, pulse 2 is second, and pulse 3 is third. The excitation pulses are centered at times τ_1, τ_2 and τ_3. Since the time axis can be taken with respect to the specific rule, we can choose $\tau_3 = 0$. The third order signal at time t is then an integral over contributions from all possible combinations of delays t_1, t_2 and t_3 such that are positive. Notice that this setup of pulses is the so-called gedanken or thought experiment. The pulses 1, 2, and 3 here are the sources of interactions constructing the response function. Later, for various experiments, we associate these virtual pulses with various real pulses.

enters the third order polarization (13.16). Each of them is a sum of three optical pulses. The whole product will thus be a sum of $3 \times 3 \times 3 = 27$ terms. However, notice that in the FWM scheme we only have several distinct output wavevectors, which can be distinguished by a specific outgoing direction (wavevector). These are (the total oscillatory frequency of the configuration is also given):

$\boldsymbol{k}_1 \Leftrightarrow \Omega$	$2\boldsymbol{k}_1 - \boldsymbol{k}_2 \Leftrightarrow \Omega$	$2\boldsymbol{k}_2 - \boldsymbol{k}_3 \Leftrightarrow \Omega$
$3\boldsymbol{k}_1 \Leftrightarrow 3\Omega$	$2\boldsymbol{k}_1 + \boldsymbol{k}_2 \Leftrightarrow 3\Omega$	$2\boldsymbol{k}_2 + \boldsymbol{k}_3 \Leftrightarrow 3\Omega$
$\boldsymbol{k}_2 \Leftrightarrow \Omega$	$2\boldsymbol{k}_1 - \boldsymbol{k}_3 \Leftrightarrow \Omega$	$2\boldsymbol{k}_3 - \boldsymbol{k}_1 \Leftrightarrow \Omega$
$3\boldsymbol{k}_2 \Leftrightarrow 3\Omega$	$2\boldsymbol{k}_1 + \boldsymbol{k}_3 \Leftrightarrow 3\Omega$	$2\boldsymbol{k}_3 + \boldsymbol{k}_1 \Leftrightarrow 3\Omega$
$\boldsymbol{k}_3 \Leftrightarrow \Omega$	$2\boldsymbol{k}_2 - \boldsymbol{k}_1 \Leftrightarrow \Omega$	$2\boldsymbol{k}_3 - \boldsymbol{k}_2 \Leftrightarrow \Omega$
$3\boldsymbol{k}_3 \Leftrightarrow 3\Omega$	$2\boldsymbol{k}_2 + \boldsymbol{k}_1 \Leftrightarrow 3\Omega$	$2\boldsymbol{k}_3 + \boldsymbol{k}_2 \Leftrightarrow 3\Omega$

$-\boldsymbol{k}_1 + \boldsymbol{k}_2 + \boldsymbol{k}_3 \Leftrightarrow \Omega$
$\boldsymbol{k}_1 - \boldsymbol{k}_2 + \boldsymbol{k}_3 \Leftrightarrow \Omega$
$\boldsymbol{k}_1 + \boldsymbol{k}_2 - \boldsymbol{k}_3 \Leftrightarrow \Omega$
$\boldsymbol{k}_1 + \boldsymbol{k}_2 + \boldsymbol{k}_3 \Leftrightarrow 3\Omega$

The set of wavevectors with opposite directions generates conjugate responses. The first thing that we observe is that the configurations that have frequencies 3Ω are off-resonant with our form of the response function (since we do not include this type of triple-frequency optical transition). The second observation is that the wavevector configuration $2\boldsymbol{k}_1 - \boldsymbol{k}_2$ can be obtained from $\boldsymbol{k}_1 - \boldsymbol{k}_2 + \boldsymbol{k}_3$ if we take $\boldsymbol{k}_1 = \boldsymbol{k}_3$. It thus follows that the most general third order independent signals with our response function are

$$k_{\mathrm{I}} \equiv -\boldsymbol{k}_1 + \boldsymbol{k}_2 + \boldsymbol{k}_3 \,, \tag{15.22}$$

$$k_{\mathrm{II}} \equiv \boldsymbol{k}_1 - \boldsymbol{k}_2 + \boldsymbol{k}_3 \,, \tag{15.23}$$

$$k_{III} \equiv \boldsymbol{k}_1 + \boldsymbol{k}_2 - \boldsymbol{k}_3 \tag{15.24}$$

and others will be given by these three.

Let us consider $-\boldsymbol{k}_1 + \boldsymbol{k}_2 + \boldsymbol{k}_3$. This configuration appears from six types of field permutations. We may have

$$E_1(\boldsymbol{r}, t - t_3 - t_2 - t_1) E_2(\boldsymbol{r}, t - t_3 - t_2) E_3(\boldsymbol{r}, t - t_3) , \tag{15.25}$$

or

$$E_1(\boldsymbol{r}, t - t_3 - t_2 - t_1) E_3(\boldsymbol{r}, t - t_3 - t_2) E_2(\boldsymbol{r}, t - t_3) , \tag{15.26}$$

$$E_2(\boldsymbol{r}, t - t_3 - t_2 - t_1) E_3(\boldsymbol{r}, t - t_3 - t_2) E_1(\boldsymbol{r}, t - t_3) , \tag{15.27}$$

$$E_2(\boldsymbol{r}, t - t_3 - t_2 - t_1) E_1(\boldsymbol{r}, t - t_3 - t_2) E_3(\boldsymbol{r}, t - t_3) , \tag{15.28}$$

$$E_3(\boldsymbol{r}, t - t_3 - t_2 - t_1) E_2(\boldsymbol{r}, t - t_3 - t_2) E_1(\boldsymbol{r}, t - t_3) , \tag{15.29}$$

and finally

$$E_3(\boldsymbol{r}, t - t_3 - t_2 - t_1) E_1(\boldsymbol{r}, t - t_3 - t_2) E_2(\boldsymbol{r}, t - t_3) . \tag{15.30}$$

The corresponding phase factors containing t_1, t_2, t_3 (over which we integrate) ordered in the same way are

$$\exp(\mathrm{i}\Omega(-t_1 + t_3)) , \tag{15.31}$$

$$\exp(\mathrm{i}\Omega(-t_1 + t_3)) , \tag{15.32}$$

$$\exp(\mathrm{i}\Omega(t_3 + 2t_2 + t_1)) , \tag{15.33}$$

$$\exp(\mathrm{i}\Omega(t_1 + t_3)) , \tag{15.34}$$

$$\exp(\mathrm{i}\Omega(t_3 + 2t_2 + t_1)) , \tag{15.35}$$

$$\exp(\mathrm{i}\Omega(t_1 + t_3)) . \tag{15.36}$$

The Liouville pathways, (15.13) to (15.20), have similar phase factors in variables t_3, t_2 and t_1. If a given Liouville pathway does not have a phase factor which cancels the phase factor of the field, it brings only a very small contribution to the integral. We will, therefore, take only such contributions where the cancelation occurs and the rest will be neglected.

The electric field enters (13.16) three times with three different time arguments. Let us define the delays between incoming pulses as $T_1 = \tau_2 - \tau_1$, $T_2 = \tau_3 - \tau_2$ and abbreviate the expressions as follows:

$$E_j^{(1)}(\vartheta) \equiv \mathcal{E}_j(t + \vartheta - t_3 - t_2 - t_1) , \tag{15.37}$$

$$E_j^{(2)}(\vartheta) \equiv \mathcal{E}_j(t + \vartheta - t_3 - t_2) , \tag{15.38}$$

and

$$E_j^{(3)}(\vartheta) \equiv \mathcal{E}_j(t + \vartheta - t_3) . \tag{15.39}$$

For the third order polarization in RWA we can now write one integral expression

$$P^{(3)}_{\mathrm{RWA}}(t, T_2, T_1) \approx \mathrm{e}^{-\mathrm{i}\Omega(t-T_1)} \iiint_0^\infty \mathrm{d}t_3\mathrm{d}t_2\mathrm{d}t_1[S_\mathrm{R}(t_3, t_2, t_1)A_\mathrm{R} + S_\mathrm{NR}(t_3, t_2, t_1)A_\mathrm{NR} + S_\mathrm{DC}(t_3, t_2, t_1)A_\mathrm{DC}] , \quad (15.40)$$

where

$$A \equiv A(t_3, t_2, t_1; t, T_2, T_1)$$

$$A_\mathrm{R} = \mathrm{e}^{\mathrm{i}\Omega t_3 - \mathrm{i}\Omega t_1}\left[E_1^{(1)*}(T_2 + T_1)E_2^{(2)}(T_2)E_3^{(3)}(0) + E_1^{(1)*}(T_2 + T_1)E_2^{(3)}(T_2)E_3^{(2)}(0)\right] , \quad (15.41)$$

$$A_\mathrm{NR} = \mathrm{e}^{\mathrm{i}\Omega t_3 + \mathrm{i}\Omega t_1}\left[E_1^{(2)*}(T_2 + T_1)E_2^{(1)}(T_2)E_3^{(3)}(0) + E_1^{(2)*}(T_2 + T_1)E_2^{(3)}(T_2)E_3^{(1)}(0)\right] \quad (15.42)$$

and

$$A_\mathrm{DC} = \mathrm{e}^{\mathrm{i}\Omega t_3 + 2\mathrm{i}\Omega t_2 + \mathrm{i}\Omega t_1}\left[E_1^{(3)*}(T_2 + T_1)E_2^{(1)}(T_2)E_3^{(2)}(0) + E_1^{(3)*}(T_2 + T_1)E_2^{(2)}(T_2)E_3^{(1)}(0)\right] . \quad (15.43)$$

Here, we have collected the Liouville pathways into groups according to their characteristic oscillating phase factors. Thus, we have the *rephasing*

$$S_\mathrm{R}(t_3, t_2, t_1) = R_{2g}(t_3, t_2, t_1) + R_{3g}(t_3, t_2, t_1) - R^*_{1f}(t_3, t_2, t_1) , \quad (15.44)$$

the *nonrephasing*

$$S_\mathrm{NR}(t_3, t_2, t_1) = R_{1g}(t_3, t_2, t_1) + R_{4g}(t_3, t_2, t_1) - R^*_{2f}(t_3, t_2, t_1) , \quad (15.45)$$

and the *double quantum coherence*

$$S_\mathrm{DC}(t_3, t_2, t_1) = R_{4f}(t_3, t_2, t_1) - R^*_{3f}(t_3, t_2, t_1) , \quad (15.46)$$

pathway groups.

Equation (15.40) enables us to calculate the signal for any third order nonlinear experiment employing a three-pulse sequence with arbitrary pulse shapes for arbitrary pulse ordering. The reason why the complex conjugated pathways with the minus sign appear in (15.44)–(15.46) is that their particular phase factors cancel with the fields. One can notice that the pathway R_{1g} is nonrephasing, while the pathway R^*_{1f} is rephasing. This is easily explained by considering the involved transitions. While in R_{1g} we have $\mathrm{e}^{-\mathrm{i}\omega_{eg}t_1 - \mathrm{i}\omega_{eg}t_3}$, the R_{1f} pathway must have $\mathrm{e}^{-\mathrm{i}\omega_{eg}t_1 - \mathrm{i}\omega_{ef}t_3}$. From Figure 15.1 we have $\omega_{ef} \approx -\omega_{eg}$ and thus the phase factor turns into $\mathrm{e}^{-\mathrm{i}\omega_{eg}t_1 + \mathrm{i}\omega_{eg}t_3}$ which has to be complex conjugated in order to cancel with field factors.

The notation that we introduced in (15.37) to (15.39) and Figure 15.5 enables us to understand the origin of the individual contributions. The upper index α of $E_j^{(\alpha)}$ denotes one of the three interaction times. The value $\alpha = 1$ corresponds to the first interaction, $\alpha = 2$ to the second and so on. In (15.41) to (15.43), we insert the envelopes of the fields in an order in which the pulse centers arrive at the sample (from left to right). The order of the lower indices on the envelopes is, therefore, always $1-2-3$. The upper indices correspond to the order in which these pulses actually interact with the matter. We can see that for instance the nonrephasing contribution originates from the order $2-1-3$ and $3-1-2$ and so on.

15.1.4 Third Order Polarization in Impulsive Limit

The laser pulses usually employed to excite molecular systems in laboratory realizations of the third order spectroscopy may be very short, sometimes with just a few cycles of the optical oscillation within the pulse envelope. It is, therefore, often advantageous to assume the nonlinear response originating from excitation by such a short pulse. In order to keep the resonance properties of the pulses in place, the "short pulse" has to be understood as "slowly" varying with respect to the optical frequency, while it is short with respect to slowly varying system dynamics. It is, therefore, only allowed to have the Dirac delta function properties with respect to the slow envelopes of the response functions. Such delta function is often referred to as the *physical delta function* (see a discussion for example in [26]). Equation (15.40) can be significantly simplified by this short pulse assumption. Most importantly, one can immediately see that for different pulse orderings, different types of Liouville pathways contribute to the signal. Let us first assume that the pulses arrive in time ordered as $\boldsymbol{k}_1$, $\boldsymbol{k}_2$, $\boldsymbol{k}_3$ (we will denote this ordering as $1-2-3$). As they are short, their overlap can be neglected. This means that $T_1 > 0$ and $T_2 > 0$ in Figure 15.5. We now have six combinations of the pulse envelopes in (15.40), but only one of them yields a nonzero integration value. Let us examine the first term in (15.41), that is

$$\mathcal{E}_1^*(t+T_2+T_1-t_3-t_2-t_1)\mathcal{E}_2(t+T_2-t_3-t_2)\mathcal{E}_3(t-t_3)$$
$$\approx \delta(t+T_2+T_1-t_3-t_2-t_1)\delta(t+T_2-t_3-t_2)\delta(t-t_3)\,. \qquad (15.47)$$

This term yields the conditions:

$$t+T_2+T_1-t_3-t_2-t_1=0\,, \qquad (15.48)$$

$$t+T_2-t_3-t_2=0\,, \qquad (15.49)$$

and

$$t-t_3=0 \qquad (15.50)$$

for the integral in (15.40) to give a nonzero contribution. Equations (15.48)–(15.50) can be easily satisfied by $t_3 = t$, $t_2 = T_2$ and $t_1 = T_1$ and the contribution to the polarization thus yields:

$$P^{(3)}_{\mathrm{RWA}}(t, T_2, T_1) \approx \mathrm{e}^{-\mathrm{i}\Omega(t-T_1)} S_{\mathrm{R}}(t, T_2, T_1) \,. \tag{15.51}$$

The second term in (15.40) which corresponds to the role of the second and third pulses switched, similarly yields the conditions $t + T_2 + T_1 - t_3 - t_2 - t_1 = 0$, $t - t_3 - t_2 = 0$ and $t + T_2 - t_3 = 0$. This can be satisfied by $t_3 = t + T_2$, $t_2 = -T_2$ and $t_1 = T_2 + T_1$, and t_2 has to be negative. For negative t_2, however, the response functions are zero. Similar conclusions will be reached for all other integrals.

Switching the order of pulses into $(1 - 3 - 2)$ will yield the rephasing signal as well. We can, therefore, conclude that for both orderings $(1-2-3)$ and $(1-3-2)$ only the rephasing Liouville pathways contribute to the signal. This has an important implication in the description of the experiment. If we measure a signal emerging into the phase-matched direction $\boldsymbol{k}_s = -\boldsymbol{k}_1 + \boldsymbol{k}_2 + \boldsymbol{k}_3$, and if the pulse with the wavevector $\boldsymbol{k}_1$ is guaranteed to arrive at the sample first, the response function responsible for the signal will be always of the rephasing type.

Let us now switch the order of the first two pulses to yield ordering $(2 - 1 - 3)$. Following the same arguments the reader can check that this ordering yields the nonrephasing contribution. The very same conclusion is reached for the pulse order $(3-1-2)$. This means that in the direction $\boldsymbol{k}_s = -\boldsymbol{k}_1+\boldsymbol{k}_2+\boldsymbol{k}_3 \equiv \boldsymbol{k}_2-\boldsymbol{k}_1+\boldsymbol{k}_3$, if the pulse $\boldsymbol{k}_1$ is guarantied to arrive as second, the corresponding signal is always of the nonrephasing type. For a given direction we therefore conclude that it is possible to selectively probe the rephasing or nonrephasing group of Liouville pathways.

In an experimental arrangement when the pulse $\boldsymbol{k}_1$ precedes $\boldsymbol{k}_2$ we measure the response S_{R} in the direction $-\boldsymbol{k}_1 + \boldsymbol{k}_2 + \boldsymbol{k}_3$. One can of course redefine the indices of the directions as well. If we denote the original direction $\boldsymbol{k}_1$ by $\boldsymbol{k}'_2$ and vice versa ($\boldsymbol{k}'_1 = \boldsymbol{k}_2, \boldsymbol{k}'_2 = \boldsymbol{k}_1, \boldsymbol{k}'_3 = \boldsymbol{k}_3$), and keep the order of pulses in time intact, we realize that in the direction $-\boldsymbol{k}'_1 + \boldsymbol{k}'_2 + \boldsymbol{k}'_3 = \boldsymbol{k}_1 - \boldsymbol{k}_2 + \boldsymbol{k}_3$ we measure the non-rephasing response S_{NR}.

The third pathway S_{DC} contributes to the signal only when the order of the pulses is $(2-3-1)$ or $(3-2-1)$. The S_{DC} pathways contain a fast oscillating term canceling the 2Ω term in the field factor which appears in the orderings $(2 - 3 - 1)$ and $(3-2-1)$. Although the oscillations during the population interval t_2 are very fast, they also survive RWA. This signal is emitted into $-\boldsymbol{k}_1 + \boldsymbol{k}_2 + \boldsymbol{k}_3$ only if the last of the interacting pulses is $\boldsymbol{k}_1$. It is also a basis of certain spectroscopic techniques aiming at exclusively studying the properties of the higher lying excited states.

Summarizing these conclusions we thus find that we can request the strict pulse ordering such that pulse 1 is always first, pulse 2 is always second, and pulse 3 is always third. Then the rephasing signal will be measured in the phase matching direction $k_{\mathrm{I}} = -\boldsymbol{k}_1 + \boldsymbol{k}_2 + \boldsymbol{k}_3$, the nonrephasing signal will be measured in $k_{\mathrm{II}} = \boldsymbol{k}_1-\boldsymbol{k}_2+\boldsymbol{k}_3$ and the double quantum coherence will be measured in $k_{\mathrm{III}} = \boldsymbol{k}_1+\boldsymbol{k}_2-\boldsymbol{k}_3$. These are three independent types of measurements in the FWM experiment and all other FWM experiments can be given in terms of these three techniques. The natural variables of the measurement are the delay times between the pulses. The induced polarization amplitude is given by the corresponding contribution to the response function and the variables of the response function t_1, t_2, t_3 become equivalent to the delay times between the ordered pulses.

15.2
Multilevel System in Contact with the Bath

Now that we have completed the discussion of the relation between the response function, excitation fields and the generated induced polarization (the signal in the heterodyne detection scheme), we can turn our attention to the effects of bath degrees of freedom and the measurement of relaxation phenomena. The formulae for the response function contain evolution superoperators which we could easily represent in the basis of electronic states. However, the evolution superoperator matrix elements would in this case still be operators on the Hilbert space of the bath degrees of freedom. Only after tracing the response functions over the bath, we obtain the observable response function.

First, we will concentrate on the response in the so-called adiabatic limit, that is, when the bath induces fluctuations of electronic energies [100]. In this case we have only the pure dephasing effect and the bath-induced transitions between electronic states are forbidden. For the evolution superoperators we then have either

$$\tilde{\mathcal{U}}^{(ee)}_{klmn}(t_2) = \delta_{kl}\delta_{nm}\delta_{km}\tilde{\mathcal{U}}^{(ee)}_{kkkk}(t_2) , \tag{15.52}$$

or

$$\tilde{\mathcal{U}}^{(ee)}_{klmn}(t_2) = \delta_{km}\delta_{ln}(1-\delta_{kl})\tilde{\mathcal{U}}^{(ee)}_{klkl}(t_2) . \tag{15.53}$$

The former elements describe the evolution of populations, while the latter describe the evolution of coherence. One would be tempted to conclude that

$$\tilde{\mathcal{U}}^{(ee)}_{kkkk}(t_2) = 1 , \tag{15.54}$$

because we expect the populations to remain constant in this limit. However, one should not forget that $\tilde{\mathcal{U}}^{(ee)}_{kkkk}(t_2)$ is an operator describing the evolution of the bath degrees of freedom. The bath can undergo reorganization after the system was excited, without the population of the excited state changing due to relaxation. Assuming also that

$$\tilde{\mathcal{U}}^{(eg)}_{kglg}(t) = \delta_{kl}\tilde{\mathcal{U}}^{(eg)}_{kgkg}(t) \tag{15.55}$$

for the evolution of the optical coherence elements, we find that the total response of the multilevel system in the adiabatic limit splits into two contributions, where one corresponds to the evolution of populations in the t_2 interval, (15.52), and the other to the evolution of coherences, (15.53). For instance, for the pathway R_{2g} we thus get:

$$R_{2g}(t_3, t_2, t_1) = R^{(\text{pop})}_{2g}(t_3, t_2, t_1) + R^{(\text{coh})}_{2g}(t_3, t_2, t_1) , \tag{15.56}$$

where

$$R_{2g}^{(\mathrm{pop})}(t_3,t_2,t_1) = \mathrm{i}^3 \sum_k \mu_{gk}\mu_{kg}\mu_{gk}\mu_{kg} \times \mathrm{Tr}\left\{\tilde{\mathcal{U}}^{(eg)}_{kgkg}(t_3)\tilde{\mathcal{U}}^{(ee)}_{kkkk}(t_2)\tilde{\mathcal{U}}^{(ge)}_{gkgk}(t_1)\,W_{eq}\right\} \mathrm{e}^{-\mathrm{i}\omega_{kg}(t_3-t_1)}\,, \tag{15.57}$$

and

$$R_{2g}^{(\mathrm{coh})}(t_3,t_2,t_1) = \mathrm{i}^3 \sum_{k\neq l} \mu_{gl}\mu_{gk}\mu_{lg}\mu_{kg} \times \mathrm{Tr}\left\{\tilde{\mathcal{U}}^{(eg)}_{lglg}(t_3)\tilde{\mathcal{U}}^{(ee)}_{lklk}(t_2)\tilde{\mathcal{U}}^{(ge)}_{gkgk}(t_1)\,W_{eq}\right\} \mathrm{e}^{-\mathrm{i}\omega_{lg}t_3+\mathrm{i}\omega_{kg}t_1}\,. \tag{15.58}$$

This distinction will become very significant when considering separately energy and phase relaxation.

15.2.1
Energy Fluctuations of the General Multilevel System

Let the system be given by a set of the energy levels: $|1\rangle, |2\rangle, \ldots, |N\rangle$. The system Hamiltonian is thus

$$\hat{H}_{\mathrm{mol}} = \sum_{k=1}^{N} \varepsilon_k |k\rangle\langle k|\,, \tag{15.59}$$

and we assume the bath Hamiltonian H_{B} to represent a bath of harmonic oscillators, which induce fluctuations of system energies. The system–bath interaction in general is given by the same linear coupling energy gap operators as in (14.11), and we define the fluctuations of all states with respect to one specific state – the ground state as the reference

$$\Delta\hat{V}_{kk}(t) = U_g^{\dagger}(t)\left(\hat{H}_k - \hat{H}_g\right)U_g(t) - \hbar\omega_{kg}\,, \tag{15.60}$$

$$\hat{H}_{\mathrm{SB}} = \sum_k \Delta V_{kk}|k\rangle\langle k|\,. \tag{15.61}$$

We additionally take into account that the bath is in equilibrium with respect to the electronic ground state. The ground state energy thus becomes constant. Since different states are not coupled to each other, the evolution of the system in a given state is adiabatic, i.e. the potential surface of different states do not intersect. It turns out that the optical response function for this system can be exactly calculated.

The system third order response functions $R_1(t_3,t_2,t_1), \ldots, R_4(t_3,t_2,t_1)$ (13.24)–(13.27) can be written as four-point dipole correlation functions. They contain four interactions at specific times, and consequently three delays between interaction times. We can express them using a single function

$$R_1(t_3, t_2, t_1) = C^{(4)}(t_1, t_1 + t_2, t_1 + t_2 + t_3, 0) , \quad (15.62)$$

$$R_2(t_3, t_2, t_1) = C^{(4)}(0, t_1 + t_2, t_1 + t_2 + t_3, t_1) , \quad (15.63)$$

$$R_3(t_3, t_2, t_1) = C^{(4)}(0, t_1, t_1 + t_2 + t_3, t_1 + t_2) , \quad (15.64)$$

$$R_4(t_3, t_2, t_1) = C^{(4)}(t_1 + t_2 + t_3, t_1 + t_2, t_1, 0) , \quad (15.65)$$

where the function

$$C^{(4)}(\tau_4, \tau_3, \tau_2, \tau_1) = \mathrm{Tr}\left(\hat{\mu}(\tau_4)\hat{\mu}(\tau_3)\hat{\mu}(\tau_2)\hat{\mu}(\tau_1)\right) \quad (15.66)$$

is the four-point correlation function. This case is described in detail in Appendix A.8. For the dipole operator

$$\hat{\mu} = \sum_{ab}^{a \neq b} \mu_{ab} |a\rangle\langle b| \quad (15.67)$$

and for an isolated ground state $|g\rangle$ we obtain the general cumulant expression for the four-point correlation function

$$\begin{aligned} C^{(4)}(\tau_4, \tau_3, \tau_2, \tau_1) = & \sum_{bcd} \mu_{gd}\mu_{dc}\mu_{cb}\mu_{bg} \\ & \times \exp[-i\varepsilon_d \tau_{43} - i\varepsilon_c \tau_{32} - i\varepsilon_b \tau_{21} \\ & + f_{dcbg}(\tau_4, \tau_3, \tau_2, \tau_1)] , \end{aligned} \quad (15.68)$$

where $\tau_{ij} = \tau_i - \tau_j$ and

$$\begin{aligned} f_{dcbg}(\tau_4, \tau_3, \tau_2, \tau_1) = & -g_{dd}(\tau_{43}) - g_{cc}(\tau_{32}) - g_{bb}(\tau_{21}) \\ & + g_{dc}(\tau_{32}) + g_{dc}(\tau_{43}) - g_{dc}(\tau_{42}) \\ & - g_{db}(\tau_{32}) + g_{db}(\tau_{31}) + g_{db}(\tau_{42}) - g_{db}(\tau_{41}) \\ & + g_{cb}(\tau_{21}) + g_{cb}(\tau_{32}) - g_{cb}(\tau_{31}) . \end{aligned} \quad (15.69)$$

Notice that the ground state in this expression has zero energy, it is not fluctuating and all g-functions including the ground state vanish. For the three-band system the summation indices run over g, e, f bands. Compared to the lineshape functions of Section 11.5 we now keep only two indices for the lineshape function. So $g_{aa,bb} \equiv g_{ab}$.

We can now associate all distinct Liouville space pathways to the function $C^{(4)}$ by adjusting its time variables and various contributions can be calculated for the given lineshape function $g(t)$.

15.2.2 Off-Diagonal Fluctuations and Energy Relaxation

Consider now additional off-diagonal fluctuations. Due to off diagonal elements of the Hamiltonian the evolution in time mixes different states. Hence, the system dynamics now includes population transport, and the system cannot be described using only the adiabatic evolution operators. We have two types of dynamics now: the coherence evolution and the population transport. From the beginning we make two approximations: First is that the off diagonal fluctuations are independent of each other. Second, we adopt the secular approximation. In this case for the off diagonal elements of the density matrix – the coherences – the off-diagonal fluctuations induce only lifetime-induced dephasing as in Section 14.5. We can, therefore, continue using the cumulant expressions for the evolution of density matrix coherences. They will only be amended by additional dephasing terms originating from the off-diagonal fluctuations. For the diagonal terms of the density matrix – populations – the off-diagonal fluctuations induce population transport which can be described using the Pauli master equation, with rates calculated by some of the methods introduced in Chapter 11.

Consider the Liouville space pathway (or the Feynman diagram) when starting from the ground state $|g\rangle\langle g|$ the system first interacts on the left side and the electronic inter-band coherence $|e_i\rangle\langle g|$ is created and it propagates during t_1. As we are interested in population, let the second interaction happen on the right creating electronic population $|e_i\rangle\langle e_i|$. As the population transfer is now allowed, in the second interval t_2 this population can be transferred to the state $|e_f\rangle\langle e_f|$ with the probability $G_{e_f e_i}(t_2)$. We can assume the next transition to occur on either side of the density matrix. Hence, we can set the resulting inter-band coherence to be of the general form $|b\rangle\langle c|$, which can be later assigned to one of $|e\rangle\langle g|$, $|g\rangle\langle e|$, $|f\rangle\langle e|$, or $|e\rangle\langle f|$. The contribution to the response function of such population-transport pathway is then given by [101]

$$T(t_3, t_2, t_1) = -(\mathrm{i})^3 \sum_{cbe'e} \mu_{cb}\mu_{\nu\nu'}\mu_{e_i g}\mu_{e_i g} G_{e_f e_i}(t_2) \mathcal{F}^{(\mathrm{I})}_{cbe_f e_i}(t_3, t_2, t_1) , \quad (15.70)$$

where $\nu\nu'$ has to be changed to $e_f b$ when the signal is generated on the left side of the diagram, and to ce_f, when it is generated on the right. $G_{e_f e_i}(t_2)$ when $e_f = e_i$ is the population survival probability. The phase function is determined by the cumulant expansion with respect to the diagonal fluctuations:

$$\mathcal{F}^{(\mathrm{I})}_{cbe_f e_i}(t_3, t_2, t_1) = \exp\Big[\mathrm{i}\omega_{cb}t_3 - \mathrm{i}\omega_{e_i g}t_1 - (\gamma_c + \gamma_b)t_3 - \gamma_{e_i}t_1 + f^{(\mathrm{I})}_{cbe_i}(t_3, t_2, t_1)\Big] , \quad (15.71)$$

where

$$\begin{aligned} f^{(\mathrm{I})}_{cbe_i}(t_3, t_2, t_1) = &-g_{e_i e_i}(t_1) - g_{bb}(t_3) - g^*_{cc}(t_3) \\ &- g_{be_i}(t_1 + t_2 + t_3) + g_{be_i}(t_1 + t_2) + g_{be_i}(t_2 + t_3) \\ &+ g_{ce_i}(t_1 + t_2 + t_3) - g_{ce_i}(t_1 + t_2) - g_{ce_i}(t_2 + t_3) \\ &+ g_{cb}(t_3) + g^*_{bc}(t_3) + g_{ce_i}(t_2) - g_{be_i}(t_2) \\ &+ 2\mathrm{i}\,\mathrm{Im}\,[g_{ce_f}(t_2 + t_3) - g_{ce_f}(t_2) - g_{ce_f}(t_3) \\ &- g_{be_f}(t_2 + t_3) + g_{be_f}(t_2) + g_{be_f}(t_3)] \,. \end{aligned} \tag{15.72}$$

γ_ν is the dephasing constant due to the state lifetime

$$\gamma_\nu = \frac{|K_{\nu\nu}|}{2} \,. \tag{15.73}$$

The population Green's function is a solution of the Pauli master equation

$$\dot{G}_{e'e}(t) = \sum_{j \neq e'} K_{e'j} G_{je} - \left(\sum_{j \neq e'} K_{je'} \right) G_{e'e} \,, \tag{15.74}$$

where K_{ij} are the population transport rates. This equation can be represented in the matrix form

$$\frac{\partial}{\partial t} \hat{G}(t) = -\hat{\tilde{K}} \hat{G}(t) \,, \tag{15.75}$$

where the population transport rate matrix is constructed as: $\tilde{K}_{ab} = -K_{ab} + \delta_{ab} \sum_j K_{jb}$.

Combining the adiabatic model with the transport expressions thus allows a complete approximate description of the dynamic relaxation in optical measurements. The final expressions for the three – rephasing, non rephasing and double quantum coherence – the response functions using this approach are given in Appendix A.10.

15.2.3
Fluctuations in a Coupled Multichromophore System

In previous subsections we have described how diagonal and off-diagonal fluctuations can be incorporated into the response functions. In this subsection we will consider a general many-body system consisting of N resonantly interacting two-level systems representing, for example, aggregates as discussed in Chapter 5. Nonlinear optical properties of such complexes of coupled chromophores (for example, molecular aggregates, proteins and so on) are described using a Frenkel exciton model

$$\hat{H}_\mathrm{S} = \sum_{m=1}^{N} \varepsilon_m \hat{B}^\dagger_m \hat{B}_m + \sum_{mn}^{N} J_{nm} \hat{B}^\dagger_m \hat{B}_n \,, \tag{15.76}$$

where $\hat{B}_m^\dagger$ and $\hat{B}_m$ are excitation creation and annihilation operators on a molecule m, ε_m and J_{mn} are site energy of the mth chromophore and resonant coupling between nth and mth chromophores, respectively. In the exciton representation we obtain the multilevel system as follows. It contains the single ground state $|g\rangle$ and one- and two-exciton bands (manifolds). In the one-exciton manifold each state is denoted as $|e_j\rangle$ ($j = 1, \ldots, N$). In the double-exciton manifold the number of two-exciton states is $N(N-1)/2$ and they are denoted as $|f_k\rangle$.

The difference of this system from the previous subsection is that now the system Hamiltonian is essentially a nondiagonal matrix. Thus, for an arbitrary type of coupling with the bath, the dynamics in principle is not adiabatic and processes such as population transport and coherence decays are necessarily important.

The one-exciton Hamiltonian matrix is $h_{jk}^{(1)} = \delta_{jk}\varepsilon_j + \zeta_{jk} J_{jk}$, where $\zeta_{jk} = 1 - \delta_{jk}$ describes the single excitation band. The two-exciton Hamiltonian block is $h_{(kl),(mn)}^{(2)} \equiv (\varepsilon_k + \varepsilon_l)\delta_{km}\delta_{ln} + J_{km}\delta_{ln}\zeta_{km} + J_{ln}\delta_{km}\zeta_{ln}$ where $k > l$ and $m > n$ and the K couplings of the aggregate (see Chapter 5) have been neglected for simplicity. Transition from the molecular excitation representation to the exciton (eigenstate) basis is obtained using a unitary transformations

$$\sum_{mn} c_{em} c_{en} h_{mn}^{(1)} = E_e \,, \tag{15.77}$$

$$\sum_{mnkl} C_{f,mn} C_{f,kl} h_{mn,kl}^{(2)} = E_f \,. \tag{15.78}$$

The interaction with the optical field is described by the following dipole operator

$$\hat{\mu} = \sum_m \boldsymbol{d}_m \left(\hat{B}_m^\dagger + \hat{B}_m^\dagger \right) , \tag{15.79}$$

where $\boldsymbol{d}_m$ is a molecular transition dipole. Applying the unitary transformation the dipoles of inter-band transitions (eigen-dipoles) are obtained:

$$\begin{aligned} \boldsymbol{\mu}_{ge} &= \sum_m c_{em} \boldsymbol{d}_m \\ \boldsymbol{\mu}_{ef} &= \sum_{mn} C_{f,mn} (c_{en} \boldsymbol{d}_m + c_{em} \boldsymbol{d}_n) \,. \end{aligned} \tag{15.80}$$

Here we extend indices of C so that $C_{f,mn} = C_{f,nm}$ and $C_{f,mm} = 0$.

We next assume the diagonal molecular transition energy fluctuations of the form:

$$\hat{H}_{\mathrm{SB}} = \sum_{m\alpha} z_{m\alpha} \hat{q}_\alpha \hat{B}_m^\dagger \hat{B}_m \,, \tag{15.81}$$

also, each molecule has its own independent set of fluctuating coordinates uncorrelated to the other molecules, that is the correlation function matrix is diagonal, $\sum_\alpha z_{m\alpha} z_{n\alpha} = 0$ and only one correlation function is then relevant

$$\sum_\alpha |z_{m\alpha}|^2 \langle \hat{q}_\alpha(t) \hat{q}_\alpha(0) \rangle = C(t) \,. \tag{15.82}$$

In the previous subsection we were able to define correlation functions and lineshape functions for all pairs of energy levels. Now we define the fluctuations of each site: these fluctuations translate into the eigenbasis through transformations of the type (15.77) and (15.78). It is convenient to use the spectral density (see Section 8.6), which can be given by:

$$\mathcal{C}''(\omega) = \frac{1}{2}\int_0^\infty \mathrm{d}t \exp(\mathrm{i}\omega t)(\mathcal{C}(t) - \mathcal{C}(-t)) . \tag{15.83}$$

In the exciton basis, we obtain fluctuating transition energies and couplings between the eigenstates. These fluctuations are characterized by spectral densities

$$\begin{aligned}
C''_{e_1e_2,e_3e_4}(\omega) &= \left(\sum_m c_{e_1m}c_{e_2m}c_{e_3m}c_{e_4m}\right)\mathcal{C}''(\omega) ,\\
C''_{e_1e_2,f_3f_4}(\omega) &= \left(\sum_m c_{e_1m}c_{e_2m}\sum_k^{k\neq m} C_{f_3,mk}C_{f_4,mk}\right)\mathcal{C}''(\omega) ,\\
C''_{f_1f_2,e_3e_4}(\omega) &= C''_{e_3e_4,f_1f_2}(\omega) ,\\
C''_{f_1f_2,f_3f_4}(\omega) &= \left[\sum_m\left(\sum_k^{k\neq m} C_{f_1,mk}C_{f_2,mk}\right)\left(\sum_l^{l\neq m} C_{f_3,ml}C_{f_4,ml}\right)\right]\mathcal{C}''(\omega) .
\end{aligned} \tag{15.84}$$

Given the correlation functions for fluctuations we are now able to calculate all lineshape functions and population relaxation rates. This information is sufficient to use the expressions of the previous subsection of the lineshape functions and the linear and third order response functions can now be calculated. However, it should be noted that all conditions of (15.68) and (15.71) are not satisfied by the excitonic model: the diagonal and off diagonal fluctuations of excitons are essentially correlated as they originate from the same molecular fluctuations. Hence the response function expressions of Appendix A.10 should be used with caution.

15.2.4
Inter-Band Fluctuations: Relaxation to the Electronic Ground State

So far, the only transitions between the electronic ground state and the first electronic excited states we considered were the transitions stimulated by the radiation (absorption and stimulated emission). To account for nonradiative relaxation of the system to the ground state, or for spontaneous emission, we have to extend the range of processes that our response functions describe. In particular we have to include pathways in which the system is transferred to the excited state by the first two interactions with the incident fields, but relaxes to the ground state before the third interaction when the second interval is relatively long. The relaxation processes ensure that the induced polarization to third order and all related signals will always vanish for very long time $t_2 \to \infty$.

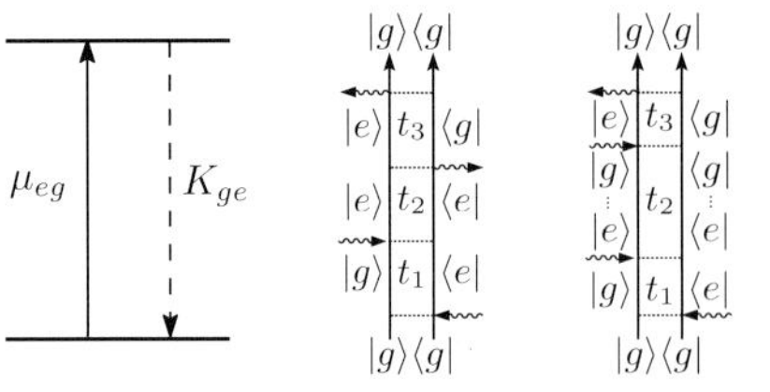

Figure 15.6 Two-level system with relaxation. The two Feynman diagrams correspond to the part of the signal from unrelaxed molecules and the molecules that have undergone relaxation to the electronic ground state.

If the relaxation process can be described by the one-way relaxation rate K_{ge} (see Figure 15.6), then every Liouville pathway which passes through the excited state in the population interval t_2 (say R_{2g}) splits into two contributions. First, with the probability $e^{-K_{ge}t_2}$ the system remains excited, and its contribution to the signal is the same as in nonrelaxing system. The corresponding Feynman diagram is presented in Figure 15.6. Second, the system has relaxed to the ground state some time during the population interval with the probability $(1-e^{-K_{ge}t})$. The probability that it has happened in the time interval $(t_2', t_2' + dt)$, $0 < t_2' < t_2$ is $K_{ge}dt$ and all such contributions have to be integrated.

The Feynman diagram of the relaxed contribution is depicted in Figure 15.6. The R_{2g} pathway, (15.14), for a two-level system can be simplified to

$$R_{2g}(t_3, t_2, t_1) = i^3|d_{eg}|^4 e^{i\Omega(t_1-t_3)}\text{Tr}\left\{\tilde{\mathcal{U}}^{(eg)}(t_3)\tilde{\mathcal{U}}^{(ee)}(t_2)\tilde{\mathcal{U}}^{(ge)}(t_1)\,W_{eq}\right\} . \quad (15.85)$$

In the presence of the relaxation, the evolution superoperator element reads

$$\tilde{\mathcal{U}}^{(ee)}(t_2)|_{\text{relax}} = e^{-K_{ge}t}\tilde{\mathcal{U}}^{(ee)}(t_2) , \quad (15.86)$$

and correspondingly the Liouville space pathway's amplitude decays with the relaxation rate as

$$R_{2g}(t_3, t_2, t_1)|_{\text{relax}} = e^{-K_{ge}t_2} R_{2g}(t_3, t_2, t_1) . \quad (15.87)$$

Analogously, we write the response function for the relaxation diagram by identifying the evolution superoperator term $\tilde{\mathcal{K}}^{(ge)}(t_2)$ which transfers the population from the excited state to the ground state. As discussed above, we have to sum contributions corresponding to the system relaxation at all times t_2' between 0 and t_2. We assume that the system propagated from 0 to t_2' in the excited state $|e\rangle$ and then is transferred to $|g\rangle$ and propagated further in the ground state. This can be expressed as

$$\tilde{\mathcal{K}}^{(ge)}(t_2) = K_{ge}\int_0^{t_2} dt_2'\tilde{\mathcal{U}}^{(gg)}(t - t_2')\tilde{\mathcal{U}}^{(ee)}(t_2') . \quad (15.88)$$

In order to obtain a nonzero (rephasing) contribution the interactions with electric field after the relaxation have to be such as depicted in Figure 15.6. This corresponds not to the R_2 diagram, but to a complex conjugation of the R_1 diagram as can be easily verified in Figure 15.2.

$$R^*_{1g\leftarrow e}(t_3, t_2, t_1) = \mathrm{i}^3|d_{eg}|^4 \mathrm{Tr}\left\{\tilde{\mathcal{U}}^{(eg)}(t_3)\tilde{\mathcal{K}}^{(ge)}(t_2)\tilde{\mathcal{U}}^{(ge)}(t_1)\, W_{eq}\right\}$$
$$= \mathrm{i}^3|d_{eg}|^4 K_{ge} \int_0^{t_2} \mathrm{d}t_2'$$
$$\times \mathrm{Tr}\left\{\tilde{\mathcal{U}}^{(eg)}(t_3)\tilde{\mathcal{U}}^{(gg)}(t - t_2')\tilde{\mathcal{U}}^{(ee)}(t_2')\tilde{\mathcal{U}}^{(ge)}(t_1)\, W_{eq}\right\} \quad (15.89)$$

Because there is only one interaction from the right side of the diagram the diagram contributes a minus sign, and the total third order response is

$$S^{(3)}(t_3, t_2, t_1) = R_{3g}(t_3, t_2, t_1) + \mathrm{e}^{-K_{ge} t_2} R_{2g}(t_3, t_2, t_1) - R^*_{1g\leftarrow e}(t_3, t_2, t_1)\,. \quad (15.90)$$

Applying the cumulant expansion to (15.89) results in

$$R^*_{1g\leftarrow e}(t_3, t_2, t_1) = K_{ge} R_{3g}(t_3, t_2, t_1)\mathrm{e}^{-g_{ee}(t_2+t_3)+g_{ee}(t_2)+g^*_{ee}(t_2+t_3)-g^*_{ee}(t_2)}$$
$$\times \int_0^{t_2} \mathrm{d}t_2' \mathrm{e}^{-K_{ge} t_2'} \mathrm{e}^{g_{ee}(t_2+t_3-t_2')-g_{ee}(t_2-t_2')-g^*_{ee}(t_2+t_3-t_2')+g^*_{ee}(t_2-t_2')}$$
$$(15.91)$$

Instead of going into further details, we can now prove that at $t_2 \to \infty$ the total signal $S^{(3)}$ goes to zero. At long times, the imaginary part of the line broadening function corresponding to thermodynamic bath behaves linearly and thus

$$g_{ee}\left(t_2 + t_3 - t_2'\right) - g_{ee}\left(t_2 - t_2'\right) - g^*_{ee}\left(t_2 + t_3 - t_2'\right) + g^*_{ee}\left(t_2 - t_2'\right) =$$
$$2\mathrm{i}\,\mathrm{Im}\left\{g_{ee}\left(t_2 + t_3 - t_2'\right) - g_{ee}\left(t_2 - t_2'\right)\right\} \approx 2\mathrm{i}\,\mathrm{Im}\, g_{ee}(t_3) = \mathrm{i}\alpha t_3\,,$$
$$(15.92)$$

where α is some real constant. Equation (15.91) thus yields

$$R^*_{1g\leftarrow e}(t_3, t_2, t_1) \approx -K_{ge} R_{3g}(t_3, t_2, t_1)\frac{\mathrm{e}^{-K_{ge} t_2} - 1}{K_{ge}} \underset{t_2\to\infty}{=} R_{3g}(t_3, t_2, t_1)\,,$$
$$(15.93)$$

and correspondingly the total response $S^{(3)}$, (15.90), converges to zero at large t_2 despite the fact that the R_{3g} contribution remains unaffected by the relaxation.

A similar procedure can be used for derivation of the response function of a system with the energy transfer between the donor and the acceptor. The approach is limited to problems with the one-way relaxation or energy transfer in which (15.88) holds. In the case of two states separated by energy comparable to $k_B T$, that is with the significant back transfer rate K_{eg}, one cannot write $\tilde{\mathcal{K}}^{(ge)}$ in a simple form, and thus the treatment of electronic energy relaxation processes within the response function formalism beyond Markov approximation is limited to several special cases.

15.2.5 Energetic Disorder in Four-Wave Mixing

The Liouville pathways represent responses of certain combinations of energy levels connected through optical transitions. In the above sections we have always assumed that we work with a single representative molecular system. In the real experiment, the cuvette usually contains a solution of systems of interest. We thus have to deal with an ensemble of systems. The consequence of this setup for the linear response was described in Section 14.6.

Let us again assume that we have the ensemble of identical copies of the systems but their energy levels are scattered randomly around the mean values. These static, or more generally, slow, fluctuations are then independent of the fast fluctuating modes that cause the homogeneous broadening effects. Because of the independence of the two types of fluctuations, as the lowest-level approach, we can assume that the response functions could be written as product of the slow and fast parts

$$R_n(t_3, t_2, t_1) = R_n^{(\text{slow})}(t_3, t_2, t_1) R_n^{(\text{fast})}(t_3, t_2, t_1) , \tag{15.94}$$

and the fast and slow response functions differ only in their corresponding sets of lineshape functions.

Consider a two-level system with one ground and one excited state. The slow lineshape function as described in Section 14.6 is a parabolic function of time

$$g_{ee}^{(\text{slow})}(t) = \frac{1}{2}\Delta^2 t^2 . \tag{15.95}$$

We then find that the slow part of the response is the same for all rephasing pathways

$$R_{2g}^{(\text{slow})}(t_3, t_2, t_1) = R_{3g}^{(\text{slow})}(t_3, t_2, t_1) = I_{\text{inh}}(t_3 - t_1) = \mathrm{e}^{-\frac{\Delta^2}{2}(t_3 - t_1)^2} , \tag{15.96}$$

and all nonrephasing pathways

$$R_{1g}^{(\text{slow})}(t_3, t_2, t_1) = R_{4g}^{(\text{slow})}(t_3, t_2, t_1) = I_{\text{inh}}(t_3 + t_1) = \mathrm{e}^{-\frac{\Delta^2}{2}(t_3 + t_1)^2} . \tag{15.97}$$

The inhomogeneity factor $I_{\text{inh}}(t)$ thus enters as a Gaussian function that is convoluted with the homogeneous response function. This property comes due to delicate interference of wavefunction phase rotations in the t_1 and t_3 intervals, where in the rephasing pathways the phases rotate in opposite directions while they continue rotation in the same direction in the nonrephasing pathway. This property becomes very important in Section 15.3.1.

We have just derived the famous *photon echo effect*. The rephasing nonlinear response is equal to its fast (homogeneous) part at $t_1 = t_3$, whereas the nonrephasing response decays with both t_1 and t_3 times. Thus, if we design an experiment that probes only the rephasing part of the FWM measurement and we keep $t_1 = t_3$ we can probe only homogeneous part of the response. This is never possible in the linear response. Let us now discuss the photon echo (PE) effect and its observation in more detail. We will relax the requirement $t_1 = t_3$ and take $t_2 = 0$. The PE measurement is as follows: first we excite by a short laser pulse; we then wait for time t_1 and send in the other two pulses (from different directions) at the same time. This generates the nonlinear polarization, and we record the rephasing polarization as a function of t_3. As $t_3 \ll t_1$, the intensity is proportional to $\exp(-(\Delta^2/2)t_1^2)$, which considerably. However, as t_3 increases the intensity grows, peaks when $t_1 = t_3$ according to (15.96). This peak is termed optical echo. The signal intensity as a function of t_1 at $t_3 = t_1$ will recover from the decay caused by the inhomogeneous effects according to (15.94).

The same type of echo phenomenon occurs also for a three-level system where we must assume that the static fluctuations of the f state are equal to $\times 2$ the fluctuation of state e. This property can be satisfied for a weakly anharmonic oscillator described by the Hamiltoanian

$$\hat{H} = (\hbar\omega_0 + \delta)\,\hat{a}^\dagger\hat{a} + d\hat{a}^\dagger\hat{a}^\dagger\hat{a}\hat{a}\,, \tag{15.98}$$

where d is the energy anharmonicity and δ is the static fluctuation. Additionally, for this type of system we get the double quantum coherence contributions to the full response function. These R_{3f} pathways are convoluted with

$$R_{3f}^{(\mathrm{slow})}(t_3, t_2, t_1) = R_{4f}^{(\mathrm{slow})}(t_3, t_2, t_1) = I_{\mathrm{inh}}(t_3 + 2t_2 + t_1) = \mathrm{e}^{-\frac{\Delta^2}{2}(t_3+2t_2+t_1)^2}\,. \tag{15.99}$$

However, for more complicated systems the echo effect becomes challenging to observe. It still holds for individual Liouville space pathways, but when we have many states in the e and f bands the delicate interference of phase rotations can be masked after adding up all relevant Liouville space pathways.

15.2.6
Random Orientations of Molecules

The amplitude of an optical transition depends on the *projection* of the molecular dipole moment onto the polarization vector of the excitation optical field. In multi-state systems a single response function can include dipole moments of different molecules that may have various orientations. It is clear that the projections will depend on the orientation of the molecule in the laboratory frame of reference. It is not easy to align molecules in solution, and unless we have some very special experiment setup, we can expect them to be oriented isotropically in all directions. Our task in this section is to average the molecular response over these orientations. It means that we cannot ignore the tensor character of the response function any more. Actually, we can use it to our advantage as we do in Section 16.5

We have labeled the interaction order in the response functions in (15.2)–(15.5). This interaction order now becomes important when the acting optical fields have distinct electric field polarizations. Let us denote the polarization vectors of the incoming fields and the emitted field (in four-wave mixing) by $\boldsymbol{e}_1, \boldsymbol{e}_2, \boldsymbol{e}_3$ and $\boldsymbol{e}_4$. Let us also denote $\boldsymbol{n}_1, \boldsymbol{n}_2, \boldsymbol{n}_3$ and $\boldsymbol{n}_4$ the normalized orientations of the dipole moments in the third order response function of a single Liouville space pathway. For a given arrangement of the excitation and detection polarizations, and for a given orientation of the molecular aggregate under consideration, the contribution of the Liouville space pathway in FWM experiment will have in general the following prefactor $(\boldsymbol{n}_1 \cdot \boldsymbol{e}_1)(\boldsymbol{n}_2 \cdot \boldsymbol{e}_2)(\boldsymbol{n}_3 \cdot \boldsymbol{e}_3)(\boldsymbol{n}_4 \cdot \boldsymbol{e}_4)$. Let us denote averaging over the isotropic distribution of orientations as $\langle \ldots \rangle_O$. The averaged amplitude of the contribution is thus

$$\langle (\boldsymbol{n}_4 \cdot \boldsymbol{e}_4)(\boldsymbol{n}_3 \cdot \boldsymbol{e}_3)(\boldsymbol{n}_2 \cdot \boldsymbol{e}_1)(\boldsymbol{n}_1 \cdot \boldsymbol{e}_1) \rangle_O \, . \tag{15.100}$$

Because orientations of the four dipole moments in the response can be in principle all different, individual Liouville pathways will have different prefactors. Assuming that molecular rotation due to Brownian motion in solution occurs on a timescale much slower than the time scale of the nonlinear experiment, this type of averaging can be treated independently of the time dependence of the response functions.

The orientational average of dipoles can be calculated by applying Euler transformation expressions. Let us take the molecular frame where the molecular transition dipoles are defined. The optical fields are defined in the laboratory frame. We can denote the transformation operation from the molecular frame to the lab frame as

$$\boldsymbol{d}^{(\mathrm{lab})} = \boldsymbol{T}\boldsymbol{d}^{(\mathrm{mol})} \, , \tag{15.101}$$

where the matrix $\boldsymbol{T}$ can be given in terms of three rotations. In matrix form we have

$$\boldsymbol{T} = \begin{pmatrix} \cos(\psi) & -\sin(\psi) & 0 \\ \sin(\psi) & \cos(\psi) & 0 \\ 0 & 0 & 1 \end{pmatrix} \begin{pmatrix} \cos(\theta) & 0 & -\sin(\theta) \\ 0 & 1 & 0 \\ \sin(\theta) & 0 & \cos(\theta) \end{pmatrix} \times \begin{pmatrix} 1 & 0 & 0 \\ 0 & \cos(\phi) & -\sin(\phi) \\ 0 & \sin(\phi) & \cos(\phi) \end{pmatrix} , \tag{15.102}$$

where (ψ, θ, ϕ) are the Euler rotation angles ranging in the intervals ϕ and ψ – $(0, 2\pi)$, θ – $(0, \pi)$. In terms of the vector components we can write

$$\boldsymbol{d}_\nu^{(\mathrm{lab})} = \boldsymbol{T}_{\nu\mu}\boldsymbol{d}_\mu^{(\mathrm{mol})} \, , \tag{15.103}$$

where summation over $\mu = x, y, z$ is implied. For the product of transition dipoles we then have

$$(\boldsymbol{n}_4 \cdot \boldsymbol{e}_4)(\boldsymbol{n}_3 \cdot \boldsymbol{e}_3)(\boldsymbol{n}_2 \cdot \boldsymbol{e}_1)(\boldsymbol{n}_1 \cdot \boldsymbol{e}_1) = \sum_{\nu_4\nu_3\nu_2\nu_1} (\boldsymbol{e}_4)_{\nu_4}(\boldsymbol{e}_3)_{\nu_3}(\boldsymbol{e}_2)_{\nu_2}(\boldsymbol{e}_1)_{\nu_1} \times (\boldsymbol{n}_4)_{\nu_4}(\boldsymbol{n}_3)_{\nu_3}(\boldsymbol{n}_2)_{\nu_2}(\boldsymbol{n}_1)_{\nu_1} \, . \tag{15.104}$$

Now using (15.103) we can write the molecular dipoles in the lab frame

$$(\boldsymbol{n}_4 \cdot \boldsymbol{e}_4)(\boldsymbol{n}_3 \cdot \boldsymbol{e}_3)(\boldsymbol{n}_2 \cdot \boldsymbol{e}_1)(\boldsymbol{n}_1 \cdot \boldsymbol{e}_1) = \sum_{\nu_4\nu_3\nu_2\nu_1} \sum_{\mu_4\mu_3\mu_2\mu_1} (\boldsymbol{e}_4)_{\nu_4}(\boldsymbol{e}_3)_{\nu_3}(\boldsymbol{e}_2)_{\nu_2}(\boldsymbol{e}_1)_{\nu_1}$$
$$\times \boldsymbol{T}_{\nu_4\mu_4}\boldsymbol{T}_{\nu_3\mu_3}\boldsymbol{T}_{\nu_2\mu_2}\boldsymbol{T}_{\nu_1\mu_1}(\boldsymbol{n}_4)_{\mu_4}(\boldsymbol{n}_3)_{\mu_3}(\boldsymbol{n}_2)_{\mu_2}(\boldsymbol{n}_1)_{\mu_1} . \tag{15.105}$$

Now the vectors $\boldsymbol{e}$ are defined in the lab frame, the molecular dipoles in the molecular frame and $\boldsymbol{T}$-s contain the transformation. The orientational averaging thus amounts to the integration of the product $\boldsymbol{T}_{\nu_4\mu_4}\boldsymbol{T}_{\nu_3\mu_3}\boldsymbol{T}_{\nu_2\mu_2}\boldsymbol{T}_{\nu_1\mu_1}$ over the whole range of the Euler angles. This can be given in the matrix form [102]

$$\langle(\boldsymbol{n}_4 \cdot \boldsymbol{e}_4)(\boldsymbol{n}_3 \cdot \boldsymbol{e}_3)(\boldsymbol{n}_2 \cdot \boldsymbol{e}_1)(\boldsymbol{n}_1 \cdot \boldsymbol{e}_1)\rangle_O = F^{(4)T}(\boldsymbol{e})M^{(4)}F^{(4)}(\boldsymbol{n}) , \tag{15.106}$$

where

$$F^{(4)}(\boldsymbol{x}) = \begin{pmatrix} (\boldsymbol{x}_4 \cdot \boldsymbol{x}_3)(\boldsymbol{x}_2 \cdot \boldsymbol{x}_1) \\ (\boldsymbol{x}_4 \cdot \boldsymbol{x}_2)(\boldsymbol{x}_3 \cdot \boldsymbol{x}_1) \\ (\boldsymbol{x}_4 \cdot \boldsymbol{x}_1)(\boldsymbol{x}_3 \cdot \boldsymbol{x}_2) \end{pmatrix} , \tag{15.107}$$

and

$$M^{(4)} = \frac{1}{30}\begin{pmatrix} 4 & -1 & -1 \\ -1 & 4 & -1 \\ -1 & -1 & 4 \end{pmatrix} . \tag{15.108}$$

For a given setup of the external fields and for the defined molecular dipoles, the amplitudes of all Liouville space pathways for FWM setup can be orientationally averaged using this procedure (15.77)–(15.78).

The same formalism applies to the linear response as well. In the linear response we have the product of two transition amplitudes, one is the excitation, the other is the detection in the form of

$$\langle(\boldsymbol{n}_2 \cdot \boldsymbol{e}_2)(\boldsymbol{n}_1 \cdot \boldsymbol{e}_1)\rangle_O = F^{(2)\mathrm{T}}(\boldsymbol{e})M^{(2)}F^{(2)}(\boldsymbol{n}) , \tag{15.109}$$

where now

$$F^{(2)}(\boldsymbol{x}) = (\boldsymbol{x}_2 \cdot \boldsymbol{x}_1) , \tag{15.110}$$

and

$$M^{(2)} = \frac{1}{3} . \tag{15.111}$$

However, due to the trace property of the linear response function we usually have the same transition involved in the absorption of the electromagnetic quantum and in the emission into the *same* electric field. The amplitude that enters the linear polarization thus becomes

$$\mu_{eg}^2 E^2 \langle(\boldsymbol{n} \cdot \boldsymbol{e})(\boldsymbol{n} \cdot \boldsymbol{e})\rangle_O \equiv \frac{1}{3}|\boldsymbol{\mu}_{eg}|^2|\boldsymbol{E}|^2 . \tag{15.112}$$

The tensorial properties of the linear response function can thus be often ignored.

Special attention should be paid to the orientational averaging of three scalar products of vectors

$$\langle (\boldsymbol{n}_3 \cdot \boldsymbol{e}_3)(\boldsymbol{n}_2 \cdot \boldsymbol{e}_1)(\boldsymbol{n}_1 \cdot \boldsymbol{e}_1) \rangle_O = \frac{1}{6}[\boldsymbol{e}_3 \cdot (\boldsymbol{e}_2 \times \boldsymbol{e}_1)][\boldsymbol{n}_3 \cdot (\boldsymbol{n}_2 \times \boldsymbol{n}_1)] \,. \tag{15.113}$$

This type of averaging would emerge is the second order spectroscopy (three-wave mixing). First notice that for such products to be nonzero all laser fields must be perpendicular, which yields a very weak response since the field-overlapping space would be small. Second, if the system is a single dipole, the product of material transition dipole is always zero. Thus the second order response in the dipole approximation is zero. However, going beyond dipole approximation, this product survives and the second order response even in the second order to the field could, in principle, be measured. It should be noted that the system must contain inherent chirality in its structure for the response to be non vanishing as can be inspected by performing space inversion operation of the system.

15.3
Application of the Response Functions to Simple FWM Experiments

15.3.1
Photon Echo Peakshift: Learning About System–Bath Interactions

Linear and nonlinear spectra provide us with information about both the electronic properties of the studied molecules (positions of the absorption band) and the properties of the molecular environment (lineshapes) as well as of the relaxation dynamics. In a two-level system (a single molecule) the bath in most of the cases is characterized by the single correlation function $C(t)$ of the energy gap. The so-called *photon echo peakshift* (PEPS) experiment is one from which energy gap correlation function can be obtained. It enables us to gain insight into the effects of bath reorganization that are otherwise very hard to attain with any other formulation of the theory.

For the PEPS measurement, the third order nonlinear signal in the direction $k_{\text{I}} = -\boldsymbol{k}_1 + \boldsymbol{k}_2 + \boldsymbol{k}_2$ (photon echo) is measured with a slow detector. The delays between pulses are denoted as $T_1 \equiv \tau$ and $T_2 \equiv T$: these are expected to be always positive, $\tau \geq 0$ and $T \geq 0$. Thus, only the rephasing part of the response function is therefore detected. The detector integrates the signal intensity over the time t. Thus, we measure

$$S_{\text{PEPS}}(T, \tau) = \int_0^\infty \mathrm{d}t |S^{(3)}_{k_{\text{I}}}(t, T, \tau)|^2 \,. \tag{15.114}$$

This signal at a given time T rises from the initial finite value and then decays. In the S_{PEPS} we look for the value $\tau^*(T)$ of τ for which the S_{PEPS} is maximal. This value is called the *photon echo peakshift*, as it is shifted from zero value, and can be formally defined by the relation

$$\frac{\partial}{\partial \tau} S_{\text{PEPS}}(T, \tau)|_{\tau=\tau^*(T)} = 0 , \tag{15.115}$$

of course by assuming that there are no oscillations in the measured amplitude.

In the impulsive limit, the third order signal reads

$$|S_{k_I}^{(3)}(t, T, \tau)|^2 \approx I_{\text{inh}}^2(t - \tau)|R_{3g}(t, T, \tau) + R_{2g}(t, T, \tau)|^2 , \tag{15.116}$$

where $I_{\text{inh}}^2(t) = \mathrm{e}^{-\Delta_{\text{in}}^2 t^2}$ is the inhomogeneity factor introduced in Section 15.2.5. If we denote the line shape function by $g(t)$ and, according to previous sections for a two-level molecule, we can write:

$$\begin{aligned} |R_{3g}(t, T, \tau) + R_{2g}(t, T, \tau)|^2 &= \mathrm{e}^{-2\,\text{Re}\,[g(\tau)-g(T)+g(t)+g(\tau+T)+g(\tau+t)-g(\tau+T+t)]} \\ &\quad \times \cos^2[\text{Im}\,(g(T) + g(t) - g(T + t))] \end{aligned} \tag{15.117}$$

Because the signal is expected to decay fast with t an expansion of the exponent in t, up to the second order could yield a reasonable approximation. We will follow derivation of Cho *et al.* [30, 31] to reveal striking insights into relation of the photon echo peakshift and the energy gap correlation function. To this end we expand the $g(t)$ function as $g(a + t) \approx g(a) + \dot{g}(a)t + \ddot{g}(a)t^2/2$, where the dot denotes a time derivative. Since

$$\text{Re}\,\ddot{g}(t) = \int_0^\infty \mathrm{d}\omega\, C''(\omega) \coth\left(\frac{\hbar\omega}{2k_B T}\right) \cos\omega t , \tag{15.118}$$

where $C''(\omega)$ is the spectral density and

$$\text{Im}\,\dot{g}(t) = \int_0^\infty \mathrm{d}\omega \frac{C''(\omega)}{\omega} (\cos\omega t - 1) , \tag{15.119}$$

the cosine part of (15.117) can be expressed in the second cumulant expansion as $\cos^2[\text{Im}\,(g(T) + g(t) - g(T + t))] \approx \exp[-(\text{Im}\,\dot{g}(T))^2 t]$, and the exponent of the exponential part could also be expanded to the second order in t. Equation (15.114) thus becomes

$$S_{\text{PEPS}}(T, \tau) \approx \int_0^\infty \mathrm{d}t \mathrm{e}^{B(T,\tau)t - A(T,\tau)t^2 - 2P(\tau) - \Delta_{\text{in}}^2 \tau^2} , \tag{15.120}$$

where

$$A(T, \tau) = \Delta_{\text{in}}^2 + (\text{Im}\,\dot{g}(T))^2 + \text{Re}\,\{\ddot{g}(0) - \ddot{g}(T) - \ddot{g}(\tau + T)\} , \tag{15.121}$$

$$B(T,\tau) = 2\Delta_{in}^2\tau - 2\mathrm{Re}\,\{\dot{g}(T) - \dot{g}(\tau + T)\}\,. \tag{15.122}$$

The integration of (15.120) gives

$$S_{\mathrm{PEPS}}(\tau,T) \approx \frac{\sqrt{\pi}}{2\sqrt{A(T,\tau)}}\exp\left[-2\mathrm{Re}\,g(\tau) - \Delta_{in}^2\tau^2 + \frac{B(T,\tau)^2}{4A(T,\tau)}\right] \times\left[1 + \mathrm{erf}\left(\frac{B(\tau,T)}{2\sqrt{A(T,\tau)}}\right)\right]. \tag{15.123}$$

Equation (15.123) represents an approximate result for the photon echo signal and can be used for calculating the photon echo peakshift.

If the inhomogeneous broadening is much larger than the homogeneous broadening, that is $\mathrm{Re}\,\ddot{g}(0) \ll \Delta_{in}^2$ the integrated photon echo is approximately given by $S_{\mathrm{PEPS}}(T,\tau) \approx \exp\{-2\mathrm{Re}\,g(\tau)\}$ for $\tau > \Delta_{in}^{-1}$. This can be verified by noticing that in this limit $B \approx 2\Delta_{in}^2\tau$ and $A \approx \Delta_{in}^2$, and the error function $\mathrm{erf}(x) \approx 1$ for $x > 1$. The integrated photon echo signal does not depend on T here, and the time dependent photon echo signal has its maximum at time $t = \tau$. The peakshift is given by the condition $\mathrm{Re}\,\dot{g}(\tau)|_{\tau=\tau^*(T)} = 0$ which leads to $\tau^*(T) = 0$ at $\tau > \Delta_{in}^{-1}$.

If the inhomogeneous and homogeneous line widths Δ_{in}^2 and $\mathrm{Re}\,\ddot{g}(0)$ are comparable, one can apply another expansion of the signal, this time in τ. We expand all quantities in the second order of τ which leads to $B \approx 2(\Delta_{in}^2 + \mathrm{Re}\,\ddot{g}(T))\tau$, $A \approx \mathrm{Re}\,\ddot{g}(0) + \Delta_{\mathrm{in}} + (\mathrm{Im}\,\dot{g}(T))^2$, $\mathrm{Re}\,g(\tau) \approx \ddot{g}(0)\tau^2/2$, and then expand (15.123) to the second order in τ. Taking the derivative of such an expanded integrated signal S_{PEPS} according to τ, and solving the resulting linear equation for τ, lead to the following expression for the peakshift

$$\tau^*(T) = \frac{1}{\sqrt{\pi}}\frac{(\Delta_{in}^2 + \mathrm{Re}\,\ddot{g}(T))\sqrt{\ddot{g}(0) + \Delta_{in}^2 + (\mathrm{Im}\,\dot{g}(T))^2}}{\ddot{g}(0)\left[\ddot{g}(0) + 2\Delta_{in}^2 + (\mathrm{Im}\,\dot{g}(T))^2\right] + \Delta_{in}^2(\mathrm{Im}\,\dot{g}(T))^2}\,. \tag{15.124}$$

In the case that the inhomogeneous broadening can be completely ignored (for example, in liquids), the peakshift expression simplifies dramatically to

$$\tau^*(T) = \frac{1}{\sqrt{\pi}}\frac{\mathrm{Re}\,\ddot{g}(T)}{\ddot{g}(0)\sqrt{\ddot{g}(0) + (\mathrm{Im}\,\dot{g}(T))^2}}\,. \tag{15.125}$$

Often $\mathrm{Im}\,\dot{g}(T)$ is small compared to $\ddot{g}(0)$ and can be neglected, and the peakshift therefore directly reveals the real part of the energy gap correlation function $\mathrm{Re}\,\ddot{g}(t) = \mathrm{Re}\,C(t)$. The long T time peakshift is correspondingly equal to zero, $\tau^*(T \to \infty) = 0$.

Most interestingly, in the limit of long population times $T \to \infty$ in (15.124), that is, for Δ_{in}^2 comparable to $\ddot{g}(0)$, the long time peakshift remains nonzero. We have $\mathrm{Im}\,\dot{g}(T \to \infty) \approx \lambda$, $\mathrm{Re}\,\ddot{g}(T \to \infty) \approx 0$ and thus

$$\tau^*(T \to \infty) = \frac{1}{\sqrt{\pi}}\frac{\Delta_{in}^2\sqrt{\ddot{g}(0) + \Delta_{in}^2 + \lambda^2}}{\ddot{g}(0)(\ddot{g}(0) + 2\Delta_{in}^2 + \lambda^2) + \Delta_{in}^2\lambda^2}\,. \tag{15.126}$$

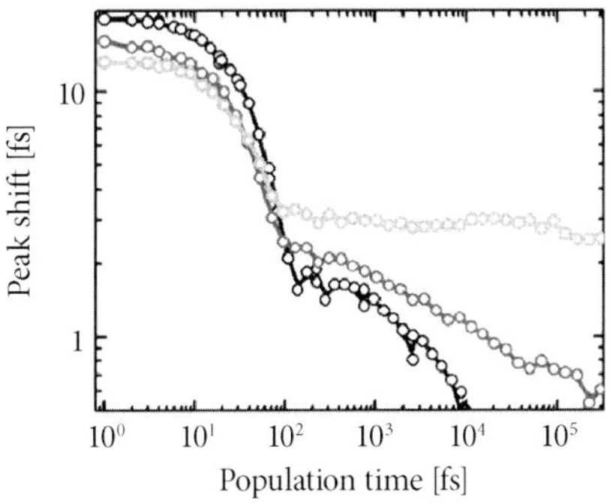

Figure 15.7 Peakshift measured on a chromophore captured by a mouse antibody at different level of antibody development [103]. The disorder in the environment of the chromophore as revealed by the long time value of the peakshift decreases as the antibody matures (from light gray via dark gray to black), demonstrating how its structure becomes more specific for binding the chromophore. (Figure taken from [103]; © 2006, National Academy of Sciences, USA.)

Consequently, the nonzero long time value of the peakshift is the evidence of a finite inhomogeneity.

An interesting example of the peakshift application is presented in Figure 15.7. In a study reported in [103] a chromophore molecule (fluorescein) was injected into a mouse, and the evolution of the photon echo peakshift of the chromophore bound by the antibody was monitored as the antibody matured. Over time, the antibody becomes more specifically designed for binding of the chromophore. One can imagine the antibodies having their protein "trap" designed more specifically to the chromophore molecule, rather than using some generic design. This results in a decrease of the heterogeneity of the environment of the chromophore as it gets trapped by more antibodies specifically tailored to it.

According to (15.124) and (15.125), the peakshift can be used to estimate the bath correlation function, that is, it can be used to investigate fluctuations of the energy gap. If a complex of two or more molecules is measured, several molecules may contribute to a single electronic transition and the total energy gap correlation function is built from contributions of individual molecules. For the case of two identical molecules with energy gap fluctuations uncorrelated with each other, a certain combination of one color peakshifts (measured with all three pulses of the same frequency) and two-color peakshifts (measured with the first two pulses of one frequency and the third pulse of another frequency) allows ones to estimate the excitonic mixing and thus the value of excitonic coupling between the two molecules. For two different molecules, one can even estimate the difference between the correlation functions on the two molecules.

Simulation of such an experimental determination of the coupling coefficient and the difference between the monomeric energy gap correlation functions in a heterodimer has been done in [104]. Figure 15.8 represents a comparison of the coupling coefficient and the difference of the energy gap correlation functions on the two molecules forming a dimer calculated from the photon echo peakshift, with the actual values that had entered the simulation. The overall good agreement between the calculated and expected values for the waiting time T demonstrates that despite the many approximations in the derivation of the peakshift expression, (15.126), it enables us to retrieve a quantitative information about molecular system interaction with its environment.

15.3.2 Revisiting Pump-Probe

We have described the pump probe experiment in Section 13.3.2 where we introduced it in terms of the induced polarization. Here we relate it explicitly with the response function formalism.

The pump probe technique for the three-band systems can be understood as the preparation of the excited population, or more precisely – the nonequilibrium state by the pump pulse, propagation with some delay, and then probing of the resultant non equilibrium density matrix by the probe pulse. In the FWM setup according to previous sections, the populations ρ_{ee} are prepared by the first two $\boldsymbol{k}_1$ and $\boldsymbol{k}_2$ pulses in the rephasing and nonrephasing signals by interaction configurations $\boldsymbol{k}_1 - \boldsymbol{k}_2$ and $-\boldsymbol{k}_1 + \boldsymbol{k}_2$. Let the $\boldsymbol{k}_1 = \boldsymbol{k}_2 = \boldsymbol{k}_{\text{pu}}$. In this case the pump probe intensity will be detected in the probe direction $-\boldsymbol{k}_1 + \boldsymbol{k}_2 + \boldsymbol{k}_3 = \boldsymbol{k}_{\text{pr}}$ and $\boldsymbol{k}_1 - \boldsymbol{k}_2 + \boldsymbol{k}_3 = \boldsymbol{k}_{\text{pr}}$ when $\boldsymbol{k}_3 = \boldsymbol{k}_{\text{pr}}$. The pump probe signal can thus be represented by the sum of the rephasing and nonrephasing signals with the six Feynman diagrams as shown in Figure 15.9.

Following Section 15.1.3 we can thus write the induced polarization by taking $T_1 = 0$ and $T_2 = T$ as the delay time between the pump and the probe

$$P^{(3)}_{\text{PP}}(t,T) \approx \mathrm{e}^{-\mathrm{i}\Omega t} \iiint\limits_0^\infty \mathrm{d}t_3 \mathrm{d}t_2 \mathrm{d}t_1 \left[S_{\text{R}}(t_3,t_2,t_1) A_{\text{R}} + S_{\text{NR}}(t_3,t_2,t_1) A_{\text{NR}} \right] ,$$

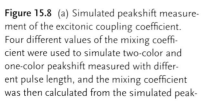

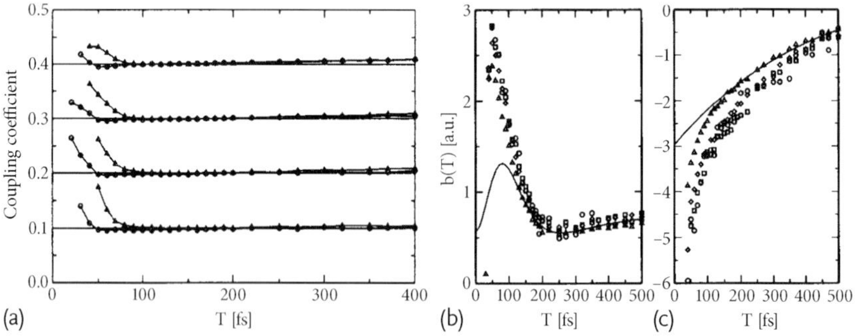

Figure 15.8 (a) Simulated peakshift measurement of the excitonic coupling coefficient. Four different values of the mixing coefficient were used to simulate two-color and one-color peakshift measured with different pulse length, and the mixing coefficient was then calculated from the simulated peakshift by the formula from [104]. (b,c) The difference $b(t) = C_{\text{A}}(t) - C_{\text{B}}(t)$ of the two molecules A and B of the dimer reconstructed from the simulated peakshift, and compared with the actual value. (Figure adapted from [104]; Copyright © 2004, American Institute of Physics.)

where

$$A_{\mathrm{R}} = e^{i\Omega(t_3-t_1)} \mathcal{E}_{\mathrm{pu}}^*(t+T-t_3-t_2-t_1)\mathcal{E}_{\mathrm{pu}}(t+T-t_3-t_2)\mathcal{E}_{\mathrm{pr}}(t-t_3) ,$$

$$A_{\mathrm{NR}} = e^{i\Omega(t_3+t_1)} \mathcal{E}_{\mathrm{pu}}(t+T-t_3-t_2-t_1)\mathcal{E}_{\mathrm{pu}}^*(t+T-t_3-t_2)\mathcal{E}_{\mathrm{pr}}(t-t_3) .$$

according to Section 13.3.2 the self-heterodyning leads to the following expression of the pump probe intensity

$$I_{\mathrm{PP}}(T) = \int \mathrm{d}t\, P_{\mathrm{PP}}^{(3)}(t,T)\mathcal{E}_{\mathrm{pr}}^*(t) . \tag{15.127}$$

It is convenient to use the impulsive limit, as t_2 is usually much larger than t_1 or t_3. This is because T is the physical delay between pulses, while the pulse duration controls t_1, t_3 and t. For this purpose we can choose $t_2 = T$, $t_1 = 0$, $t_3 = t$. Altthough we use ultra-short pulse, we can assume the frequency resolution of the pump pulse to be sufficient to create a specific excitation in the e band (see the discussion of thc physical delta-function in Section 15.1.4). We can therefore choose a specific initial state in the time-evolution operator for T propagation. The detected amplitude at a specific frequency is determined by the frequency variable conjugate with t_3 delay. The frequency-resolved pump probe spectrum is thus given by the Fourier transform of the response functions:

$$I_{\mathrm{PP}}(\omega_{\mathrm{pr}}, T) \approx S_{\mathrm{R}}(\omega_{\mathrm{pr}}, T, t_1 = 0) + S_{\mathrm{NR}}(\omega_{\mathrm{pr}}, T, t_1 = 0) .$$

We can also assume the so-called two-dimensional pump probe spectrum where the specific excitation frequency is associated with the specific emission frequency. We then have

$$I_{\mathrm{PP}}(\omega_{\mathrm{pr}}, T, \omega_{\mathrm{pu}}) \approx S_{\mathrm{R}}(\omega_3, T, -\omega_{\mathrm{pu}}) + S_{\mathrm{NR}}(\omega_3, T, \omega_{\mathrm{pu}}) .$$

The "minus" sign of ω_{pu} in the rephasing term denotes the specific phase rotation of the time-evolution in the rephasing response function.

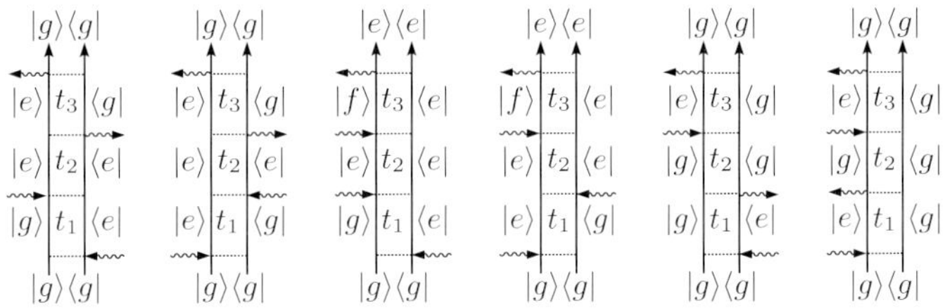

Figure 15.9 Double sided Feynman diagrams corresponding to the pump probe experiment. The g, e, f denote the bands where the excitation propagates. The delay times t_1, t_2, t_3 of the response function are associated with the pump probe experiments by performing Fourier transforms with respect to t_1 and t_3 and taking $t_2 \equiv T_2 \equiv \tau_{\mathrm{D}}$ as the delay between pump and probe.

15.3.3
Time-Resolved Fluorescence

We will briefly describe one more "theoretical" experiment here based on the response function expressions. Consider the experiment where an optical pulse excites the system which is then left to propagate on its own. As it propagates it has the probability to perform a spontaneous emission of the field as in fluorescence. If we were able to record the emission at a given delay time τ_D after the excitation, we could imagine the *time-resolved fluorescence* (TRF) experiment.

A theoretical description of the TRF can be based on the third order response function theory and involves the system relaxation dynamics due to the coupling with the bath. For the TRF calculations we proceed along the same ideas as for the pump probe, but we can make few more restrictions. Again, the first two interactions are performed with the excitation pulse, which resonantly excites a single exciton state e and creates its population. Time t_1 is thus limited within the excitation pulse and $\hat{\mathcal{U}}(t_1)$ is responsible for the preparation process. The second delay between the interactions t_2 is a waiting time when the field is off. During the third time t_3 the emission takes place by a transition from some exciton state e' to the ground state, and the dynamics of the corresponding $e'g$ optical quantum coherence generates the outgoing field, which is detected by a detector. Within these restrictions only two terms in the response function are related to the TRF [26]:

$$R_{\mathrm{TRF}}(t_3, t_2, t_1) = \mathrm{i}^3 \left[\mathrm{Tr}\langle \hat{\boldsymbol{\mu}} | \mathcal{U}^{(eg)}(t_3) \mathcal{V}^{(egee)} \mathcal{U}^{(ee)}(t_2) \mathcal{V}^{(eeeg)} \mathcal{U}^{(eg)}(t_1) \mathcal{V}^{(egee)} | \rho_{gg}\rangle \right.$$
$$\left. + \mathrm{Tr}\langle \hat{\boldsymbol{\mu}} | \mathcal{U}^{(eg)}(t_3) \mathcal{V}^{(egee)} \mathcal{U}^{(ee)}(t_2) \mathcal{V}^{(eege)} \mathcal{U}^{(ge)}(t_1) \mathcal{V}^{(geee)} | \rho_{gg}\rangle \right] . \quad (15.128)$$

They are represented by a single Feynman diagram in Figure 15.10: the two terms are obtained by interchanging the order of first two interactions. We next assume the impulsive limit for the excitation and full frequency resolution for detection, and we denote the free-field propagation time as $\tau_D \equiv t_2$. The TRF is then given as

$$F(\omega, \tau_D) = \mathrm{Re} \int_0^\infty \mathrm{d}\tau \mathrm{e}^{\mathrm{i}\omega t_3} \mathrm{i} R_{\mathrm{TRF}}(t_3, \tau_D, t_1 \to 0) . \quad (15.129)$$

Since we have assumed that during the excitation only one excited state population is resonantly excited (that can be achieved when the incoming field frequency is tuned to a specific inter-band energy gap in the system), we set

$$\mathcal{V}^{(eeeg)} \mathcal{U}^{(eg)}(0) \mathcal{V}^{(egee)} | \rho_{gg}\rangle = \mathcal{V}^{(eege)} \mathcal{U}^{(ge)}(0) \mathcal{V}^{(geee)} | \rho_{gg}\rangle \approx - \left| \rho_{ee}^{(g)} \right\rangle , \quad (15.130)$$

where $|\rho_{ee}^{(g)}\rangle$ denotes the excited state e population with the bath equilibrium corresponding to the ground state. The final expression for the TRF kernel is

$$R_{\mathrm{TRF}}(t_3, \tau_D) = -\mathrm{i}^3 \mathrm{tr} \langle \hat{\boldsymbol{\mu}} | \mathcal{U}^{(eg)}(t_3) \mathcal{V}^{(egee)} \mathcal{U}^{(ee)}(\tau_D) \left| \rho_{ee}^{(g)} \right\rangle \quad (15.131)$$

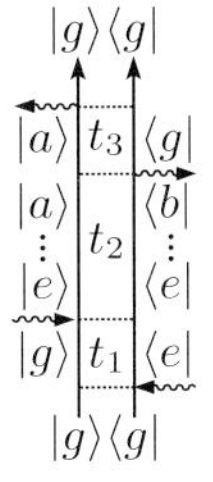

Figure 15.10 Double sided Feynman diagram corresponding to the time-resolved fluorescence.

(we skipped $t_1 = 0$).

This type of response function contribution to the TRF (15.131) is illustrated by the Feynman diagram presented in Figure 15.10. The system starts in its ground state population $|g\rangle\langle g|$. After two successive interactions (both happen within the single laser excitation pulse) the excited state population $|e\rangle\langle e|$ is created. The system then evolves in the excited state during the waiting time t (population transfer and population to coherence transfer events are possible), leaving the system in another state configuration $|a\rangle\langle b|$. Time τ separates two interactions after which the system returns to the ground state $|g\rangle\langle g|$. The evolution during this last interval determines the emission spectrum, which is obtained by the Fourier transform.

The calculation of the fluorescence dynamics requires that the propagator of the full density matrix be calculated, so all information about the system and the bath is fully determined. This is easily accomplished for a single two-level system coupled with the bath using the cumulant expansion, which gives

$$\begin{aligned} R_{\text{TRF}}(t_3, \tau_{\text{D}}) \sim\ & \mathrm{i}^3 \exp(-\mathrm{i}\omega_{eg} t_3) \\ & \times \exp\left(-g^*(t_3) + 2\mathrm{i}\,\mathrm{Im}\left[g(\tau_{\text{D}}) - g(\tau_{\text{D}} + t_3)\right]\right) . \end{aligned} \tag{15.132}$$

For more complicated systems the propagators can be calculated using methods of the dissipation theory and the TRF can be calculated.

16
Coherent Two-Dimensional Spectroscopy

Two-dimensional coherent spectroscopy which will be introduced in this chapter is a relatively new addition to the family of nonlinear optical spectroscopies. It uses the full potential of the photo-induced third order nonlinear response, because it aims at a full resolution of the response. In Section 13.3.3 we have shown that with the heterodyne detection scheme, it is possible to detect the electric field of the signal itself, rather than just its intensity, and it is even possible to phase that field so that we can identify its in-phase and out-of-phase component. This was possible by comparing the heterodyne detected field with the results of for instance pump probe spectrum, a method which is in a certain sense a more simple scheme for generation of the similar signal. Comparison with the pump probe allows us to interpret the signal, or more precisely, the response, in terms of absorptive and refractive parts. In the following sections we will show how this interpretation is related to the structure of the third order response derived in Section 13.1.

16.1
Two-Dimensional Representation of the Response Functions

Let us start by comparing a general structure of the first order and the third order responses. We compare the first order response function of (13.20) which we now write as:

$$J(t) = \mathrm{i}\,\mathrm{Tr}\left\{\hat{\mu}^{(ge)}\mathcal{J}^{(eggg)}(t)\,W^{(gg)}\right\}\mathrm{e}^{-\mathrm{i}\Omega t}\,, \tag{16.1}$$

with $\mathcal{J}^{(eggg)}(t) = \tilde{\mathcal{U}}^{(eg)}(t)\mathcal{V}^{(eggg)}$, and some representative third order response function, for example, R_{2g} of (15.14), which can be written as:

$$R_{2g}(t_3, t_2, t_1) = -\mathrm{i}\mathrm{Tr}\left\{\hat{\mu}^{(ge)}\mathcal{J}^{(egee)}(t_3)\tilde{\mathcal{U}}^{(ee)}(t_2)\right.$$
$$\left.\times\;\hat{\mu}^{(eg)}\mathcal{J}^{(gegg)}(t_1)\,W^{(gg)}\right\}\mathrm{e}^{-\mathrm{i}\Omega(t_3-t_1)}\,. \tag{16.2}$$

In (16.2) we can recognize a repeating pattern consisting of the superoperator $\mathcal{J}$ and the transition dipole operator, which also appears in (16.1). The comparison of (16.1) and (16.2) suggests an analogy between the absorption spectrum given by the Fourier transform of the response function $J(t)$ and a possible nonlinear spectrum given by Fourier transforms along the times t_3 and t_1. In this analogy the frequency ω_1 dependence obtained by the Fourier transform of the third order response in time t_1 would be the same as the frequency dependence of an absorption spectrum:

$$\int_0^\infty dt_1 e^{i\omega_1 t_1} \int_0^\infty dt_3 e^{i\omega_3 t_3} R_{2g}(t_3, t_2, t_1) \approx \mathcal{W}(\omega_3)\tilde{\mathcal{U}}^{(ee)}(t_2)\mathcal{D}(\omega_1) . \tag{16.3}$$

Because of the phase factor $e^{i\Omega t_1}$ the inverse Fourier transform is sometimes used to obtain absorption that peaks at positive frequencies, $\omega_1 > 0$. However, the sign of frequency is a feature of mutual agreement. We choose to take the negative frequency sign for the rephasing and the positive frequency for the nonrephasing signal. We thus always use the same type of the Fourier transforms for both $t_1 \to \omega_1$ and $t_3 \to \omega_3$. The frequency ω_3 dependence obtained by the Fourier transform in time t_3 would be the same as that one of the "absorption" of a system out of equilibrium, whose statistical operator is $\hat{W}^{(ee)}$. In this particular case, $\hat{W}^{(ee)}$ corresponds to the system in the excited state, and thus the nonequilibrium "absorption" in fact also includes the stimulated emission.

The $\mathcal{D}(\omega)$ function in (16.3) is sometimes denoted as the *doorway* function and $\mathcal{W}(\omega_3)$ is the *window* function [91] and this representation of the response function is sometimes denoted as the *doorway-window* representation. As described in the previous paragraph the doorway function reflects the absorption spectrum. The window function of the full rephasing or nonrephasing response function consists of several terms. One part of the term reflects the change of ground state population, what is denoted as the *ground state bleach*. The other part describes the light emission from the excited state: this is the *excited state emission* or the *stimulated emission*. The last part corresponds to the absorption of quantum that takes the single excited state into the double excited state. This is usually denoted as the *excited state absorption* or the *induced absorption*.

The evolution superoperator $\tilde{\mathcal{U}}^{(ee)}(t_2)$ resides between the two generalized doorway and window spectra in (16.3). The two-dimensional Fourier transformed spectrum, therefore, evolves with the delay t_2. From Section 15.1.3 we know that we can record the full time dependence of the third order response in t_1, and t_2 using a specific pulse setup. Formally, we can construct two-dimensional plots correlating the absorption in ω_1 and the nonequilibrium absorption-emission spectra in ω_3. The t_2 evolution of the spectra may involve various relaxation phenomena: features that reveal relaxation of energy among several excited states, or coherence between given pairs of levels. From Section 15.1.4 we know that for $t_1 > 0$, that is, when the pulse $\boldsymbol{k}_1$ preceded the pulse $\boldsymbol{k}_2$, only the rephasing pathways contribute to the signal in the direction $-\boldsymbol{k}_1 + \boldsymbol{k}_2 + \boldsymbol{k}_3$ (here we neglect pulse overlap effects). Thus the we can define the *rephasing* 2D spectrum as

$$S_R(\omega_3, t_2, \omega_1) = \int_0^\infty dt_3 \int_0^\infty dt_1\, S_R^{(3)}(t_3, t_2, t_1) e^{i\omega_3 t_3 + i\omega_1 t_1} . \tag{16.4}$$

At the same time, we can also define the *nonrephasing* 2D spectrum and construct a similar quantity as

$$S_{NR}(\omega_3, t_2, \omega_1) = \int_0^\infty dt_3 \int_0^\infty dt_1\, S_{NR}^{(3)}(t_3, t_2, t_1) e^{i\omega_3 t_3 + i\omega_1 t_1} . \tag{16.5}$$

The limits of the integration are set naturally, because for $t_3 < 0$ and $t_1 < 0$ the response is zero. The quantities defined in (16.4) and (16.5) can be directly measured experimentally. The heterodyne detection scheme of Section 13.3.3 enables us to measure the third order signal electric field, directly proportional in the impulsive limit to $S^{(3)}$.

Notice that if the explicit pulse ordering is imposed in Section 15.1.3 in the impulsive limit (Section 15.1.4) the rephasing signal is associated with the wavavector configuration $k_I = -\boldsymbol{k}_1 + \boldsymbol{k}_2 + \boldsymbol{k}_3$, the nonrephasing is then related to $k_{II} = +\boldsymbol{k}_1 - \boldsymbol{k}_2 + \boldsymbol{k}_3$. The rephasing and nonrephasing signals are thus associated with the linearly independent wavevector configurations. We can thus also write

$$S_R(\omega_3, t_2, \omega_1) \equiv S_{k_I}(\omega_3, t_2, \omega_1) \tag{16.6}$$

$$S_{NR}(\omega_3, t_2, \omega_1) \equiv S_{k_{II}}(\omega_3, t_2, \omega_1) . \tag{16.7}$$

As described in Section 15.3.2 the pump probe signal is given by the rephasing and nonrephasing parts. We thus can define the sum

$$S_{k_I}(\omega_3, t_2, -\omega_1) + S_{k_{II}}(\omega_3, t_2, \omega_1) = S_{PP}(\omega_3, t_2, \omega_1) \tag{16.8}$$

as the *two-dimensional pump probe* signal. Alternatively, this sum is referred to as the *total* signal; however, we prefer the pump probe since it does not include the double quantum coherence signal, and in this sense it is not "total".

There is a third part to the response – the double quantum coherence – that comes from $k_{III} = \boldsymbol{k}_1 + \boldsymbol{k}_2 - \boldsymbol{k}_3$ configuration. We can thus define the two-dimensional spectrum associated with that direction. Let us inspect the characteristic time evolutions of the corresponding response function. It has two contributions given by (15.19) and (15.20). The time evolution is implied by factor

$$\exp(-i\Omega(t_1 + 2t_2 + t_3)) , \tag{16.9}$$

where Ω is the fundamental resonance - the inter-band gap. One quantity of interest is thus the t_2 interval since it oscillates with double frequency and, therefore, shows resonances of double excitations. The second interval can be chosen to be t_1 or t_2. Two representations of the *two-dimensional double quantum coherence spectrum* can be suggested

$$S_{k_{III}}(t_3, t_2, t_1) \rightarrow S_{k_{III}}(t_3, \omega_2, \omega_1) \tag{16.10}$$

and

$$S_{k_{III}}(t_3, t_2, t_1) \rightarrow S_{k_{III}}(\omega_3, \omega_2, t_1) . \tag{16.11}$$

The representation $S_{k_{III}}(t_3, \omega_2, \omega_1)$ can be advantageous since it correlates the absorption resonances along ω_1 with double-resonances along ω_2. So it can indirectly resolve the double exciton wavefunction mapping onto the single excitons [105].

Let us go back to the rephasing signal. The actual relation between absorption spectra and the 2D spectra defined above can be found by considering a simple case of a two-level system. For long population times t_2 the dependence of $g(t)$ function on t_2 becomes linear (see Appendix A.8). Both contributing rephasing response functions then have a form

$$R_{2g}(t_3, t_2, t_1) = R_{3g}(t_3, t_2, t_1) \approx e^{-g(t_3)-g^*(t_1)-i\Omega(t_3-t_1)} . \tag{16.12}$$

The Fourier transform, (16.4) leads to

$$S_{k_1}(\omega_3, t_2, \omega_1) \approx G(\omega_3 - \Omega)G^*(\omega_1 - \Omega) , \tag{16.13}$$

where we defined

$$G(\omega) = \int_0^\infty dt e^{-g(t)-i\omega t} . \tag{16.14}$$

The function G is related to the susceptibility χ and determines the absorption spectrum $\kappa_a(\omega) \approx \text{Re}\, G(\omega - \Omega)$ (see (14.19)). To simplify (16.14) we can use the homogenous limit form of the energy gap function $g(t) = \Gamma t$ with some real dephasing rate Γ [26] and we obtain:

$$G(\omega) \approx \frac{\Gamma}{\Gamma^2 + \omega^2} + i\frac{\omega}{\Gamma^2 + \omega^2} . \tag{16.15}$$

The first term on the right hand side corresponds to the Lorenzian absorption spectrum, while in the second term we meet a dispersive lineshape corresponding to the refraction index. We can see that the rephasing 2D spectrum is not strictly proportional to the absorption spectrum, but it is a mixture of the absorptive and dispersive contributions.

Let us now study the simple profile of a peak in the 2D spectrum that can be obtained analytically for an arbitrary multilevel adiabatic model. The 2D signals are given as sums over Liouville space pathways, hence, each pathway can be transformed accordingly:

$$S(\omega_3, t_2, \omega_1) = \sum_n S_n(\omega_3, t_2, \omega_1) . \tag{16.16}$$

These have the following form

$$S_n(\omega_3, t_2, \omega_1) = A_{(n)} \iint dt_1 dt_3 e^{i\omega_3 t_3 + i\omega_1 t_1} \times \exp(-i\varepsilon_3 t_3 - i\varepsilon_2 t_2 \pm i\varepsilon_1 t_1)_{(n)} , \tag{16.17}$$

where the subscript n denotes different terms of the summation. $A_{(n)}$ is a complex prefactor, given by the transition dipoles and excitation fields, while the exponent on the second row is the propagator of the density matrix. Here ε_j coincides with the energy gap ω_j between the *left* and *right* states of the system density matrix relevant to the time interval t_j. Stimulated emission and ground state bleach carry "+" sign, while induced absorption has "−" overall sign. The Fourier transforms in (16.17) map the contributions to the frequency-frequency plot $(t_1, t_3) \to (\omega_1, \omega_3) \sim (\mp|\varepsilon_1|, \varepsilon_3)$ (the upper sign is for k_{I}, the lower for k_{II}). Diagonal peaks at $\omega_1 = \mp\omega_3$ are usually distinguished, while the anti-diagonal line is defined as $\mp\omega_1 + \omega_3 = \text{const}$. The whole 2D spectrum becomes a function of t_2: either oscillatory for density matrix coherences $|a\rangle\langle b|$ with the characteristic oscillation energy $\varepsilon_2 = \omega_{ab} \neq 0$, or static for populations $|a\rangle\langle a|$ ($\varepsilon_2 = 0$).

Equation (16.17) can be analytically integrated by adding phenomenological dephasing $\exp(-\text{i}\varepsilon_j t_j - \gamma_j t_j)$. For a single contribution $\boldsymbol{S}_n$ giving rise to a peak at $(\omega_1, \omega_3) = (\mp|\varepsilon_1|, \varepsilon_3)$ we shift the origin of (ω_1, ω_3) plot to the peak center by introducing the displacements ($\omega_1 + \varepsilon_1 = -s_1$, $\omega_3 - \varepsilon_3 = s_3$ for the rephasing pathways, while $\omega_1 - \varepsilon_1 = s_1$, $\omega_3 - \varepsilon_3 = s_3$ for the nonrephasing). For $\gamma \approx \gamma_1 \approx \gamma_3$ we get the peak profile

$$\boldsymbol{S}_n(s_3, t_2, s_1) = A_n L(s_1, s_3) \mathrm{e}^{-\gamma_2 t_2} \cos(|\varepsilon_2| t_2 + \phi(s_1, s_3)) , \tag{16.18}$$

where the lineshape and phase for the k_{I} (upper sign) and k_{II} (lower sign) signals are

$$L(s_1, s_3) = \frac{\sqrt{[\gamma^2 \pm s_1 s_3]^2 + \gamma^2 (s_3 \mp s_1)^2}}{\left(s_1^2 + \gamma^2\right)\left(s_3^2 + \gamma^2\right)} , \tag{16.19}$$

$$\phi(s_1, s_3) = \text{sgn}(\varepsilon_2) \arctan\left(\frac{\gamma(s_3 \mp s_1)}{(\mp s_1 s_3 - \gamma^2)}\right) . \tag{16.20}$$

The phase ϕ and the full profile for $A_n = 1$ and $t_2 = 0$ are shown in Figure 16.1. The rephasing and nonrephasing configurations are obtained by flipping

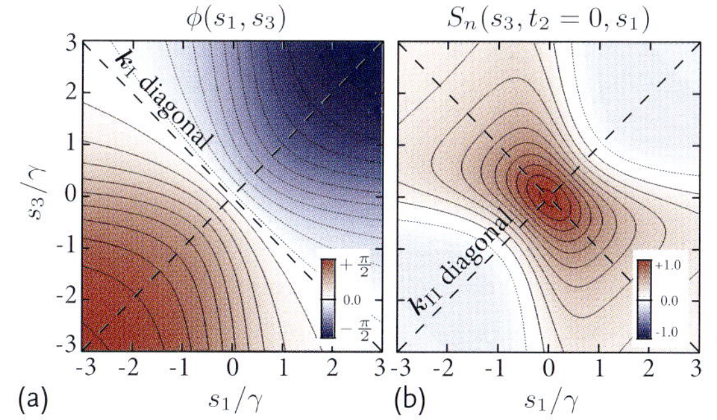

Figure 16.1 Phase ϕ of the contribution (16.18) (a) and peak profile S_n (b) as a function of the shift from the peak center ($s_1 = s_3 = 0$) using relative coordinates. The diagonal lines of the $\boldsymbol{k}_{\text{I}}$ and $\boldsymbol{k}_{\text{II}}$ contributions to the 2D spectra are shown by dashed lines.

the direction of the s_1 axis. At the center of the peak ($s_1 = s_3 = 0$), we have and $\phi = 0$, leading to $S_n \propto \cos(|\varepsilon_2|t_2)$. However, for ($s_1 \neq 0, s_3 \neq 0$) we find $S_n \propto \cos(|\varepsilon_2|t_2 + \phi(s_1, s_3))$ with nonzero phase $\phi(s_1, s_3) \neq 0$. Note that the sign of the phase ϕ is opposite for the peaks above ($\varepsilon_2 < 0$) and below ($\varepsilon_2 > 0$) the diagonal line, and this applies to all contributions.

The whole two dimensional spectrum is a sum of all relevant contributions. Assuming that all dephasings are similar, different contributions to the same peak will have the same spectral shape and they may be summed. We can then simplify the two dimensional spectrum by writing the signal as a sum of peaks $\bar{\sum}$, which have static (from populations) and oscillatory (from coherences) parts:

$$S(\omega_3, t_2, \omega_1) = e^{-\gamma_2 t_2} \bar{\sum_{i,j}} L_{ij}(\omega_1, \omega_3) \times \left[A^{\mathrm{p}}_{ij} + A^{\mathrm{c}}_{ij} \cdot \cos(|\omega_{ij}|t_2 + \phi_{ij}(\omega_1, \omega_3)) \right] . \qquad (16.21)$$

Here ω_{ij} is the characteristic oscillatory frequency of a peak (ij), $A^{\mathrm{p}}_{ij}(t_2)$ and $A^{\mathrm{c}}_{ij}(t_2)$ are the real parts of orientationally-averaged prefactors. The spectral lineshape is given by $L_{ij}(\omega_1, \omega_3)$.

We can thus quantify the system's fine structure of the excited states, as well as coherent dynamics using the two dimensional spectrum and by carefully inspecting the spectral lineshapes. We apply these ideas below while studying simple model systems.

Interpretation of many types of two dimensional spectra can be based on this simple shape. In the sections that follow, more involved spectral shapes that take into account the sometimes complicated evolution of the line broadening function $g(t)$ and the excitation transfer dynamics will be studied. Let us notice in short some of the most important features of the homogenous lineshape. First, the rephasing and nonrephasing spectra have certain characteristic orientations (Figure 16.2), and both their real and imaginary parts contain positive and negative contributions. These features survive even in spectra of more complicated systems. The contribution of the R_2 pathway, (16.2), which contains evolution in the excited state during the waiting time t_2 could be readily interpreted as a stimulated emission (SE). Combined with the corresponding nonrephasing pathway, they would lead to a decrease in absorption if they were measured in a pump probe ($\boldsymbol{k}_1 = \boldsymbol{k}_2$) configuration. This is also a somewhat general feature, because the real part of the total spectrum should represent absorption. On the other hand, we can see that this interpretation is only approximatively valid. It will later be seen that even for a real part of a two-level system (which excludes any absorption to the higher excited states) the 2D spectrum is not purely positive. The R_3 pathway contains propagation in the ground state and it has the same sign as R_2. It cannot, therefore, represent absorption. Rather, it has to stand for the ground state bleaching (GSB), that is the loss of absorption due to the decreased population of the ground state. The nonrephasing counterpart of this signal can be classified in the very same way. For higher lying excited states, one more pathway with a negative sign containing propagation in the excited state during the waiting time t_2 contributes to both

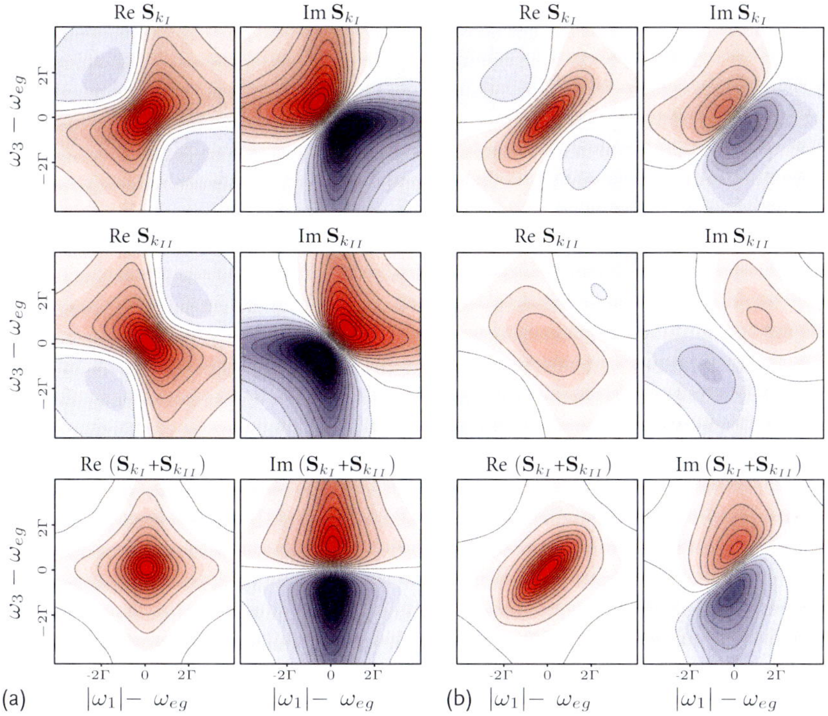

Figure 16.2 Basic homogeneous and inhomogeneous lineshapes of the 2D spectroscopy. (a) The homogeneous system. (b) The inhomogeneously broadened system.

the rephasing and nonrephasing signal. Their combined negative signal is the one corresponding to the so-called excited state absorption (ESA).

So far, we have considered all molecules in the sample to be identical. Now, let us introduce inhomogeneity in the form of diagonal (energy) disorder. We will see that with respect to disorder there is more to the 2D spectrum than just a product of absorption spectra. The response function formalism can handle a simple disorder by including it into the line shape function $g(t)$ as we have seen in Section 15.2.5. The product doorway-window expression of the 2D spectrum, (16.3), is no longer valid because the response

$$R_{2g}(t_3, t_2, t_1) = R_{3g}(t_3, t_2, t_1) \approx e^{-g(t_3)-g^*(t_1)-\Delta(t_3-t_1)^2-i\Omega(t_3-t_1)} , \qquad (16.22)$$

cannot be split into a product of t_3 and t_1 dependent functions. The double Fourier transform analytically leads to Erf functions. The result of the analytical calculation is depicted in Figure 16.2 as well. This 2D lineshape has to correspond to a sum of the contributions of individual molecules of the ensemble, because the 2D spectra of noninteracting sub-ensembles is additive. The inhomogeneity can thus be represented as an averaging over 2D spectra of species with different transition frequencies Ω. Because all such spectra differ only by their position at the diagonal of the spectrum, the lineshape has to be elongated as is shown in Figure 16.2.

Two-dimensional spectra, therefore, broaden only along the diagonal, and both the homogeneous and the inhomogeneous line widths can be in principle estimated from a measurement on the disordered system.

As we will show later, the 2D spectrum not only directly reveals the homogeneous and inhomogeneous line shapes it can additionally allow us to estimate the resonance coupling by the presence of the crosspeaks, and it enables an observation of energy relaxation processes between electronic levels by build-up of the crosspeaks in time t_2.

16.2 Molecular System with Few Excited States

We have already introduced basics of two-dimensional lineshapes in the previous section. Now we will add even more details, and we will concentrate on properties of systems with more than two levels, including their time evolution.

16.2.1 Two-State System

In this part we briefly consider an ideal quantum system of two energy levels: the ground state $|g\rangle$ and the excited state $|e\rangle$. This model effectively represents an isolated resonant transition of, for example, an atom. Since there is always some spectral line broadening phenomenon involved, we add phenomenological decay of interband coherences. The total Hamiltonian in the system eigenstate basis consists of the material part and the coupling with the electric field

$$\hat{H} = \varepsilon_g |g\rangle\langle g| + \varepsilon_e |e\rangle\langle e| - \hat{\mu} E(t) .$$

As usual we set the energy of the ground state $\varepsilon_g = 0$ and the linear response function (13.18) together with $\hat{\rho}_{eq} = |g\rangle\langle g|$ results in

$$S^{(1)}(t_1) = -\frac{2}{\hbar}\theta(t_1)|\mu_{ge}|^2 e^{-\gamma t_1} \sin\left(\omega_{eg} t_1\right) .$$

The Fourier-transformed linear response function is

$$S^{(1)}(\omega_1) = \frac{i}{\hbar}|\mu_{eg}|^2 \left[\frac{1}{\gamma + i(\omega_1 - \omega_{eg})} - \frac{1}{\gamma + i(\omega_1 + \omega_{eg})}\right] . \tag{16.23}$$

This function defines the optical susceptibility and thus the absorption spectrum. For an isolated two-level system it is a Lorentzian-shaped function, centered at ω_{eg} with the linewidth γ.

In a third order photon echo response function of a two-level system we have only two contributions, ground state bleaching and stimulated emission which, for the rephasing part, is

$$S^{\text{GSB}}(t_3, t_2, t_1) \equiv R_2(t_3, t_2, t_1) = \mu^4 e^{i\omega_{eg}(t_1 - t_3) - \gamma(t_1 + t_3)} ,$$
$$S^{\text{SE}}(t_3, t_2, t_1) \equiv R_3(t_3, t_2, t_1) = \mu^4 e^{i\omega_{eg}(t_1 - t_3) - \gamma(t_1 + t_3)} , \tag{16.24}$$

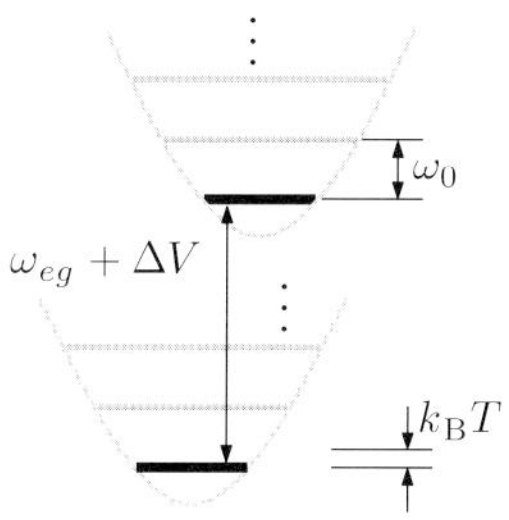

Figure 16.3 Model system of displaced harmonic oscillators.

while $R_1(t_3, t_2, t_1)$ and $R_4(t_3, t_2, t_1)$ contributions are zero. Thus, the corresponding response function is

$$S^{(3)}(t_3, t_2, t_1) = \left(\frac{i}{\hbar}\right)^3 \theta(t_1)\theta(t_2)\theta(t_3)\mu^4[R_2(t_3, t_2, t_1) + R_3(t_3, t_2, t_1)]$$
$$= 2\left(\frac{i}{\hbar}\right)^3 \theta(t_1)\theta(t_2)\theta(t_3)\mu^4 e^{i\omega_{eg}(t_1-t_3)-\gamma(t_1+t_3)} \qquad (16.25)$$

The difference in the nonrephasing response function is just the sign in front of $i\omega_{eg}t_1$. By applying the Fourier transform we then obtain the rephasing and non rephasing two dimensional spectra as follows:

$$S_{k_I}(\omega_3, t_2, \omega_1) = 2\left(\frac{i}{\hbar}\right)^3 \mu^4 \frac{1}{\gamma - i(\omega_3 - \omega_{eg})} \cdot \frac{1}{\gamma - i(\omega_1 + \omega_{eg})},$$
$$S_{k_{II}}(\omega_3, t_2, \omega_1) = 2\left(\frac{i}{\hbar}\right)^3 \mu^4 \frac{1}{\gamma - i(\omega_3 - \omega_{eg})} \cdot \frac{1}{\gamma - i(\omega_1 - \omega_{eg})}. \qquad (16.26)$$

From these analytical expressions we observe that the nonrephasing 2D spectra of the real and imaginary parts are just mirrored images of the rephasing spectra with respect to ω_1 axis. This is a consequence of the simple damping model as described in the previous section.

16.2.2
Damped Vibronic System – Two-Level Molecule

Let us now consider the system of Section 8.6. We repeat the energy level diagram of the system in Figure 16.3. In one dimension the electronic potential of the ground state is $V_g(q) = m\omega_0 q^2/2$ and the displaced electronic excited state is described by the potential $V_e(q) = \omega_{eg} + m\omega_0(q-d)^2/2$. Here ω_0 is the vibrational frequency, ω_{eg} is the energy gap between the minima of two potentials and d is a displacement parameter.

Vibrational dynamics in the harmonic potentials results in an infinite set of wavefunctions ψ_m with quantum numbers $m = 1, \ldots, \infty$ and corresponding energies $E_m = \hbar\omega(m + 1/2)$ with respect to the bottom of the corresponding potential surface. Transitions between the ladder of the electronic ground state and the one of the electronic excited state determine the vibronic progression in the absorption spectrum. The intensity of each vibronic peak is scaled by the overlap of vibrational wavefunctions in the ground and excited state potentials. Following these assumptions the absorption spectrum of this Franck–Condon molecule can be given by

$$\kappa_{\mathrm{abs}}^{\mathrm{FC}}(\omega) \propto \omega \sum_{m,n=1}^{\infty} \mathrm{e}^{-\frac{n\omega_0}{k_B T}} |F_{mn}|^2 \mathrm{Re} \int_0^{\infty} \mathrm{d}t \mathrm{e}^{\mathrm{i}(\omega-\omega_{eg})t-\mathrm{i}\omega_0(n-m)t-\gamma t} , \tag{16.27}$$

where the line-broadening parameter γ is introduced and F_{mn} is the Franck–Condon wavefunction overlap integral for the $m - n$ transition [98]. It is the matrix element of the displacement operator:

$$F_{mn} = \langle m|\hat{D}(d)|n\rangle . \tag{16.28}$$

Here $\hat{D} = \exp(-1/2d^2 + d(\hat{a}^\dagger - \hat{a}))$ which is here given in terms of bosonic (vibrational) creation $\hat{a}^\dagger$ and annihilation $\hat{a}$ operators, d^2 is due to normalization.

The above description considers the electronic+vibrational system as a closed system. However, the line-broadening parameter γ includes the dephasing phenomenologically without a more detailed physical insight. A more general model of a realistic molecule is needed to capture both the vibrational-type ladder of the energy spectrum and the spectral broadening. The cumulant expansion technique allows us to describe various types of vibrational baths and includes these effects explicitly.

As described in Section 8.6 the nuclear dynamics of such a system can be described by various forms of spectral density. Fast-decaying modes of molecular vibrations will result in homogeneous broadening and strong coupling to the high-frequency vibrations will result in vibrational progression in the absorption. By taking various limits with respect to vibrational frequency, vibrational damping, different damping regimes can be achieved representing different baths. The cases of undamped, damped, and overdamped regimes are discussed in the following.

The spectral line shape function $g(t)$ describes the spectral lineshapes. Following Section 8.6 the *overdamped semi-classical bath* is described by the spectral density

$$C''_{\text{o-sc}}(\omega) = 2\lambda \frac{\gamma\omega}{\omega^2 + \gamma^2} , \tag{16.29}$$

and in the high-temperature limit we get the corresponding lineshape function

$$g_{\text{o-sc}}(t) = \frac{\lambda}{\gamma}\left(\frac{2}{\beta\gamma} - \mathrm{i}\right)(\mathrm{e}^{-\gamma t} + \gamma t - 1) . \tag{16.30}$$

Another form of the overdamped bath is given by the *quantum-overdamped* model of the density

$$C''_{\text{o}-\text{q}}(\omega) = \frac{4\lambda\omega\gamma^3}{(\omega^2 + \gamma^2)^2} . \tag{16.31}$$

In the high-temperature limit the corresponding lineshape function is somewhat more complicated [106]

$$g_{\text{o-q}}(t) = \frac{2\lambda}{\beta\gamma^2}\left(e^{-\gamma t}\gamma t + 3e^{-\gamma t} + 2\gamma t - 3\right) - i\frac{\lambda}{\gamma}\left(e^{-\gamma t}\gamma t + 2e^{-\gamma t} + \gamma t - 2\right) \tag{16.32}$$

The damped regime where the spectral density still maintains a resonance will be described by the spectral density

$$C''_{\text{d}}(\omega) = \frac{4\lambda\omega\omega_0^2\gamma}{\left(\omega^2 - \omega_0^2\right)^2 + \gamma^2\omega^2}, \tag{16.33}$$

however, the spectral lineshape should be calculated numerically.

Let us first consider the overdamped vibrational modes. Comparison of the quantum and semiclassical models of the overdamped bath is presented in Figure 16.4. The spectral densities are similar but the quantum model has sharper cutoff at high frequencies. The real parts of the lineshape functions $g(t)$ (Figure 16.4b) determine the absorption linewidth. The corresponding absorption spectra are presented in Figure 16.4c and have only minor differences in their absorption lineshapes. The distortions in the lineshapes are mainly determined by the real part of the corresponding lineshape function. For the *semi-classical* bath, it is smaller by λ/γ compared to the *quantum* bath at $\gamma t \gg 1$. The slopes of the imaginary parts are, however, both equal at $\gamma t \gg 1$.

The lineshape functions of the *overdamped* system correspond to *fast*, $2\lambda k_B T\gamma^{-2} \gg 1$, or *slow*, $2\lambda k_B T\gamma^{-2} \ll 1$, decay regimes. In the former case we get the Lorentzian absorption lineshape and in the latter case the Gaussian. The same form of the lineshapes is obtained in the 2D spectrum as well; however,

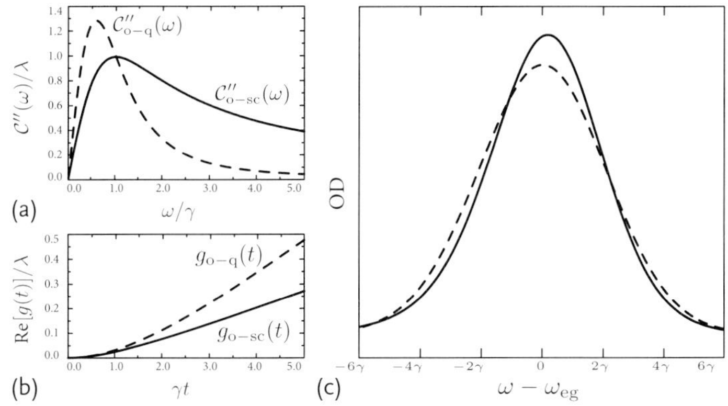

Figure 16.4 (a) Quantum (dashed line) and semi-classical (solid line) spectral density functions of an overdamped two-level system and corresponding (b) real parts of the lineshape functions and (c) absorption spectra.

modes of slow decay of phonon bath fluctuations reflect to the inhomogeneous lineshape broadening rather than homogeneous [106].

Let us now relax the requirement of the overdamped regime and consider the damped vs. undamped vibrations. The spectral densities shown in Figure 16.5a now reflect the resonance at vibronic frequency ω_0. The corresponding lineshape functions in Figure 16.5b reflect the oscillatory character: in the case of undamped mode $g(t)$ follows a shifted cosine function, while for the damped mode the oscillations decay into a straight line showing decay time of vibronic coherences. Absorption spectrum of such system is presented in Figure 16.5c for two values of the damping strength γ. The spectrum demonstrates three well resolved peaks of vibrational progression at frequencies $\omega = \omega_{eg}$, $\omega_{eg} + \omega_0$ and $\omega_{eg} + 2\omega_0$. In the case of *undamped* vibrations (dashed line in Figure 16.5c), the progression is well resolved since all peaks of the progression have the same shape. The damped vibrations cause nonuniform broadening of peaks in the progression. Peaks that are at higher energies are broadened more. As a result, the higher-energy shoulder of the vibrational progression will be reduced due to damped vibrations. This is essentially the consequence of the fact that the higher-quantum vibrational states experience larger decay rates as found in (8.63) of Section 8.4.

The realistic electronic two-state system is usually coupled to the set of vibrational modes where some of them are coherent, some overdamped. This can be modeled using the compound bath model with the spectral density consisting of several parts. It should be noted that the lineshape function is the linear transformation of the spectral density. Thus, utilization of the lineshape function $g_{\text{o-sc}}(t) + g_{\text{d}}(t)$, composed of the *overdamped semi-classical* bath and *damped* vibrations, gives

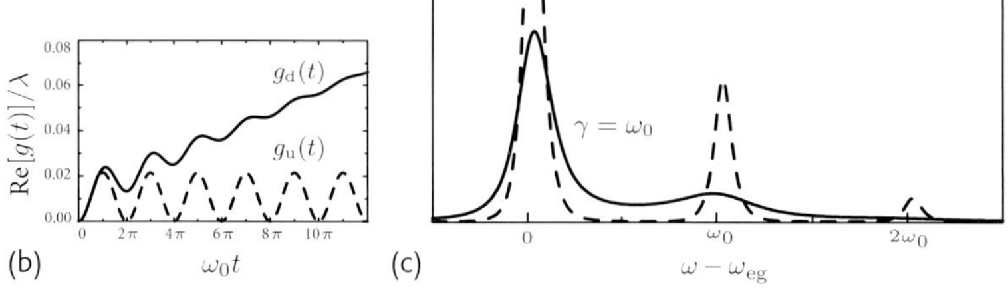

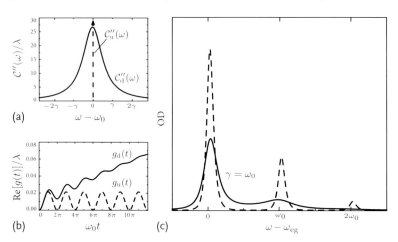

Figure 16.5 (a) Spectral densities of *undamped* (dashed line) and *damped* (solid line) baths; (b) the real part of the lineshape functions corresponding to the different vibronic bath models corresponding to (a). (c) Absorption spectra of the displaced harmonic oscillator (the Huang–Rhys factor $s = 0.3$) using $\gamma = \omega_0$ of the *damped* bath (solid line) and $\gamma \ll \omega_0$ for the *undamped* bath (dashed line). In the undamped case the progression peaks are additionally broadened for clarity.

the broadened vibrational progression in the spectrum. The 2D spectra obtained by using this composition of the system-bath correlation is the commonly used approach to reflect different types of fluctuations and is discussed in detail in the literature [43, 44, 94].

Expressions for the third order response function in terms of the lineshape functions can be obtained using the second order cumulant expansion. The two-dimensional signals, contributing to the rephasing and nonrephasing spectrum can be easily separated. The complex two-dimensional signal of a molecule is then given by

$$\begin{aligned} S_{k_{\mathrm{I}}}(t_3, t_2, t_1) &= \frac{1}{2}\mathrm{e}^{-\mathrm{i}\omega_0(t_3-t_1)} \\ &\times \mathrm{e}^{-g^*(t_1+t_2)-g^*(t_1)+g^*(t_1+t_2+t_3)} \\ &\times \mathrm{e}^{\mathrm{Re}\left[g(t_2+t_3)+g(t_2)-g(t_3)\right]} \\ &\times \cos\left\{\mathrm{Im}\left[g(t_2+t_3)+g(t_2)+g(t_3)\right]\right\} \end{aligned} \tag{16.34}$$

for the rephasing signal and

$$\begin{aligned} S_{k_{\mathrm{II}}}(t_3, t_2, t_1) &= \frac{1}{2}\mathrm{e}^{-\mathrm{i}\omega_0(t_3+t_1)} \\ &\times \mathrm{e}^{g(t_1+t_2)-g(t_1)-g(t_1+t_2+t_3)} \\ &\times \mathrm{e}^{\mathrm{Re}\left[g(t_2+t_3)+g(t_2)-g(t_3)\right]} \\ &\times \cos\left\{\mathrm{Im}\left[g(t_2+t_3)+g(t_2)-g(t_3)\right]\right\} \end{aligned} \tag{16.35}$$

for the nonrephasing. Notice that since we associate the two dimensional spectra directly with the response functions in the impulsive limit, we use time delays $t_1 \equiv T_1$, $t_2 \equiv T_2$ and $t_3 \equiv t$.

The rephasing 2D spectrum and its time-resolved peak intensities are plotted in Figure 16.6. For the vibronic bath, two cases are illustrated: *damped* vibrations ($\gamma = \omega_0/4$, Figure 16.6a,b) and *undamped* vibrations ($\gamma \to 0$, Figure 16.6c,d). As described above in both cases we use the compound spectral density by adding together the lineshape function representing the semiclassical *overdamped* bath and the damped vibrational part. In the figures, 2D spectra at population times $t_2 = 0$, $2\pi\omega_0^{-1}$ and $5\pi\omega_0^{-1}$ are plotted. Other parameters are such that the reorganization energy in all cases was the same ($2s\omega_0$).

In the 2D plots we observe two strong diagonal peaks (1-1 and 2-2) reflecting the vibrational progression of the absorption. The two off-diagonal crosspeaks (1-2 and 2-1) show coherent quantum interplay of vibronic inner structure. The peak lineshapes show slightly larger broadenings in spectra obtained by using the *damped* spectral density compared to those of *undamped* vibrations. The nonuniform broadening, as introduced in the absorption simulations previously, is observed in the 2D spectrum as well. The time-resolved peak intensities display the coherent nature of the vibrational system: the peaks oscillate coherently in time with the vibrational frequency. Decay of these coherences in the case of *damped* vibrations results in vanishing of cross-peaks: as all spectra are normalized to the maximum of the

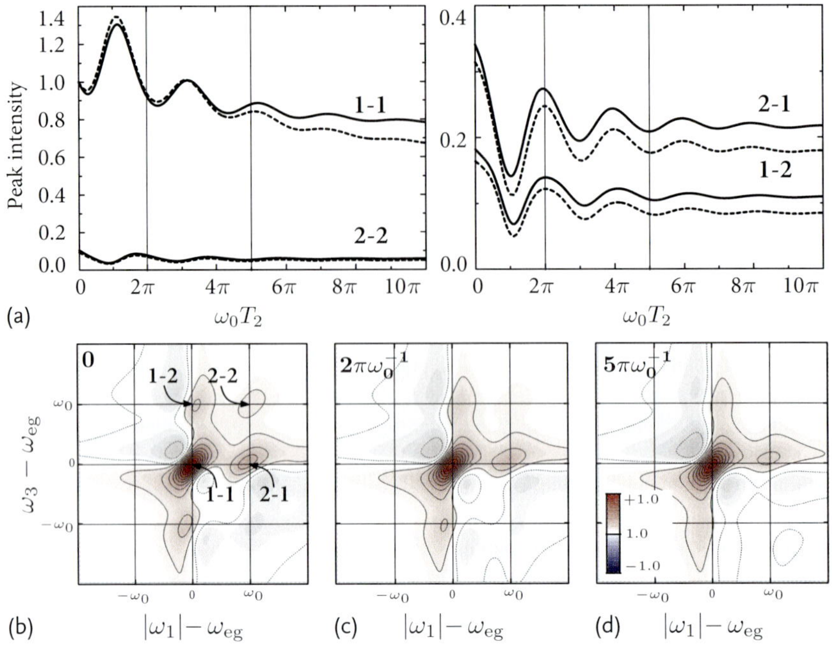

Figure 16.6 Time-resolved 2D spectra of vibrations. (a) Intensities of peaks in the real parts of the rephrasing 2D spectra as functions of the delay time t_2 in the case of damped vibrations with semi-classical (solid lines) and quantum (dashed lines) overdamped bath. The corresponding 2D spectra with semiclassical bath at $T_2 \equiv t_2 = 0$, $2\pi\omega_0^{-1}$ and $5\pi\omega_0^{-1}$ are depicted in (b–d). All values are normalized to the maximum of the rephasing spectrum at $T_2 \equiv t_2 = 0$.

rephasing signal at $t_2 = 0$, the intensities of the crosspeaks and the upper diagonal peak are notably lower than those of the main peak at $(\omega_1, \omega_3) = (\omega_{eg}, \omega_{eg})$. Also, the negative features are more pronounced in spectra of the monomer with undamped vibrations.

The main differences in oscillatory dynamics of peaks in 2D spectra is the damping-induced decay of coherences in the case of *damped* vibrations. Such a decay can be easily related to the value of the damping strength of the vibrational mode γ. The shape of functions of the spectral peak dynamics obtained by using the *quantum* model of the overdamped bath coincides with the *semi-classical* bath simulations at short population times.

16.3 Electronic Dimer

An excitonically coupled dimer is an archetypical molecular system describing excitation properties in molecular aggregates. It has been extensively described in Section 5.3. The general scheme of a realistic molecular heterodimer as well as the corresponding exciton band structure are presented in Figure 16.7a–c. The dimer consists of two coupled chromophores, represented by two dipoles $\boldsymbol{d}_1$ and $\boldsymbol{d}_2$ with the interdipole distance vector $\boldsymbol{R}_{12}$ and angle ϕ. In the Frenkel exciton Hamiltonian the chromophore energies are denoted as ϵ_1 and ϵ_2 and the coupling constant is J, thus the Hamiltonian has four terms

$$\hat{H}_{\text{mol}} = \epsilon_1 \hat{a}_1^\dagger \hat{a}_1 + \epsilon_2 \hat{a}_2^\dagger \hat{a}_2 + J\left(\hat{a}_1^\dagger \hat{a}_2 + \hat{a}_2^\dagger \hat{a}_1\right) . \tag{16.36}$$

The eigenenergies and wave functions of this Hamiltonian can be explicitly calculated, as was done in Section 5.3.

Let us consider the features of the two-dimensional spectrum of the electronic dimer, that can be extended to the aggregates of more molecules. In these systems there are two additional effects compared to the single molecule. These include the excitation energy transfer, which should be understood as the energy relaxation between the exciton states, what is the incoherent effect, and the resonance coupling between chromophores in the site basis, which is the coherent effect causing the exciton delocalization in the eigenstate basis. These two properties induce three types of peaks into the 2D rephasing and nonrephasing spectra.

The 2D spectra are constructed by pathways R_1 to R_4 (15.13)–(15.20). The pathways R_1 and R_2 contain time evolutions of the system in the excited state, $\tilde{\mathcal{U}}^{(ee)}(t_2)$. An arbitrary changes in the excited state propagation will be displayed in the two-dimensional rephasing or nonrephasing spectrum. (R_3 and R_4 pathways have different character.) A complete population transfer from one excited state to another, therefore, exhibits itself in 2D spectrum as a transfer of the original peak ampli-

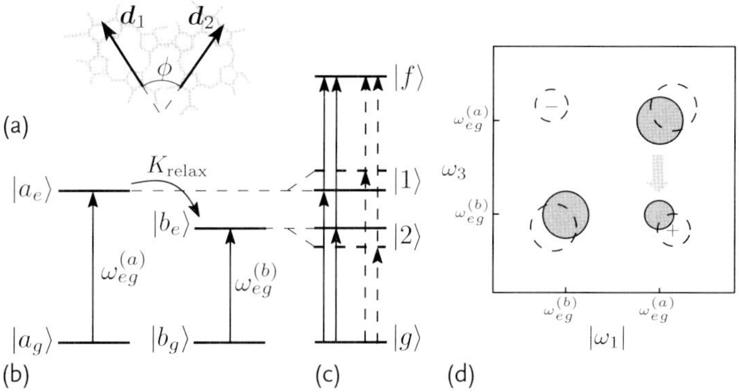

Figure 16.7 Schemes of transition dipoles, energy levels and 2D spectrum of a weakly coupled dimer. (a) Transition dipole vectors in real space associated with chromophore molecules; (b) representation of the dimer by the energy states of its monomers; (c) representation by collective states - excitons $|g\rangle = |a_g\rangle|b_g\rangle$, $|1\rangle = |a_e\rangle|b_g\rangle + c|a_g\rangle|b_e\rangle$, $|2\rangle = c|a_e\rangle|b_g\rangle - |a_g\rangle|b_e\rangle$ (these are not normalized states, c is a constant), and $|f\rangle = |a_e\rangle|b_e\rangle$. (d) The features of the corresponding 2D spectra: the full circles represent diagonal peaks of a weakly coupled dimer including the energy relaxation peaks. Dashed circles represent the peaks of a coupled dimer.

tude to a new spectra position. This results in a crosspeak between two frequency regions as in Figure 16.7b–d.

Population transfer is one way for an off-diagonal feature or a *crosspeak* to appear in the 2D spectrum. In the dimer from Figure 16.7a one can imagine the R_3 pathway (15.15) starting by the transition from the ground state of the monomer A to its excited state, followed by its de-excitation by the second pulse, propagation in the ground state during t_2, and further followed by the excitation of the monomer B by the third pulse. Such R_3 pathway raises a static crosspeak (with respect to t_2) at the same spectral position as the one previously identified to be due to population transfer. Notice that such transferless crosspeaks (second type) could also be easily constructed for systems which are not coupled, because all transitions were done by light alone. However, such crosspeaks survive only in the system of coupled chromophores.

The third type of crosspeak also appears due to coupling between molecules. For example, in the R_2 pathway (15.14) we have the following pattern of time evolutions $\tilde{\mathcal{U}}^{(eg)}(t_3)\tilde{\mathcal{U}}^{(ee)}(t_2)\tilde{\mathcal{U}}^{(ge)}(t_1)$. Notice that superscript e here denotes the band of states. In the secular approximation we may have the following explicit term $\tilde{\mathcal{U}}^{(eg)}_{e_2g}(t_3)\tilde{\mathcal{U}}^{(ee)}_{e_2e_1}(t_2)\tilde{\mathcal{U}}^{(ge)}_{ge_1}(t_1)$ where the subscripts label specific energy levels. After Fourier transforms of $t_1 \rightarrow \omega_1$ and $t_3 \rightarrow \omega_3$ this contribution will result in the crosspeak at $(\omega_3\omega_1) = (\omega_{e_2g}\omega_{ge_1})$. However, the t_2 evolution of this crosspeak is of type $\tilde{\mathcal{U}}^{(ee)}_{e_2e_1}(t_2) \propto \exp(-i\omega_{e_2e_1}t)$. We thus get the t_2 dependant crosspeak that has a coherent oscillatory character. These elements of the density matrix in the t_2 interval $\rho_{e_2e_1}$ are the so-called one-exciton coherences and they map onto the 2D spectrum as oscillatory crosspeaks. This results in a complicated time evolution of the crosspeaks during the time t_2 and provides the possibility to observe negative features in a spectrum that would otherwise be expected to represent a purely absorptive features.

As we find, the system with several energy levels in the e band shows a complicated picture for the spectroscopy. In real samples we always deal with the huge ensemble of molecules. The question might arise, why then do uncoupled two-level molecules yield the spectrum without crosspeaks? To answer this question we need to count all pathways that the perturbation theory requires for the aggregate (or the dimer is sufficient). We can do that with the help of Figure 16.7b,c. They represent the dimer in terms of collective states introduced in Chapter 5. The two-level schemes are equivalent to an uncoupled dimer. Let us now count the rephasing pathways, (15.44), that can contribute to the crosspeak, that is, those that have different first and third interval frequencies. They are presented in Figure 16.8, where one can immediately notice that there are four positive (R_{2g} and R_{3g}) and four negative (R_{1f}) contributions to the signal. Although the doubly excited state $|f\rangle$ is reached in the R_{1f} part of the rephasing response, all transitions can be viewed as normal transitions from the ground state to the excited state in one of the monomers. For example, $|1\rangle \rightarrow |f\rangle$ is the $|b_g\rangle \rightarrow |b_e\rangle$ transition, while the monomer a remains in its excited state $|a_e\rangle$. In fact, for each positive signal we

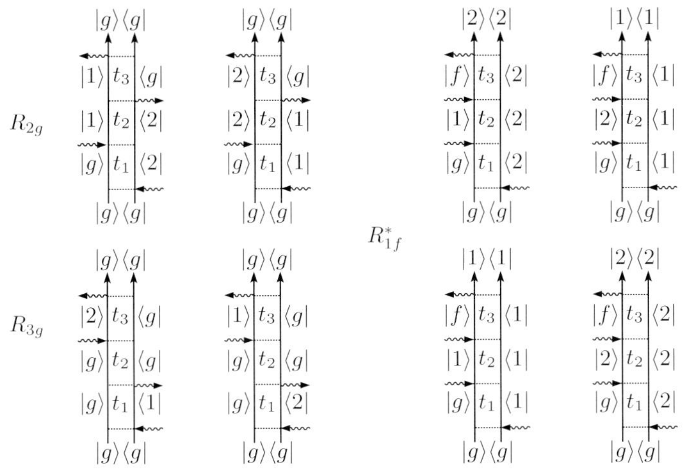

Figure 16.8 Rephasing Liouville pathways for crosspeak in a dimer 2D spectrum. The pathways denoted by the same lower case letter cancel exactly for an uncoupled dimer. Using the definitions from Figure 16.7 the R_{1f}^* pathways can be shown to contain the same transitions as those of the R_{2g} and R_{3g} type.

can find a corresponding negative one, and all crosspeaks cancel. For complexes of uncoupled monomers therefore no crosspeak can appear.

In the case of the molecular aggregates, the main message of the present analysis is that one must not neglect the presence of higher lying two-exciton states. The omission of two-exciton states in calculating the 2D spectrum would lead to unphysical crosspeaks between all energy levels of the one-exciton band.

These effects can be demonstrated on a molecular dimer. To include a finite spectral broadening and energy relaxation effects let the dimer interacts with the environment represented by two overdamped Brownian oscillator coordinates, fast and slow, with relaxation rates Λ_F and Λ_S, respectively. The semi-classical model of spectral density as described in Section 8.6 is

$$C''(\omega) = 2 \sum_{l=S,F} \lambda_l \frac{\omega \Lambda_l}{\omega^2 + \Lambda_l^2} . \tag{16.37}$$

The two modes are independent of each other. Here Λ denotes the relaxation rate of the bath. The chromophores have their individual environments, which induce the fluctuations of molecular transition energies, as described by (16.37). The corresponding lineshape function in the high-temperature limit is

$$g(t) = \sum_{l=S,F} \lambda_l \frac{2k_B T - i\Lambda_l}{\Lambda_l^2} \left(e^{-\Lambda_l t} + \Lambda_l t - 1\right) . \tag{16.38}$$

The parameters can be chosen typical of pigment molecules in photosynthetic proteins [98, 99]. The dimer is defined by $\epsilon_1 = 11\,800\,\text{cm}^{-1}$, $\epsilon_2 = 12\,200\,\text{cm}^{-1}$,

$J = 100\,\mathrm{cm}^{-1}$, $\phi = \pi/6$. Bath induced parameters are $\lambda_F = 30\,\mathrm{cm}^{-1}$, $\lambda_S = 60\,\mathrm{cm}^{-1}$, $\Lambda_F^{-1} = 50\,\mathrm{fs}$, $\Lambda_S^{-1} = 10^5\,\mathrm{ps}$. The slow bath is used to model the static fluctuations with timescale $\Lambda_S \to 0$. The population transfer rates are obtained using the secular Redfield theory: for the exciton eigenstates we have (downward) $K_{1\leftarrow 2}/J = 2.69$ and (upward) $K_{2\leftarrow 1}/J = 2.54 \times 10^{-3}$. The 2D signals were calculated as described in Section 15.2.

As a reference we first present the ideal impulsive 2D photon echo rephasing signal corresponding to the case of the short laser pulses when their spectral bandwidth is much larger than the width of the spectral region under consideration, and time delays between pulses are equivalent with the time delays between interactions in the response functions. The rephasing spectrum at two delay times is presented in the first row of Figure 16.9. The dissection of the spectra to the induced absorption (IA), stimulated emission (SE) and the ground state bleach (GSB) components corresponding to different Liouville space pathways are presented in Figure 16.10 for the real part of the rephasing signal. The population transport diagrams are merged together with the coherent (no transport) diagrams in these figures. The spectra contain both diagonal and off-diagonal elements. Across the diagonal the peaks are broadened due to the homogeneous broadening caused by the fast term of bath oscillations. The lineshapes are extensively elongated along the diagonal due to the slow term of bath oscillations. The limit $T_2 \ll \Lambda_S^{-1}$ ensures that the diagonal elongation remains for all delay times (such approach is very efficient to model the inhomogeneous broadening, and represents the static

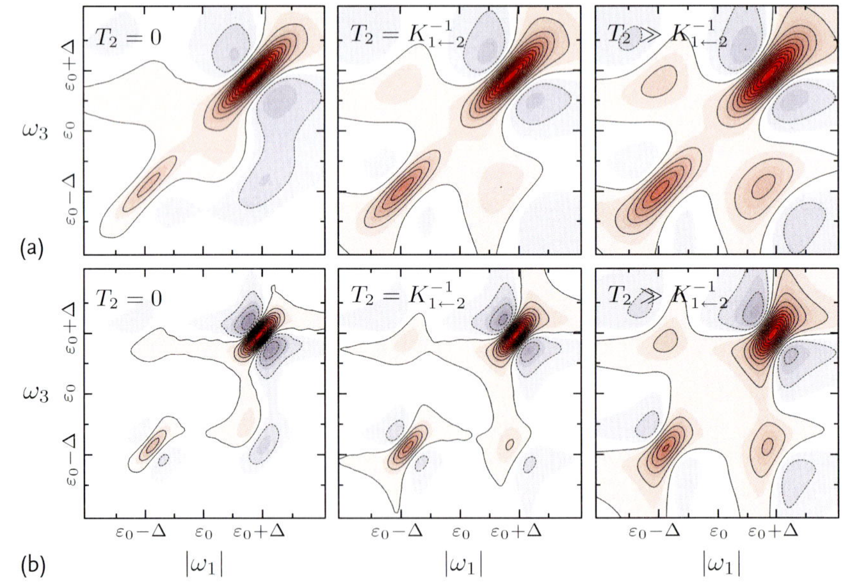

Figure 16.9 2D photon echo broad-bandwidth pulse signal at three delay times as indicated on each subplot. First row: Broad-bandwidth ideal signal. Second row: full signal reconstructed using the set of narrow-bandwidth simulations. See text for simulation parameters.

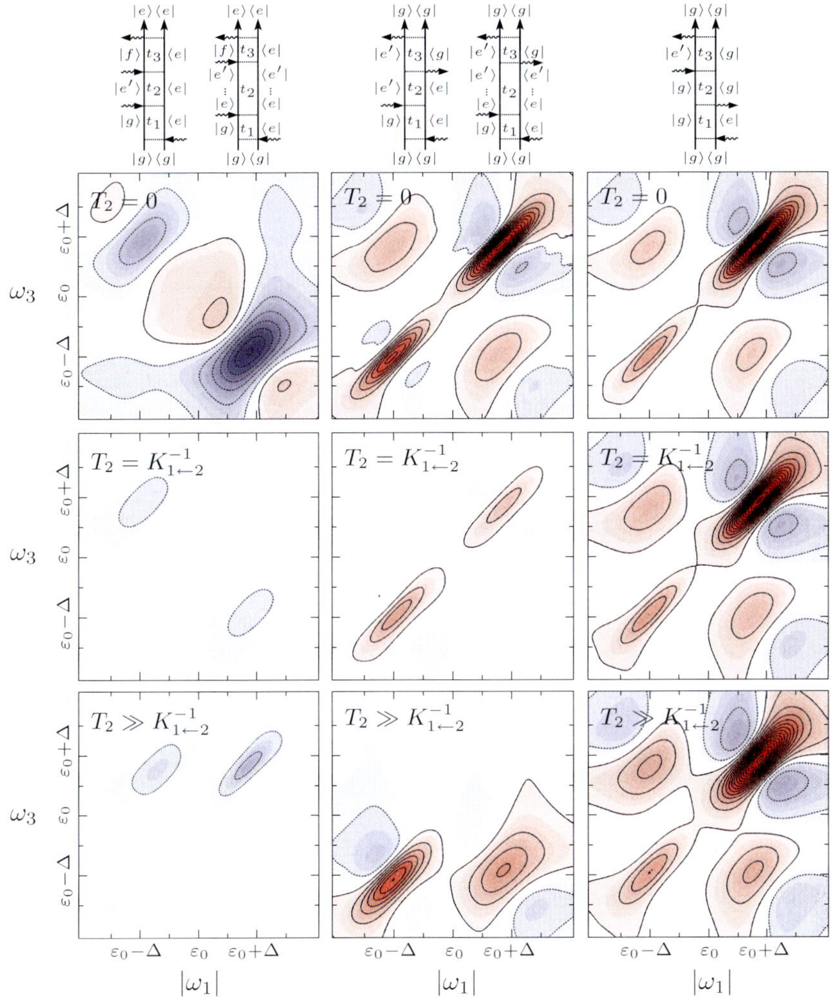

Figure 16.10 The contributions of different Liouville space pathways to the real (absorptive) part of rephasing 2D spectra of excitonically coupled dimer at the impulsive limit. First column – excited state absorption, second – excited state emission, third – ground state bleach. Corresponding constributions are drawn above the 2D plots. All graphs are normalized to the maximum of the most intensive contribution.

disorder effect). Across the diagonal the peaks are broadened due to the homogeneous broadening caused by the fast term of bath oscillations.

At the short delay times ($T_2 = 0$) the population transport is negligible and the diagonal peaks consist solely of the SE and GSB contributions. These two diagonal peaks represent two single-exciton eigenstates and are created when $e = e'$, while the off-diagonal peaks correspond to $e \neq e'$. At zero delay, the crosspeaks are created by the superposition of negative IA and positive GSB and SE contributions.

At longer delay times we see the rise of the lower-energy peaks at $\omega_1 = \omega_3 = \varepsilon_{e_1}$ demonstrating the downhill population transfer in the excitonic system. Only IA and SE contributions change over the population time T_2. GSB is conserved since there is no dynamics in the ground state contrary to the IA and SE diagrams, which depend on the population dynamics and on the coherence, and dephasing terms. At long T_2 the diagonal IA and off-diagonal SE peaks come from population transport.

Realistic optical pulses are not infinitely short. Pulses with the finite length simulate the experiment more realistically, while the impulsive limit simulations are better for the purely phenomenological understanding of different features. Because of the finite pulse duration additional effects of pulse overlap may arise [107]. The narrow-bandwidth pulses also act as band-pass filters of 2D spectra [108] since changing the length of all pulses tunes their spectral bandwidths $\sigma_\omega^{-1} = [\sigma_t]_j$. The realistic pulses can be modeled using Gaussian shapes. They have two additional parameters: the carrier frequencies Ω_j and pulse lengths $[\sigma_t]_j$ (or bandwidths). Such pulses can be included as described in Section 15.1.3. Although in this section we have described the case of the same pulse carrier frequency; however, extension to the different carrier frequencies is trivial [107].

The wavelengths of laser pulses can be tuned independently to select certain resonances in the exciton system. By comparing the pulse bandwidths to the linewidth of a single peak in the spectra, we can obtain certain detection regimes. We can assume that the pulse width is narrower than the whole exciton bandwidth ($\sigma_\omega < \Delta_e$), but broader than the width of a single peak ($\sigma_\omega > \gamma_g$) by setting $\sigma_\omega = 1.2\gamma_e \approx 0.16\Delta_e$. The model dimer has two single-exciton states with energies ε_{e_1} and ε_{e_2}; the double-exciton state energy is $\varepsilon_f = \varepsilon_{e_1} + \varepsilon_{e_2}$. The transition energies are $\omega_{fe_1} = \varepsilon_{e_2}$ and $\omega_{fe_2} = \varepsilon_{e_1}$. Therefore, only two resonant pulse frequencies have to be considered. By considering all possible configurations of the carrier frequencies of the incoming three pulses, we obtain $2^3 = 8$ possible permutations of the pulse frequencies, for example $[\Omega_1, \Omega_2, \Omega_3] = [\varepsilon_1, \varepsilon_1, \varepsilon_1], [\varepsilon_1, \varepsilon_1, \varepsilon_2] \ldots$, and so on. This laser pulse wavelength tuning scheme is sketched in Figure 16.11. However, once we select the resonant contributions, we find only six resonant configurations, the four most significant of which are presented in Figure 16.12 (by selecting the resonant pathways we have also considered population transport at nonzero delays T_2).

Appearance of specific spectral elements in manipulated spectra is controlled by laser pulse frequencies. The first laser pulse "controls" selection of the spectral element at ω_1. For instance, in configurations with $\Omega_1 = \varepsilon_{e_1}$, only spectral elements for $\omega_1 = \omega_{e_1 g}$ do not vanish. The second pulse determines the state, which further evolves in the range of T_2.

It is remarkable that pulses select the distinct Liouville space pathways with high resolution. Various diagonal peaks and the crosspeaks now can be separately characterized including their shape and amplitude. Their time evolution follows density matrix dynamics at corresponding Liouville space pathways. The spectra in Figure 16.9b, reconstructed by summing up all the signals of different laser pulse configurations, resemble the broad-bandwidth signals (the second row in Figure 16.9)

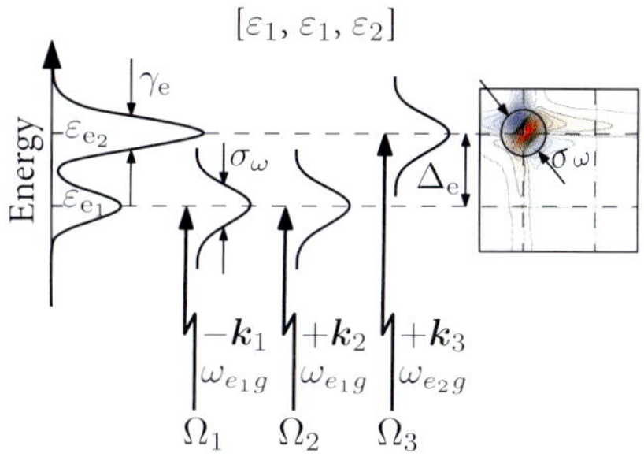

Figure 16.11 Laser pulse wavelength tuning scheme for the FWM experiment. The wavelengths of laser pulses are tuned independently to select certain resonances in the exciton system. In this example, the first ($-\boldsymbol{k}_1$) and second ($+\boldsymbol{k}_2$) pulses have wavelength resonant to $\omega_{e_1 g} = \omega_{f e_2}$ transition, while the third pulse ($+\boldsymbol{k}_3$) is tuned to $\omega_{e_2 g} = \omega_{f e_1}$. We use notation $[\varepsilon_1, \varepsilon_1, \varepsilon_2]$ for this configuration of laser frequencies.

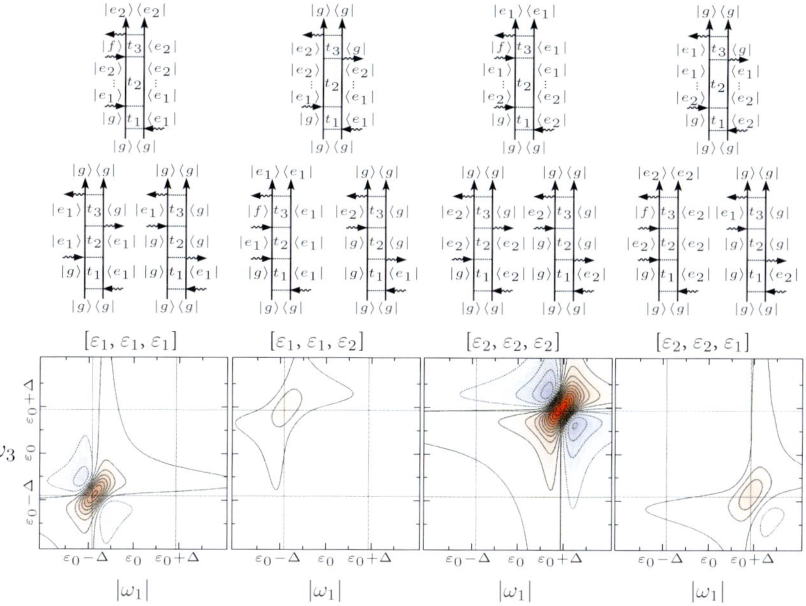

Figure 16.12 Four most intensive narrow-bandwidth signals leading to resonant selection of Feynman diagrams in the signal at delay time $T_2 = 10$ ps. The signals were simulated by varying central pulse frequencies while keeping the pulse bandwidths unchanged. Utilization of other possible laser pulse configurations gives negligible signals at $T_2 \gg K_{1 \leftarrow 2}^{-1}$. All graphs are normalized to the global maximum ($[\varepsilon_2, \varepsilon_2, \varepsilon_2]$ contribution).

very closely. By comparing the spectra it is noticeable that peaks of reconstructed broad-bandwidth spectra are slightly narrower due to the finite bandwidth of the pulses. So this pathway selection phenomenon can be denoted as the *density matrix tomography*.

Extending the phenomenon to the broader class of systems the laser pulse overlap effect can be quantitatively characterized as follows. The response function of an excitonic system is characterized by two parameters: the splitting of the single-exciton states, Δ_e, and the characteristic linewidth of each single-exciton resonance γ. The time-domain response functions then experiences the splitting-related oscillations with frequency Δ_e and the decay with timescale γ_e^{-1}. In our case $\Delta_e > \gamma_e$ and we observe well-separated exciton resonances. The ideal impulsive conditions are fulfilled when $\sigma_\omega \gg \Delta_e$ and $\sigma_\omega \gg \gamma_e$. This corresponds to the *impulsive* regime of ultrashort pulses, when their overlaps can be neglected. For realistic finite-bandwidth Gaussian pulses we need to consider pulse durations, σ_ω^{-1}. In two dimensions of time the whole area where the response function is not zero is γ_e^{-2}. The area, where pulses overlap, is σ_ω^{-2}. The ratio $\eta = \sigma_\omega^{-2}/\gamma_e^{-2}$ thus characterizes the relative pulse-overlap magnitude. Our finite-bandwidth simulations are in the regime $\gamma_e < \sigma_\omega < \Delta_e$. We have $\eta < 1$ and the pulses can thus specifically select resonant peaks, the response function decays slowly compared to the pulse duration and, therefore, the pulse-overlaps make a very small contribution. We call this regime *quasi-impulsive*. That is the ideal regime to be used for selection of specific pathways, i.e. for tomography, of the system with well-separated peaks. The ideal impulsive experiment can then be reconstructed from a set of narrow-bandwidth measurements. Both the impulsive and quasi-impulsive measurements yield very similar results and it is instructive to consider only the impulsive regime, which is mostly reviewed in this book.

The other narrow-bandwidth limit is when $\sigma_\omega < \gamma_e$ irrespective of Δ_e. In this case, the response function decays much faster than the pulse duration and the experiment approaches the frequency domain regime. Varying central frequencies can select certain resonances, but their characterization is more complicated since pulse overlaps dominate due to poor time resolution. This is the *nonimpulsive* regime of the experiment.

16.4
Dimer of Three-Level Chromophores – Vibrational Dimer

Let us now extend the previous model system and consider a homodimer of three-level chromophores. This model essentially represents the two coupled high-frequency (*frequency* $\gg k_B T$) vibrations tackled by the two dimensional measurement. The Hamiltonian of the system is

$$\hat{H}_{\text{mol}} = \omega_0 \sum_{m=1}^{2} \hat{a}_m^\dagger \hat{a}_m + J \sum_{m \neq n}^{2} \hat{a}_m^\dagger \hat{a}_n + \frac{\Delta}{2} \sum_{m=1}^{2} \hat{a}_m^\dagger \hat{a}_m^\dagger \hat{a}_m \hat{a}_m \,, \qquad (16.39)$$

where now operators $\hat{a}$ are of bosonic nature, J describes the resonance transfer of excitations and Δ is the anharmonicity parameter. The one-exciton block is identical to a homodimer of two-level systems

$$\hat{h}^{(1)} = \begin{pmatrix} \omega_0 & J \\ J & \omega_0 \end{pmatrix} \tag{16.40}$$

and corresponding exciton energies and eigenvectors are identical to the ones of the electronic dimer.

In site representation the two-exciton block can be built from the electronic dimer by adding states $\hat{a}_1^\dagger \hat{a}_1^\dagger |0\rangle$ and $\hat{a}_2^\dagger \hat{a}_2^\dagger |0\rangle$ with energies $2\omega_0 + \Delta$ which are denoted as the overtone states of single chromophores. Together with the combination state $\hat{a}_2^\dagger \hat{a}_1^\dagger |0\rangle$ we get three double-exciton states in total. The coupling constant between the overtone states and multiexciton state with energy $2\omega_0$ is now equal to $\sqrt{2}J$ as comes from the Hamiltonian matrix elements:

$$\hat{h}^{(2)} = \begin{pmatrix} 2\omega_0 + \Delta & \sqrt{2}J & 0 \\ \sqrt{2}J & 2\omega_0 & \sqrt{2}J \\ 0 & \sqrt{2}J & 2\omega_0 + \Delta \end{pmatrix} . \tag{16.41}$$

Diagonalization of one- and two-exciton blocks allows us to define the exciton basis

$$|e_j\rangle = \sum_{m=1}^{N} c_{jm} \hat{a}_m^\dagger |0\rangle , \tag{16.42}$$

$$|f_k\rangle = \sum_{m=1}^{N} \sum_{n=m}^{N} C_{mn}^{(k)} \left(\zeta_{mn} + \frac{1}{\sqrt{2}} \delta_{mn} \right) \hat{a}_m^\dagger \hat{a}_n^\dagger |0\rangle . \tag{16.43}$$

The one-exciton eigenvector matrix is identical to the one given in Section 5.3. The two-exciton eigenvector matrix in terms of the mixing angle is now

$$\frac{1}{\sqrt{2}} \begin{pmatrix} \frac{\sin\vartheta}{2\cos\vartheta/2} & \frac{-\cos\vartheta - 1}{\sqrt{2}\cos\vartheta/2} & \frac{\sin\vartheta}{2\cos\vartheta/2} \\ 1 & 0 & -1 \\ \frac{\sin\vartheta}{2\sin\vartheta/2} & \frac{-\cos\vartheta + 1}{\sqrt{2}\sin\vartheta/2} & \frac{\sin\vartheta}{2\sin\vartheta/2} \end{pmatrix} , \tag{16.44}$$

where $\vartheta = \arctan(4J/\Delta)$ and eigenvectors are given in rows. The eigenenergies of one- and two-exciton states are

$$\varepsilon_{e_1} = \omega_0 + J , \tag{16.45}$$

$$\varepsilon_{e_2} = \omega_0 - J \tag{16.46}$$

and

$$\varepsilon_{f_1} = 2\omega_0 + 2J \frac{\cos\vartheta - 1}{\sin\vartheta} , \tag{16.47}$$

$$\varepsilon_{f_2} = 2\omega_0 + \Delta , \tag{16.48}$$

$$\varepsilon_{f_3} = 2\omega_0 + 2J \frac{\cos\vartheta + 1}{\sin\vartheta} . \tag{16.49}$$

Dipole moments for the ground to one- and two-exciton states are

$$\begin{pmatrix} \boldsymbol{\mu}_{ge_1} \\ \boldsymbol{\mu}_{ge_2} \end{pmatrix} = \frac{1}{\sqrt{2}} \begin{pmatrix} -1 & 1 \\ 1 & 1 \end{pmatrix} \begin{pmatrix} \boldsymbol{d}_1 \\ \boldsymbol{d}_2 \end{pmatrix} \tag{16.50}$$

and

$$\begin{pmatrix} \boldsymbol{\mu}_{e_1 f_1} \\ \boldsymbol{\mu}_{e_1 f_2} \\ \boldsymbol{\mu}_{e_1 f_3} \end{pmatrix} = \frac{1}{\sqrt{2}} \begin{pmatrix} -\frac{\sin\vartheta+\cos\vartheta+1}{2\cos\vartheta/2} & \frac{\sin\vartheta+\cos\vartheta+1}{2\cos\vartheta/2} \\ -1 & -1 \\ -\frac{\sin\vartheta+\cos\vartheta-1}{2\sin\vartheta/2} & \frac{\sin\vartheta+\cos\vartheta-1}{2\sin\vartheta/2} \end{pmatrix} \begin{pmatrix} \boldsymbol{d}_1 \\ \boldsymbol{d}_2 \end{pmatrix} \tag{16.51}$$

and

$$\begin{pmatrix} \boldsymbol{\mu}_{e_2 f_1} \\ \boldsymbol{\mu}_{e_2 f_2} \\ \boldsymbol{\mu}_{e_2 f_3} \end{pmatrix} = \frac{1}{\sqrt{2}} \begin{pmatrix} \frac{\sin\vartheta-\cos\vartheta-1}{2\cos\vartheta/2} & \frac{\sin\vartheta-\cos\vartheta-1}{2\cos\vartheta/2} \\ 1 & -1 \\ \frac{\sin\vartheta-\cos\vartheta+1}{2\sin\vartheta/2} & \frac{\sin\vartheta-\cos\vartheta+1}{2\sin\vartheta/2} \end{pmatrix} \begin{pmatrix} \boldsymbol{d}_1 \\ \boldsymbol{d}_2 \end{pmatrix} . \tag{16.52}$$

Having transition dipole moments defined one can construct analytic expressions of the total 2D signal from individual Feynman diagrams. Notice that we now have the same as for the electronic dimer but we have more induced absorption Liouville space pathways. In Figure 16.13 the spectra of the dimer of three-level molecules using different dephasing rates are presented. Coherent induced absorption diagrams produce spectral elements on the diagonal and give oscillations of corresponding diagonal peaks, that is, perceptible in 2D spectra in the limit of small γ. However, if the dephasing rate is $\gamma \gg J$, the inter-state coherence dynamics cannot be separated from diagonal elements and the whole spectra resemble a single anharmonic oscillator.

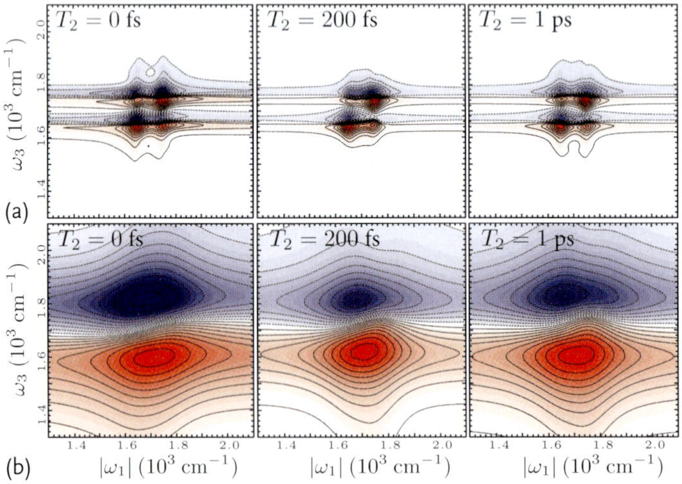

Figure 16.13 The real part of complete 2D spectra of the dimer of a three-level system with dephasing constant $\gamma = J$ (a) and $\gamma = 5J$ (b). $J = 50\,\text{cm}^{-1}$ and $\Delta = 15\,\text{cm}^{-1}$ in both cases.

16.5 Interferences of the 2D Signals: General Discussion Based on an Electronic Dimer

Considerable effort has been devoted to developing methods for relating the two-dimensional signals to dynamical properties in electronic aggregates. It is well established that the diagonal ($|\omega_1| = \omega_3$) peak positions correspond to excitation energies and linewidths show lifetimes and couplings to the environment. The crosspeaks ($|\omega_1| \neq \omega_3$) carry additional information about couplings and correlations of different states, which determine the energy flow pathways and timescales.

The induced polarization vector, which is measured in the spectra, reflects dynamical properties of the system density matrix during each of the time delays t_1, t_2 and t_3 (in the impulsive limit). These properties can be deduced with the help of the Feynman diagrams given in Figure 16.14. Let us consider the rephasing signal. The signal is proportional to the response function $S^{(3)}_{k_1}(t_3 t_2 t_1)$ given by a sum over all resonant pathways of the density matrix. These pathways represent sequences of intermediate states, which show up between the various interactions with the laser pulses. Let us label different pathways by η. The contribution of each pathway to the signal may be represented as $S^{(3)}_\eta(t_3 t_2 t_1) = M_\eta \cdot G_\eta(t_3 t_2 t_1)$, where M_η is a geometry-dependent amplitude of the pathway given by the transition dipole configuration and G represents the time propagators of the density matrix which is independent of the excitation polarizations. $G_\eta(t_3, t_2, t_1)$ contains all relevant system dynamics in the pathway η, as can be deduced from the diagrams in Fig-

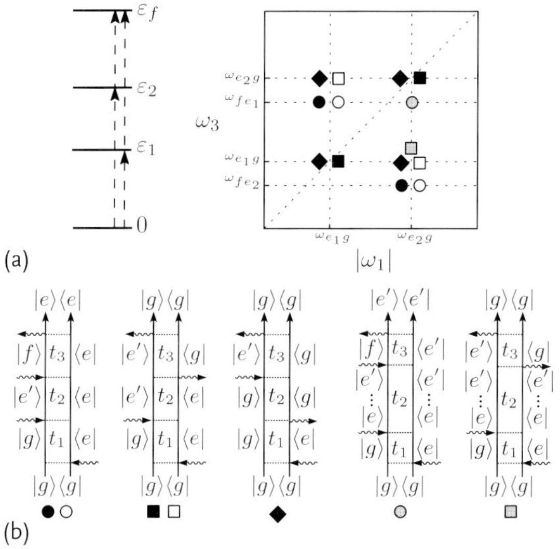

Figure 16.14 Five types of Liouville space pathways of an electronic dimer. (a) Scheme of dimer energy levels, (b) Feynman diagrams for the rephasing spectrum; symbols relate pathways to the 2D spectrum. Open symbols denote electronic coherences during t_2 delay and full symbols the populations during t_2.

ure 16.14: during t_1 the system experiences oscillations between the ground and the singly-excited states e with inter-band optical frequency. During the delay t_2 only intra-band excitonic properties show up. These correspond to either coherent oscillations between two singly-excited states or populations. Population relaxation and transport takes place as well. Inter-band oscillations are again observed during t_3 (these also include inter-band one- to two-exciton frequencies). Fourier transforms with respect to t_1 and t_3 translate the inter-band oscillations into resonant peaks along ω_1 and ω_3 axes respectively. Excited state evolution can then be directly followed by comparing the 2D peak patterns for various t_2.

To gain some insights in Figure 16.14 we sketch the rephasing spectrum of a simplified model dimer of two-level chromophores which has two single exciton energy levels and a single double exciton level. The spectrum at finite t_2 contains contributions from all diagrams, as indicated by the graphic symbols. One specific feature that is observed purely from a symbol pattern on the 2D plot is the absence of open symbols on the diagonal line. The full symbols reflect the populations during the interval t_2. These come from the stimulated emission and the ground state bleach. Thus, the diagonal line of the rephasing spectrum always contains only populations and in the secular density matrix theories these are either static (no relaxation) or exponential functions of time (incoherent exciton relaxation). The first conclusion that comes out is that the diagonal peaks within this level of theory are never oscillatory. The second conclusion comes from simple inspection that for one bleach contribution always comes one stimulated emission contribution, what corresponds to the effective two-level system. The two-level system shows photon echo phenomenon in the rephasing experiment configuration. Thus, the diagonal line of the 2D rephasing spectrum always includes the photon echo and thus shows diagonally elongated lineshapes. These two conclusions, as they can be associated with the independent two-level systems, are not related to the dimer, this applies for any (electronic and vibrational) excitonic aggregates.

So now how about crosspeaks? The crosspeaks in Figure 16.14 are composed of both open and solid symbols. The open symbols must show oscillatory t_2 evolution, while solid symbols are slowly-varying. The open symbols originate from electronic coherences during t_2, thus the place to look for electronic coherent effects is the off-diagonal region in the rephasing spectrum. This again applies not only for the dimer but for an arbitrary excitonic aggregate.

The amplitude M_η of a pathway is a crucial parameter that determines how strong a specific feature emerges in the total spectrum. It turns out, the amplitude can be controlled by polarization configuration of the laser pulses in the experiment design.

The amplitude of a pathway has fundamental symmetry properties, that is, it is invariant to certain permutations of pulse polarization configurations (PPC). First we notice that when the delay time t_2 in Figure 16.14 is set to 0, the signal must be invariant to exchange of the second and third pulses since they are indistinguishable. This implies the permutation symmetry of wavevectors and polarizations of these two pulses at $t_2 = 0$. As quantum dynamics develops during $t_2 > 0$ this symmetry breaks down.

Three independent pulse polarization configurations which exist for isotropic systems in the dipole approximation [102] can be inferred from (15.107) (Section 15.2.6). These are obtained when

$$F^{(4)}(e) = \begin{pmatrix} (e_4 \cdot e_3)(e_2 \cdot e_1) \\ (e_4 \cdot e_2)(e_3 \cdot e_1) \\ (e_4 \cdot e_1)(e_3 \cdot e_2) \end{pmatrix} \tag{16.53}$$

is equal to one of

$$\begin{pmatrix} 1 \\ 0 \\ 0 \end{pmatrix}, \quad \begin{pmatrix} 0 \\ 1 \\ 0 \end{pmatrix}, \quad \begin{pmatrix} 0 \\ 0 \\ 1 \end{pmatrix}. \tag{16.54}$$

These three independent unit vectors correspond to $(e_4 e_3 e_2 e_1) = (x\,x\,y\,y), (x\,y\,x\,y)$, $(x\,y\,y\,x)$. All other configurations can then be obtained from these three fundamental ones as their linear superpositions. Based on this property and using the idea of the previous paragraph we can write a differential signal

$$A \equiv S^{(3)}_{xxyy} - S^{(3)}_{xyxy}\,. \tag{16.55}$$

Here the subscript indicates the pulse polarization configuration as pulse polarization directions $\nu_4\nu_3\nu_2\nu_1$. This signal must vanish identically at $t_2 = 0$ due to the above described symmetry and will gradually build up, with t_2 showing coherent evolution, dephasing and transport. The signal A thus highlights dynamical features of the spectra. However, these features are mixed in A.

For our model dimer in Figure 16.14 the signal A with t_2 will show only oscillating crosspeaks due to open symbols, later these will decay due to dephasing and the stationary crosspeaks will develop due to the ground state bleach and the induced absorption. The higher energy diagonal peak at ε_e will develop due to population relaxation from state e to e' together with down-pointing triangles (the filled circle with filled up pointing triangle both corresponding to the higher energy state e, will disappear).

Other properties of the pathways lead to more symmetries. The pathways η may be classified according to whether the state of the system during t_2 is a population (population pathways, η_{pp}) or a coherence (coherent pathways, η_{cp}). These are represented, respectively, by diagonal (ρ_{ee}) and off-diagonal ($\rho_{ee'}$, $e \neq e'$) exciton density matrix elements. Filled symbols in Figure 16.14 indicate contributions from population pathways and open symbols – from coherence pathways. As can be seen, a common feature of all population pathways is that the first two interactions induce the same transition, $\alpha = \text{d}$ (doorway). The last two interactions induce the same transition too, $\alpha = \text{w}$ (window). The amplitudes of this set of interactions for isotropic systems for three independent pulse polarization configurations are as follows:

$$M_{\eta_{\text{pp}}}(x\,x\,y\,y) = \frac{2}{30}\left[2\boldsymbol{\mu}_{\text{w}}^2\boldsymbol{\mu}_{\text{d}}^2 - (\boldsymbol{\mu}_{\text{w}} \cdot \boldsymbol{\mu}_{\text{d}})^2\right] \tag{16.56}$$

$$M_{\eta_{\rm pp}}(xyxy) = M_{\eta_{\rm pp}}(xyyx) = \frac{1}{30}\left[-\boldsymbol{\mu}_{\rm w}^2\boldsymbol{\mu}_{\rm d}^2 + 3(\boldsymbol{\mu}_{\rm w}\cdot\boldsymbol{\mu}_{\rm d})^2\right] . \tag{16.57}$$

We find that the population pathways have the symmetry for an arbitrary delay t_2 (here the PPC is denoted by $M_\eta(\nu_4\nu_3\nu_2\nu_1)$). Based on this symmetry it is possible to use the combination of PPC

$$B \equiv \boldsymbol{S}^{(3)}_{xyyx} - \boldsymbol{S}^{(3)}_{xyxy} \tag{16.58}$$

to cancel population signatures and to display coherent quantum dynamics. The B signal can be interpreted as follows: only coherent t_2 dynamics will be seen by the coherence pathways. For our model dimer in Figure 16.14 all filled symbols will thus be eliminated in B and the open symbols from excited state absorption and stimulated emission with density matrix coherences will remain. This signal is thus very promising for inspection of electronic coherences in molecular aggregates. Additionally, as we described above the diagonal line in the 2D rephasing spectrum is covered by population involving Liouville space pathways. The B signal thus *eliminates the diagonal features in the rephasing 2D spectrum.* Only off-diagonal (open symbol) contribution will thus remain. This conclusion again applies to a general excitonic aggregate.

Another set of pathways can be isolated by inspecting the coherent contribution to stimulated emission pathway in Figure 16.14b. From the SE diagram we find that these pathways are characterized by only two optical transitions: the first and third transitions on the right side of the diagram are identical $\alpha = \rm r$, while the second and fourth transitions on the left side of the diagram are also identical $\alpha = \rm l$ (we notice that this property is valid for density matrix coherences $\rho_{ee'}$ and populations ρ_{ee} during t_2). By calculating the pathway amplitude we find

$$M^{(\rm SE)}_{\eta_{\rm cp}}(xxyy) = M^{(\rm SE)}_{\eta_{\rm cp}}(xyyx) = \frac{1}{30}\left[-\boldsymbol{\mu}_{\rm l}^2\boldsymbol{\mu}_{\rm r}^2 + 3(\boldsymbol{\mu}_{\rm l}\cdot\boldsymbol{\mu}_{\rm r})^2\right] , \tag{16.59}$$

$$M^{(\rm SE)}_{\eta_{\rm cp}}(xyxy) = \frac{2}{30}\left[2\boldsymbol{\mu}_{\rm l}^2\boldsymbol{\mu}_{\rm r}^2 - (\boldsymbol{\mu}_{\rm l}\cdot\boldsymbol{\mu}_{\rm r})^2\right] . \tag{16.60}$$

This suggests an additional symmetry and an additional signal which cancels excitonic coherences in SE pathway

$$C \equiv \boldsymbol{S}^{(3)}_{xyyx} - \boldsymbol{S}^{(3)}_{xxyy} . \tag{16.61}$$

The C signal will eliminate the stimulated emission coherent pathways: these overlap with population pathways (induced absorption coherent peaks along ω_3 which reflect double-exciton resonances are shifted from these crosspeaks when the two-exciton state energy is different from $\varepsilon_e + \varepsilon_{e'}$). Population relaxation during t_2 can then be monitored through the redistribution of crosspeak amplitudes. In Figure 16.14 the signal C eliminates the open and filled squares from the second diagram at all delay times. By checking the GSB diagram we find that diamonds at the diagonal will also be eliminated. Thus, the diagonal peaks will be also eliminated in the C signal the population transport contribution (shaded symbols) will be enhanced.

It should also be noted that at $t_2 = 0$ we find $B = C$, which is a good reference point and $A \equiv B + C$ at any delay time. These signals can thus be used for realistic electronic systems which can lead to better resolution of coherent and dissipative exciton dynamics. We demonstrate the application of these signals in Section 17.2.2.

16.6 Vibrational vs. Electronic Coherences in 2D Spectrum of Molecular Systems

In previous sections we have analyzed distinct systems: electronic dimers and monomers coupled to vibrations. A few simple electronic level systems with well defined high-frequency vibrations present an important study case of the general phenomena of electron-phonon interaction in the molecular system. In this section we show that the two systems can show a considerable similarity.

In molecules and their aggregates, electronic transitions are coupled to various intra- and intermolecular vibrational modes. Vibrational energies of these are of the order of 100–3000 cm^{-1}, while the magnitudes of the resonant couplings, J, in excitonic aggregates (for example, in photosynthetic pigment-protein complexes or in J-aggregates) are in the same range. Thus, vibronic and excitonic systems show considerable spectroscopic similarities. Because of excitonic or vibrational coherences, electronic and/or vibrational beats in the 2D spectrum are expected as described in Sections 16.2.2 and 16.5 (consider B signal). In complexes of coupled molecules with transitions possibly modulated by vibrational modes it might be difficult to decide what is the origin of oscillations in the 2D spectrum. Indeed, similar spectral beats originating entirely from a high-energy vibrational wavepacket motion have been observed [109, 110].

We can compare two generic model systems (Figure 16.15b,c) which exhibit distinct internal coherent dynamics. The simplest model of an isolated molecular electronic excitation is the vibronic system represented by two electronic states, $|g\rangle$ and $|e\rangle$, which are coupled to a one-dimensional nuclear coordinate q. This is the celebrated *displaced oscillator* (DO) system (Figure 16.15c). Taking $\hbar = 1$, the vibronic potential energy surface of the $|e\rangle$ state is shifted up by the electronic transition energy ω_{eg} and its minimum is shifted by d with respect to the ground state $|g\rangle$; d is the dimensionless displacement. This setup results in two vibrational ladders of quantum sub-states $|g_m\rangle$ and $|e_n\rangle$, $m, n = 0, \ldots, \infty$, characterized by the Huang–Rhys (HR) factor $\text{HR} = d^2/2$ as described in Section 16.2.2.

The other model system, which shows similar spectroscopic properties but has completely different coherent internal dynamics, is the *excitonic homodimer* (ED) without vibrations. It consists of two two-level chromophores (sites) with identical transition energies ϵ. The two sites are coupled by the inter-site resonance coupling J. As a result, the dimer has one ground state $|g\rangle$, two single-exciton states ($|e_1\rangle$ and $|e_2\rangle$ with energies $\varepsilon_{e_1,e_2} = \epsilon \pm J$, respectively), and a single double-exciton state $|f\rangle$ with energy $\varepsilon_f = 2\epsilon - \Delta$, where Δ is the bi-exciton binding energy.

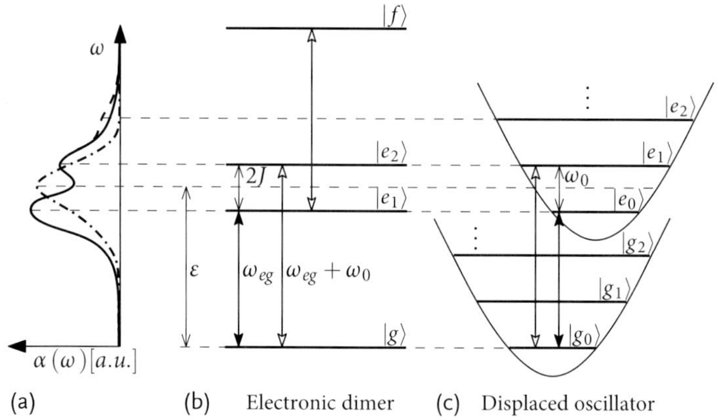

Figure 16.15 Energy level structure of the electronic dimer (b) and displaced oscillator (c) and corresponding linear absorption spectra (a). Dash-dotted line in (a) reflects the typical laser bandwith.

The absorption spectrum of both systems has been described, but it is constructive to revisit its main features here. It is as follows. The absorption of the displaced oscillator is determined by transitions from the $|g_m\rangle$ vibrational ladder into $|e_n\rangle$ scaled by the Franck–Condon (FC) vibrational wavefunction overlaps. Choosing HR $= 0.3$ and $k_B T \approx \frac{1}{3}\omega_0$ and assuming Lorentzian lineshapes with the linewidth γ, we get the vibrational progression in the absorption spectrum (dashed line in Figure 16.15). Here ω_0 is the vibrational energy. The most intensive peaks at ω_{eg} and $\omega_{eg} + \omega_0$ correspond to 0-0 and 0-1 vibronic transitions.

Qualitatively similar peak structure is featured in the absorption of the electronic dimer, where the spectrum shows two optical transitions $|g\rangle \to |e_1\rangle$ and $|g\rangle \to |e_2\rangle$, assuming both are allowed. Choosing $J = \omega_0/2$ and the angle φ between the chromophore transition dipoles equal to $\pi/6$, and using adequate linewidth parameters, we get absorption peaks (solid line in Figure 16.15) that exactly match the strongest two peaks of the displaced oscillator. The absorption spectrum thus does not show the difference between these two distinct systems.

Let us now consider the two-dimensional rephasing spectrum. As described in Section 16.1 the Liouville space pathways can raise as static as oscillatory peaks. We can then write the 2DPE plot by expressing the signal as a sum of peaks $\sum$, which have static (from populations) and oscillatory (from coherences) parts:

$$S_R(\omega_3, t_2, \omega_1) = e^{-\bar{\gamma}_2 t_2} \sum_{i,j} L_{ij}(\omega_1, \omega_3) \times \left[A^{p}_{ij} + A^{c}_{ij} \cdot \cos(|\omega_{ij}| t_2 + \phi_{ij}(\omega_1, \omega_3)) \right] . \tag{16.62}$$

Here ω_{ij} is the characteristic oscillatory frequency of a peak (ij), $A^{p}_{ij}(t_2)$ and $A^{c}_{ij}(t_2)$ are the real parts of orientationally averaged prefactors of population and coherence (electronic or vibronic) contributions, respectively. The spectral lineshape is given by $L_{ij}(\omega_1, \omega_3)$.

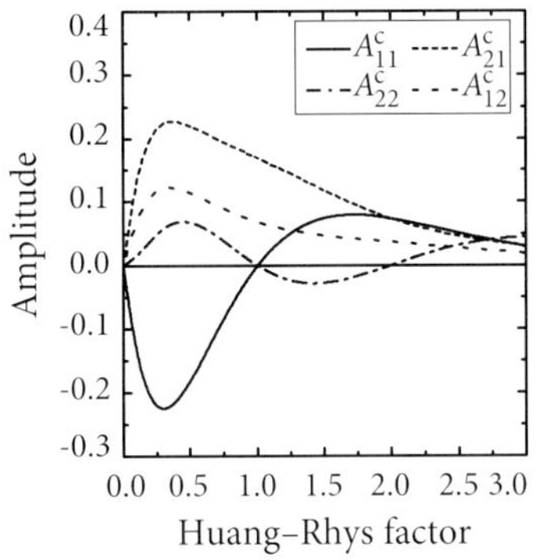

Figure 16.16 The amplitudes of oscillatory peaks of 2D spectra of the DO model for $\boldsymbol{k}_{\mathrm{I}}$ signal. Note that the negative amplitude denotes a phase shift of π of the oscillation.

To apply this expression to the two model systems, we assume a typical situation where the carrier frequencies of the laser pulses is tuned to the center of the absorption spectrum and their bandwidths select the two strongest absorption peaks. In the 2D spectra two diagonal and two off-diagonal peaks for ED and DO are observed. Indices i and j in (16.62) run over the positions of the peaks and thus can be (1,1), (1,2), (2,1), and (2,2). For clarity we study the spectral dynamics with t_2 at the short delays.

The transition dipole properties of the ED result in the picture where all static amplitudes of the ED are positive and the crosspeaks have equal amplitudes $A^{\mathrm{p}}_{12} = A^{\mathrm{p}}_{21}$. The oscillatory amplitudes of the crosspeaks are equal as well: $A^{\mathrm{c}}_{12} = A^{\mathrm{c}}_{21}$. The spectral beats with t_2 can thus only have the same phases in the 2D spectrum when measured at peak centers. Additionally, the oscillatory ESE and ESA parts in ED cancel each other if $\Delta = 0$ and their broadenings are equal. As these relationships do not depend on coupling J and transition dipole orientations, all ED systems should behave similarly.

The amplitude relationships, however, are different for the displaced oscillator system. The amplitudes A^{c}_{ij} of the oscillatory peaks now depend on the Huang–Rhys factor and are plotted in Figure 16.16. The amplitudes A^{c}_{11} and A^{c}_{22} maintain the opposite sign when HR < 2 and are both positive when $2 <$ HR < 3. The oscillation amplitudes A^{c}_{11} and A^{c}_{22} change the sign at HR $= 1$. Amplitudes A^{c}_{12} and A^{c}_{21} are always positive. The amplitudes of static contributions are positive in the whole range of parameters.

We thus find very different behavior of oscillatory peaks of DO and ED systems. A realistic 2D spectra for both DO and ED systems calculated by including phenomenological relaxation and Gaussian laser pulse shapes [101, 107] are plotted in Figure 16.17. The structure and the t_2 evolution of the spectra illustrate the dynamics discussed above and clearly shows the distinctive spectral properties of the vibronic vs. electronic system: (i) diagonal peaks in the $\boldsymbol{k}_{\mathrm{I}}$ signal oscillate in DO, but only exponentially decay in ED (the oscillatory traces come from the overlapping tails of off-diagonal peaks), (ii) the relative amplitude of oscillations is much

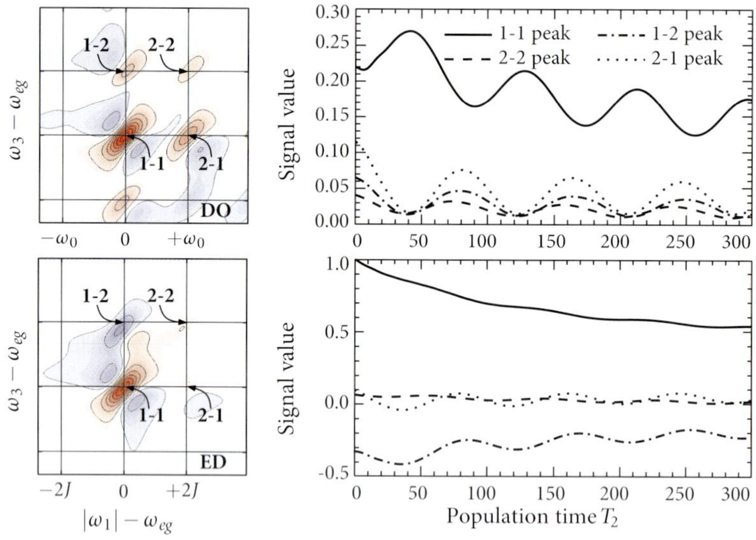

Figure 16.17 2DPE spectra and peak values of the DO and ED as a function of population time t_2, of the $\boldsymbol{k}_{\text{I}}$ and $\boldsymbol{k}_{\text{II}}$ signals. Spectra are normalized to the maxima of the total spectra of the DO and ED.

stronger in DO as compared to ED, where the ESA and ESE cancelation suppresses the oscillations, (iii) opposite oscillation phases are observed in DO, while all peaks oscillate in-phase in ED.

Weakly damped electronic and vibronic coherent wavepackets in molecular systems can thus be discriminated based on fundamental theoretical considerations. Dynamics of diagonal peaks and crosspeaks as well as the relative phase between them in the rephasing signal can now be classified for vibrational and excitonic systems as follows. (i) Static diagonal peaks and oscillatory off-diagonal peaks signify pure electronic coherences, not involved in the energy transport. (ii) Oscillatory diagonal peaks in accord with off-diagonal peaks (0 or π phase relationships) signify vibronic origins. The oscillation phase is 0 for electronic coherences and 0 or π for vibronic coherences.

17
Two Dimensional Spectroscopy Applications for Photosynthetic Excitons

The model systems described in previous chapters are useful for physical understanding and interpretations of the spectroscopy measurements. The spectra and dynamic features related to the transient system evolution in the nonequilibrium excited states can thus be easily understood. Realistic systems can show much more complicated dynamics which we describe next. In this chapter we discuss applications of this theory to the photosynthetic pigment–protein complexes. For most of the simulations presented in this chapter the Frenkel exciton model has been used including the secular population transport model unless indicated separately. The spectral response functions and the two dimensional spectra have been calculated using the expressions with the lineshape functions. In the following only the rephasing 2D spectra are discussed as they display the photon echo effect mapped onto the diagonally elongated lineshapes of various peaks. As this book is about theoretical approaches, we do not describe all experimental details. Instead, in this chapter we review some applications of theoretical approaches to the simulations of the two-dimensional rephasing spectra of few photosynthetic aggregates.

17.1
Photosynthetic Molecular Aggregates

The harvesting of solar energy and its conversion to chemical energy are essential for all forms of life. The primary events that start the whole process of photosynthesis include the photon absorption, transport and charge separation events. The charge separation in the core of pigment–protein reaction center complexes is the first energy conversion step in photosynthesis. The subsequent electron transfer across a thylakoid membrane of chloroplasts triggers a proton transfer reaction, creating a charge gradient that drives a chain of chemical reactions leading eventually to the stable storage of solar energy [98, 99]. This complicated photophyscal/chemical process takes place in membrane-bound photosynthetic complexes. Photosynthetic apparatuses of bacteria and higher plants are the subjects of investigation for scientists from diverse areas trying to understand the structure and underlying mechanisms of highly effective natural solar energy conversion. In order to model the photosynthetic complexes of living organisms on a microscop-

Molecular Excitation Dynamics and Relaxation, First Edition. L. Valkunas, D. Abramavicius, and T. Mančal.

ic scale the accumulated knowledge of different fields of physics, chemistry, and biology needs to be applied [24, 98, 99, 113]. Despite numerous experimental and theoretical publications over the last 20 years revealing structural and spectroscopic properties of photosynthetic complexes there are still open questions in connecting some of their structural peculiarities with the functional properties. As more precise structures of photosynthetic complexes become available there appear more possibilities to explore their spectral properties and excitation dynamics obtained directly from the structural data. These results reveal the roles of different parts of the complexes that they perform in order to optimize the efficiency of the light harvesting and excitation energy transfer within the particular complex and between the complexes to the so-called reaction center, where the energy conversion takes place.

Microscopic understanding of these processes and how they may be tuned could be used to engineer artificial solar cells which mimic the high efficiency of natural organisms.

Below in this section we describe a few well studied photosynthetic aggregates.

17.1.1
Fenna–Matthews–Olson Complex

The Fenna–Matthews–Olson (FMO) complex [111, 112] in green sulfur bacteria is the first light-harvesting system whose X-ray structure has been determined (Figure 17.1). It is known that the FMO complex mediates the transfer of excitation energy from the light-harvesting antennae, chlorosomes, to the so-called reaction center [113], where energy conversion from molecular exciton into the charge pair occurs. The FMO protein is a trimer made of identical subunits, each containing seven bacteriochlorophyll pigments. *Chlorobium tepidum* (C.t.) and *Prosthecochloris aestuarii* (P.a.) are the most thoroughly investigated species with known structures [99, 111, 112, 114, 115]. The two structures are virtually identical, with minor differences in the positions and orientations of the bacteriochlorophylls. However, differences in the local protein environment significantly affect the bacteriochlorophyll site energies. The nature of these interactions is not fully understood. A recent electrostatic computation of the electrochromic shifts of the FMO site energies in both species found that the major contributions to the shifts were from the charged amino acids and the ligands [116].

Interactions of the bacteriochlorophyll *a* molecules in the FMO complexes with the local environment are thus responsible for the differences in the site energies of these pigments. These variations determine the delocalization of the collective electronic states (excitons) as well as their temporal and spatial energy relaxation dynamics.

This system has been extensively studied by linear spectroscopy such as absorption as well as linear and circular dichroism [113]. The linear absorption shows clearly a few peaks reflecting excitonic coherent transitions. These peaks have been attributed to delocalized excitons over specific pairs of molecules. The Frenkel exci-

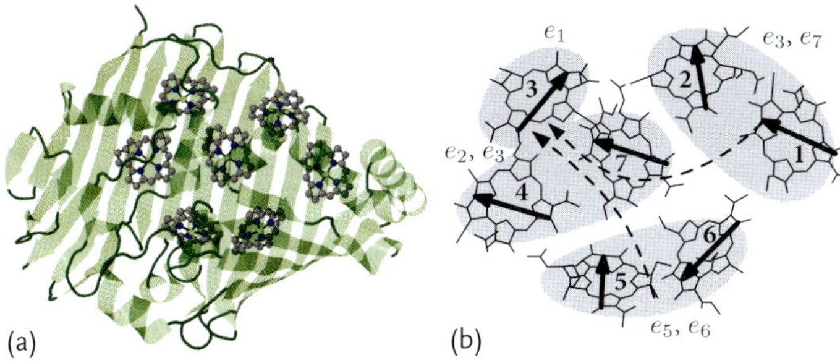

Figure 17.1 Structure of the FMO aggregate: one monomer including the protein (a) and "undressed" pigments (b). Shaded regions represent delocalized excitons, black solid arrows are the molecular transition dipoles, dashed arrows the exciton relaxation pathways.

ton model has been extensively used. Its parameters (site energies and interactions) were fitted to a number of experiments.

Two-dimensional optical spectroscopy [62, 117–119] has revealed that as the excitation energy is transferred towards the reaction center it proceeds in a coherent fashion – it is oscillatory, very fast, and one-directional. It has been suggested that this may be due to correlated chromophore transition energy fluctuations caused by protein thermal motion [120] or by assistance from the coherent protein vibrations [110]. The FMO is thus an ideal test system where coherent vs. dissipative processes can be studied by spectroscopic means with high resolution.

17.1.2
LH2 Aggregate of Bacterial Complexes

In general, bacterial photosynthetic apparatuses are simpler than those of the higher plants and often serve as trial models. One such outstanding system is the photosynthetic light harvesting complex 2 (LH2) of purple bacteria, which is distinguished by an extremely redshifted absorption band and a highly symmetric structure displayed in Figure 17.2 [121–123]. The absorption of the LH2 complex of *Rhodopseudomonas acidophila* (now *Rhodoblastus acidophilus*) strain 10050 is due to Q_y transitions of the bacteriochlorophyll *a* molecules. These molecules are organized in two rings, called B800 and B850 by their lowest absorption wavelength. The overall structure of the complex is cylindrical, characterized by C_9 symmetry with two transmembrane proteins α and β, both sequenced [124], three bacteriochlorophyll *a* and either one or two carotenoid molecules per symmetric unit [125, 126]. Nine bacteriochlorophyll molecules are distributed rather sparsely composing the B800 ring close to the cytoplasmic surface of the membrane, whereas 18 bacteriochlorophylls of the B850 ring on the periplasmic side are made of densely packed dimeric units.

The absorption spectrum of the LH2 complex in the region of Q_y transitions of the bacteriochlorophylls at 77 K has two clear sharp distinct peaks [127] with maxima positioned at 801 nm (B800 band) and 867 nm (B850 band). The contribution

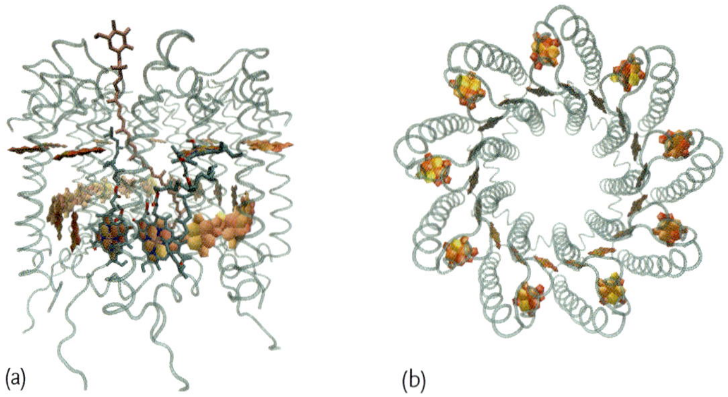

Figure 17.2 Schematic structure of the LH2 aggregate. Turquoise – proteins, gold – bacteriochlorophylls. There are a few caroteinoid chromophores as well. (a) Side view, (b) top view.

of the particular groups of pigments to producing this spectrum is defined rather uniquely with respect to the ring structure. Recent femtosecond coherent experiments reveal more details on the excitonic interactions in the LH2 and similar LH3 [60, 61] aggregates. The experiments reveal the evidence of quantum mechanical interference which represents a previously undescribed strategy for control of excitonic dynamics as revealed by the phase map of quantum beating signals in the two-dimensional signals of the LH2.

Since the first crystallographic structure of the LH2 complex of *Rhodopseudomonas acidophila* was published [125], there have been numerous studies performed, often combining experiments with theoretical modeling, seeking to connect structural properties with functions of the complex (see [24, 127, 128]). Several microscopic models based on the X-ray structure have been suggested to reproduce the absorption spectrum of the LH2 aggregate [129].

17.1.3 Photosystem I (PS-I)

Photosynthetic complex photosystem I (PS-I) is a pigment–protein apparatus shared by bacteria and plants that converts the photon into electrical energy [130]. It exists in trimeric and monomeric forms, but the trimeric species have the same optical properties as monomers: in both the absorbed light at room temperature has > 95% probability to induce charge separation [131]. This suggests that the energy exchange between monomers is negligible and a single monomer can be used to model the spectroscopy measurements.

A recently reported high resolution structure of a cyanobacteria *Thermosynechococcus elongatus* monomer [132] revealed 96 bacteriochlorophylls, and 22 carotenoids embedded in the protein frame. The whole PS-I can be divided into two parts (Figure 17.3): an outer nonsymmetric, 90 antenna bacteriochlorophyll array surrounds a central six bacteriochlorophyll core, which is identified as the *reaction center*, where charge separation takes place. The structure of the PS-I monomer

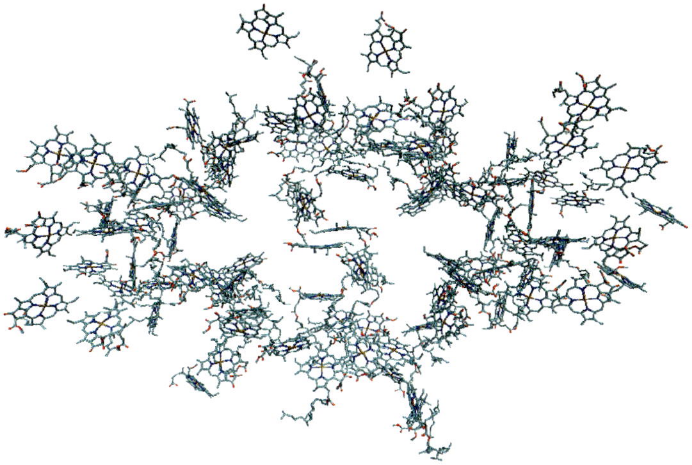

Figure 17.3 Spatial distribution of the 96 bacteriochlorophylls in photosynthetic complex PS-I. The reaction center can be identified in the center. The protein scaffold has been skipped.

(see Figure 17.3) is apparently optimized for the efficient energy conversion, as demonstrated by experiments and supported by numerical simulations [133, 134].

The room temperature PS-I absorption spectrum consists of a broad (700–645 nm) main antenna absorption band and a shoulder to the red from 715 mn [130]. The lowest-energy band of the reaction center is at 700 nm. The "red" absorption band which extends below the RC absorption is a unique feature of the PS-I complex. While the PS-I complexes from different species have a very similar main absorption band, they mostly differ in the red absorption region. Features of charge transfer (CT) states [135] were attributed to a dimeric pattern of bacteriochlorophylls responsible for this red shoulder in both *Synechococcus* and *Synechocystis*.

Exciton dynamics in the PS-I show both coherent and incoherent components which reflect the interplay of localized and delocalized excitons [136]. One- and two-color photon echo peak-shift (3PEPS) measurements performed by Vaswani *et al.* indicated strong excitonic couplings between pigments absorbing at different energies, while the red chromophores show fast decay of the 3PEPS signal due to strong coupling with the protein [137].

Microscopic exciton dynamics simulations for PS-I have been carried out right after high-resolution structural information became available [133, 136, 138, 139]. An effective Frenkel exciton Hamiltonian has been constructed using semi-empirical INDO/S electronic structure calculations combined with the experimental structure [139, 140]. Efficient numerical optimization algorithms have also been applied to search for the transition energy and dipole orientation of each BChl [135, 141]. The resulting parameters provide a good fit to experimental absorption, circular dichroism (CD), and time-resolved fluorescence spectra.

17.1.4
Photosystem II (PS-II)

Photosystem II (PS-II), which is the most abundant photosynthetic complex in nature [142], is responsible for multiple phenomena, such as photon energy harvesting, delivery of excitons to the reaction center positioned in the middle, charge separation, and most importantly, initiation of the water splitting chemistry. Its high resolution structure shows two branches of pigments, D1 and D2, each made up of two chlorophyll molecules and one pheophytin, and other pigments that are separated from these six core pigments either energetically or spatially [143] (see Figure 17.4). Six chromophores of the reaction center are tightly packed within a $\sim 30^3$ Å^3 volume and show strong resonant exciton interactions.

The PS-II aggregate allows direct study of the charge separation process because the reaction center can be physically isolated from the surrounding antennae in a solution. The reaction center is a small aggregate containing only eight chromophores. However, different from the FMO, which is of similar size, the absorption spectrum of the PS-II reaction center is poorly featured [144]. It contains a single broad band from 660 to 690 nm, which at low temperature slightly splits into a double-peak structure. Because of this poor resolubility the simulation studies of the aggregate become very valuable.

The Frenkel exciton model and the system-bath coupling have been parametrized for PS-II by Raszewski *et al.* using a numerical optimization algorithm, which yields good agreement with linear optical properties [145]. A more elaborate spectral density of the system-bath coupling was used by Novoderezhkin *et al.* [146], by employing 48 vibrational bath modes extracted from low-temperature fluorescence line-narrowing data. Despite extensive studies of electron separation and transfer

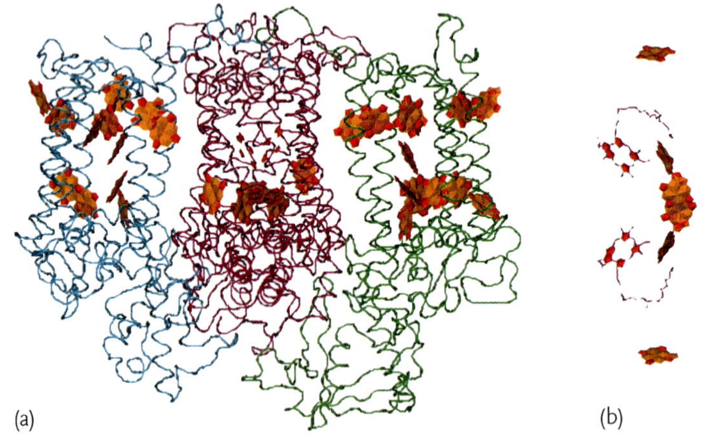

Figure 17.4 (a) Full PS-II complex of plants. Chromophores are displayed by gold pellets, twisted tubes represent protein backbones. Pink – reaction center. (b) Chlorophylls (gold) and pheophytins (pink) of the reaction center (twisted view compared to (a)).

timescales, the primary charge separation site in PS-II RC is still not clearly identified: several radical pair (RP) evolution scenarios fit the existing experiments [147].

The absorption spectrum is not very sensitive to the CT states. This is to be expected since the isolated CT states carry no oscillator strength from the ground state. This selection rule is broken by mixing the CT with the Frenkel exciton states, leading to a weak absorption of the CT state. However, for the same reason, CT states are strongly coupled to the medium causing large broadening, which makes them harder to resolve.

Recently, the experimental 2D spectra of the PS-II RC were recorded [148]. These spectra showed rapid energy equilibration among the PS-II RC pigments and later slow decay of the signal. The 2D studies provide a detailed picture of the excitation frequency dependent spectral signatures of charge separation.

17.2 Simulations of 2D Spectroscopy of Photosynthetic Aggregates

Over the last five years a huge leap, fueled by the invention of two-dimensional electronic spectroscopy, was achieved in understanding coherent and dissipatory effects in the photosynthetic light harvesting pigment–protein antenna complexes that reside inside specific membranes in bacteria and plants. As the main function of the peripheral chromophores is funneling of energy to the reaction center, this process has been extensively probed by time-resolved two-dimensional optical spectra. In this section we present some modeling of two-dimensional spectroscopy in the photosynthetic aggregates described above. For calculations the theory described in previous chapters is being used. Also notice that further on we use the impulsive limit (unless separately indicated) so the interval of the response function t_2 is equivalent to the time delay between optical pulses T_2. In the two-dimensional plots we use color scale where blue peaks indicate the positive peak amplitude, while yellow represents the negative. Green color reflects zero values.

17.2.1 Energy Relaxation in FMO Aggregate

Spectroscopic studies of light harvesting and subsequent energy conversion in photosynthesis can track quantum dynamics happening at the microscopic level. The Q_y band of FMO was described using the Hamiltonian of Aartsma [24, 149] as refined by Brixner *et al.* [117]. Atomic coordinates were taken from [114]. The electric transition dipoles were assumed to pass through the nitrogen *b* and *d* atoms according to the crystallographic nomenclature. The monomer electric dipole strength was 5.4 D [114]. The coordinates of the magnesium atoms located at the center of BChl a molecules were used as the reference points of chromophore coordinates when necessary.

The commonly accepted model for the coupling to the environment is the assumption that each chromophore experiences identical, however, uncorrelated fluctuations. The semiclassical overdamped spectral density given by (8.105) is then often used to represent the fluctuations of the chromophores. The bath relaxation time $\gamma^{-1} = 100$ fs and the reorganization energy $\lambda = 55\,\text{cm}^{-1}$ for each bacteriochlorophyll then properly describe the homogeneous linewidth in the absorption spectrum. However, it should be noticed that these bath parameters are not unique as the spectrum weakly depends on the fine structure of the *overdamped* bath since the overall spectral density is a broad featureless function. This part of the bath represents dynamical fluctuating environment. The static disorder can be included explicitly on top of the fluctuations using the Monte Carlo sampling and ensemble averaging. The typical disorder for FMO site energies is distributed according to Gaussian shape (width is $20\,\text{cm}^{-1}$ as proposed by Brixner *et al.* [117]). The orientational averaging is performed as described in Section 15.2.6. The absorption calculated using this procedure is presented in Figure 17.5. There we clearly identify three low-energy peaks and one shoulder on the higher energy side. While this shape is created by seven exciton states, some of them have small transition dipoles and poorly contribute to the spectrum.

The two-dimensional rephasing spectrum was calculated by summing over all possible Liouville space pathways. It is easy to check that for seven chromophore molecules we have seven single exciton states and $7 \times 6/2 = 21$ double exciton states. In that case we have 7×7 ground state bleach pathways, 7×6 stimulated emission pathways that include only exciton coherences during t_2 interval, 7×7 stimulated emission pathways that include exciton populations and their energy relaxation during the same interval. Additionally, we have $7 \times 6 \times 21$ induced absorption pathways with coherences during t_2 and $7 \times 7 \times 21$ induced absorption pathways with populations during T_2. So, for the FMO aggregate we need to sum up 2051 pathways (in general for N electronic two-level sites we will have $N(3N-1) + N^2(N-1)^2(1+1/2(N-1))$ number of pathways). For the homogeneous line broadenings we use the cumulant expansion as given in Appendix A.10.

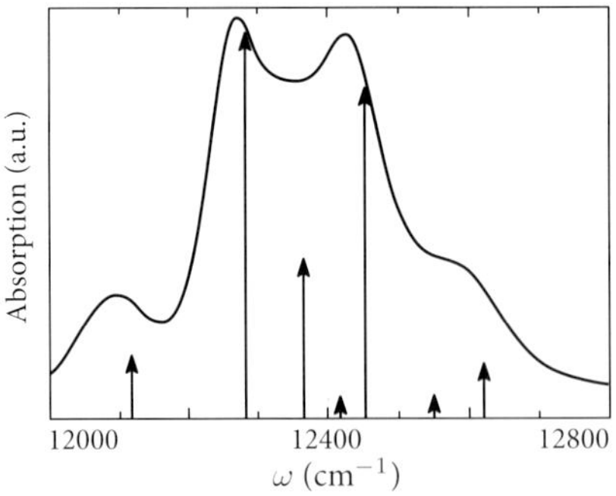

Figure 17.5 Simulated absorption spectrum of the FMO aggregate [150]. The vertical arrows represent the excitonic energy levels and the length of the arrow is the oscillator strength $|\mu_{eg}|^2$.

The simulations thus require secular exciton transport rates that are calculated using the secular Redfield equation (Section 11.4). It is used solely in the interval t_2 where the population dynamics reduces to the master equation for diagonal density matrix elements.

For simplicity we choose the regular laser setup where all laser polarizations are parallel. Since the whole spectrum is orientationally averaged with respect to the single molecule, wa can take $xxxx$ polarization configuration.

The simulated 2D rephasing spectra at different delay times are shown in Figure 17.6. It shows a set of diagonal peaks that reflect main excitonic resonances. These diagonal peaks are slightly elongated along the diagonal reflecting the inhomogeneity of the system. The green-to-yellow features above the diagonal at $T_2 = 0$ represent the induced absorption. As time T_2 evolves the blue crosspeaks below the diagonal becomes visible; these reflect the energy relaxation from the higher-energy diagonal peaks into the lower-energy crosspeaks. The 2D spectrum thus reveals the energy relaxation pathways with high resolution. However, there are no well resolved crosspeaks that would oscillate, so we do not observe excitonic coherent effects with this approach (see Section 16.5 on static and oscillatory peak patterns).

The effects of coherent dynamics have drawn considerable interest in FMO since experiments of Engel [118], Fleming [151], Scholes [152] and Kauffmann [153] groups observed some coherent beats (up to 1 ps) of various peaks in the two-dimensional spectra. One reason behind these oscillations is the weak exciton intra-band dephasing, what could keep coherences "alive".

Long-range correlated chromophore transition energy fluctuations is one such possible mechanism as suggested by Lee *et al.* [64]. Assuming that energy levels of all chromophores fluctuate in-phase, the exciton wavefunction phases are not affected by this motion; the entire exciton band energy is modulated together. This effect can be easily described using the theory presented in previous chapters.

As the molecular transition energies are linearly coupled to the bath, all bath-induced properties are determined by the following matrix of correlation functions or spectral densities $C''_{mn}(\omega)$. This matrix describes how fluctuation of energy of the molecule m is connected to that of the molecule n. The transformation of spectral

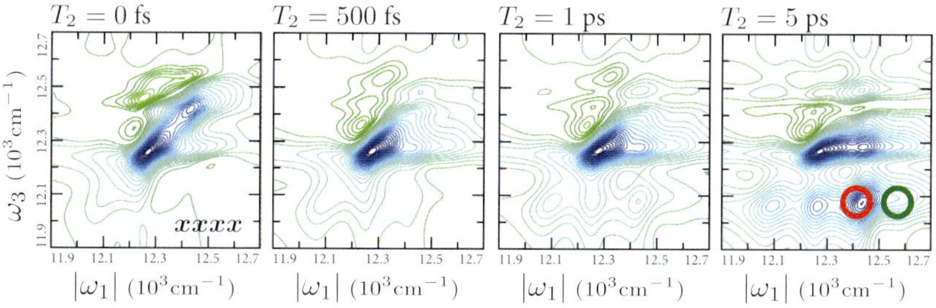

Figure 17.6 Time evolution of the two-dimensional rephasing spectra of the FMO aggregate simulated using $xxxx$ pulse polarizations [150]. Red and green circles mark the crosspeaks reflecting the energy relaxation.

densities between the molecular and the exciton basis is then given by

$$C''_{e_4e_3,e_2e_1}(\omega) = \sum_{mn} C''_{mn}(\omega) c_{me_4} c^*_{me_3} c^*_{ne_2} c_{ne_1} \,. \tag{17.1}$$

The common approach that has been described above is that the transition energy fluctuations of all chromophores are uncorrelated. This case is given by $C''_{mn}(\omega) = \delta_{mn} C''_u(\omega)$. The dynamical properties that are observed in the spectra of excitons depend on fluctuations *in the exciton basis* set, where

$$C''_{e_4e_3,e_2e_1}(\omega) = C''_u(\omega) \sum_m c_{me_4} c^*_{me_3} c^*_{me_2} c_{me_1} \equiv \xi_{e_4e_3,e_2e_1} C''_u(\omega) \,. \tag{17.2}$$

$\xi_{e_4e_3,e_2e_1}$ is the exciton overlap matrix. Since in general $\xi_{e_4e_3,e_2e_1}$ is finite for all combinations of exciton indices, we find that uncorrelated fluctuations contribute to both exciton transport (via off-diagonal fluctuations, $C''_{ee',e'e'}$) and pure dephasing (via diagonal fluctuations, $C''_{ee,ee}$). These cases have been discussed with respect to relaxation in a harmonic oscillator in Chapter 8.

The opposite extreme case is when the transition energy fluctuations of all chromophores are fully positively correlated. This case is given by $C''_{mn}(\omega) = C''_c(\omega)$ and in the exciton eigenstate basis the spectral density assumes the form

$$C''_{e_4e_3,e_2e_1}(\omega) = \delta_{e_4e_3} \delta_{e_2e_1} C''_c(\omega) \,. \tag{17.3}$$

The correlated fluctuations of molecular transition energies thus lead to correlated diagonal fluctuations of exciton transition energies. Off-diagonal fluctuations of excitons, which could lead to exciton transport, are absent and *the population transport vanishes*. As the transport is observed, the case of fully correlated fluctuations is thus not realistic in FMO aggregate.

Let us include these correlation effects in the simulation of two-dimensional spectra [120] and let us label, using (i), the model which neglects all correlations so that $C_{mn}(t) = 0$ $(m \neq n)$. As the intermediate case we can assume (model ii) exponentially decaying inter-chromophore correlations

$$C^{(\text{ii})}_{mn}(t) = e^{-|r_m - r_n|/l} C(t) \,, \tag{17.4}$$

where l is the spatial correlation distance. We denote this case by the *exponential model*. Let us also denote the highly asymmetic case (iii), denoted by *a cut-off* model that makes a sharp cut-off with the distance, that is

$$C^{(\text{iii})}_{mn}(t) = \theta\left(\frac{|\boldsymbol{r}_m - \boldsymbol{r}_n|}{l} - 1\right) C(t) \,, \tag{17.5}$$

where $\theta(x)$ is the step function: the correlation vanishes when the distance between chromophores is larger than the correlation distance l.

The strength of the intermediate correlations was chosen as follows. The FMO active region size (largest distance between central Mg atoms of chromophores in FMO) is $\sim$ 27 Å, thus in the exponential model we take $l = 30$ Å. For the cut-off

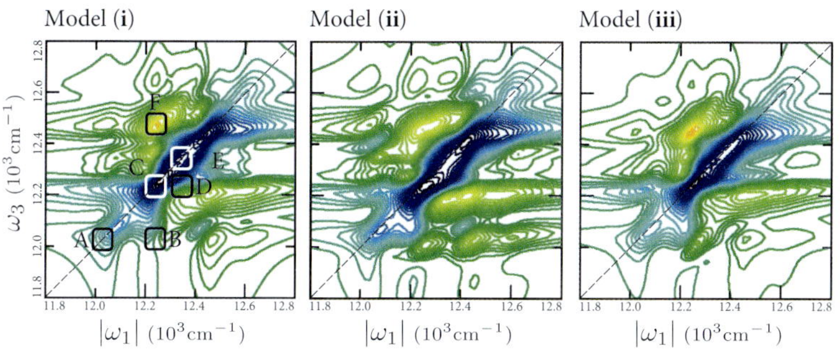

Figure 17.7 2D spectrum of FMO calculated using the three models of correlated fluctuations for $T_2 = 0$ as indicated [120]. Regions A to F (each $50 \times 50\,\text{cm}^{-1}$) were selected for further examination of the time T_2 dependence. This is shown in Figure 17.8. Notice that the diagonal disorder is not included in these calculations.

model we take $l = 12\,\text{Å}$ that makes sites 3 and 4 correlated as well as 5 and 6 correlated.

How do the exciton dynamics in these correlated conditions show up in spectroscopic signals? The absorption spectrum weakly depends on the spatial correlations so it is not considered. The single-exciton dynamics is more directly mapped by the 2D rephasing spectra. On the off diagonal regions in the 2D spectra the peaks belong to populations and coherences. At $T_2 = 0$ populations of all models are identical, only spectral broadening along ω_1 and ω_3 axes is affected by correlations. So the 2D spectra shown in Figure 17.7 for the three fluctuation models at $T_2 = 0$ are similar on the diagonal region, but show differences in the off-diagonal area below the diagonal.

This dependence on correlations changes dramatically in the time-resolved 2D spectra. We have selected six areas of the spectrum, which correspond to the most prominent peaks in the 2D spectrum and their crosspeaks to monitor their variation with the population delay time T_2. In Figure 17.8 diagonal peaks A, C and E demonstrate that the population relaxation timescales depend on these correlations. There is a clear difference between the three models in peaks A, and especially C. Population redistribution in the interval 200–1000 fs is much weaker in the cut-off model for these excitons (1 and 2). The most notable difference is the highly-oscillatory crosspeak dynamics for correlated fluctuation models. These simulations thus show that spatial correlations of fluctuations dramatically reshape the crosspeak region and the time-dependence of various crosspeaks in the 2D rephasing spectra.

17.2.2
Energy Relaxation Pathways in PS-I

As described in Section 16.5 some general principles for the design of two-dimensional rephasing spectra by controlling pulse polarizations can be developed. These have been applied to FMO photosynthetic aggregate in [150, 154]. The same strategy has been used to study exciton dynamics in a "big" photosynthetic complexe such as PS-I [155]. Here, we briefly revisit this study. One special interest in PS-I is that it contains three sets of special chromophores. There are so-called red chromophores, whose energies are lower than of the reaction center chromophores; these are A2, A3, A4, A19, A20, A21, A31, A32, A38, A39, B6, B7, B11, B31, B32 and B33 [132]. The other set in the center of the structure determines the reaction center (S1–S6). The third set is the main antenna region that surrounds the reaction center.

The exciton model parameters for the PS-I complex were taken as refined by Vaitekonis *et al.* [135]. The coupling of the chromophores to the bath is included using the model of spatially uncorrelated fluctuations. We are then left with the spectral densities. However, now the bath spectral densities depend on the type of the chromophore [141]:

$$C_n''(\omega) = 2\lambda \frac{\Lambda\omega}{\omega^2 + \Lambda^2} + \sum_j^5 \eta_{j,n} \frac{\omega^4}{\Omega_j^3} \exp\left(-\frac{\omega}{\Omega_j}\right) . \tag{17.6}$$

The first overdamped Brownian oscillator mode with the coupling strength $\lambda = 16\,\text{cm}^{-1}$ and the relaxation rate $\Lambda = 32\,\text{cm}^{-1}$ controls the homogeneous linewidth of the spectra. The remaining, Ohmic, part of the spectral density determines the population transport rates and was tuned by fitting the time resolved fluorescence. The Ohmic frequencies (in cm^{-1}) are: $\Omega_1 = 10.5$, $\Omega_2 = 25$, $\Omega_3 = 50$, $\Omega_4 = 120$ and $\Omega_5 = 350$; for the red chromophores we use $\eta_{1,r} = 0.0792$, $\eta_{2,r} = 0.0792$, $\eta_{3,r} = 0.24$, $\eta_{4,r} = 1.2$, $\eta_{5,r} = 0.096$; for the remaining chromophores $\eta_{j,n\neq r} = 0.024$ for all j.

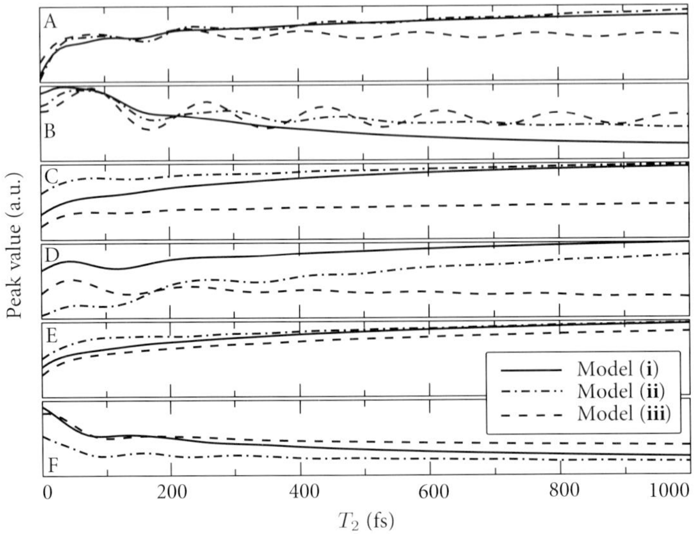

Figure 17.8 Time dependent amplitudes of the A–F regions of Figure 17.7 for the three fluctuation models [120].

The PS-I aggregate contains a lot of chromophores and the full cumulant function expressions were reduced by approximating all spectral lineshapes to Lorentzians with the predefined dephasing constant γ: each Liouville space pathway with pattern

$$\exp(-i\omega_3 t_3 - i\omega_2 t_2 + i\omega_1 t_1) \tag{17.7}$$

($|\omega_2| > 0$) was appended by exponential damping:

$$\exp(-i\omega_3 t_3 + i\omega_1 t_1 - \gamma_3 t_3 - \gamma_2 t_2 - \gamma_1 t_1)\,, \tag{17.8}$$

while for population-involving pathways ($\omega_2 = 0$) the secular Redfield equation with respect to t_2 has been solved similar to Section 16.2.1. The damping constants for all excitons γ_e have been calculated using the Redfield model (Section 11.4). Additionally, the double-exciton states were represented in the one-exciton product basis [105] to simplify parametrization of double-exciton dephasings. Uncorrelated static diagonal Gaussian fluctuations of all pigment transition energies with $90\,\mathrm{cm}^{-1}$ variance were added by statistical sampling to account for the inhomogeneous spectral linewidth.

The exciton state energies ε_e and wavefunctions c_{ne}, obtained by diagonalizing the single-exciton block of the Hamiltonian matrix, describe the involvement of different pigments in the spectroscopy regions. The single-exciton wavefunction coefficients c_{ne} denote how the eigenstate e projects into the nth chromophore. The single-exciton states have been classified according to their participation in certain structural patterns [155]. The excitons belonging to the red states, the reaction center states, have been identified. There are four exciton states which link the reaction center with the peripheral antenna: these have been denoted as linker states. They are expected to be responsible for exciton delivery to the RC. The linker states have energies close to the lower energy edge of the bulk antenna band ($14\,600$–$14\,700\,\mathrm{cm}^{-1}$). In Figure 17.9 we show this assignment in real space. The

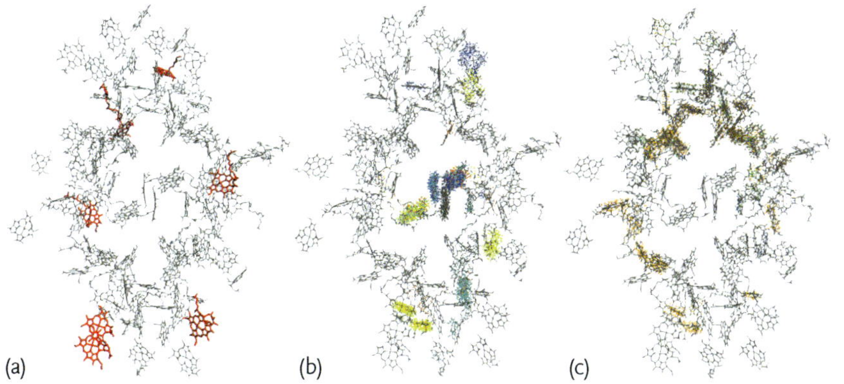

Figure 17.9 Exciton probability distribution of groups of exciton states shown in real space (density of dots represents $|c_{ne}|^2$). (a) Red states, (b) RC and linker states, (c) delocalized antenna states [155].

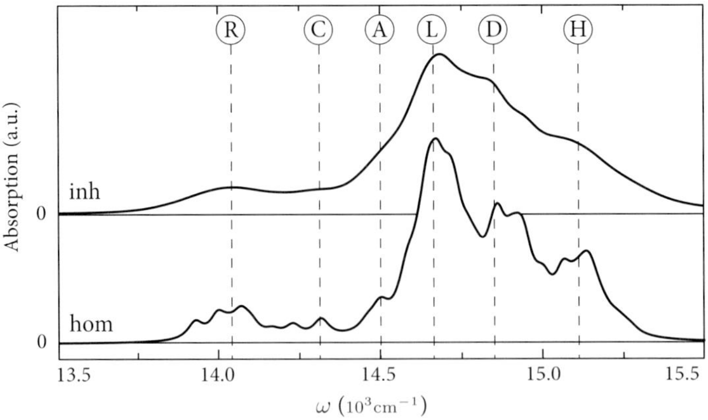

Figure 17.10 Simulated absorption of the PS-I complex [155]; "inh" denotes full simulations averaged over 1000 diagonal disorder configurations, "hom" – homogeneous model. R – red excitons, C – RC peak, A – bulk antenna lower-edge, L – linker states, D – delocalized states over most of the antenna, H – bulk antenna higher-edge.

red states are mostly localized on few chromophores scattered throughout the PS-I, the RC and linker states cover the reaction center and show long tails extending to the edges of the antenna. Figure 17.9 shows that the delocalized antenna states completely surround the RC and overlap with the tails of the linker states.

The absorption spectrum shown in Figure 17.10 can be clearly separated into the antenna region between 14 500 and 15 300 cm^{-1} and the red absorption region 13 700–14 300 cm^{-1}. The absorption is inhomogeneously broadened with a limited structure. The six main peaks are marked by vertical lines. For better peak assignment we also show the homogeneous spectra calculated without inhomogeneous broadening and using constant small homogeneous 30 cm^{-1} linewidth.

We can thus introduce six types of transitions: the red states (R) at 14 000 cm^{-1}, the RC transition (C) at 14 300 cm^{-1}, the linker transitions (L) at 14 670 cm^{-1}, the delocalized antenna region (D) at 14850 cm^{-1}. The bulk antenna starts at 14 500 cm^{-1}, which is the low-energy edge (A), up to the upper-energy edge (H) of the bulk antenna at 15 110 cm^{-1} and covers A, L, D, and H features. The R, C, L, D, and H states can be identified in the absorption.

Two-dimensional rephasing spectra show exciton correlations as crosspeaks in 2D correlation plots. Let us take one of the primary laser pulse polarization configurations $e_4 e_3 e_2 e_1 = xxyy$. In Figure 17.11 we display the rephasing 2D signal at $t_2 = 0$. Its primary diagonally-elongated feature originates from single-exciton contributions (bleaching and stimulated emission) associated with R, C, and overlapping L and D states as marked by dotted circles. Some weak broad off-diagonal features extending from the diagonal can be observed. The diagonal line mimics the absorption. The R peak is well-separated from the rest peaks whereas other diagonal transitions strongly overlap. The C diagonal peak can be identified as an extended shoulder of the antenna band. The strongest peak corresponds to the L

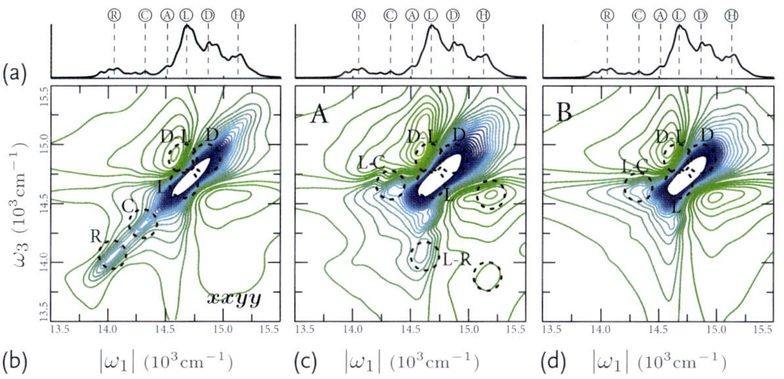

Figure 17.11 (a) The homogeneous absorption spectrum. (b) $xxyy$ tensor component of the 2D rephasing spectrum [155]. (c) Signal $A(\omega_3, t_2 = 1\text{ fs}, \omega_1)$. (d) Signal $B(\omega_3, t_2 = 0, \omega_1)$. The 2D plots are normalized to their maximum.

region. However, off-diagonal regions around the bulk antenna bands are mostly featureless.

Let us now employ the optimized pulse polarization configurations, as described in Section 16.5. It has been shown that the A signal vanishes at $t_2 = 0$. At short delay ($t_2 = 1$ fs in simulations) it gives essentially the zero-time derivative of the 2D spectrum. In Figure 17.11 it shows mainly as two overlapping diagonal peaks, which can be identified with the L and D features. A set of crosspeaks are observed: D–L, L–C are well-resolved, while others may be related to H–L, H–R. Diagonal features of the red states and of the reaction center are absent. Spectral features of this signal indicate ultrafast exciton dynamics within the bulk antenna and demonstrate that the delocalized states are very active at short times. Strong L–C features on both sides of the diagonal show that the excitons reach the RC at very short times.

The A signal in Figure 17.11 contributes to very short t_2 delays right after the excitation. We find that the exciton dynamics is very significant shaping the spectrum. Let us now consider the dynamical properties at later times. The complete t_2 evolution of the elementary tensor component $xxyy$ is depicted in Figure 17.12. Exciton equilibration in the bulk antenna region within 1 ps is seen as change of the 2D pattern around $(\omega_3, |\omega_1|) = (14\,750, 14\,750)\text{ cm}^{-1}$. The population is subsequently trapped by the red states within 5 ps. This signal shows an extended crosspeak lineshape at $\omega_3 = 14\,000\text{ cm}^{-1}$, close to the red state exciton energy indicating that the red states act as the energy sink as the reaction center is not active (charge separation is not included in the model).

However, the density matrix coherences are hindered in the $xxyy$ configuration. The coherent evolution can be monitored by the $B(\omega_3, t_2, \omega_1)$ spectrum. The spectrum B at $t_2 = 0$ as shown in Figure 17.11 (and similarly at $t_2 = 50$ fs in Figure 17.13) mostly reveals the bulk antenna region. The pattern has a very strong diagonally-elongated peak which covers L and D states. Weaker C–L crosspeaks in-

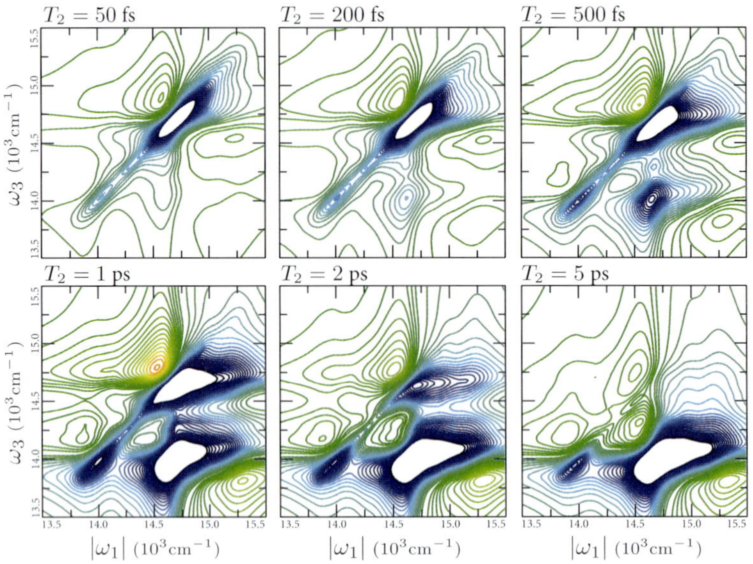

Figure 17.12 Variation of the 2D rephasing spectrum with t_2 as indicated [155]. The 2D plots are normalized to their maximum. Signals are averaged over 500 diagonal disorder configurations.

dicate strong correlation with the reaction center. These features show that strong excitonic correlations are mostly active within the linker and the delocalized antenna states and they extend to the RC. Subsequent evolution with t_2 in Figure 17.13 shows dynamics of various crosspeaks, including well-resolved L–C and D–L. 100–200 fs period oscillations cause the peak sign alternations. Since this signal probes exciton coherences, it naturally decays with the coherence decay timescale: its amplitude drops by a factor of 10 between 0 and $\sim$ 150 fs, and by a factor of 100 at 300 fs. The red states do not contribute to this signal, whereas the RC shows strong crosspeaks through the linker states.

These simulations indicate that the reaction center is clearly visible in the coherent signals and is not masked by the bulk antenna contributions. The predicted RC-related crosspeaks demonstrate a high degree of organization of the PS-I complex: while the RC is spatially separated from the antenna, the linker exciton states participating in the RC penetrate into the outer antenna, making exciton transport to the antenna very robust. This provides RC signatures in 2D signals. Since the coherences decay within 150 fs, only exciton populations determine the spectra in Figure 17.12 at later delay time. Hence, the separate analysis of the C signal as defined in Section 16.5 is not necessary.

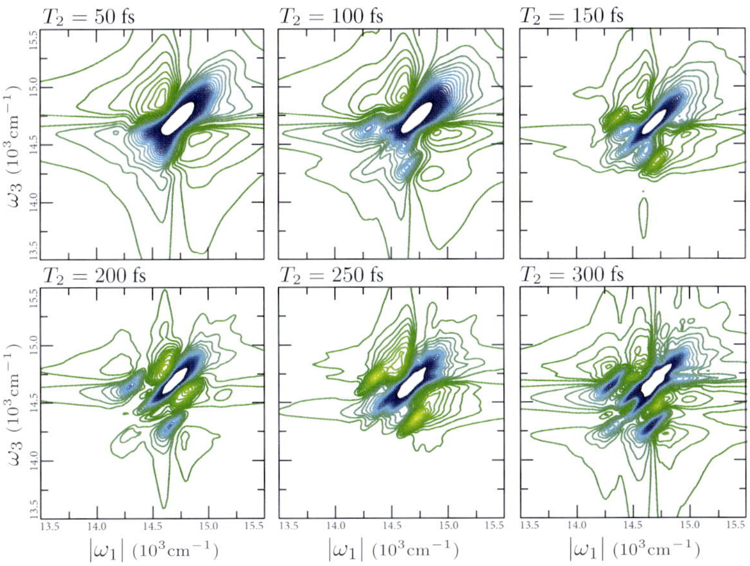

Figure 17.13 $B(\omega_3, t_2, \omega_1)$ spectrum which targets exciton coherence dynamics [155].

17.2.3 Quantum Transport in PS-II Reaction Center

The secular energy relaxation picture described in previous subsections describes purely classical energy transfer. This is because in the secular description the quantum coherences of the density matrix in the off-diagonal areas are not mixed with the rate equation of classical populations – the density matrix diagonal elements. The density matrix diagonal elements thus grow or decay monotonously. The coherences display oscillatory damped motion. The full Redfield equation, however, mixes all these dynamics as was described in Section 11.10.

Whether quantum effects, stemming from entanglement of chromophores, persist in the energy transport at room temperature, despite the rapid decoherence effects caused by environment fluctuations in photosynthetic aggregates may be questioned [79]. Let us study the photosynthetic reaction center of the photosystem II that is relatively small system but has strong couplings within. Let us consider how quantum transport may be observed by two-dimensional coherent rephasing spectroscopy.

We focus on the reaction center (RC) of the photosystem II (PS-II). Its core consists of two, D1 and D2, branches of pigments: the special pair, P_{D1} and P_{D2}, accessory Chl_{D1} and Chl_{D2}, and pheophytins $Pheo_{D1}$ and $Pheo_{D2}$ (shown in Figure 17.14a). These together with two additional pigments $Chlz_{D1}$ and $Chlz_{D2}$ (which are further away) form the primary excitonic system [145, 146, 156]. For simplicity we further include only the central four chlorophyll pigments, P_{D1}, P_{D2} and Acc_{D1}, Acc_{D2} which are closely packed in the RC core.

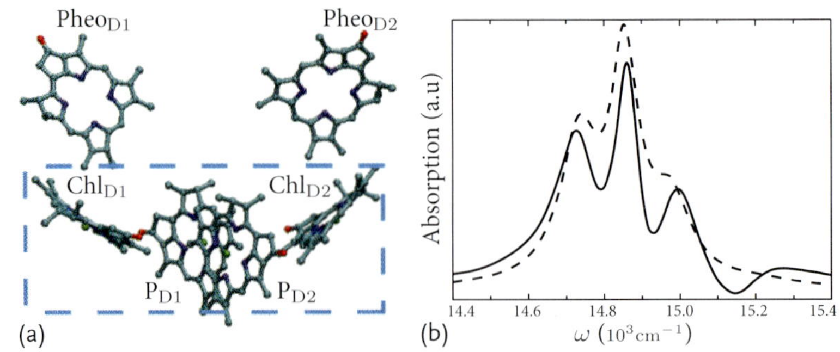

Figure 17.14 (a) Structure of the RC part of PS-I's aggregate. Only the core is shown. Blue dashed lines indicate chlorophylls. (b) Absorption spectrum of the selected region using classical transport model (dashed line) and the quantum transport model (solid line).

The surrounding proteins cause decoherence and energy relaxation through fluctuating transition energies of pigments which are characterized by the multimode spectral density similar to the PS-I aggregate described in the previous section. Energy relaxation and transport are usually described by coupling the exciton system to a phonon bath and deriving equations of motion for the reduced exciton density matrix ρ. The Redfield equation for the density matrix reads (see Section 11.2)

$$\dot{\rho}_{ab} = -\frac{\mathrm{i}}{\hbar}\left[\hat{H}_S, \rho\right]_{ab} + \sum_{cd} K_{ab,cd}\rho_{cd} \,. \tag{17.9}$$

The first term represents the free exciton system, and the tetradic relaxation superoperator K represents dephasing and transport rates. A drawback of this approach is that it was derived using the second order theory and thus it only works in a limited parameter regime. It may yield unphysical density matrix: populations may become negative or diverge [157]. Additional secular approximation stabilizes the relaxation dynamics. K is then reduced to a population block $K_{e_1e_1,e_2e_2}$, that yields a well-behaving classical master equation for populations. Dephasing rates $K_{e_1e_2,e_1e_2}$ erase quantum coherences over the delay time. This level of theory, which have been used in previous subsections, should be understood as the *classical transport* (CT) regime [79].

An alternative Lindblad equation approach has been described in Section 11.7, where the relaxation superoperator is given by:

$$K\rho = \sum_{\alpha} \hat{L}_\alpha \rho \hat{L}_\alpha^\dagger - \frac{1}{2}\rho \hat{L}_\alpha^\dagger \hat{L}_\alpha - \frac{1}{2}\hat{L}_\alpha^\dagger \hat{L}_\alpha \rho \,. \tag{17.10}$$

Here $\hat{L}_\alpha$ is a set of system operators which represent the coupling of the exciton system to the environment. Since the Lindblad equation is not limited to the secular approximation, it can couple populations and coherences in a balanced way so that the simulated density matrix dynamics is physically reasonable. As the populations become entangled with the quantum coherences, this transport regime is denoted as the *quantum transport* (QT) [79].

The heuristic approach to parametrize the Lindblad equation can be acceptable, where for excitons the Lindblad operators are defined using

$$\hat{L}_\alpha = \sum_{mn} u^\alpha_{mn} \hat{B}^\dagger_m \hat{B}_n \, . \tag{17.11}$$

The matrix elements u^α_{mn} are some complex numbers. Now the matrix $Z_{mn,m'n'} = \sum_\alpha u^{\alpha*}_{mn} u^\alpha_{m'n'} \equiv \langle u^*_{mn} u_{m'n'} \rangle$ contains the complete information required to construct the relaxation superoperator.

In the spectroscopy simulations we are dealing with the eigenstate basis so we have to transform the correlation matrix

$$Z_{e_4e_3,e_2e_1} = \sum_{mnkl} Z_{mn,kl} c_{me_4} c^*_{ne_3} c^*_{ke_2} c_{le_1} \, . \tag{17.12}$$

In the secular approximation, only autocorrelations are retained. Then $Z_{ee',ee'} = K_{ee,e'e'}$ is the population relaxation rate from state e' to state e and $Z_{ee,ee} = 2|K_{eg,eg}| - |K_{ee,ee}|$ is twice the pure dephasing of the eg coherence. These can be calculated by using a microscopic bath model from the predefined spectral density using either Redfield and modified Redfield, or Förster theory. The other elements of Z matrix need additional assumptions. The general statistics of the correlation coefficients suggests

$$Z_{e_4e_3,e_2e_1} = \sqrt{Z_{e_4e_3,e_4e_3} Z_{e_2e_1,e_2e_1}} \cos(\varphi_{e_4e_3,e_2e_1}) \, , \tag{17.13}$$

where the cosine function can be an arbitrary number between -1 and 1. The cosines can be related to the exciton spatial overlap factors $\xi_{ee'} = \sum_m |c_{me}||c_{me'}|$. Hence, only Lindblad operators involving overlapping excitons will be correlated. In a simple model one can calculate three products of overlap factors, $\xi_{e_4e_3} \cdot \xi_{e_2e_1}$, $\xi_{e_4e_2} \cdot \xi_{e_3e_1}$, $\xi_{e_4e_1} \cdot \xi_{e_3e_2}$, and set $\cos(\varphi_{e_4e_3,e_2e_1}) = 1$ if the largest product is greater than a cutoff parameter $0 < \eta < 1$. Otherwise $\cos(\varphi_{e_4e_3,e_2e_1}) = 0$. The simulation methods and parameters for calculating the entire relaxation superoperator are then calculated as described in Section 11.7.

The two-dimensional rephasing spectra using QT and CT simulations for short (0 ps) and at long (10 ps) t_2 delay are compared in Figure 17.15. The signal has two main diagonal peaks D1 and D2 (blue–negative) corresponding to the excitons e_1 and e_2, whose strength depends on their populations $\rho_{e_1e_1}$ and $\rho_{e_2e_2}$. The main crosspeak C1 is related to population transfer from e_2 to e_1. The weaker diagonal peak D3 represents the e_3 exciton. The other (yellow – positive) crosspeaks C4–C6 reflect double-exciton resonances. At $T_2 = 10$ ps delay, C1 becomes the strongest signifying the exciton transfer. The overall spectral pattern of QT and CT is similar but details (some peaks, spectral linewidths and peak amplitudes) are different.

The QT and CT dynamics is markedly different in the time evolution of diagonal peaks (D1, D2) and crosspeaks (C1, C2, C3) with t_2, as depicted in Figure 17.15b. QT shows strong oscillations of D1 and D2 lasting for over 600 fs. These reflect the non equilibrium populations and are correlated to the beating of C2 and C1. The CT simulations also show rapidly decaying ($\sim$ 300 fs) oscillations of C1 and

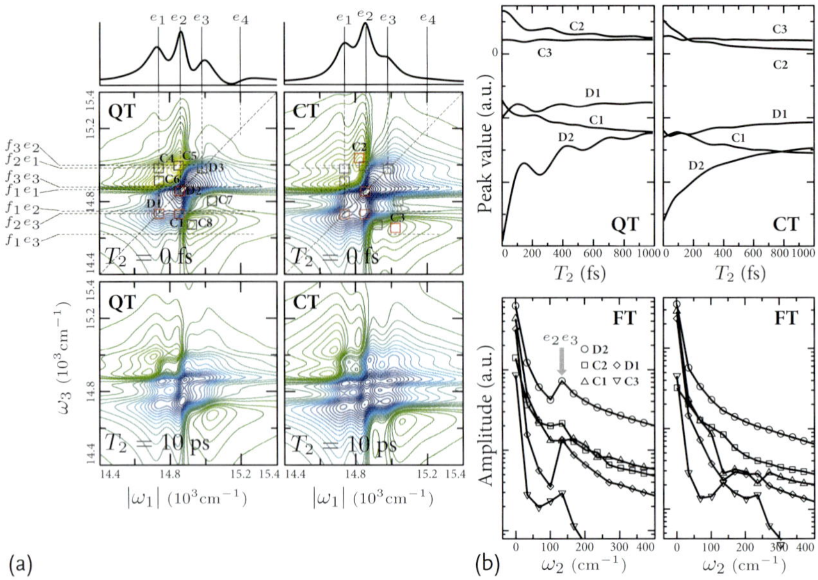

Figure 17.15 (a) 2D spectrum of the RC with (QT) and without (CT) quantum transport in the initial moment ($T_2 = 0$) and at the equilibrium in the excited state ($T_2 = 10$ ps); (b) time evolution of diagonal peaks and crosspeaks using CT and QT models.

C2 (these are related to coherences showing quantum beats), the population peaks D1 and D2 are nonoscillatory.

Our simulations thus reveal that population relaxation acquires oscillatory components due to its coupling with coherence oscillations which are displayed in the 2D spectrum. The corresponding diagonal peaks do not show oscillations in CT. Oscillatory diagonal peaks are thus a clear signature of QT. This conclusion holds for the rephasing 2D signal calculated here. The quantum beats of the combined rephasing and non rephasing signals or of off-diagonal peaks in [62, 118] do not necessarily imply QT since non rephasing 2D diagonal peaks include stimulated emission coherences which oscillate even in the CT case.

18
Single Molecule Spectroscopy

A new spectroscopic method based on the detection of the spectroscopic signals of single molecules or multichromophore systems bypasses the ambiguities concerning the ensemble averaging and provides direct experimental information for individual molecules or molecular complexes. Such a spectroscopic method allows the researchers to resolve effects caused by decoherence of excitation with higher resolution and also to trace the spectral changes due to environment fluctuations. The main purpose of this chapter is to present the theoretical basis suitable for the analysis of observations obtained using the single molecular spectroscopy.

18.1
Historical Overview

Rapid development of new spectroscopic equipment in the mid-1980s has led to the emerging of novel advanced techniques which have helped to uncover many interesting spectroscopic properties of various chemical compounds. In 1989, the absorption spectra of single molecules were measured for the first time [158], which has marked the dawn of an era of single-molecule spectroscopy (SMS). Since then, SMS has proven to be a valuable tool to inspect the subtle properties of individual molecules not obscured by the ensemble average. Indeed, traditional spectroscopic methods allow one to measure only some statistically averaged quantities, describing the whole ensemble as the observable system. In this way, the probability distribution of these quantities, their dynamic and/or static variations caused by heterogeneity of the system, as well as fundamental interactions between distinct molecules and their environment remain undetermined. The ability of SMS measurements to reveal such uncertainties and, therefore, to provide much more new information has resulted in growing interest in applying SMS techniques not only to simple fluorophores such as dye molecules [159, 160] or colloidal semiconductor quantum dots (QDs) [161], but also to complex biophysical systems, such as green fluorescent proteins [162] or pigment–protein light-harvesting complexes (LHCs) from plants [163–165] and photosynthetic bacteria [166–170]. In the latter applications, SMS has been successfully applied in the labeling experiments, when simple fluorophores are attached to complex macromolecules and provide some valuable

Molecular Excitation Dynamics and Relaxation, First Edition. L. Valkunas, D. Abramavicius, and T. Mančal.

information on molecular interactions, reaction kinetics, conformational dynamics or molecular motion [171, 172].

SMS measurements have also revealed several unexpected properties of single-molecule systems, for example, spectral diffusion, the phenomenon that occurs when the absorption frequencies of a molecule change due to some variations of its local surroundings. Another intriguing effect discovered by the SMS is the so-called fluorescence intermittency, or blinking. In virtually all fluorescing systems studied to date at the single-molecule level the measured fluorescence intensity fluctuates rapidly and abruptly despite continuous illumination [173, 174]. The sudden and uncorrelated fluctuations occur mostly between two well-defined strongly- and non- or weakly-emitting levels (the corresponding states are commonly referred to as on and off states, respectively) and usually serve as a simple signature of single emitters. However, such unpredictable behavior limits the application of simple fluorophores as fluorescent probes; therefore, much effort has been made in order to understand the underlying mechanisms responsible for fluorescence blinking and to find out how it could be controlled or even eliminated.

In the early studies of fluorescence blinking of single molecules in molecular crystals [175] it was found that the probability of the times the system spent in the on and off states (on- and off-times) to a great extent can be described by single-exponential distribution, as predicted by the quantum jump theory of transitions between singlet and metastable triplet states [176]. Later, the blinking effect with much longer off-times that could not arise due to intersystem crossing was discovered for various fluorescing systems [174]. In most of these, off-times vary across almost all experimentally accessible time scales, typically spanning over 4 orders of magnitude or even from microseconds to several hours in the case of semiconducting QDs [177]. Moreover, in almost all these very varied systems the dwell times of both on and off states are not exponentially distributed, but follow an inverse power-law or its simple modifications, with the exponent m typically lying between 1 and 2 [174]. Despite much research in this field, the explanation for probably one of the most intriguing riddles of SMS – why such diverse systems of various complexity exhibit very similar blinking statistics leading to the absence of typical time scale and even to the weak ergodicity breaking [176] – is still under discussion.

In order to resolve (at least partially) this problem, several models describing fluorescence blinking in semiconducting QDs have been proposed so far. In these models the dark state of QD is associated with the photo-ejection of an electron. According to the so-called trap models [173, 178, 179], it is assumed that the electron can tunnel through a barrier to a trap located nearby, and the dark period ends when the trapped electron hops back. Alternatively, power-law blinking statistics naturally arise if one considers one- or two-dimension random walk involving a first-passage problem. In 2005, Tang and Marcus [180] suggested a diffusion-controlled electron-transfer model, where a light-induced one-dimensional diffusion in energy space is considered. Additionally, more models of power-law statistics have been proposed, but none of the existing theories can explain all the experimentally observed issues of the blinking phenomenon. Moreover, no or very limited theoretical background regarding fluorescence blinking in other systems exists.

Recently, new data on fluorescence intermittency in the single complexes of major light-harvesting complexes (LHCII) of green plants have been collected [164, 165]. It was found that these complex biomolecular structures exhibit similar blinking behavior as structurally much simpler fluorescent dyes or QDs. Furthermore, in contrast to the dyes, in vivo LHCII particles, being subunits of large photosystem II, perform important physiological functions of very efficient light harvesting as well as excitation energy transfer and regulation via nonphotochemical quenching (NPQ). Thus, the fluorescence blinking of the single pigment–protein complex evidently represents the (quantum) electronic transition of the emitting pigment reflecting its position in the potential configuration of the protein. The latter evidently behaves according to the classical physics laws. So the whole fluorescence blinking phenomenon tightly relates to the complex quantum/classical dynamics.

18.2
How Photosynthetic Proteins Switch

It is generally accepted that proteins are mobile entities, undergoing a variety of structural deformations on different time scales. They move in their conformational landscape probing different conformational sub-states [181]. The simplest model demonstrating the switching ability via the fluorescence intermittency in LHCII complexes [165] assumes that the LHCII trimer can be found in two states depending on the protein structural arrangement: either a bright (*on*) state, when the fluorescence signal from the irradiated LHCII trimer is clearly detected, or a dark (*off*) state, when the fluorescence is almost switched off. If one projects the manifold of all rapid molecular vibrations in the LHCII onto a single reaction coordinate x, then the *on* and *off* states of the LHCII trimer mentioned above would correspond to two minima on the configurational potential energy surface of the protein [182]. Possible transitions between these two states can be attributed to another generalized coordinate y reflecting some specific slow protein conformational change which disturbs the energy balance between different pigments involved in the light-harvesting and quenching process. To characterize the transition rates between the two (*on* and *off*) states, the potential energy surfaces attributed to each of them can be separately defined as independent potential wells (see Figure 18.1), which in the harmonic approximation are given by:

$$U_1(x, y) = \frac{1}{2}\lambda_1 x^2 + \frac{1}{2}\gamma_1 y^2 ,$$
$$U_2(x, y) = \frac{1}{2}\lambda_2 (x - x_0)^2 + \frac{1}{2}\gamma_2 (y - y_0)^2 + U_0 , \tag{18.1}$$

where indices "1" and "2" denote *on* and *off* states, respectively; λ_i and γ_i determine the reorganization energies in the ith potential along the coordinates x and y, respectively; x_0 and y_0 indicate the equilibrium position of the second potential; and U_0 is the vertical difference between the potential minima. According to this description, the system resides mainly in the vicinity of the U_1 potential well when the y values are small, and the transition into the minimum of the U_2 potential well occurs by increasing the y value (when $y_0 > 0$). Since the LHCII trimer can be found only in one of these two states at any given time, a random walk in the phase space of the coordinates x and y will lead to a random switching between the *on* and *off* states. The dynamics of these transitions should then resemble the experimentally observed dynamics of fluorescence intermittency [164, 165]. Changes in the environmental conditions might induce some variation of the potential surface, which will result in a shift of the dynamic equilibrium to either *on* or *off* state.

We can safely assume that transitions between the *on* and *off* potential surfaces occur strictly vertically, meaning that the coordinates x and y do not change during the transition, as shown in Figure 18.1. The rate of the downward transition from the point A on the *on* potential surface to the point B with the same coordinates on the *off* potential surface is equal to k_1. Similarly, the rate of the downward transition $C \rightarrow D$ from the *off* to the *on* potential surface is denoted as k_2. In addition, both relaxation rates should contain the factor $\exp(-\alpha|\Delta U|/(\hbar\omega_0))$ reflecting the so-called energy gap law [45]. Here ω_0 is the dominant frequency responsible for the transitions between the points on the energy surfaces under consideration, and α is some function, weakly (logarithmically) dependent on the potential energy difference $|\Delta U|$ between those points, so that we can treat it as some constant parameter ($\alpha \cong 1 \div 3$). The ratio of the upward and downward transition rates is defined by the detailed balance relationship via the corresponding Boltzmann factor: $k_i^{(\mathrm{up})}/k_i = \exp(-|\Delta U|/(k_\mathrm{B} T))$, where k_B is the Boltzmann constant and T denotes the temperature.

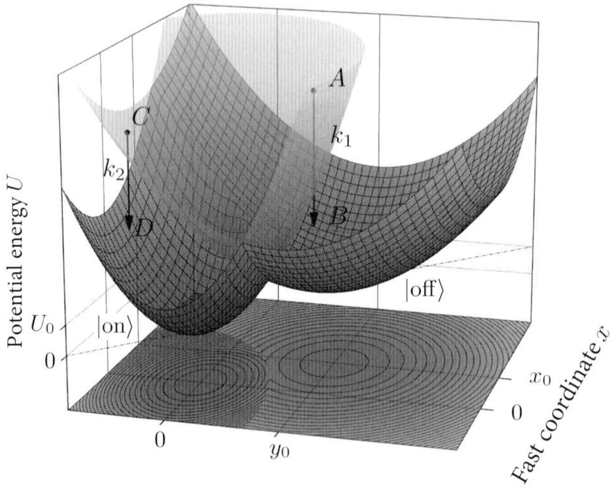

Figure 18.1 Potential surfaces of the *on* and *off* states in the phase space of the x and y coordinates. k_1 and k_2 denote the relaxation rates of the *on* → *off* and *off* → *on* transitions, respectively.

The time-dependent probability density $\rho_i(x, y, t)$ for finding the system at the point with coordinates x and y at time t, when the system is either in the *on* ($i = 1$) or *off* ($i = 2$) potential surface, obeys the following Fokker–Planck [183, 184]

$$\frac{\partial \rho_i}{\partial t} = \left[D_{ix}\mathcal{L}_x + D_{iy}\mathcal{L}_y - k_i H_i(x, y)\right]\rho_i(x, y, t) , \tag{18.2}$$

where D_{ix} and D_{iy} are the diffusion coefficients in the ith potential along the x and y directions, respectively; $\mathcal{L}_x$ and $\mathcal{L}_y$ are the corresponding diffusion operators:

$$\mathcal{L}_z \rho_i(x, y, t) = \left[\frac{\partial^2}{\partial z^2} + \frac{1}{k_B T}\frac{\partial}{\partial z}\frac{\partial U_i(x, y)}{\partial z}\right]\rho_i(x, y, t) ,$$
$$i = 1, 2 ; \quad z = x, y ; \tag{18.3}$$

and

$$H_1(x, y) = e^{-\alpha\frac{|\Delta U|}{\hbar\omega_0}} \cdot \min\left\{1, e^{\frac{U_1 - U_2}{k_B T}}\right\} ,$$
$$H_2(x, y) = e^{-\alpha\frac{|\Delta U|}{\hbar\omega_0}} \cdot \min\left\{1, e^{\frac{U_2 - U_1}{k_B T}}\right\} . \tag{18.4}$$

Assuming that the diffusion along the x coordinate is much faster than along the y coordinate, the terms determining fast dynamics can be adiabatically eliminated from (18.2). In this case the x-dependence of the probability densities approaches the stationary (Gaussian) distribution exponentially fast, thus integration of (18.2) yields

$$\frac{\partial \bar{\rho}_i(y, t)}{\partial t} = \frac{\partial}{\partial t}\int dx \rho_i = (D_{iy}\mathcal{L}_y - \kappa_i(y))\bar{\rho}_i(y, t) , \tag{18.5}$$

where

$$\kappa_1(y) = k_1\sqrt{\frac{\lambda_1}{2\pi k_B T}}\int dx e^{-\frac{1}{2k_B T}\lambda_1 x^2} H_1(x, y) , \tag{18.6}$$

$$\kappa_2(y) = k_2\sqrt{\frac{\lambda_2}{2\pi k_B T}}\int dx e^{-\frac{1}{2k_B T}\lambda_2(x - x_0)^2} H_2(x, y) . \tag{18.7}$$

The initial conditions of (18.5) can be chosen as follows. First we define the stationary solution $\bar{\rho}_1^{(st)}$ of (18.5) when transition to the *off* state is inactive. Then, we multiply the obtained steady-state solution (the Gaussian distribution) by the effective rate $\kappa_1(y)$ given by (18.6). This function determines the initial distribution of the population of the *off* state. Similarly, the initial probability density for the population of the *on* state is given by the product $\bar{\rho}_2^{(st)}(y)\kappa_2(y)$. It is noteworthy that after substituting expressions for $H_i(x, y)$ (see (18.4)) and normalizing, both initial distributions coincide:

$$\bar{\rho}_i(y, t = 0) \propto \bar{\rho}_i^{(st)}(y)\kappa_i(y) \propto \int dx \exp\left(-\alpha\frac{|\Delta U|}{\hbar\omega_0}\right)$$
$$\times \min\left(\exp\left(-\frac{U_1(x, y)}{k_B T}\right), \exp\left(-\frac{U_2(x, y)}{k_B T}\right)\right) . \tag{18.8}$$

Table 18.1 Fitted model parameters.

Model parameter	Value pH 6	Value pH 8	Model parameter	Value pH 6	Value pH 8
$\lambda = \lambda_2/\lambda_1$	0.3	0.2	k_1^{-1}	190 ms	430 ms
$\gamma = \gamma_2/\gamma_1$	0.68	0.72	k_2^{-1}	3.6 ms	4.8 ms
$x_0\sqrt{\lambda_1/(k_B T)}$	1.0	1.4	$(D_{1y}\gamma_1/(k_B T))^{-1}$	2.4 s	3.8 s
$y_0\sqrt{\gamma_1/(k_B T)}$	8.59	8.57	$(D_{2y}\gamma_1/(k_B T))^{-1}$	1 s	1.4 s
$U_0/(k_B T)$	0.5	1.5	$\hbar\omega_0/(\alpha k_B T)$	0.4	1.0

A more detailed numerical analysis reveals that (18.8) defines a very sharp distribution with the maximum located near the intersection point $y^{(0)}$ of the one-dimensional functions $U_1(x = 0, y)$ and $U_2(x = x_0, y)$, so that it might be well approximated as $\delta(y - y^{(0)})$.

Solutions of (18.5) allow us to determine the survival probabilities on the *on* and *off* potential surfaces, $S_i(t)$, by integrating $\bar{\rho}_i(y, t)$ over the y coordinate:

$$S_i(t) = \int dy \bar{\rho}_i(y, t) . \tag{18.9}$$

Finally, the quantity corresponding to the experimentally collected blinking statistics [165] determining the probability $P_i(t)$ that a transition from one state to another occurs within the time interval $(t; t + dt)$, is defined as

$$P_i(t) = -\frac{dS_i(t)}{dt} , \quad i = 1, 2 . \tag{18.10}$$

The presented model contains several parameters, which will be used for fitting the experimental data. Upon introducing relative representations of the coordinates determining both potential surfaces, five of these characterize the *on* and *off* potential surfaces ($\lambda = \lambda_2/\lambda_1$, $\gamma = \gamma_2/\gamma_1$, x_0, y_0 and U_0), while the other three determine the dynamics of the transitions between the *on* and *off* states (k_1, k_2 and ω_0). All these parameters were varied while fitting the experimental data of blinking statistics at various pH values [165]. We note that the diffusion coefficients D_{iy} only determine the time scale of the protein conformational changes, so they do not change the shape of the $P_i(t)$ distributions on the logarithmic scale but only shift them along the time and probability density axes. From the magnitude of those shifts the diffusion coefficients were determined. The fitting results for data collected under two particular conditions of the environment, namely at pH 6 and pH 8, are demonstrated in Figure 18.2, and the corresponding fitting parameters are presented in Table 18.1. The latter pH value corresponds to natural physiological conditions ensuring strong fluorescence of isolated trimers, while the former is similar to the one usually observed under NPQ conditions.

It is clear that a description of all the possible conformational changes of the LHCII trimer using only two generalized coordinates and simple harmonic potential wells cannot reveal all the subtle details of the dynamic spectral properties of

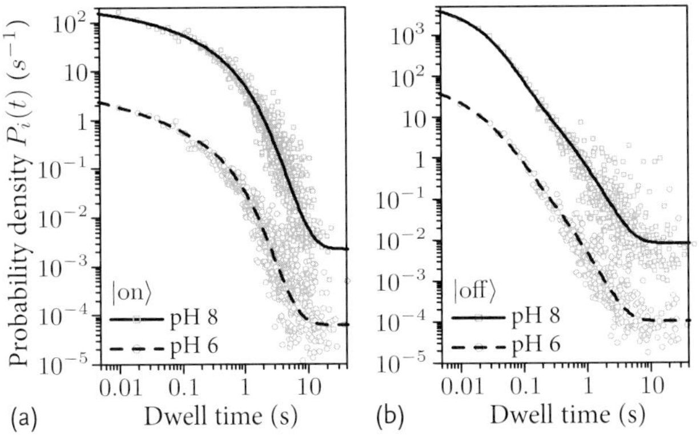

Figure 18.2 Experimentally obtained (circles) and simulated (lines) probability densities of the dwell times in bright (a) and dark (b) states for two different acidity levels of the environment. For visual clarity, the upper data corresponding to pH 8 were multiplied by a factor of 100.

such a photosynthetic pigment protein. Nevertheless, this simplified model with properly chosen parameters can very well reproduce the experimentally observed fluorescence intermittency on the whole experimentally accessible time-scale, as demonstrated in Figure 18.2. From the obtained model parameters (Table 18.1) several interesting properties of the potential surfaces of the *on* and *off* states can be outlined. First of all, the potential energy surface of the *off* state is less steep than that of the *on* state, and its minimum is located slightly above the minimum of the *on* state. Furthermore, a clear dependence of the parameters on the environmental acidity is revealed. Under more adverse environmental conditions at a low pH level the potential surfaces of both *on* and *off* states exhibit notable deformations. The potential well of the *off* state becomes slightly steeper (along the fast x coordinate), and its minimum approaches the minimum of the *on* state, which results in an increased probability for the system to switch to the dark, nonfluorescing state.

While analyzing the model parameters that determine the transitions between the *on* and *off* states, we first notice that the energy $\hbar\omega_0$ of the dominating phonon mode taking part in the transitions is of the order of the thermal energy $k_B T$. The lower pH level determines a higher protonation state of the system, which results in the increased effective mass of the vibrating molecules and, therefore, an almost twofold drop of the frequency ω_0 of their vibrations. The relatively high value of $\hbar\omega_0$ can explain the flattening of the calculated probability densities at longer dwell times (Figure 18.2), a somewhat different behavior compared to other existing models dealing with the power-law blinking statistics [173, 183].

An interesting outcome is the notable difference between the diffusion and transition rates in the *on* and *off* states. The rate of the transition from *on* to *off* state is $\sim$ 50–90 times slower than that of the backward transition. Such a high ratio of the transition rates in opposite directions reveals why the population of the *off* states decreases in time much faster than that of the *on* states. After the transition to the dark state the system usually remains in this state for a very short time so that the measured fluorescence intermittency should resemble very short blinking events, rather than short flashes.

A somewhat similar fluorescence blinking behavior was observed in many other fluorescing complexes, ranging from simple single dye molecules [159, 160, 185] and semiconductor quantum dots [161, 173, 178, 179, 186, 187], to the diverse range of more complex fluorescing systems [162, 163, 188]. If the exponential switching behavior of single dye molecules observed on shorter time scales can be attributed to the quantum transitions between singlet and metastable triplet states [175, 176], the power-law blinking statistics observed in various other systems still does not have a proper explanation. For disordered biological systems the power-law exponents strongly depend on the environmental conditions. The blinking effect observed to take place in single LHCIIs seems to be even more outstanding if one takes into account the complex internal structure of these units containing more than 40 distinct pigments, each with its own spectral properties. If the fluorescence intermittency were attributed to individual uncoupled chromophores, due to the stochastic nature of the blinking phenomenon, the averaged signal from all pigments would almost completely lack any noticeable blinking events. In contrast, experimental observations of fluorescence blinking support the significance of the protein scaffold binding all the pigments together and enforcing them to act as a whole quantum unit. It seems that during evolution the plants 'have learned' to take advantage of the blinking of simple emitters and implemented it at a slightly more macroscopic level, when the protein's motion and deformation influence the inter-pigment couplings, molecular fluctuations, and possible pathways for the excitation energy transfer. As a result, the switching behavior of the LHCIIs has become their intrinsic property governed by the lability and adaptability of the protein scaffold. The latter property not only determines the system evolution, but also manifests itself as system adaptation to the varying environmental conditions, such as acidity, illumination level, and so on.

Such switching ability between bright and dark states implies that the mechanisms responsible for NPQ should be closely related to the phenomenon of fluorescence intermittency.

18.3 Dichotomous Exciton Model

The fluorescence spectral changes and intensity blinking behavior have also been observed by the SMS of the peripheral light-harvesting complexes LH2 from photosynthetic bacteria [166–168]. To determine possible structural changes of the LH2 complex and the time scale of these changes the experimentally measured spectral profile must also be associated with the microscopic structural parameters. Since the B850 antenna ring is arranged by excitonically coupled pigment molecules (bacteriochlorophylls), the static disorder of pigment site energies, and the coupling of the pigment electronic excitations to phonons, which gives rise to the so-called dynamic disorder, have to be taken into account explicitly. The fluorescence spectral profile of the B850 is known to be sensitive to the structural fluctuations (static disorder) and to the dynamic disorder determined by the electronic excitation interaction with the vibrational modes of the molecules and protein scaffold (see Chapter 5). Therefore, the experimentally observable differences of the SM spectral line shape and the peak wavelength should be associated with different realizations of the static disorder. Transitions between different spectral states occur due to the changes of the static disorder, which in their turn are induced by the conformational changes of the protein surrounding of the pigment molecules. The latter occur either spontaneously or are light-induced due to the nonradiative relaxation of the absorbed excitation energy. The simplest possible disorder model in which each pigment can switch between the two states of different electronic excitation energy already demonstrates the spectral switching behavior of the fluorescence spectra [169, 170]. Because of this assumption let us consider the B850 band by describing the exciton energy spectrum using the following Hamiltonian [98]:

$$H = \sum_{n=1}^{N} \left(\varepsilon_n^{(0)} + q_n \right) |n\rangle\langle n| + \sum_{n,m=1}^{N} V_{nm} |n\rangle\langle m| + H_{\text{ph}} \tag{18.11}$$

where $\varepsilon_n = \varepsilon_n^{(0)} + q_n$ is the excitation energy of the nth pigment molecule modulated by the collective coordinate of the thermal bath q_n, $|n\rangle$ and $\langle n|$ are the ket and bra vectors for the excitation to be localized on the nth molecule in the aggregate, respectively. The matrix element V_{nm} denotes the resonance interaction between the nth and the mth pigment molecules. H_{ph} denotes the phonon bath composed of the intra-molecular and protein vibrations.

Because of slow bath degrees of freedom, $q^{(\text{slow})}$, the excitation energies ε_n are stochastic parameters, characterized by their distribution function. For the dichotomous model it consists of two parts: the Gaussian disorder and the dichotomous protein conformational disorder. We denote the Gaussian disorder as the inhomogeneous part. It is characterized by the inhomogeneous distribution function (IDF), which can be usually represented as a Gaussian function with a mean value ε_{h} (the subscript h denotes the helix binding the pigment, $\text{h} = \alpha, \beta$) with the full-width at half-maximum Γ^{inh}:

$$f_{\text{inh}} \left(\varepsilon_n^{\text{inh}} \right) \propto \exp \left(- \frac{4 \ln(2) \left(\varepsilon_n^{\text{inh}} - \varepsilon_{\text{h}} \right)^2}{(\Gamma^{\text{inh}})^2} \right) \tag{18.12}$$

The dichotomous exciton model assumes that protein fluctuations introduce two conformational states for each BChl pigment (see Figure 18.3). These states are characterized by their population probabilities, p_j, $j = 1, 2$. The two conformational states shift the mean excitation energies of a pigment by $+\Delta E$ for $j = 2$ or $-\Delta E$ for $j = 1$. Thus, the total electronic transition energy of each bacteriochlorophyll molecule can be expressed as

$$\varepsilon_n = \varepsilon_n^{\text{inh}} \pm \Delta E \, , \tag{18.13}$$

where the case of plus corresponds to dichotomous state $j = 2$ and the case of minus to $j = 1$.

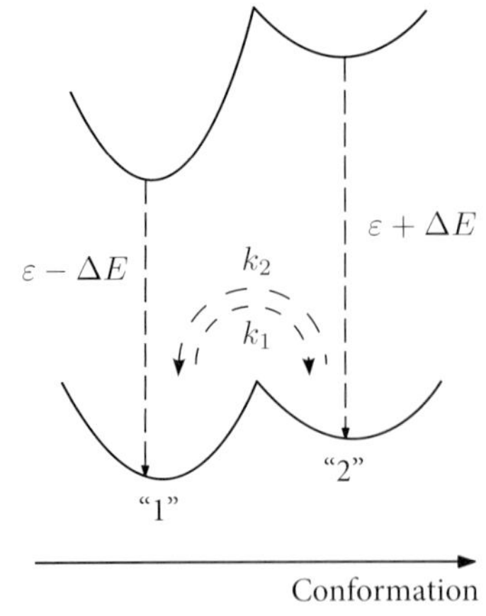

Figure 18.3 Potential surfaces in the ground- and excited-states of a BChl molecule from the B850 ring of the LH2 complex. The potentials are plotted against the conformational coordinate of the protein surrounding. It is postulated that each BChl molecule can reside in either of the two conformational states, denoted "1" and "2" and the transition rates between these two states as indicated as k_1 and k_2, respectively.

The remaining bath degrees of freedom constitute fast fluctuations, $q^{(\text{fast})}$, which are treated perturbatively. These fluctuations are characterized by a spectral density, which for the model of uncorrelated fluctuation for different molecules is defined as

$$C''_{n,n}(\omega) = \frac{1}{2} \int_{-\infty}^{\infty} \mathrm{d}t \exp(\mathrm{i}\omega t) \left\langle \left[q_n^{(\text{fast})}(t), q_n^{(\text{fast})}(0) \right] \right\rangle \tag{18.14}$$

while the bath average is taken with respect to the equilibrium phonon Hamiltonian H_{ph}.

Transition between the two conformational states occurs due to the thermally induced potential energy barrier crossing, or it can also be light-induced as a result of dissipation into the surrounding protein scaffold. In the first case, the protein temperature is that of the ambient whereas in the case of the light-induced changes the protein might be locally heated and/or the characteristic energy barrier separating the potential energy minima can differ from that of the spontaneous transition. Therefore, this rate is sensitive to the temperature and is linearly dependent on the small changes of temperature. In a typical SMS experiment [168] the LH2 complex is excited approximately 10^7 times per second. More than 90% of the absorbed photons are dissipated in the protein scaffold and further to the surrounding of the complex. The calculated temperature change of the complex in such a dynamic equilibrium is negligible (approximately 10^{-4} for 6 μW of the excitation power), which is reached in about 60 ps. If the complex is adiabatically isolated, the expected temperature increase is $\sim$ 2 K. Such a temperature increase would not be associated with a noticeable enhancement of the probability to overcome the energy barrier. However, the probability of the transition might be enhanced at early times after the internal conversion before the thermalization is reached, that is, when the pigment–protein is heated locally. In this case the frequency of conformational transitions should be linearly dependent on the excitation intensity as was observed experimentally. The thermally induced barrier crossing mechanism is also consistent with the observed fast jumps between long-lasting spectrally distinguishable states. Spontaneous transitions are also taking place as demonstrated experimentally. Thus, the total rate of the transition can be defined phenomenologically as:

$$k_{i,n} = k_i^{\mathrm{s}} + k_i^0 \sum_k P_k \left| c_n^k \right|^2 , \qquad (18.15)$$

where k_i^{s} is the rate of the spontaneous transition and k_i^0 is the fitting parameter of the light-induced transition rate, which is assumed to be site independent. $k_{i,n}$ is defined as the transition rate of the nth pigment in the ring from state "1" to state "2". Thus, the second term determines the rate of the light-induced change of the protein conformation and, therefore, it is proportional to the sum of the contributions of the excited pigment to the exciton states weighted with the thermodynamic population of those states. It is also dependent on the frequency of the excitation of the complex, which is accounted for by k_i^0. The rate of spontaneous changes, k_i^{s}, is determined by the height of the energy barrier separating states "1" and "2". Transition rates determined in this way are defined for a particular realization of the static disorder. However, the barrier crossing is a stochastic process, and it should be considered when calculating the time-span in a particular conformational state.

As this model takes into account the switching probability that each pigment molecule in the antenna ring may be in two possible energy states, it reproduces the bulk fluorescence spectrum with the spectrum of the single LH2 averaged in time as well as the statistics of the fluorescence peak distribution [169, 170]. It also explains the fluorescence changes attributing them to conformational motions of the protein. In general the conformational changes of the protein are represented as a diffusive motion between the local minima in the multidimensional energy landscape, therefore, the two-state model should be considered as an evident simplification. However, it provides an intuitively clear picture of possible protein motions as attributing the movement of the protein surrounding between the two equilibrium positions to the potential energy barrier in between. Moreover, it also resembles the two-level model used to describe the hole burning and spectral diffusion. The exponential kinetics evidently neglecting the details of the spectral diffusion was assumed to characterize the transition between two states.

Appendix

A.1
Elements of the Field Theory

In the classical theory of fields several basic concepts are useful for manipulation of field-related values [190]. We first denote a scalar field $\phi(\boldsymbol{r}) \equiv \phi(x, y, z)$, which is a single-valued function of space. A vector field $\boldsymbol{A}(\boldsymbol{r}) \equiv \boldsymbol{x} A_x(x, y, z) + \boldsymbol{y} A_y(x, y, z) + \boldsymbol{z} A_z(x, y, z)$ assigns a vector to each space point; A_x, A_y and A_z are scalar fields. A gradient operation of the scalar field is denoted as

$$\operatorname{grad}\phi = \frac{\partial \phi}{\partial x}\boldsymbol{x} + \frac{\partial \phi}{\partial y}\boldsymbol{y} + \frac{\partial \phi}{\partial z}\boldsymbol{z} , \tag{A1}$$

where $\boldsymbol{x}$, $\boldsymbol{y}$ and $\boldsymbol{z}$ are the unit vectors of the Cartesian coordinate system. The gradient of a scalar field is thus the vector field which describes the rate of change of the field. The variation of the vector field within the space is described by two types of differential operations. These are the divergence and the curl or rotor of the vector field

$$\operatorname{div}\boldsymbol{A} = \frac{\partial A_x}{\partial x} + \frac{\partial A_y}{\partial y} + \frac{\partial A_z}{\partial z} . \tag{A2}$$

This operator describes the source of the field at a given point. The curl or rotor of the vector field describes the rotary nature of the vector field at a point

$$\operatorname{curl}\boldsymbol{A} = \begin{vmatrix} \boldsymbol{x} & \boldsymbol{y} & \boldsymbol{z} \\ \frac{\partial}{\partial x} & \frac{\partial}{\partial y} & \frac{\partial}{\partial z} \\ A_x & A_y & A_z \end{vmatrix} . \tag{A3}$$

Notice that the divergence creates a scalar field, while the curl creates the vector field.

These differentiation operations can be easily denoted using the *nabla* operator

$$\nabla = \boldsymbol{x}\frac{\partial}{\partial x} + \boldsymbol{y}\frac{\partial}{\partial y} + \boldsymbol{x}\frac{\partial}{\partial z} . \tag{A4}$$

Molecular Excitation Dynamics and Relaxation, First Edition. L. Valkunas, D. Abramavicius, and T. Mančal.

We thus have

$$\text{grad}\phi = \nabla\phi \,, \tag{A5}$$

$$\text{div}\boldsymbol{A} = \nabla \cdot \boldsymbol{A} \tag{A6}$$

and

$$\text{curl}\boldsymbol{A} = \nabla \times \boldsymbol{A} \,. \tag{A7}$$

Often we face the problem of differentiation of some field products. In that case the general differentiation rules apply, that is for instance

$$\nabla \cdot (\phi \boldsymbol{A}) = (\nabla\phi) \cdot \boldsymbol{A} + \phi(\nabla \cdot \boldsymbol{A}) \,, \tag{A8}$$

$$\nabla(\boldsymbol{A} \cdot \boldsymbol{B}) = \nabla_{\text{A}}(\boldsymbol{A} \cdot \boldsymbol{B}) + \nabla_{\text{B}}(\boldsymbol{A} \cdot \boldsymbol{B}) \,, \tag{A9}$$

where ∇_A operates only on field $\boldsymbol{A}$. Slightly more complicated are operations with the curl:

$$\nabla \cdot (\boldsymbol{A} \times \boldsymbol{B}) = \boldsymbol{B} \cdot (\nabla \times \boldsymbol{A}) - \boldsymbol{A} \cdot (\nabla \times \boldsymbol{B}) \,, \tag{A10}$$

$$\nabla \times (\phi \boldsymbol{A}) = \phi \cdot (\nabla \times \boldsymbol{A}) + (\nabla\phi) \times \boldsymbol{A} \,, \tag{A11}$$

$$\nabla \times (\boldsymbol{A} \times \boldsymbol{B}) = \boldsymbol{A}(\nabla \cdot \boldsymbol{B}) - \boldsymbol{B}(\nabla \cdot \boldsymbol{A}) + (\boldsymbol{B} \cdot \nabla)\boldsymbol{A} - (\boldsymbol{A} \cdot \nabla)\boldsymbol{B} \,. \tag{A12}$$

Some useful relations of the fields operators for electrodynamics are

$$\nabla \times \nabla \times \boldsymbol{A} = \nabla(\nabla \cdot \boldsymbol{A}) - (\nabla \cdot \nabla)\boldsymbol{A} \,. \tag{A13}$$

Here

$$(\nabla \cdot \nabla)\boldsymbol{A} \equiv \nabla^2 \boldsymbol{A} \equiv \Delta \boldsymbol{A} = \left(\frac{\partial^2}{\partial x^2} + \frac{\partial^2}{\partial y^2} + \frac{\partial^2}{\partial z^2} \right) \boldsymbol{A} \,. \tag{A14}$$

Operator $\Delta = \partial^2/\partial x^2 + \partial^2/\partial y^2 + \partial^2/\partial z^2$ is known as the Laplace operator. Some identities are

$$\nabla \times (\nabla\phi) = 0 \,, \tag{A15}$$

$$\nabla \cdot (\nabla \times \boldsymbol{A}) = 0 \,. \tag{A16}$$

These operations can be also represented using the Levi-Civita symbol ϵ_{ijk} which is equal to 1 for cyclic configuration of indices $i\,j\,k = 123, 312, 231$, equal to -1 for anti-cyclic configuration 132, 321, 213, otherwise at least two indices are equal and then the Levi-Civita symbol is equal to 0. The vector differential calculus can be described using the Levi-Civita symbol since

$$\boldsymbol{A} \times \boldsymbol{B} = \epsilon_{ijk} e_i A_j B_k \,, \tag{A17}$$

where triple summation over $i\,j\,k$ is implied, $e_1 \equiv x$ $e_2 \equiv y$ $e_3 \equiv z$. Then the curl operation is

$$(\nabla \times \boldsymbol{A})_i = \epsilon_{ijk} e_i \frac{\partial}{\partial e_j} A_k \,. \tag{A18}$$

A.2
Characteristic Function and Cumulants

In statistics for a stochastic variable x characterized by the probability density $p(x)$ the moments are averages $\bar{\ldots}$ (or $\langle\ldots\rangle$) and are calculated as

$$\bar{x^n} \equiv \langle x^n \rangle = \int dx\, x^n p(x) \,. \tag{A19}$$

We define the characteristic function

$$G(k) = \langle \exp(ikx) \rangle \,. \tag{A20}$$

If we expand the exponent we have

$$G(k) = \sum_n \frac{(ik)^n}{n!} \langle x^n \rangle \,. \tag{A21}$$

We can thus observe that the derivatives of the characteristic function generate moments

$$\langle x^n \rangle = i^{-n} \frac{d^n G(k)}{dk^n}|_{k=0} \,. \tag{A22}$$

We can also define

$$G(k) = \exp(g(k)) \tag{A23}$$

and use the expansion

$$g(k) = \sum_n \frac{(ik)^n}{n!} c_n \,. \tag{A24}$$

c_n are denoted as cumulants. Using the derivatives of the characteristic function the cumulants can be given by superpositions of moments. For instance

$$c_1 = \langle x \rangle \tag{A25}$$

$$c_2 = \langle x^2 \rangle - \langle x_1 \rangle^2 \tag{A26}$$

$$c_3 = \langle x^3 \rangle - 3\langle x \rangle \langle x^2 \rangle + 2\langle x \rangle^3 \tag{A27}$$

$$\vdots \tag{A28}$$

Let us assume the Gaussian probability density

$$p(x) = \frac{1}{\sigma\sqrt{2\pi}} \exp\left(-\frac{(x - \langle x \rangle)^2}{2\sigma^2}\right) \,. \tag{A29}$$

The characteristic function for these *Gaussian* statistics can be explicitly calculated

$$G(k) = \exp\left(i\langle x \rangle k - \frac{\sigma^2}{2} k^2\right) \,. \tag{A30}$$

So it turns out that

$$g(x) = \mathrm{i}\langle x\rangle k - \frac{\sigma^2}{2}k^2 \tag{A31}$$

and so the cumulant expansion is only up to k^2 term. We also get two cumulants: the mean $c_1 = \langle x\rangle$ and the variance $c_2 = \sigma^2$.

Similar manipulation can be used for a time-ordered set of a stochastic variable or the *stochastic process* $x(t)$. In the case of the stationary process the characteristics of the process do not depend on time explicitly; however, they depend only on the time differences. We can thus introduce the correlation functions. The characteristic functional

$$G(k(t)) = \left\langle \exp\left(\mathrm{i}\int \mathrm{d}t k(t) x(t)\right)\right\rangle \tag{A32}$$

can be defined for an arbitrary real-valued function $k(t)$ [52]. This functional then generates the correlation functions

$$\left.\frac{\delta^n G(k(t))}{\delta k(t_1)\delta k(t_2)\dots\delta k(t_n)}\right|_{k(t)=0} = \mathrm{i}^n\langle x(t_1)x(t_2)\dots x(t_n)\rangle\ . \tag{A33}$$

The functional can then be expanded as

$$G(k(t)) = 1 + \sum_n \frac{\mathrm{i}^n}{n!}\int \mathrm{d}t_1 \int \mathrm{d}t_2 \dots \int \mathrm{d}t_n \langle x(t_1)x(t_2)\dots x(t_n)\rangle\ . \tag{A34}$$

The function $x(t)$ is a random Gaussian process with zero mean when its correlation functions factorize as

$$\langle x(t_1)x(t_2)\dots x(t_{2l+1})\rangle = 0\ , \quad l = 0, 1, 2, \dots \tag{A35}$$

$$\langle x(t_1)x(t_2)\dots x(t_{2l})\rangle = \sum_{\text{pairs}} \prod_{\alpha=1}^{l} \langle x(t_{i\alpha})x(t_{j\alpha})\rangle\ , \quad l = 1, 2, 3, \dots \tag{A36}$$

where the sum over "*pairs*" runs over all the different ways in which the $2l$ indices can be subdivided into l unordered pairs. For example,

$$\begin{aligned}\langle x(t_1)x(t_2)x(t_3)x(t_4)\rangle &= \langle x(t_1)x(t_2)\rangle\langle x(t_3)x(t_4)\rangle \\ &+ \langle x(t_1)x(t_3)\rangle\langle x(t_2)x(t_4)\rangle + \langle x(t_1)x(t_4)\rangle\langle x(t_2)x(t_3)\rangle\ .\end{aligned} \tag{A37}$$

For the characteristic function this yields

$$G(k(t)) = \exp\left(-\frac{1}{2}\int \mathrm{d}t_2 \int \mathrm{d}t_1 k(t_2)\langle x(t_2)x(t_1)\rangle k(t_1)\right)\ . \tag{A38}$$

The Gaussian stochastic trajectory is, therefore, fully characterized by the two-point correlation functions.

A.3 Weyl Formula

Suppose that we have operators $\hat{A}$ and $\hat{B}$ and their commutator $[\hat{A}, \hat{B}] = c$ is a number. Then the commutator

$$[\hat{A}, \exp(\alpha \hat{B})] = \sum_n \frac{\alpha^n}{n!}[\hat{A}, \hat{B}^n] . \tag{A39}$$

Now

$$[\hat{A}, \hat{B}^n] = \hat{A}\hat{B}^n - \hat{B}^n\hat{A} = (c + \hat{B}\hat{A})\hat{B}^{n-1} - \hat{B}^n\hat{A} = \ldots = nc\hat{B}^{n-1} . \tag{A40}$$

This yields

$$[\hat{A}, \exp(\alpha \hat{B})] = \alpha c \exp(\alpha \hat{B}) . \tag{A41}$$

By using these expressions we can write

$$\exp(-\alpha \hat{B})[\hat{A}, \exp(\alpha \hat{B})] = \alpha c \exp(-\alpha \hat{B}) \exp(\alpha \hat{B}) \tag{A42}$$

or

$$\exp(-\alpha \hat{B})\hat{A}\exp(\alpha \hat{B}) = \hat{A} + \alpha c . \tag{A43}$$

We next denote

$$f(x) = \exp(\hat{A}x)\exp(\hat{B}x) , \tag{A44}$$

where x is a number, and $\hat{A}$ and $\hat{B}$ are any operators. Differentiation of f leads to

$$\begin{aligned} \frac{\mathrm{d}}{\mathrm{d}x} f(x) &= f(x)\exp(-\hat{B}x)\hat{A}\exp(\hat{B}x) + f(x)\hat{B} \\ &= f(x)(\hat{A} + \hat{B} + xc) . \end{aligned} \tag{A45}$$

Putting $\hat{A} + \hat{B} = \hat{Z}$ gives the differential equation

$$f'(x) - f(x)\cdot(\hat{Z} + xc) = 0 , \tag{A46}$$

the solution of which is given by

$$f(x) = \exp\left(\hat{Z}x + c\frac{x^2}{2}\right) \tag{A47}$$

or taking $x = 1$ we get the *Weyl formula*

$$\exp(\hat{A})\exp(\hat{B}) = \exp(\hat{A} + \hat{B})\exp\left(\frac{1}{2}[\hat{A}, \hat{B}]\right) . \tag{A48}$$

A.4
Thermodynamic Potentials and the Partition Function

The energy and entropy of an isolated system can be characterized by thermodynamic potentials. The natural parameters of the system in this case are the internal energy U, the entropy S, the volume V and the number of particles N. They completely define the state of the system, so if we take a function

$$U = U(S, V, N) \tag{A49}$$

then

$$\mathrm{d}U = \frac{\partial U}{\partial S}\Big|_{V,N}\mathrm{d}S + \frac{\partial U}{\partial V}\Big|_{S,N}\mathrm{d}V + \frac{\partial U}{\partial N}\Big|_{S,V}\mathrm{d}N \tag{A50}$$

where

$$\frac{\partial U}{\partial S}\Big|_{V,N} = T\,, \quad \frac{\partial U}{\partial V}\Big|_{S,N} = -p\,, \quad \frac{\partial U}{\partial N}\Big|_{S,V} = \mu\,. \tag{A51}$$

T determines the temperature, p is a pressure, μ defines the chemical potential. We can similarly write

$$\mathrm{d}S = \frac{1}{T}\mathrm{d}U + \frac{p}{T}\mathrm{d}V - \frac{\mu}{T}\mathrm{d}N\,, \tag{A52}$$

where

$$\frac{\partial S}{\partial U}\Big|_{V,N} = \frac{1}{T} \quad \frac{\partial S}{\partial V}\Big|_{U,N} = \frac{p}{T} \quad \frac{\partial S}{\partial N}\Big|_{V,U} = -\frac{\mu}{T}\,. \tag{A53}$$

Additional set of thermodynamic potentials can be easily created as superpositions of the other potentials. Thus, the free energy is given by

$$F = U - TS = -pV + \mu N\,. \tag{A54}$$

The free energy is a function of T, V, N. The enthalpy

$$H = U + pV \tag{A55}$$

is a function of S, p, N. The Gibbs free energy

$$G = U - TS + pV \tag{A56}$$

is a function of T, p, N. The grand thermodynamic potential

$$\Phi = F - \mu N \tag{A57}$$

is considered as a function of T, V, μ. Maxwell relations are derived from the relations between the second order partial derivatives of the potentials. These are described in any textbook of thermodynamics, see, for instance, [37].

Thermodynamic potentials are tightly related with the partition functions. The canonical ensemble describes a closed system. The partition function of the canonical ensemble directly leads to the free energy

$$F = -k_B T \ln Z \,. \tag{A58}$$

Other properties follow from thermodynamic relations. For instance, the internal energy

$$U = -\frac{\partial \ln Z}{\partial [1/(k_B T)]} \,, \tag{A59}$$

and the entropy

$$S = \frac{\partial}{\partial T}(k_B T \ln Z) \,. \tag{A60}$$

The grand canonical ensemble describes the open system and its partition function directly provides the grand potential

$$\Phi = -k_B T \ln Z_g \,. \tag{A61}$$

Thermodynamic properties of the open system can thus be described using the partition function through thermodynamic relations.

A.5 Fourier Transformation

The symmetric integral form of the Fourier transformation is defined as follows

$$\tilde{f}(\lambda) = \int_{-\infty}^{\infty} f(x) e^{-2\pi i \lambda x} dx \tag{A62}$$

and the similar expression is given for the inverse transformation

$$f(x) = \int_{-\infty}^{\infty} \tilde{f}(\lambda) e^{2\pi i \lambda x} d\lambda \,. \tag{A63}$$

In mathematics all variables, x and λ, are dimensionless. However, in a physical context we always add dimensional meaning to the quantities: if x is time in seconds, then λ is the frequency in Hertz, or if x is coordinate in meters, then λ relates to wavenumbers.

The following symmetry identities follow from definitions given by (A62) and (A63):

$$f(x) = \delta(x) \,; \quad \tilde{f}(\lambda) = 1 \tag{A64}$$

$$f(x) = 1\,; \quad \tilde{f}(\lambda) = \delta(\lambda)\,, \tag{A65}$$

$$\int_{-\infty}^{\infty} e^{\pm 2\pi i y x} dx = \delta(y)\,. \tag{A66}$$

In realistic applications it is necessary to use the sampling theorem, which states that for a function of frequency defined from $-\Lambda/2$ to $\Lambda/2$ and sampled by N points with resolution $d\lambda = \Lambda/N$, the original function of time should be sampled at intervals $dx = \Lambda^{-1}$. The number of points N must be the same in the original and in the transformed function. Thus, the length of time trajectory $X = N\Lambda^{-1}$ gives frequency resolution $d\lambda = \Lambda/N$. If these conditions are satisfied, the numerical representation of relations, given by (A64) and (A66), is automatically satisfied. This gives dimensionless relationship:

$$dx d\lambda = \frac{1}{N}\,. \tag{A67}$$

In physical applications it is more convenient to use the cyclic frequency

$$\omega = 2\pi\lambda_t \tag{A68}$$

as a conjugate time variable, and a wavevector

$$k = 2\pi\lambda_x \tag{A69}$$

as a conjugate coordinate variable. Also, due to wave properties of the electromagnetic field we will use the following form of time-space transformations and we skip overbars; instead we explicitly denote the argument:

$$f(k,\omega) = \int_{-\infty}^{\infty} f(x,t) e^{-ikx + i\omega t} dx dt \tag{A70}$$

and

$$f(x,t) = \int_{-\infty}^{\infty} \frac{dk}{2\pi} \int_{-\infty}^{\infty} \frac{d\omega}{2\pi} f(k,\omega) e^{ikx - i\omega t}\,. \tag{A71}$$

or equivalently we use transformation for time or coordinate only. For the cyclic values we have nonsymmetric relations

$$f(t) = \delta(t)\,; \quad f(\omega) = 1\,, \tag{A72}$$

$$f(t) = 1\,; \quad f(\omega) = 2\pi\delta(\omega)\,, \tag{A73}$$

which gives

$$\int_{-\infty}^{\infty} dx e^{\pm ixy} = 2\pi\delta(y)\,. \tag{A74}$$

In quantum mechanics the momentum-wavevector relation $p = \hbar k$ implies that the momentum can be used as a conjugate variable to a coordinate in the Fourier transform meaning. In the momentum integration it is useful to invoke the interval defined

$$\frac{\mathrm{d}p}{2\pi\hbar} \tag{A75}$$

which has a dimension of inverse coordinate. Thus, $\mathrm{d}x \cdot \mathrm{d}p/(2\pi\hbar)$ is a dimensionless quantity.

A.6 Born Rule

Here we show that $|\langle\psi|a_k\rangle|^2$ is the probability for the system to be observed in the state $|a_k\rangle$. Let us have an operator $\hat{\mathcal{F}}_i^{(N)}$ which gives us the frequency $\nu_i^{(N)}$ of the occurrence of the state $|a_i\rangle$ among the states constructed out of the eigenstates $|a_i\rangle$ of an ensemble of N identical systems. We first construct the following vector

$$\left|\Phi_k^{(N)}\right\rangle = \left(\mathcal{F}_k^{(N)} - |\langle\psi|a_k\rangle|^2\right)|\Psi^{(N)}\rangle\,. \tag{A76}$$

The vector $|\Psi\rangle$ is an eigenvector of $\mathcal{F}_k$ with the eigenvalue $|\langle\psi|a_k\rangle|^2$ if

$$\lim_{N\to\infty}\left\langle\Phi_k^{(N)}\right|\left.\Phi_k^{(N)}\right\rangle = 0\,. \tag{A77}$$

This is not difficult to show. First we write out the scalar product $\langle\Phi_k^{(N)}|\Phi_k^{(N)}\rangle$ explicitly

$$\left\langle\Phi_k^{(N)}\right|\left.\Phi_k^{(N)}\right\rangle = I_1^{(N)} - 2I_2^{(N)} + |\langle\psi|a_k\rangle|^4\,, \tag{A78}$$

where

$$\begin{aligned} I_1^{(N)} &= \langle\Psi^{(N)}|\mathcal{F}_k^{(N)}\mathcal{F}_k^{(N)}|\Psi^{(N)}\rangle\,,\\ I_2^{(N)} &= |\langle\psi|a_k\rangle|^2\,\mathrm{Re}\,\langle\Psi^{(N)}|\mathcal{F}_k^{(N)}|\Psi^{(N)}\rangle\,. \end{aligned} \tag{A79}$$

We start with evaluation of the term $I_2^{(N)}$. First, using (6.83) and (6.84), we write it as

$$I_2^{(N)} = \frac{1}{N}\sum_{\alpha}\sum_{i_1,i_2,\ldots,i_N}\delta_{k i_\alpha}|\langle\psi|a_{i_1}\rangle_1|^2\ldots|\langle\psi|a_{i_N}\rangle_N|^2\,. \tag{A80}$$

For each α value we obtain

$$\frac{1}{N}\sum_{i_1,i_2,\ldots,i_N}\delta_{k i_\alpha}|\langle\psi|a_{i_1}\rangle_1|^2\ldots|\langle\psi|a_{i_N}\rangle_N|^2 = \frac{|\langle\psi|a_k\rangle|^2}{N}\,, \tag{A81}$$

where the fact that $\sum_i |a_i\rangle\langle a_i|$ represents a unity operator is taken into account. This summation can be done for all systems in the ensemble, except of the one with index equal to a given α. The term $I_2^{(N)}$ comprises N of such terms and consequently

$$I_2^{(N)} = |\langle\psi|a_k\rangle|^4 \,. \tag{A82}$$

The term $I_1^{(N)}$ can be calculated in an analogical manner. We can write it as

$$I_1^{(N)} = \frac{1}{N^2}\sum_{\alpha\beta}\sum_{i_1,i_2,\ldots,i_N} \delta_{ki_\alpha}\delta_{ki_\beta}|\langle\psi|a_{i_1}\rangle_1|^2 \ldots |\langle\psi|a_{i_N}\rangle_N|^2 \,. \tag{A83}$$

We have N cases where $\alpha = \beta$ and it allows us to evaluate $|\langle\psi|a_k\rangle|^2/N$, and $N(N-1)$ terms where $\alpha \neq \beta$, thus, providing the following result: $|\langle\psi|a_k\rangle|^4/N^2$. In total we get

$$I_1^{(N)} = \frac{1}{N}|\langle\psi|a_k\rangle|^2 + \left(1 - \frac{1}{N}\right)|\langle\psi|a_k\rangle|^4 \,. \tag{A84}$$

Inserting (A82) and (A84) into (A78) we obtain

$$\lim_{N\to\infty}\left\langle \Phi_k^{(N)} \middle| \Phi_k^{(N)} \right\rangle = \frac{1}{N}\left(|\langle\psi|a_k\rangle|^2 + |\langle\psi|a_k\rangle|^4\right) = 0 \,, \tag{A85}$$

and any vector $|\Psi\rangle$ is therefore an eigenvector of the operator $\mathcal{F}_k$ with the corresponding eigenvalue $|\langle\psi|a_k\rangle|^2$.

A.7
Green's Function of a Harmonic Oscillator

The equation for a driven harmonic oscillator is

$$\ddot{x} + \omega_0^2 x = f(t) \,. \tag{A86}$$

We make a Fourier transform and get the solution

$$x(\omega) = -\frac{f(\omega)}{\omega^2 - \omega_0^2} \,. \tag{A87}$$

The inverse Fourier transform will be also performed. First we rewrite

$$\frac{1}{\omega^2 - \omega_0^2} = \frac{1}{2\omega_0}\left(\frac{1}{\omega - \omega_0} - \frac{1}{\omega + \omega_0}\right) \,. \tag{A88}$$

Then we can write a convolution expression

$$x(t) = \int_0^\infty \mathrm{d}t'\, G(t - t') f(t') \,, \tag{A89}$$

where

$$G(\tau) = \int_{-\infty}^{\infty} -\frac{1}{2\omega_0} \int \frac{d\omega}{2\pi} e^{-i\omega\tau} \left(\frac{1}{\omega - \omega_0} - \frac{1}{\omega + \omega_0} \right) \qquad \text{(A90)}$$

is called a Green's function. For calculation of the integral over the frequency we apply the Cauchy integration formula. However, we want to construct a causal function that reflects $f(t)$ causal relation to the $x(t)$. We should then have $t' < t$ in (A90). Specific selection of the integration contour and poles give the following. For poles we have to check the exponent $\exp(-i\omega\tau)$ at $\tau > 0$; if we take the complex frequency $\omega = \pm\omega + i\eta$ with $\eta \to +0$, the exponent $\exp(\mp i\omega\tau + \eta\tau)$ diverges for positive τ. When we take $\omega = \pm\omega_0 - i\eta$, the exponent $\exp(\mp i\omega_0\tau - \eta\tau)$ decays with $\tau > 0$. So the poles must be taken at the lower complex half plane and this will ensure that $G(\tau)$ is nonzero only for positive times. The integration then yields:

$$G(\tau) = -\frac{\theta(\tau)}{\omega_0} \sin(\omega_0\tau) \,. \qquad \text{(A91)}$$

To emphasize the complex nature of the causal Green's function in the frequency domain we write

$$G(\omega) = -\lim_{\eta\to 0} \frac{1}{2\omega_0} \left(\frac{1}{\omega - \omega_0 + i\eta} - \frac{1}{\omega + \omega_0 + i\eta} \right) \qquad \text{(A92)}$$

or

$$G_x(\omega) = -\lim_{\eta\to 0} \frac{1}{\omega^2 - \omega_0^2 + i\omega\eta} \,. \qquad \text{(A93)}$$

Here we rescaled $2\eta \to \eta$ as it approaches zero.

A.8 Cumulant Expansion in Quantum Mechanics

The wavefunction propagator for a time-dependent Hamiltonian in quantum mechanics is given by:

$$\hat{U}(t) = \exp_+ \left(-i \int_0^t d\tau \hat{H}(\tau) \right) \,. \qquad \text{(A94)}$$

It is reminiscent of the characteristic functional, and thus, we can apply the cumulant expansion. When the time-dependence is due to Gaussian fluctuations of energy values (the adiabatic approximation) in the eigenstate basis we have

$$\langle a|\hat{U}(t)|a\rangle = \exp_+ \left(-i \int_0^t d\tau \varepsilon_a(\tau) \right) \,. \qquad \text{(A95)}$$

for the state $|a\rangle$. The propagator is given by series of the expansion and its statistical averaging gives

$$\langle \hat{U}_{aa}(t)\rangle = e^{-i\varepsilon_a t}\left\langle 1 - i\int_0^t d\tau\tilde{\varepsilon}_a(\tau) - \int_0^t d\tau_2 \int_0^{\tau_2} d\tau_1\tilde{\varepsilon}_a(\tau_2)\tilde{\varepsilon}_a(\tau_1) + \ldots\right\rangle . \tag{A96}$$

The mean value of the fluctuation has been included into ε_a, and $\tilde{\varepsilon}_a$ is a zero-mean Gaussian fluctuation. The first integral vanishes as averaging is performed. Taking the cumulant expansion we find the exact expression

$$\langle \hat{U}_{aa}(t)\rangle = e^{-i\varepsilon_a t - g_{aa}(t)} , \tag{A97}$$

where

$$g_{aa}(t) = \int_0^t d\tau_2 \int_0^{\tau_2} d\tau_1 \langle\tilde{\varepsilon}_a(\tau_2)\tilde{\varepsilon}_a(\tau_1)\rangle \tag{A98}$$

is a linear transformation of the correlation function and it is denoted as the lineshape function. Here we use two indices for the lineshape function. Sometimes we use four indices for the lineshape function. That form correlates two elements of a general fluctuating hamiltonian matrix. For instance

$$g_{ab,cd}(t) = \int_0^t d\tau_2 \int_0^{\tau_2} d\tau_1 \langle\tilde{h}_{ab}(\tau_2)\tilde{h}_{cd}(\tau_1)\rangle \tag{A99}$$

and $g_{aa,bb}(t) \equiv g_{ab}(t)$.

Some useful properties related to the lineshape functions are

$$g_{aa}(0) = 0 , \tag{A100}$$

$$\int_a^b d\tau_2 \int_c^d d\tau_1 C_{ab}(\tau_2 - \tau_1) = g_{ab}(a-d) - g_{ab}(a-c) - g_{ab}(b-d) + g_{ab}(b-c) , \tag{A101}$$

which become apparent by considering the geometry of triangular integration areas; moreover

$$\int_0^t d\tau C_{ab}(\tau) = \dot{g}_{ab}(t) , \tag{A102}$$

$$C_{ab}(t) = \ddot{g}_{ab}(t) , \tag{A103}$$

as well as

$$g_{ab}(t) = g^*_{ba}(-t)\ . \tag{A104}$$

As the correlation function is defined using the spectral density $C''(\omega)$, the line-shape function can be conveniently expressed using the Fourier transform of the correlation function

$$g(t) = \int_{-\infty}^{\infty} \frac{\mathrm{d}\omega}{2\pi} \frac{C(\omega)}{\omega^2} (1 - \mathrm{e}^{-\mathrm{i}\omega t} - \mathrm{i}\omega t)\ , \tag{A105}$$

or using the spectral density

$$g(t) = \int_{-\infty}^{\infty} \frac{\mathrm{d}\omega}{2\pi} \frac{C''(\omega)}{\omega^2} \left[1 + \coth\left(\frac{\beta\omega}{2}\right)\right] \left(1 - \mathrm{e}^{-\mathrm{i}\omega t} - \mathrm{i}\omega t\right)\ . \tag{A106}$$

A.8.1
Application to the Double Slit Experiment

The recipe described above applies to the dephasing. Let us consider (6.34). We expand the exponential and evaluate individual terms

$$\begin{aligned}\langle \eta_1 | \eta_2 \rangle \approx 1 &- \frac{\mathrm{i}}{\hbar} \int_0^t \mathrm{d}\tau \langle \eta_0 | \hat{U}_1^\dagger(\tau) \Delta\Xi\, \hat{U}_1(\tau) | \eta_0 \rangle \\ &- \frac{1}{\hbar^2} \int_0^t \mathrm{d}\tau \int_0^\tau \mathrm{d}\tau' \langle \eta_0 | \hat{U}_1^\dagger(\tau) \Delta\Xi\, \hat{U}_1(\tau - \tau') \Delta\Xi\, \hat{U}_1(\tau') | \eta_0 \rangle + \ldots\end{aligned} \tag{A107}$$

We will make an assumption that $\langle \eta_1 | \Delta\Xi | \eta_1 \rangle = 0$, which can always be enforced by redefinition, $\Delta\Xi \to \Delta\Xi - \langle \eta_1 | \Delta\Xi | \eta_1 \rangle$, when it not satisfied. The first nonzero term is, therefore, the one of the second order. We define functions

$$C(\tau, \tau') = \langle \eta_0 | U_1^\dagger(\tau) \Delta\Xi\, U_1(\tau - \tau') \Delta\Xi\, U_1(\tau') | \eta_0 \rangle\ , \tag{A108}$$

and

$$g(t) = \frac{1}{\hbar^2} \int_0^t \mathrm{d}\tau \int_0^\tau \mathrm{d}\tau' C(\tau, \tau')\ . \tag{A109}$$

In terms of the cumulant expression the overlap can be written as an exponential

$$\langle \eta_1 | \eta_2 \rangle = \mathrm{e}^{-g(t)}\ . \tag{A110}$$

This expression is exact if the bath effect is Gaussian.

A.8.2
Application to Linear Optical Response

The linear optical response is given by the correlation function of the dipole operator

$$C^{(2)}(t) = \mathrm{Tr}_{\mathrm{B}}\{\hat{\mu}(t)\hat{\mu}(0)\,W(0)\}\,. \tag{A111}$$

Here $W(0)$ is the full density matrix at zero time. The dipole operator can be easily obtained if we take system eigenstate basis, and assume that the bath adds diagonal fluctuations to the system energies. The dipole operator in the presence of the bath is then

$$\begin{aligned}\hat{\mu}(t) &= \mathrm{e}^{\mathrm{i}\hat{H}t}\hat{\mu}\mathrm{e}^{-\mathrm{i}\hat{H}t}\\ &= \sum_{ab}\mu_{ab}\mathrm{e}^{\mathrm{i}\omega_{ab}t}|a\rangle\langle b|\exp_{+}\left(-\mathrm{i}\int_t^0 \mathrm{d}\tau\varepsilon_a(\tau) - \mathrm{i}\int_0^t \mathrm{d}\tau\varepsilon_b(\tau)\right)\,.\end{aligned} \tag{A112}$$

For the density matrix we assume the canonical distribution in the observable system and bath subsystems. We use the cumulant expansion by assuming that the bath is arranged by a set of harmonic oscillators and thus the fluctuations are Gaussian. The correlation function is then given by cumulant expansion

$$\begin{aligned}C^{(2)}(t) &= \sum_{ab}\frac{\mu_{ab}\mu_{ba}}{Z}\mathrm{e}^{-\beta\varepsilon_a}\mathrm{e}^{-\mathrm{i}\omega_{ba}t}\\ &\quad \mathrm{Tr}_{\mathrm{B}}\left\{\exp_{+}\left(-\mathrm{i}\int_t^0 \mathrm{d}\tau\varepsilon_a(\tau) - i\int_0^t \mathrm{d}\tau\varepsilon_b(\tau)\right)\rho_{\mathrm{B}}\right\}\,.\end{aligned} \tag{A113}$$

The exponent denotes the expansion

$$\begin{aligned}\exp_{+}&\left(-\mathrm{i}\int_t^0 \mathrm{d}\tau\varepsilon_a(\tau) - \mathrm{i}\int_0^t \mathrm{d}\tau\varepsilon_b(\tau)\right) = 1 - \mathrm{i}\int_t^0 \mathrm{d}\tau\varepsilon_a(\tau) - \mathrm{i}\int_0^t \mathrm{d}\tau\varepsilon_b(\tau)\\ &- \int_t^0 \mathrm{d}\tau_2\int_t^{\tau_2} \mathrm{d}\tau_1\varepsilon_a(\tau_2)\varepsilon_a(\tau_1) - \int_0^t \mathrm{d}\tau_1\int_0^{t_1} \mathrm{d}\tau_2\varepsilon_b(\tau_2)\varepsilon_b(\tau_1)\\ &- \mathrm{i}\int_t^0 \mathrm{d}\tau_1\int_0^t \mathrm{d}\tau_2\varepsilon_a(\tau_1)\varepsilon_b(\tau_2) + \ldots\end{aligned} \tag{A114}$$

Taking the statistical average yields

$$\begin{aligned}&\left\langle\exp_{+}\left(-\mathrm{i}\int_t^0 \mathrm{d}\tau\varepsilon_a(\tau) - \mathrm{i}\int_0^t \mathrm{d}\tau\varepsilon_b(\tau)\right)\right\rangle\\ &= 1 - g_{aa}(-t) - g_{bb}(t) - g_{ab}(t) - g_{ab}(-t)\,.\end{aligned} \tag{A115}$$

For the correlation function we then have

$$C^{(2)}(t) = \sum_{ab} \frac{\mu_{ab}\mu_{ba}}{Z} e^{-\beta\varepsilon_a} \exp(-i\omega_{ba}t - g^*_{aa}(t) - g_{bb}(t) - g_{ab}(t) - g^*_{ba}(t)) , \tag{A116}$$

where

$$Z = \sum_a e^{-\beta\varepsilon_a} \tag{A117}$$

and is the partition function and $\beta = (k_B T)^{-1}$.

A.8.3
Application to Third Order Nonlinear Response

The third order optical response is given by the correlation function of the dipole operator

$$C^{(4)}(t_4, t_3, t_2, t_1) = \mathrm{Tr}_B\{\hat{\mu}(t_4)\hat{\mu}(t_3)\hat{\mu}(t_2)\hat{\mu}(t_1) W(0)\} . \tag{A118}$$

It is mathematically more involved than the two-point correlation function of the previous subsection, but it is straightforward to proceed with the cumulant expansion. The result is given in a following form:

$$\begin{aligned} C^{(4)}(t_4, t_3, t_2, t_1) = \sum_{abcd} & \frac{\mu_{ad}\mu_{dc}\mu_{cb}\mu_{ba}}{Z} e^{-\beta\varepsilon_a} \\ & \times \exp(-i\omega_{da}t_{43} - i\omega_{ca}t_{32} - i\omega_{ba}t_{21} \\ & + f_{dcba}(t_4, t_3, t_2, t_1)) , \end{aligned} \tag{A119}$$

where $t_{ij} = t_i - t_j$. To calculate f_{dcba} we have to consider the propagator

$$\exp_+ \left(-i \int_{t_3}^{t_4} d\tau \varepsilon_d(\tau) - i \int_{t_2}^{t_3} d\tau \varepsilon_c(\tau) - i \int_{t_1}^{t_2} d\tau \varepsilon_b(\tau) - i \int_{t_4}^{t_1} d\tau \varepsilon_a(\tau) \right) . \tag{A120}$$

The cumulant expansion finally gives

$$\begin{aligned} f_{dcba}(t_4, t_3, t_2, t_1) = & -g_{dd}(t_{43}) - g_{cc}(t_{32}) - g_{bb}(t_{21}) - g_{aa}(t_{14}) \\ & + g_{dc}(t_{32}) + g_{dc}(t_{43}) - g_{dc}(t_{42}) \\ & - g_{db}(t_{32}) + g_{db}(t_{31}) + g_{db}(t_{42}) - g_{db}(t_{41}) \\ & - g_{da}(t_{31}) + g_{da}(t_{34}) + g_{da}(t_{41}) \\ & + g_{cb}(t_{21}) + g_{cb}(t_{32}) - g_{cb}(t_{31}) \\ & - g_{ca}(t_{21}) + g_{ca}(t_{24}) + g_{ca}(t_{31}) - g_{ca}(t_{34}) \\ & + g_{ba}(t_{14}) + g_{ba}(t_{21}) - g_{ba}(t_{24}) . \end{aligned} \tag{A121}$$

This expression again holds for the adiabatic regime when observable states do change their states due to bath fluctuations.

A.9
Matching the Heterodyned FWM Signal with the Pump-Probe

Imagine two experiments: one in which we excite our system with three pulses coming from three directions $\boldsymbol{k}_1$, $\boldsymbol{k}_2$ and $\boldsymbol{k}_{\mathrm{pr}}$, and one in which the first two pulses merge into one, coming from the direction $\boldsymbol{k}_{\mathrm{pu}} \equiv \boldsymbol{k}_1$. The first and the second pulse comes at exactly the same time, while the last pulse comes with a given delay time T. The former of the two measurements gives us a background-free signal field $E_{\mathrm{s}}(t)$ into the direction $-\boldsymbol{k}_1 + \boldsymbol{k}_2 + \boldsymbol{k}_{\mathrm{pr}}$ (and to many other directions), while the latter one results in exactly the same signal, but going into the direction $-\boldsymbol{k}_{\mathrm{pu}} + \boldsymbol{k}_{\mathrm{pu}} + \boldsymbol{k}_{\mathrm{pr}} = \boldsymbol{k}_{\mathrm{pr}}$. The latter scheme clearly represents the pump probe measurement, and it enables us to determine the differential intensity

$$\Delta I_{\mathrm{pp}} \approx \mathrm{Re}\left(E_{\mathrm{s}}(t)\, E^{*}_{\mathrm{pr}}(t)\right) . \tag{A122}$$

In the former situation we can use heterodyne detection and we obtain a similar result

$$\Delta I_{\mathrm{LO}} \approx \mathrm{Re}\left(E_{\mathrm{s}}(t)\, E^{*}_{\mathrm{LO}}(t)\right) , \tag{A123}$$

which only differs by an arbitrary phase of the local oscillator.

Let us assume the three involved fields have phases ϕ_{s}, ϕ_{pr} and ϕ_{LO} so that we can write

$$E_{\mathrm{s}}(t) = \mathcal{E}_{\mathrm{s}}(t)\mathrm{e}^{-\mathrm{i}\Omega t + \mathrm{i}\phi_{\mathrm{s}}} , \quad E_{\mathrm{pr}}(t) = \mathcal{E}_{\mathrm{pr}}(t)\mathrm{e}^{-\mathrm{i}\Omega t + \mathrm{i}\phi_{\mathrm{pr}}} . \tag{A124}$$

$$E_{\mathrm{LO}}(t) = \mathcal{E}_{\mathrm{LO}}(t + t_4)\mathrm{e}^{-\mathrm{i}\Omega(t+t_4) + \mathrm{i}\phi_{\mathrm{LO}}} . \tag{A125}$$

The local oscillator comes usually from the same source as all other pulses, and their envelopes are, therefore, the same, that is $\mathcal{E}_{\mathrm{LO}}(t) = \mathcal{E}_{\mathrm{pr}}(t)$. We assume all pulses to have a central frequency Ω and the local oscillator is sent prior to all other pulses. The delay between the last pulse of the sequence, which is the probe pulse, will be denoted t_4. The two differential intensities then read:

$$\Delta I_{\mathrm{pp}} \approx \mathrm{Re}\left(\mathcal{E}_{\mathrm{s}}(t)\mathcal{E}_{\mathrm{pr}}(t)\mathrm{e}^{\mathrm{i}(\phi_{\mathrm{s}} - \phi_{\mathrm{pr}})}\right) , \tag{A126}$$

$$\Delta I_{\mathrm{LO}} \approx \mathrm{Re}\left(\mathcal{E}_{\mathrm{s}}(t)\mathcal{E}_{\mathrm{LO}}(t + t_4)\mathrm{e}^{\mathrm{i}(\phi_{\mathrm{s}} - \phi_{\mathrm{LO}}) + \mathrm{i}\Omega t_4}\right) \tag{A127}$$

The measurement can be performed equally well in frequency domain, that is by first dispersing the signal according to the frequency and then looking at the intensity at each frequency:

$$\Delta I_{\mathrm{pp}}(\omega) \approx \mathrm{Re}\left(\mathcal{E}_{\mathrm{s}}(\omega)\mathcal{E}_{\mathrm{pr}}(\omega)\mathrm{e}^{\mathrm{i}(\phi_{\mathrm{s}} - \phi_{\mathrm{pr}})}\right) , \tag{A128}$$

$$\Delta I_{\mathrm{LO}}(\omega) \approx \mathrm{Re}\left(\mathcal{E}_{\mathrm{s}}(\omega)\mathcal{E}_{\mathrm{LO}}(\omega)\mathrm{e}^{\mathrm{i}(\phi_{\mathrm{s}} - \phi_{\mathrm{LO}}) - \mathrm{i}(\omega - \Omega)t_4}\right) , \tag{A129}$$

where argument ω denotes the Fourier transform of the corresponding real envelops. A quick look at (A129) shows us that $\Delta I_{\mathrm{LO}}(\omega)$ is modulated by the factor

$e^{i\omega t_4}$, and correspondingly as a function of ω it oscillates with frequency t_4. The frequency of this oscillation is therefore given by the delay t_4 which is set experimentally.

Now we can use a numerical trick which will enable us to compensate for the arbitrary phase change between (A128) and (A129). We will Fourier transform the differential absorption measured with the local oscillator:

$$\Delta I_{\rm LO}(t) = \frac{1}{2\pi} \int_{-\infty}^{\infty} d\omega \Delta I_{\rm LO}(\omega) e^{-i\omega t} . \tag{A130}$$

Because $\Delta I_{\rm LO}(\omega)$ is a real function, its Fourier transform is symmetric, that is $\Delta I_{\rm LO}(-t) = \Delta I_{\rm LO}(t)$. Setting the negative time part of the function to zero, and Fourier transforming the function back to frequency domain we obtain a complex function, which should correspond to the function $\Delta I'_{\rm LO}(\omega) = \tilde{\mathcal{E}}_{\rm s}(\omega)\tilde{\mathcal{E}}_{\rm LO}(\omega) e^{i(\phi_{\rm s}-\phi_{\rm LO})-i(\omega-\Omega)t_4}$ of which we measured the real part:

$$\Delta I'_{\rm LO}(\omega) = 2 \int_{-\infty}^{\infty} dt\, \Theta(t) \Delta I_{\rm LO}(\omega) e^{i\omega t} , \tag{A131}$$

The last step of the procedure is to multiply $\Delta I'_{\rm LO}(\omega)$ by a phase factor $e^{i\Delta\phi}$ such that

$$\mathrm{Re}\left(\Delta I'_{\rm LO}(\omega) e^{i\Delta\phi}\right) = \Delta I_{\rm pp}(\omega) . \tag{A132}$$

In theory we can easily see that

$$\Delta\phi = -\phi_{\rm pr} + \phi_{\rm LO} + (\omega - \Omega) t_4 , \tag{A133}$$

but experimentally, the only way to determine the phase is by fitting. It is important to realize that if we would not have set a delay between the local oscillator and the signal, and if we were extremely unlucky in setting our local oscillator phase, we could have had it such that $\Delta I_{\rm LO} \approx 0$. With a considerably long delay t_4, any value of $\phi_{\rm s} - \phi_{\rm LO}$ is allowed, and any value can be compensated by fitting. It is, therefore, these "artificial" oscillations on $\Delta I_{\rm LO}$ caused by the delay t_4 which enable us to recover the signal phase.

This phasing procedure is of extreme importance for the two-dimensional coherent spectroscopy that is introduced in Chapter 16. There the phase of the spectrum in the experiment is adjusted by comparison with the pump probe spectrum just as described here. However, this is not necessary in simulations where the signals are given by response functions.

A.10
Response Functions of an Excitonic System with Diagonal and Off-Diagonal Fluctuations in the Secular Limit

Here we present the third order response function expression written in terms of resonant Liouville space pathways (or Feynman diagrams, shown in Figure 15.4).

We break the response function into the following contributions: induced absorption (IA), stimulated emission (SE) and ground state bleach (GSB). Additionally, we distinguish terms with the population (p) in the second interval t_2 or the coherence (c), and the population can be transferred from e to e' – such diagrams are indicated by *prime*.

The rephasing response function is given by the following seven terms:

$$\begin{aligned} S^{\mathrm{IA_p}} = &-\sum_e \sum_f G_{ee}(t_2)\left\langle|\mu_{eg}|^2|\mu_{ef}|^2\right\rangle \\ &\times \exp\big(\mathrm{i}\omega_{eg}t_1 - \mathrm{i}\omega_{fe}t_3 \\ &+ \mathrm{conj}\big[-g^*_{ee}(t_2) - g^*_{ff}(t_3) - g_{ee}(t_1+t_2+t_3) \\ &- g^*_{ef}(t_2+t_3) + g^*_{ef}(t_2) + g^*_{ef}(t_3) \\ &- g_{ee}(t_1) + g^*_{ee}(t_2+t_3) + g_{ee}(t_1+t_2) - g^*_{ee}(t_3) \\ &- g_{ef}(t_1+t_2) + g^*_{ef}(t_3) + g_{ef}(t_1+t_2+t_3)\big]\big) \end{aligned} \tag{A134}$$

$$\begin{aligned} S^{\mathrm{IA'_p}} = &-\sum_{e'\neq e} \sum_f \left\langle\mu_{e'}\mu_{e'f}\mu_{ef}\mu_e\right\rangle \\ &\times \exp\big(\mathrm{i}\omega_{eg}t_1 - \mathrm{i}\omega_{e'e}t_2 - \mathrm{i}\omega_{fe}t_3 \\ &+ \mathrm{conj}\big[-g^*_{e'e'}(t_2) - g^*_{ff}(t_3) - g_{ee}(t_1+t_2+t_3) \\ &- g^*_{e'f}(t_2+t_3) + g^*_{e'f}(t_2) + g^*_{e'f}(t_3) \\ &- g_{e'e}(t_1) + g^*_{e'e}(t_2+t_3) + g_{e'e}(t_1+t_2) - g^*_{e'e}(t_3) \\ &- g_{ef}(t_1+t_2) + g^*_{ef}(t_3) + g_{ef}(t_1+t_2+t_3)\big]\big) \end{aligned} \tag{A135}$$

$$\begin{aligned} S^{\mathrm{IA_c}} = &-\sum_{e'\neq e} \sum_f \left\langle\mu_{e'}\mu_{e'f}\mu_{ef}\mu_e\right\rangle \\ &\times \exp\big(\mathrm{i}\omega_{eg}t_1 - \mathrm{i}\omega_{e'e}t_2 - \mathrm{i}\omega_{fe}t_3 \\ &+ \mathrm{conj}\big[-g^*_{e'e'}(t_2) - g^*_{ff}(t_3) - g_{ee}(t_1+t_2+t_3) \\ &- g^*_{e'f}(t_2+t_3) + g^*_{e'f}(t_2) + g^*_{e'f}(t_3) \\ &- g_{e'e}(t_1) + g^*_{e'e}(t_2+t_3) + g_{e'e}(t_1+t_2) - g^*_{e'e}(t_3) \\ &- g_{ef}(t_1+t_2) + g^*_{ef}(t_3) + g_{ef}(t_1+t_2+t_3)\big]\big) \end{aligned} \tag{A136}$$

$$\begin{aligned} S^{\mathrm{SE_c}} = &\sum_{e\neq e'} \left\langle|\mu_e|^2|\mu_{e'}|^2\right\rangle \\ &\times \exp\big(\mathrm{i}\omega_{eg}t_1 + \mathrm{i}\omega_{ee'}t_2 - \mathrm{i}\omega_{e'}t_3 \\ &- g^*_{ee}(t_1+t_2) - g_{e'e'}(t_2+t_3) \\ &- g^*_{ee'}(t_1) + g^*_{ee'}(t_1+t_2+t_3) + g_{ee'}(t_2) - g^*_{ee'}(t_3)\big) \end{aligned} \tag{A137}$$

$$
\begin{aligned}
S^{\mathrm{SE_p}} = \sum_{e=e'} & G_{e'e}(t_2)\left\langle |\mu_e|^2 |\mu_{e'}|^2 \right\rangle \\
& \times \exp\big(\mathrm{i}\omega_{eg}t_1 - \mathrm{i}\omega_{e'g}t_3 \\
& - g^*_{ee}(t_1+t_2) - g_{e'e'}(t_2+t_3) \\
& - g^*_{ee'}(t_1) + g^*_{ee'}(t_1+t_2+t_3) + g_{ee'}(t_2) - g^*_{ee'}(t_3)\big)
\end{aligned} \tag{A138}
$$

$$
\begin{aligned}
S^{\mathrm{SE'_p}} = \sum_{e\neq e'} & G_{e'e}(t_2)|\mu_e|^2 |\mu_{e'}|^2 \\
& \times \exp\big(\mathrm{i}\omega_{eg}t_1 - \mathrm{i}\omega_{e'g}t_3 - \gamma_{e'}t_3 - \gamma_e t_1 \\
& - g_{ee}(t_1) - g^*_{e'e'}(t_3) + g_{e'e}(t_1+t_2+t_3) \\
& - g_{e'e}(t_1+t_2) - g_{e'e}(t_2+t_3) + g_{e'e}(t_2)\big)
\end{aligned} \tag{A139}
$$

$$
\begin{aligned}
S^{\mathrm{GSB}} = \sum_{ee'} & \left\langle |\mu_e|^2 |\mu_{e'}|^2 \right\rangle \exp\big(\mathrm{i}\omega_{eg}t_1 - \mathrm{i}\omega_{e'g}t_3 \\
& - g^*_{ee}(t_1) - g_{e'e'}(t_3) - g^*_{ee'}(t_1+t_2) \\
& + g^*_{ee'}(t_1+t_2+t_3) + g^*_{ee'}(t_2) - g^*_{ee'}(t_2+t_3)\big).
\end{aligned} \tag{A140}
$$

The nonrephasing response function is a sum of

$$
\begin{aligned}
S^{\mathrm{IA_p}} = -\sum_{e'=e}\sum_{f} & G_{e'e}(t_2)\left\langle |\mu_{e'f}|^2 |\mu_e|^2 \right\rangle \\
& \times \exp\big(-\mathrm{i}\omega_{e'g}t_1 + \mathrm{i}\omega_{ef}t_3 \\
& + \mathrm{conj}\big[-g^*_{e'e'}(t_1+t_2) - g^*_{ff}(t_3) - g_{ee}(t_2+t_3) \\
& - g^*_{e'f}(t_1+t_2+t_3) + g^*_{e'f}(t_1+t_2) + g^*_{e'f}(t_3) \\
& - g^*_{e'e}(t_1) + g^*_{e'e}(t_1+t_2+t_3) + g_{e'e}(t_2) - g^*_{e'e}(t_3) \\
& - g_{fe}(t_2) + g^*_{fe}(t_3) + g_{fe}(t_2+t_3)\big]\big)
\end{aligned} \tag{A141}
$$

$$
\begin{aligned}
S^{\mathrm{IA'_p}} = -\sum_{e'\neq e}\sum_{f} & G_{e'e}(t_2)\left\langle |\mu_{e'f}|^2 |\mu_e|^2 \right\rangle \\
& \times \exp\big(\mathrm{i}\omega_{eg}t_1 - \mathrm{i}\omega_{e'f}t_3 - \gamma_{e'}t_3 - \gamma_e t_1 \\
& + \mathrm{conj}\big[-g_{ee}(t_1) - g_{ff}(t_3) - g^*_{e'e'}(t_3) \\
& - g_{fe}(t_1+t_2+t_3) + g_{fe}(t_1+t_2) + g_{fe}(t_2+t_3) \\
& + g_{e'e}(t_1+t_2+t_3) - g_{e'e}(t_1+t_2) - g_{e'e}(t_2+t_3) \\
& + g_{e'f}(t_3) + g^*_{fe'}(t_3) + g_{e'e}(t_2) - g_{fe}(t_2)\big]\big)
\end{aligned} \tag{A142}
$$

$$S^{\mathrm{IA_c}} = -\sum_{e' \neq e} \sum_{f} \langle \mu_{e'} \mu_{e'f} \mu_{fe} \mu_{e} \rangle$$
$$\times \exp(-\mathrm{i}\omega_{e'g} t_1 + \mathrm{i}\omega_{ee'} t_2 + \mathrm{i}\omega_{ef} t_3$$
$$+ \mathrm{conj}[-g^*_{e'e'}(t_1 + t_2) - g^*_{ff}(t_3) - g_{ee}(t_2 + t_3)$$
$$- g^*_{e'f}(t_1 + t_2 + t_3) + g^*_{e'f}(t_1 + t_2) + g^*_{e'f}(t_3)$$
$$- g^*_{e'e}(t_1) + g^*_{e'e}(t_1 + t_2 + t_3) + g_{e'e}(t_2) - g^*_{e'e}(t_3)$$
$$- g_{fe}(t_2) + g^*_{fe}(t_3) + g_{fe}(t_2 + t_3)]) \tag{A143}$$

$$S^{\mathrm{SE_p}} = \sum_{e'=e} G_{e'e}(t_2) \langle |\mu_e|^2 |\mu_{e'}|^2 \rangle$$
$$\times \exp(-\mathrm{i}\omega_{eg} t_1 - \mathrm{i}\varepsilon_{e'g} t_3$$
$$- g^*_{ee}(t_2) - g_{e'e'}(t_1 + t_2 + t_3) - g_{ee'}(t_1)$$
$$+ g^*_{ee'}(t_2 + t_3) + g_{ee'}(t_1 + t_2) - g^*_{ee'}(t_3)) \tag{A144}$$

$$S^{\mathrm{SE'_p}} = \sum_{e' \neq e} G_{e'e}(t_2) \langle |\mu_e|^2 |\mu_{e'}|^2 \rangle$$
$$\times \exp(-\mathrm{i}\omega_{eg} t_1 - \mathrm{i}\omega_{e'g} t_3 - \gamma_{e'} t_3 - \gamma_e t_1$$
$$- g_{ee}(t_1) - g_{e'e'}(t_3) - g_{e'e}(t_1 + t_2 + t_3)$$
$$+ g_{e'e}(t_1 + t_2) + g_{e'e}(t_2 + t_3) - g_{e'e}(t_2)) \tag{A145}$$

$$S^{\mathrm{SE_c}} = \sum_{e' \neq e} \langle |\mu_e|^2 |\mu_{e'}|^2 \rangle$$
$$\times \exp(-\mathrm{i}\omega_{e'g} t_1 + \mathrm{i}\omega_{ee'} t_2 - \mathrm{i}\omega_{e'g} t_3$$
$$- g^*_{ee}(t_2) - g_{e'e'}(t_1 + t_2 + t_3) - g_{ee'}(t_1)$$
$$+ g^*_{ee'}(t_2 + t_3) + g_{ee'}(t_1 + t_2) - g^*_{ee'}(t_3)) \tag{A146}$$

$$S^{\mathrm{GSB}} = \sum_{ee'} \langle |\mu_e|^2 |\mu_{e'}|^2 \rangle$$
$$\times \exp(-\mathrm{i}\omega_{e'g} t_1 - \mathrm{i}\omega_{eg} t_3$$
$$- g_{ee}(t_3) - g_{e'e'}(t_1) - g_{ee'}(t_1 + t_2 + t_3)$$
$$+ g_{ee'}(t_2 + t_3) + g_{ee'}(t_1 + t_2) - g_{ee'}(t_2)). \tag{A147}$$

Finally the double-quantum coherence response function is given by

$$S^{2Q_1} = \sum_{e'e} \sum_{f} \langle \mu_{e'} \mu_{e'f} \mu_{fe} \mu_{e} \rangle$$
$$\times \exp(-i\omega_{eg} t_1 - i\omega_{fg} t_2 - i\omega_{e'g} t_3$$
$$- g_{e'e'}(t_3) - g_{ff}(t_2) - g_{ee}(t_1) - g_{e'f}(t_2 + t_3)$$
$$+ g_{e'f}(t_3) + g_{e'f}(t_2) - g_{e'e}(t_1 + t_2 + t_3)$$
$$+ g_{e'e}(t_2 + t_3) + g_{e'e}(t_1 + t_2) - g_{e'e}(t_2)$$
$$- g_{fe}(t_1 + t_2) + g_{fe}(t_2) + g_{fe}(t_1)) \quad \text{(A148)}$$

and

$$S^{2Q_2} = -\sum_{e'e} \sum_{f} \langle \mu_{e} \mu_{ef} \mu_{fe'} \mu_{e'} \rangle$$
$$\times \exp(-i\omega_{eg} t_1 - i\omega_{fg} t_2 + i\omega_{e'f} t_3$$
$$+ \mathrm{conj}[-g^*_{ee}(t_1) - g^*_{ff}(t_2 + t_3) - g^*_{ef}(t_1 + t_2 + t_3)$$
$$+ g^*_{ef}(t_1) + g^*_{ef}(t_2 + t_3) - g^*_{ee'}(t_1 + t_2) - g_{e'e'}(t_3)$$
$$+ g^*_{ee'}(t_1 + t_2 + t_3) + g^*_{ee'}(t_2) - g^*_{ee'}(t_2 + t_3)$$
$$- g^*_{fe'}(t_2) + g^*_{fe'}(t_2 + t_3) + g_{fe'}(t_3)]) \,. \quad \text{(A149)}$$

References

1 Landau, L.D. and Lifshitz, E.M. (1976) *Mechanics*, Butterworth Heinemann, Oxford.
2 Greiner, W. (2003) *Classical Mechanics: System of Particles and Hamiltonian Dynamics*, Springer, New York.
3 Jackson, J.D. (1999) *Classical Electrodynamics*, 3rd edn, John Wiley and Sons, New York.
4 Greenwood, D.T. (1997) *Classical Dynamics*, Dover Publications, New York.
5 Landau, L.D. and Lifshitz, E.M. (1979) *The Classical Theory of Fields*, Butterworth Heinemann, Oxford.
6 Craig, D.P. and Thirunamachandran, T. (1998) *Molecular Quantum Electrodynamics*, Dover Publications, New York.
7 Loundon, R. (2000) *The Quantum Theory of Light*. Oxford University Press.
8 van Kampen, N.G. (2007) *Stochastic Processes in Physics and Chemistry*, 3rd edn, North Holland.
9 Breuer, H.-P. and Petruccione, F. (2002) *The Theory of Open Quantum Systems*, Oxford University Press, New York.
10 Risken, H. and Frank, T. (1996) *The Fokker–Planck Equation: Methods of Solutions and Applications*, 2nd edn, Springer.
11 Atkins, P. and Friedman, R. (2007) *Molecular Quantum Mechanics*, 4th edn, Oxford University Press, New York.
12 Greiner, W. (2001) *Quantum Mechanics. An Introduction*, 4th edn, Springer, Berlin, Heidelberg.
13 Griffiths, D.J. (1995) *Introduction to Quantum Mechanics*, Pearson Prentice Hall, New York.
14 Phillips, A.C. (2003) *Introduction to Quantum Mechanics*, John Wiley & Sons, Ltd, Chichester.
15 Landau, L.D. and Lifshitz, E.M. (1977) *Quantum Mechanics (Non-Relativistic Theory)*, 3rd edn, Butterworth Heinemann, Amsterdam.
16 Tannor, D.J. (2007) *Introduction to Quantum Mechanics. A Time-dependent Perspective*, University Science Books, Sausalito.
17 Greiner, W. (1998) *Quantum Mechanics. Special Chapters*, Springer, Berlin, Heidelberg.
18 Fengt, D.H., Zhang, W.-M., and Gilmore, R. (1990) Coherent states: Theory and some applications. *Rev. Mod. Phys.*, **62**, 867–927.
19 Gazeau, J.-P. (2009) *Coherent States in Quantum Physics*. Wiley-VCH Verlag GmbH, Weinheim.
20 Kühn, O. and May, V. (2011) *Charge and Energy Transfer Dynamics in Molecular Systems*, Wiley-VCH Verlag GmbH, Weinheim.
21 Haken, H. (1983) *Quantum Field Theory of Solids*, North-Holland Publ., Amsterdam.
22 Broude, V.I., Rashba, E.I., and Sheka, E. F. (1985) *Spectroscopy of Molecular Excitons*, Springer Series in Chemical Physics 16, Springer, Berlin.
23 Swenberg, C.E. and Pope, M. (1999) *Electronic Processes in Organic Crystals and Polymers*, Oxford University Press, New York.
24 van Amerongen, H., Valkunas, L., and van Grondelle, R. (2000) *Photosynthetic*

Molecular Excitation Dynamics and Relaxation, First Edition. L. Valkunas, D. Abramavicius, and T. Mančal.

Excitons, World Scientific Co., Singapore.
25 Davydov, A.S. (1971) *Theory of Molecular Excitons*, Plenum Press, New York.
26 Mukamel, S. (1995) *Principles of Nonlinear Optical Spectroscopy*, Oxford University Press, New York.
27 Capek, V. and Silinsh, E.A. (1994) *Organic Molecular Crystals. Interaction, Localization and Transport Phenomena*, AIP Press, New York.
28 He, X.F. (1991) Excitons in anisotropic solids: The model of fractional-dimensional space. *Phys. Rev. B*, **43**, 2063.
29 Jorio, A., Dresselhous, M.S., and Dresselhous, G. (2008) Carbon nanotubes, in *Topics in Applied Physics*, vol. 111, Springer, Berlin, Heidelberg.
30 Itzykson, C. and Zuber, J.-B. (1980) *Quantum Field Theory*, McGraw-Hill, New York.
31 Abramavicius, D. and Mukamel, S. (2010) Energy-transfer and charge-separation pathways in the reaction center of photosystem II revealed by coherent two-dimensional optical spectroscopy. *J. Chem Phys*, **133**, 184501.
32 Davydov, A.S. (1985) *Solitons in Molecular Systems*, Reidel, Dordrecht.
33 Ziman, J.M. (1979) *Models of Disorder. Theoretical Physics of Homogeneously Disordered Systems*, Cambridge University Press.
34 Karski, M., Förster, L., Choi, J.-M., Steffen, A., Alt, W., Meschede, D., and Widera, A. (2010) Quantum walk in position space with single optically trapped atoms. *Science*, **325**, 174–177.
35 Schlosshauer, M. (2007) *Decoherence and the Quantum-to-Classical Transition*, Springer, Berlin.
36 Everett, H. (1957) "Relative state" formulation of quantum mechanics. *Rev. Mod. Phys.*, **29**, 454–462.
37 Greiner, W., Neise, L., Stöcker, H., and Rischke, D. (1995) *Thermodynamics and Statistical Mechanics (Classical Theoretical Physics)*, Springer.
38 Toda, M., Kubo, R., and Saito, N. (1983) *Statistical Physics I*, Springer, Berlin, Heidelberg, New York, Tokyo.
39 Engel, Y.A. and Boles, M.A. (2010) *Thermodynamics: An Engineering Approach*, 7th edn, McGraw-Hill Science/Engineering/Math.
40 Kubo, R., Toda, M., and Hashitsume, N. (1985) *Statistical Physics II*, Spinger, Berlin, Heidelberg, New York, Tokyo.
41 Leggett, A.J., Chakravarty, S., Dorsey, A.T., Matthew Fisher, P.A., Anupam G., and Zwerger, W. (1987) Dynamics of the dissipative two-state system. *Rev. Mod. Phys.*, **59**, 1–85.
42 Weiss, U. (2008) *Quantum Dissipative Systems*, World Scientific.
43 Mančal, T. Nemeth, A., Milota, F., Lukeš, V., Kauffmann, H.F., and Sperling, J. (2010) Vibrational wave packet induced oscillations in two-dimensional electronic spectra. II. Theory. *J. Chem. Phys.*, **132**, 184515.
44 Egorova, D. (2008) Detection of electronic and vibrational coherences in molecular systems by 2D electronic photon echo spectroscopy. *Chem. Phys.*, **347**, 166–176.
45 Fain, B. (2000) Irreversibilities in quantum mechanics, in *Fundamental Theories of Physics*, Kluwer Academic, Dordrecht, London.
46 Doll, R., Zueco, D., Wubs, M., Kohler, S., and Hanggi, P. (2008) On the conundrum of deriving exact solutions from approximate master equations. *Chem. Phys.*, **347**, 243–249.
47 Chaichian, M. and Demichev, A. (2001) *Path Integrals in Physics*, vol. I and II, IoP – Institute of Physics Publishing, Bristol and Philadelphia.
48 Feynman, R.P. and Vernon, F.L. (1963) The theory of a general quantum system interacting with a linear dissipative system. *Ann. Phys. (NY)*, **24**, 118.
49 Grabert, H., Schramm, P., and Ingold, G.-L. (1988) Quantum Brownian motion: The functional integral approach. *Phys. Rep.*, **168**, 115–207.
50 Diosi, L. and Strunz, W.T. (1997) The non-Markovian stochastic Schrödinger equation for open systems. *Phys. Lett. A*, **235**, 569–573.
51 Strunz, W.T. (1996) Linear quantum state diffusion for non-Markovian open

quantum systems. *Phys. Lett. A*, **224**, 25–30.
52 Wiegel, F.W. (1975) Path integral methods in statistical mechanics. *Phys. Rep.*, **16**(2), 57–114.
53 Tanimura, Y. (2006) Stochastic Liouville, Langevin, Fokker–Planck, and master equation approaches to quantum dissipative systems. *J. Phys. Soc. Jpn.*, **75**, 082001.
54 Ishizaki, A. and Fleming, G.R. (2009) Unified treatment of quantum coherent and incoherent hopping dynamics in electronic energy transfer: Reduced hierarchy equation approach. *J. Chem. Phys.*, **130**(23), 234111.
55 Gelzinis, A., Abramavicius, D., and Valkunas, L. (2011) Non-Markovian effects in time-resolved fluorescence spectrum of molecular aggregates: Tracing polaron formation. *Phys. Rev. B*, **84**, 245430.
56 Cho, M., Vaswani, H.M., Brixner, T., Stenger, J., and Fleming, G.R. (2005) Exciton analysis in 2D electronic spectroscopy. *J. Phys. Chem. B*, **109**, 10542–10556.
57 Mančal, T., Valkunas, L., Read, E.L., Engel, G.S., Calhoun, T.R., and Fleming, G.R. (2008) Electronic coherence transfer in photosynthetic complexes and its signatures in optical spectroscopy. *Spectroscopy*, **22**, 199–211.
58 Cho, M. (2009) *Two-Dimensional Optical Spectroscopy*, CRC Press, Boca Raton.
59 Schlau-Cohen, G.S., Ishizaki, A., and Fleming, R.G. (2011) Two-dimensional electronic spectroscopy and photosynthesis: Fundamentals and applications to photosynthetic light-harvesting. *Chem. Phys.*, **386**(1–3), 1–22.
60 Harel, E., and Gregory Engel, S. (2012) Quantum coherence spectroscopy reveals complex dynamics in bacterial light-harvesting complex 2 (LH2). *Proc. Natl. Acad. Sci. USA*, **109**(3), 706–711.
61 Zigmantas, D., Read, E.L., Mančal, T., Brixner, T., Gardiner, A.T., Cogdell, R.J., and Fleming, G.R. (2006) Two-dimensional electronic spectroscopy of the B800–B820 light-harvesting complex. *Proc. Natl. Acad. Sci. USA*, **103**, 12672–12677.
62 Engel, G.S., Calhoun, T.R., Read, E.L., Ahn, T.K., Mančal, T. Cheng, Y.C., Blankenship, R.E., and Fleming, G.R. (2007) Evidence for wavelike energy transfer through quantum coherence in photosynthetic systems. *Nature*, **446**, 782–786.
63 Schlau-Cohen, G.S., Calhoun, T.R., Ginsberg, N.S., Read, E.L., Ballottari, M., Bassi, R., van Grondelle, R., and Fleming, G.R. (2009) Pathways of energy flow in LHCII from two-dimensional electronic spectroscopy. *J. Phys. Chem. B*, **113**(46), 15352–15363.
64 Lee, H., Cheng, Y.-C., and Fleming, G.R. (2007) Coherence dynamics in photosynthesis: Protein protection of excitonic coherence. *Science*, **316**, 1462–1465.
65 Myers, J.A., Lewis, K.L.M., Franklin Fuller, D., Patrick Tekavec, F., Yocum, C.F., and Ogilvie, J.P. (2010) Two-dimensional electronic spectroscopy of the D1-D2-cyt b559 photosystem II reaction center complex. *J. Phys. Chem. Lett.*, **1**(19), 2774–2780.
66 Collini, E. and Scholes, G.D. (2009) Coherent intrachain energy migration in a conjugated polymer at room temperature. *Science*, **323**(5912), 369–373.
67 Milota, F., Sperling, J., Nemeth, A., and Kauffmann, H.F. (2009) Two-dimensional electronic photon echoes of a double band *J* aggregate: Quantum oscillatory motion versus exciton relaxation. *Chem. Phys.*, **357**(1–3), 45–53.
68 Caruso, F., Chin, A.W., Datta, A., Huelga, S.F., and Plenio, M.B. (2009) Highly efficient energy excitation transfer in light harvesting complexes: The fundamental role of noise-assisted transport. *J. Chem. Phys.*, **131**, 105106.
69 Rebentrost, P., Mohseni, M., and Aspuru-Guzik, A. (2009) Role of quantum coherence and environmental fluctuations in chromophoric energy transport. *J. Phys. Chem. B*, **113**, 9942–9947.
70 Fassioli, F., Nazir, A., and Olaya-Castro, A. (2010) Quantum state tuning of energy transfer in a correlated environment. *J. Phys. Chem. Lett.*, **1**, 2139–2143.
71 Nalbach, P., Eckel, J., and Thorwart, M. (2010) Quantum coherent biomolecular

energy transfer with spatially correlated fluctuations. *New. J. Phys.*, **12**, 065043.
72 Fleming, G.R., Huelga, S.F., and Plenio, M.B. (2010) Focus of quantum effects and noise in biomolecules. *New. J. Phys.*, **12**, 065002.
73 Sarovar, M., Ishizaki, A., and Fleming, G.R. (2010) Quantum entanglement in photosynthetic light-harvesting complexes. *Nat. Phys.*, **6**, 462–467.
74 Redfield, A.G. (1957) On the theory of relaxation processes. *IBM J. Res. Dev.*, **1**, 19–31.
75 Zhang, W.M., Meier, T., Chernyak, V., and Mukamel, S. (1998) Exciton-migration and three-pulse femtosecond optical spectroscopies of photosynthetic antenna complexes. *J. Chem. Phys.*, **108**, 7763–7774.
76 Yang, M. and Fleming, G.R. (2002) Influence of phonons on exciton transfer dynamics: comparison of the Redfield, Förster, and modified Redfield equations. *Chem. Phys.*, **282**, 163–180.
77 Mukamel, S. and Rupasov, V. (1995) Energy transfer, spectral diffusion and fluorescence of molecular aggregates: Brownian oscillator analysis. *Chem. Phys. Lett.*, **242**, 17–26.
78 Lindblad, G. (1976) On the generators of quantum dynamical semigroups. *Commun. Math. Phys.*, **48**, 119.
79 Abramavicius, D. and Mukamel, S. (2010) Quantum oscillatory exciton migration in photosynthetic reaction centers. *J. Chem. Phys.*, **133**, 064510.
80 Xu, R.X., Tian, B.L., Xu, J., Shi, Q., and Yan, Y.J. (2009) Hierarchical quantum master equation with semiclassical drude dissipation. *J. Chem. Phys.*, **131**(21), 214111.
81 Hu, J., Luo, M., Jiang, F., Xu, R.-X., and Yan, Y.J. (2011) Padé spectrum decompositions of quantum distribution functions and optimal hierarchical equations of motion construction for quantum open systems. *J. Chem. Phys.*, **134**(24), 244106.
82 Xu, R.-X., Cui, P., Li, X.-Q., Mo, Y., and Yan, Y.-J. (2005) Exact quantum master equation via the calculus on path integrals. *J. Chem. Phys.*, **122**, 041103.
83 Mančal, T., Balevičius, V., and Valkunas, L. (2011) Decoherence in weakly coupled excitonic complexes. *J. Phys. Chem. A*, **115**(16), 3845–3858.
84 Strumpfer, J. and Schulten, K. (2009) Light harvesting complex II B850 excitation dynamics. *J. Chem. Phys.*, **131**, 225101.
85 Xu, R.-X. and Yan, Y.-J. (2007) Dynamics of quantum dissipation systems interacting with bosonic canonical bath: Hierarchical equations of motion approach. *Phys. Rev. E*, **75**, 031107.
86 Guohua, T. and Miller, W.H. (2010) Semiclassical description of electronic excitation population transfer in a model photosynthetic system. *J. Phys. Chem. Lett.*, **1**, 891–894.
87 Ritschel, G., Roden, J., Strunz, W.T., and Eisfeld, A. (2011) An efficient method to calculate excitation energy transfer in light-harvesting systems: Application to the Fenna–Matthews–Olson complex. *New J. Phys.*, **13**, 113034.
88 Kreisbeck, C., Kramer, T., Rodriguez, M., and Hein, B. (2011) High-performance solution of hierarchical equations of motion for studying energy transfer in light-harvesting complexes. *J. Chem. Theory Comput.*, **7**, 2166.
89 Prior, J., Chin, A.W., Huelga, S.F., and Plenio, M.B. (2010) Efficient simulation of strong system-environment interactions. *Phys. Rev. Lett.*, **105**, 050404.
90 Valkunas, L., Abramavicius, D., and Butkus, V. (2011) Interplay of exciton coherence and dissipation in molecular aggregates, in *Semiconductors and Semimetals, v. 85, Quantum Efficiency in Complex Systems, part II*, Academic Press, San Diego, pp. 3–46.
91 Abramavicius, D., Palmieri, B., Voronine, D.V., Šanda, F., and Mukamel, S. (2009) Coherent multidimensional optical spectroscopy of excitons in molecular aggregates; quasiparticle, vs. supermolecule perspectives. *Chem. Rev.*, **109**, 2350–2408.
92 Domcke, W. Egorova, D., Gelin, M. (2007) Analysis of cross peaks in two-

dimensional electronic photon-echo spectroscopy for simple models with vibrations and dissipation. *J. Chem. Phys.*, 126, 074314.

93 Christensson, N., Kauffmann, H.F., Pullerits, T., and Mančal, T. (2012) Origin of long-lived coherences in light-harvesting complexes. *J. Phys. Chem. B*, **116**(25), 7449–7454.

94 Butkus, V., Zigmantas, D., Valkunas, L., and Abramavicius, D. (2012) Vibrational vs. electronic coherences in 2D spectrum of molecular systems. *Chem. Phys. Lett.*, **545**, 40–43.

95 Nalbach, P., Braun, D., and Thorwart, M. (2011) Exciton transfer dynamics and quantumness of energy transfer in the Fenna–Matthews–Olson complex. *Phys. Rev. E*, **84**, 041926.

96 Jelley, E.E. (1937) Molecular, nematic and crystal states of 1,1-diethylcyaninechloride. *Nature*, **139**, 631.

97 Scheibe, G. (1937) Über die Veränderlichkeit der Absorptionsspektren in Lösungen und die Nebenvalenzen als ihre Ursache. *Angew. Chem.*, **50**(11), 212–219.

98 Ruban, A. (2013) *The Photosynthetic Membrane. Molecular Mechanisms and Biophysics of Light Harvesting*, Wiley, Chichester.

99 Blankenship, R.E. (2002) *Molecular Mechanisms of Photosynthesis*, Wiley-Blackwell.

100 Mukamel, S. and Abramavicius, D. (2004) Many-body approaches for simulating coherent nonlinear spectroscopies of electronic and vibrational excitons. *Chem. Rev.*, **104**, 2073.

101 Abramavicius, D., Valkunas, L., and Mukamel, S. (2007) Transport and correlated fluctuations in the nonlinear optical response of excitons. *Europhys. Lett.*, **80**, 17005.

102 Andrews, D.L. and Thirunamachandran, T. (1977) On three-dimensional rotational averages. *J. Chem. Phys.*, **67**, 5026.

103 Zimmermann, J., Oakman, E.L., Thorpe, I.F., Shi, X., Abbyad, P., Brooks, C.L., Boxer, S.G., and Romesberg, F.E. (2006) Antibody evolution constrains conformational heterogeneity by tailoring protein dynamics. *Proc. Natl. Acad. Sci. USA*, **103**(37), 13722–13727.

104 Mančal, T. and Fleming, G.R. (2004) Probing electronic coupling in excitonically coupled heterodimer complexes by two-color three-pulse photon echoes. *J. Chem. Phys.*, **121**(21), 10556–10565.

105 Abramavicius, D., Voronine, D.V., and Mukamel, S. (2008) Double-quantum resonances and exciton-scattering in coherent 2D spectroscopy of photosynthetic complexes. *Proc. Natl. Acad. Sci. USA*, **105**, 8525.

106 Butkus, V., Valkunas, L., and Abramavicius, D. (2012) Molecular vibrations-induced quantum beats in two-dimensional electronic spectroscopy. *J. Chem. Phys.*, **137**(4), 044513.

107 Abramavicius, D., Butkus, V., Bujokas, J., and Valkunas, L. (2010) Manipulation of two-dimensional spectra of excitonically coupled molecules by narrow-bandwidth laser pulses. *Chem. Phys.*, **372**, 22–32.

108 Kjellberg, P., Brüggemann, B. and Pullerits, T. (2006) Two-dimensional electronic spectroscopy of an excitonically coupled dimer. *Phys. Rev. B*, **74**(2), 024303.

109 Nemeth, A., Milota, F., Mancal, T., Lukes, V., Hauer, J., Kauffmann, H.F., and Sperling, J. (2010) Vibrational wave packet induced oscillations in two-dimensional electronic spectra. Experiments, I. *J. Chem. Phys.*, **132**, 184514.

110 Christensson, N., Milota, F., Hauer, J., Sperling, J., Bixner, O., Nemeth, A., and Kauffmann, H.F. (2011) High frequency vibrational modulations in two-dimensional electronic spectra and their resemblance to electronic coherence signatures. *J. Phys. Chem. B*, **115**(18), 5383–5391.

111 Fenna, R.E. and Matthews, B.W. (1975) Chlorophyll arrangement in a bacteriochlorophyll protein from *Chlorobium limicola*. *Nature*, **258**, 573–577.

112 Li, Y.-F., Zhou, W., Blankenship, R.E., and Allen, J.P. (1997) Crystal structure of the bacteriochlorophyll *a* protein from *Chlorobium tepidum*. *J. Mol. Biol.*, **271**, 456–471.

113 Milder, M.T.W., Brüggemann, B., van Grondelle, R., and Herek, J.L. (2010) Revisiting the optical properties of the FMO protein. *Photosynth. Res.*, **104**, 257–274.

114 Camara-Artigas, A., Blankenship, R., and Allen, J.P. (2003) The structure of the FMO protein from chlorobium tepidum at 2.2 Å resolution. *Photosynth. Res.*, **75**, 49.

115 Tronrud, D.E., Schmid, M.F., and Matthews, B.W. (1986) Structure and X-ray amino acid sequence of a bacteriochlorophyll a protein from *prosthecochloris aestuarii* refined at 1.9 Å resolution. *J. Mol. Biol.*, **188**(3), 443–454.

116 Adolphs, J. and Renger, T. (2006) How proteins trigger excitation energy transfer in the FMO complex of green sulfur bacteria. *Biophys. J.*, **91**, 2778–2797.

117 Brixner, T., Stenger, J., Vaswani, H.M., Cho, M., Blankenship, R.E., and Fleming, G.R. (2005) Two-dimensional spectroscopy of electronic couplings in photosynthesis. *Nature*, **434**, 625–628.

118 Panitchayangkoon, G., Hayes, D., Fransted, K.A., J.Caram, R., Harel, E., Wen, J., Blankenship, R.E., and Engel, G.S. (2010) Long-lived quantum coherence in photosynthetic complexes at physiological temperature. *Proc. Natl. Acad. Sci. USA*, **107**, 12766–12770.

119 Read, E.L., Engel, G.S., Calhoun, T.R., Mančal, T., Ahn, T.K., Blankenship, R.E., and Fleming, G.R. (2007) Cross-peak-specific two-dimensional electronic spectroscopy. *Proc. Natl. Acad. Sci. USA*, **104**(36), 14203–14208.

120 Abramavicius, D. and Mukamel, S. (2011) Exciton dynamics in chromophore aggregates with correlated environment fluctuations. *J. Chem. Phys*, **134**(17), 174504.

121 Hu, X., Ritz, T., Damjanovic, A., Autenrieth, F., and Schulten, K. (2002) Photosynthetic apparatus of purple bacteria. *Q. Rev. Biophys.*, **35**, 1–62.

122 Cogdell, R.J., Gall, A., and Köhler, J. (2006) The architecture and function of the light-harvesting apparatus of purple bacteria: from single molecules to in vivo membranes. *Q. Rev. Biophys.*, **39**, 227–324.

123 Hunter, C.N., Daldal, F., Thurnauer, M.C., and Beatty, J.T. (eds) (2008) The purple phototrophic bacteria, *Advances in Photosynthesis and Respiration*, vol. 28, Springer, Dordrecht.

124 Zuber, H. and Brunisholz, R.A. (1991) Structure and function of antenna polypeptides and chlorophyll-protein complexes: Principles and variability, in *The Chlorophylls*, CRC, Boca Raton.

125 McDermott, G., Prince, S.M., Freer, A.A., Hawthornthwaite-Lawless, A.M., Papiz, M.Z., Cogdell, R.J., and Isaacs, N.W. (1995) Crystal structure of an integral membrane light-harvesting complex from photosynthetic bacteria. *Nature*, **374**, 517.

126 Papiz, M.Z., Prince, S.M., Howard, T., Cogdell, R.J., and Isaacs, N.W. (2003) The structure and thermal motion of the b800-850 LH2 complex from rps.acidophila at 2.0 Å resolution and 100 K: New structural features and functionally relevant motions. *J. Mol. Biol.*, **326**, 1523.

127 Georgakopoulou, S., Frese, R.N., Johnson, E., Koolhaas, C., Cogdell, R.J., van Grondelle, R., and van der Zwan, G. (2002) Absorption and CD spectroscopy and modeling of various LH2 complexes from purple bacteria. *Biophys. J.*, **82**, 2184.

128 Cogdell, R.J., Gall, A., and Köhler, J. (2006) The architecture and function of the light-harvesting apparatus of purple bacteria: From single molecules to in vivo membranes. *Q. Rev. Biophys.*, **39**, 227.

129 Rancova, O., Sulskus, J., and Abramavicius, D. (2012) Insight into the structure of photosynthetic LH2 aggregate from spectroscopy simulations. *J. Phys. Chem. B*, **116**(27), 7803–7814.

130 Gobets, B. and van Grondelle, R. (2001) Energy transfer and trapping in photosystem I. *Biochim. Biophys. Acta*, **1507**, 80–99.

131 Turconi, S., Kruip, J., Schweitzer, G., Rögner, M., and Holzwarth, A.R. (1996) A comparative fluorescence kinetics study of photosystem I monomers and trimers from *Synechocystis* PCC 6803. *Photosynth. Res.*, **49**, 263–268.

132 Jordan, P., Fromme, P., Witt, H.T., Klukas, O., Saenger, W., and Kraus, N. (2001) Three-dimensional structure of cyanobacterial photosystem I at 2.5 Å resolution. *Nature*, **411**, 909–917.

133 Sener, M.K., Lu, D., Ritz, T., Park, S., Fromme, P., and Schulten, K. (2002) Robustness and optimality of light harvesting in cyanobacterial photosystem I. *J. Phys. Chem. B*, **106**, 7948–7960.

134 Fromme, P. and Mathis, P. (2004) Unraveling the photosystem I reaction center: A history, or the sum of many efforts. *Photosynth. Res.*, **80**, 109–124.

135 Vaitekonis, S., Trinkunas, G., and Valkunas, L. (2005) Red chlorophylls in the exciton model of photosystem I. *Photosynth. Res.*, **86**, 185–201.

136 Byrdin, M., Jordan, P., Krauss, N., Fromme, P., Stehlik, D., and Schlodder, E. (2002) Light harvesting in photosystem I: Model based on the 2.5 Å structure of photosystem I from *Synechococcus elongatus*. *Biophys. J.*, **83**, 433–457.

137 Vaswani, H.M., Stenger, J., Fromme, P., and Fleming, G.R. (2006) One- and two-color photon echo peak shift studies of photosystem I. *J. Phys. Chem. B*, **110**, 26303–26312.

138 Renger, T., May, V., and Kühn, O. (2001) Ultrafast excitation energy transfer dynamics in photosynthetic pigment–protein complexes. *Phys. Rep.*, **343**, 137–254.

139 Damjanovic, A., Vaswani, H.M., Fromme, P., and Fleming, G.R. (2002) Chlorophyll excitations in photosystem I of *Synechococcus elongatus*. *J. Phys. Chem. B*, **106**, 10251–10262.

140 Yang, M., Damjanovic, A., Vaswani, H.M., and Fleming, G.R. (2003) Energy transfer in photosystem I of cyanobacteria *Synechococcus elongatus*: Model study with structure-based semi-empirical hamiltonian and experimental spectral density. *Biophys. J.*, **85**, 140–158.

141 Brüggemann, B., Sznee, K., Novoderezhkin, V., van Grondelle, R., and May, V. (2004) From structure to dynamics: Modeling exciton dynamics in the photosynthetic antenna PS1888. *J. Phys. Chem. B*, **108**, 13536–13546.

142 Renger, G. and Renger, T. (2008) Photosystem II: The machinery of photosynthetic water splitting. *Photosynth. Res.*, **98**, 53–80.

143 Guskov, A., Kern, J., Gabdulkhakov, A., Broser. M., Zouni, A., and Saenger, W. (2009) Cyanobacterial photosystem II at 2.9 Å resolution and the role of quinones, lipids, channels and chloride. *Nat. Struct. Molec. Biol.*, **16**, 334–342.

144 Raszewski, G., Saenger, W., and Renger, T. (2005) Theory of optical spectra of photosystem II reaction centers: Location of the triplet state and the identity of the primary electron donor. *Biophys. J.*, **88**(2), 986–998.

145 Raszewski, G., Diner, B.A., Schlodder, E., and Renger, T. (2008) Spectroscopic properties of reaction center pigments in photosystem II core complexes: Revision of the multimer model. *Biophys. J.*, **95**, 105–119.

146 Novoderezhkin, V., Dekker, J.P., van Amerongen, H., and van Grondelle, R. (2007) Mixing of exciton and charge-transfer states in photosystem II reaction centers: Modeling of stark spectra with modified redfield theory. *Biophys. J.*, **93**, 1293–1311.

147 Novoderezhkin, V., Andrizhiyevskaya, E., Dekker, J.P., and van Grondelle, R. (2005) Pathways and timescales of primary charge separation in the photosystem II reaction center as revealed by a simultaneous fit of time-resolved fluorescence and transient absorption. *Biophys. J.*, **89**, 1464–1481.

148 Myers, J.A., Lewis, K.L.M., Fuller, F.D., Tekavec, P.F., Yocum, C.F., and Ogilvie, J.P. (2010) Two-dimensional electronic spectroscopy of the D1-D2-cyt b559 photosystem II reaction center complex. *J. Phys. Chem. Lett.*, **1**(19), 2774–2780.

149 Vulto, S.I.E., de Baat, M.A., Neerken, S., Nowak, F.R., van Amerongen, H., Amesz, J., and Aartsma, T.J. (1999) Excited state dynamics in FMO antenna complexes from photosynthetic green sulfur bacteria: A kinetic model. *J. Phys. Chem. B*, **103**, 8153.

150 Abramavicius, D., Voronine, D.V., and Mukamel, S. (2008) Unraveling coher-

ent dynamics and energy dissipation in photosynthetic complexes by 2D spectroscopy. *Biophys. J.*, **94**, 3613–3619.

151 Calhoun, T.R., Ginsberg, N.S., Schlau-Cohen, G.S., Cheng, Y.-C., Ballottari, M., Bassi, R., and Fleming, G.R. (2009) Quantum coherence enabled determination of the energy landscape in light-harvesting complex II. *J. Phys. Chem. B*, **113**, 16291–16295.

152 Collini, E., Wong, C.Y., Wilk, K.E., Curmi, P.M.G., Brumer, P., and Scholes, G.D. (2010) Coherently wired light-harvesting in photosynthetic marine algae at ambient temperature. *Nature*, **463**, 644–647.

153 Nemeth, A., Milota, F., Mančal, T., Luke, V., Hauer, J., Kauffmann, H.F., and Sperling, J. (2010) Vibrational wave packet induced oscillations in two-dimensional electronic spectra. I. Experiments. *J. Chem. Phys.*, **132**, 184514.

154 Voronine, D.V., Abramavicius, D., and Mukamel, S. (2008) Chirality-based signatures of local protein environments of photosynthetic complexes of green sulfur bacteria. simulation study. *Biophys. J.*, **95**, 4896–4907.

155 Abramavicius, D. and Mukamel, S. (2009) Exciton delocalization and transport in photosystem I of cyanobacteria synechococcus elongates: Simulation study of coherent two-dimensional optical signals. *J. Phys. Chem. B*, **113**, 6097–6108.

156 Raszweski, G., Saenger, W., and Renger, T. (2005) Theory of optical spectra of photosystem II reaction centers: Location of the triplet state and the identity of the primary electron donor. *Biophys. J.*, **88**, 986–998.

157 Palmieri, B., Abramavicius, D., and Mukamel, S. (2009) Lindblad equations for strongly coupled populations and coherences in photosynthetic complexes. *J. Chem. Phys.*, **130**, 204512.

158 Moerner, W.E. and Kador, L. (1989) Optical detection and spectroscopy of single molecules in a solid. *Phys. Rev. Lett.*, **62**(21), 2535–2538.

159 Ambrose, W.P., Goodwin, P.M., Martin, J.C., and Keller, R.A. (1994) Single-molecule detection and photochemistry on a surface using near-field optical-excitation. *Phys. Rev. Lett.*, **72**(1), 160–163.

160 Xie, X.S. and Dunn, R.C. (1994) Probing single-molecule dynamics. *Science*, **265**(5170), 361–364.

161 Nirmal, M., Dabbousi, B.O., Bawendi, M.G., Macklin, J.J., Trautman, J.K., Harris, T.D., and Brus, L.E. (1996) Fluorescence intermittency in single cadmium selenide nanocrystals. *Nature*, **383**(6603), 802–804.

162 Dickson, R.M., Cubitt, A.B., Tsien, R.Y., and Moerner, W.E. (1997) On/off blinking and switching behaviour of single molecules of green fluorescent protein. *Nature*, **388**(6640), 355–358.

163 Bopp, M.A., Jia, Y.W., Li, L.Q., Cogdell, R.J., and Hochstrasser, R.M. (1997) Fluorescence and photobleaching dynamics of single light-harvesting complexes. *Proc. Natl. Acad. Sci. USA*, **94**(20), 10630–10635.

164 Krüger, T.P.J., Ilioaia, C., and van Grondelle, R. (2011) Fluorescence intermittency from the main plant light-harvesting complex: Resolving shifts between intensity levels. *J. Phys. Chem. B*, **115**(18), 5071–5082.

165 Krüger, T.P.J., Ilioaia, C., Valkunas, L., and van Grondelle, R. (2011) Fluorescence intermittency from the main plant light-harvesting complex: Sensitivity to the local environment. *J. Phys. Chem. B*, **115**(18), 5083–5095.

166 van Oijen, A.M., Ketelaars, M., Köhler, J., Aartsma, T.J., and Schmidt, J. (2000) Spectroscopy of individual light-harvesting 2 complexes of *Rhodopseudomonas acidophila*: Diagonal disorder, intercomplex heterogeneity, spectral diffusion and energy transfer in B800 band. *Biophys. J.*, **78**, 1570–1577.

167 Hofmann, C., Aartsma, T.J., Michel, H., and Köhler, J (2003) Direct observation of tiers in the energy landscape of a chromoprotein: A single-molecule study. *Proc. Natl. Acad. Sci. USA*, **100**(26), 15534–15538.

168 Rutkauskas, D., Novoderezhkin, V., Cogdell, R.J., and van Grondelle, R. (2004) Fluorescence spectral fluctuations of single LH2 complexes from

Rhodopseudomonas acidophila strain 10050. *Biochemistry*, **43**(15), 4431–4438, PMID: 15078088.

169 Valkunas, L., Janusonis, J., Rutkauskas, D., and van Grondelle, R. (2007) Protein dynamics revealed in the excitonic spectra of single LH2 complexes. *J. Lumin.*, **127**(1), 269–275.

170 Janusonis, J., Valkunas, R.L., Rutkauskas, D., and van Grondelle, R. (2008) Spectral dynamics of individual bacterial light-harvesting complexes: alternative disorder model. *Biophys. J.*, **94**, 1348–1358.

171 Weiss, S. (1999) Fluorescence spectroscopy of single biomolecules. *Science*, **283**(5408), 1676–1683.

172 Joo, C., Balci, H., Ishitsuka, Y., Buranachai, C., and Ha, T. (2008) Advances in single-molecule fluorescence methods for molecular biology. *Ann. Rev. Biochem.*, **77**(1), 51–76.

173 Kuno, M., Fromm, D.P., Johnson, S.T., Gallagher, A., and Nesbitt, D.J. (2003) Modeling distributed kinetics in isolated semiconductor quantum dots. *Phys. Rev. B*, **67**(12), 125304.

174 Cichos, F., von Borczyskowski, C., and Orrit, M. (2007) Power-law intermittency of single emitters. *Curr. Opin. Colloid Interface Sci.*, **12**(6), 272–284.

175 Basche, T., Kummer, S., and Brauchle, C. (1995) Direct spectroscopic observation of quantum jumps of a single-molecule. *Nature*, **373**(6510), 132–134.

176 Stefani, F.D., Hoogenboom, J.P., and Barkai, E. (2009) Beyond quantum jumps: Blinking nanoscale light emitters. *Phys. Today*, **62**(2), 34–39.

177 Chung, I. and Moungi Bawendi, G. (2004) Relationship between single quantum-dot intermittency and fluorescence intensity decays from collections of dots. *Phys. Rev. B*, **70**, 165304.

178 Kuno, M., Fromm, D.P., Hamann, H.F., Gallagher, A., and Nesbitt, D.J. (2000) Nonexponential "blinking" kinetics of single cdse quantum dots: A universal power law behavior. *J. Chem. Phys.*, **112**(7), 3117–3120.

179 Verberk, R., van Oijen, A.M., and Orrit, M. (2002) Simple model for the power-law blinking of single semiconductor nanocrystals. *Phys. Rev. B*, **66**(23), 233202.

180 Tang, J. and Marcus, R.A. (2005) Diffusion-controlled electron transfer processes and power-law statistics of fluorescence intermittency of nanoparticles. *Phys. Rev. Lett.*, **95**, 107401.

181 Frauenfelder, H., Sligar, S.G., and P.Wolynes, G. (1991) The energy landscapes and motions of proteins. *Science*, **254**, 1598–1603.

182 Valkunas, L., Chmeliov, J., Krüger, T.P.J., Ilioaia, C., and van Grondelle, R. (2012) How photosynthetic proteins switch. *J. Phys. Chem. Lett.*, **3**, 2779–2784.

183 Tang, J. and Marcus, R.A. (2005) Mechanisms of fluorescence blinking in semiconductor nanocrystal quantum dots. *J. Chem. Phys.*, **123**(5), 054704.

184 Agmon, N. and Rabinovich, S. (1992) Diffusive dynamics on potential-energy surfaces – nonequilibrium co binding to heme-proteins. *J. Chem. Phys.*, **97**(10), 7270–7286.

185 Zondervan, R., Kulzer, F., Orlinskii, S.B., and Orrit, M. (2003) Photoblinking of rhodamine 6G in poly(vinyl alcohol): Radical dark state formed through the triplet. *J. Phys. Chem. A*, **107**(35), 6770–6776.

186 Shimizu, K.T., Neuhauser, R.G., Leatherdale, C.A., Empedocles, S.A., Woo, W.K., and Bawendi, M.G. (2001) Blinking statistics in single semiconductor nanocrystal quantum dots. *Phys. Rev. B*, **63**(20), 205316.

187 Schlegel, G., Bohnenberger, J., Potapova, I., and Mews, A. (2002) Fluorescence decay time of single semiconductor nanocrystals. *Phys. Rev. Lett.*, **88**(13), 137401.

188 Wang, J., Chen, J., and Hochstrasser, R.M. (2006) Local structure of β-hairpin isotopomers by FTIR, 2D IR, and ab initio theory. *J. Phys. Chem. B*, **110**(14), 7545–7555.

189 Novoderezhkin, V., Palacios M.A., van Amerongen, H., and van Grondelle, R. (2005) Exciton dynamics in the LHCII complex of higher plants: Modelling based on the 2.72 Å crystal structure. *J. Phys. Chem. B*, **109**, 10493–10504.

190 Morse, P.M. and Feshbash, H. (1953) *Methods of Theoretical Physics*, McGraw-Hill, New York.

Index

Molecular Excitation Dynamics and Relaxation, First Edition. L. Valkunas, D. Abramavicius, and T. Mančal.